HUMANOID ROBOTS

Edited by
BEN CHOI

Humanoid Robots
Edited by Ben Choi

Republished by InTech
Janeza Trdine 9, 51000 Rijeka, Croatia

Additional hard copies can be obtained from orders@intechopen.com

Humanoid Robots, Edited by Ben Choi
p. cm.
ISBN 978-953-7619-44-2
1. Humanoid Robots I. Ben Choi

For EU product safety concerns:
IN TECH d.o.o., Prolaz Marije Krucifikse Kozulić 3, 51000 Rijeka, Croatia,
info@intechopen.com or visit our website at intechopen.com.

Meet the editor

Ben Choi, Ph.D. & Pilot, has a Ph.D. degree in Electrical and Computer Engineering and has a Pilot certificate for flying airplanes and helicopters. He is an Associate Professor in Computer Science at Louisiana Tech University. He was a visiting Research Scholar at DePaul University, University of Western Australia, and Hong Kong University of Science and Technology. He had worked in the computer industry as a System Performance Engineer at Lucent Technologies (Bell Labs) and as a Computer Consultant. He received Ph.D., M.S., and B.S. degrees from The Ohio State University, studied Computer Science, Computer Engineering, and Electrical Engineering. His areas of research include Humanoid Robots, Artificial Intelligence, Machine Learning, Intelligent Agents, Semantic Web, Data Mining, Fuzzy Systems, and Parallel Computing. His future research includes developing advanced software and hardware methods for building intelligent machines and theorizing the Universe as a Computer.

Preface

This book focuses on the state of the art developments of humanoid robots. It aims to facilitate building robots that resemble the human form and imitate human behaviors. Humanoid robots are developed to use the infrastructures designed for humans, to ease the interactions with humans, and to help the integrations into human societies. The developments of humanoid robots proceed from building individual robots to establishing societies of robots working alongside with humans.

For building individual robots, this book addresses the problems of constructing a humanoid body and mind. On constructing a humanoid body, it describes the designs of foot, knee, waist, arm, head, and face. On constructing a mind, it describes how to generate walk patterns and maintain balance, how to encode and specify humanoid motions, and how to control eye and head movements for focusing attention on moving objects. It provides methods for learning motor skills and for language acquisition. It describes how to generate facial movements for expressing various emotions and provides methods for decision making and planning. Finally, it discusses how to create artificial cognition.

For establishing societies of robots working for humans, this book addresses the problems of interactions between humans and robots. It describes how robots learn from humans and how humanoid robots use various facial expressions as a form of nonverbal communication.

This book accounts for the current leading researches and challenges in building humanoid robots in order to prepare for the near future when human societies will be advanced by using humanoid robots. It serves as a reference source for any researchers interested in humanoid robots and as a supplementary textbook for any courses in robotics.

BEN CHOI

PhD & Pilot, pro@BenChoi.org

Associate Professor in Computer Science

Louisiana Tech University

USA

Contents

1

Humanoid Robotic Language and Virtual Reality Simulation

Ben Choi
Louisiana Tech University
USA

1. Introduction

This chapter describes the development of a humanoid robotic language and the creation of a virtual reality system for the simulation of humanoid robots. In this chapter we propose a description language for specifying motions for humanoid robots and for allowing humanoid robots to acquire motor skills. Locomotion greatly increases our ability to interact with our environments, which in turn increases our mental abilities. This principle also applies to humanoid robots. However, there are great difficulties to specify humanoid motions and to represent motor skills, which in most cases require four-dimensional space representations. We propose a representation framework that includes the following attributes: motion description layers, egocentric reference system, progressive quantized refinement, and automatic constraint satisfaction. We also outline strategies for acquiring new motor skills by learning from trial and error, macro approach, and programming.

Then, we use our new humanoid motion description language and framework as the base to build a virtual reality system to simulate humanoid robots. Currently most humanoid simulation systems require manual manipulations of body parts or require capturing movements enacted by a person. We aim to describe humanoid motions using our high-level humanoid motion description language. We defined new motion primitives and new syntax that allows programmers to easily describe complex humanoid motions. To make the motion specification possible, we defined a humanoid framework that models humanoid robots by specifying their capacities and limitations. Furthermore, we developed a simulation system that can execute the humanoid motion description language and can automatically handle conflicting motion statements and can ensure that the described motions are within the limitations of humanoid robots. Our simulation results show that the proposed system has great future potentials.

The remaining of this chapter is organized as follows. Section 2 outlines the related research on high-level language approaches to describe and to simulate humanoid motions. Section 3 describes the motives for defining a humanoid motion description framework, which includes methods for specifying humanoid motions and methods for acquiring new motor skills. Section 4 outlines the methods for specifying humanoid motions, which include the concepts of motion description layers, egocentric reference system, progressive quantized refinement, and automatic constraint satisfaction. Section 5 outlines two methods for acquiring new motor skills: learning from trial and error and learning by macro approach. Section 6 describes the motives for developing a system to simulate humanoid robots in virtual reality environments. Section 7 defines a new humanoid motion description language called Cybele. It focuses on the syntactic aspects of the language, while Section 8 focuses on the semantic aspects of the language and defines the framework on which the language can be interpreted. Section 9 provides the implementation details of the humanoid simulation system. And, Section 10 gives the conclusion and outlines the future research.

2. Related Research

Research in describing humanoid motions begins with the works for describing human dances. Popular dance notation systems include Benesh (Causley, 1980), Labanotation (Hutchinson & Balanchine, 1987), and EW (Eshkol-Wachman, 2008). Benesh is the simplest one and is designed particularly for dance description. Labanotation is more comprehensive for describing human motion in general. EW can be applied on linkage systems other than human body. Computers are now used to aid the interpretation and visualization of these notations (Ryman et al., 1984; Adamson, 1987; Calvert et al., 1993; Schiphorst, 1992). Researchers used Labanotation as a basis to represent human motion, proposed to extract key motion primitives, and proposed architectures for digital representation of human movements (Badler et al., 1979). Another approach uses natural language; such as "Improv" system used natural language to script human behaviour interacting in virtual environments (Perlin & Gikdberg, 1996). Motion sequences can be generated by system that employs human biomechanical logic (Badler et al., 1994). This section outlines related work on the high-level language approaches to humanoid simulation (Nozawa et al., 2004; Nishimura et al., 2005). Several systems will be discussed, which including, Poser Python, VRML (Virtual Reality Modelling Language), Improv, STEP, and others. It focuses on high-level language approaches to humanoid simulation and omits other general concurrent languages such as OCCAM.

Poser Python (Python, 2008; Schrand, 2001) is an implementation of the Python interpreter that includes many commands that have been extended to recognize and execute commands not included in the standard Python language. Poser Python script language is a language combination that uses syntax and basic logic of Python and special commands tailored especially for Poser scene, manipulate them, and finally send them back to Poser. The language-controlled animation is a significant advantage of Poser-Python system.

VRML (Virtual Reality Modelling Language) is a scene-description language used widely on the internet. VRML uses TimeSensor to initiate and synchronize all the motions in the scene. It is possible that asynchronously generated events arrive at the identical time as one or more sensor-generated event. In such cases, all events generated are part of the same initial event cascade, and each event has the same timestamp. Based on this mechanism, VRML is quite suitable to visual presentations with user interactions. However, there is no direct way to describe complex motions with time overlapping.

Improv (Perlin & Goldberg 1996) is a system for the creation of real-time behaviour based on animated actors. Improv consists of two subsystems. The first subsystem is an animation engine that uses procedural techniques to enable authors to create layered, continuous, nonrepetitive motions and smooth transitions between them. The system uses an English-style scripting language so that creative experts who are not primarily programmers can create powerful interactive applications.

STEP (Huang et al., 2002) is a distributed logic program being used to define and control the hand gestures of embodied agents in virtual worlds. It works as an interface between the constructs of logic programming and the humanoid model defined in VRML. Using the STEP framework, different gesture dictionaries can be defined, and variants of a hand gesture, according to dynamically changing factors, can be generated on the fly. STEP also has two kinds of control logic definitions. One is parallel and the other is sequential. But different kinds of program blocks cannot be mixed together and must be executed one by one.

There are also other systems for animations such as Alice (Conway, 1997), Maya, Lightwave, and 3D Studio Max. They each have a beautiful user interface and have easy drag and drop functionality for the animation design and customization. However, most of them lack underlining programming languages that programmer can program to control various motions. Meanwhile, current motion description languages do not have motion synchronization at the language level. The details of the motion control make the simulation difficult to implement and debug at the language level. This also makes the motions non-reusable.

3. Humanoid Motion Description Framework

Locomotion greatly increases our ability to interact with our environments, which in turn increases our mental abilities. The principle that mental abilities can be improved by interacting with the environments is the basis for MIT Cog's project (Brooks et al., 1998; Arsenio, 2004). However, Cog robot currently lacks locomotion. On the other hand, Honda humanoid robots (Honda, 2008) possess the state of the art locomotion system, but lack the autonomy and the learning abilities. We envision the union of these two types of robots, such as Albert HUBO (Oh et al., 2006), as the basis of our investigation.

The humanoid robots of the near future will possess the abilities for locomotion, autonomy, and learning (Brooks, 2002; Arsenic, 2004; Burghart et al., 2005; Yokoi, 2007). Much research remains to be done on such autonomous humanoid robots (Brooks 1996; Scassellati 2001). In this chapter, we will focus on issues of developing a common framework for both specifying motions and for autonomously acquiring motor skills for such robots.

A unified framework to address both specifying motions and acquiring motor skills will facilitate the developments of autonomous humanoid robots. Neural Network, for example, may be a good medium for capturing and classifying motor skills. However, the resultant representation in terms of matrix of weights of connectivity is difficult to be interpreted and modified. Thus, in this investigation we choose to use symbolic approach by developing a description language.

Our humanoid motion description language, like any other languages, consists of syntactic and semantic aspects. Syntactic aspect specifies rules for combining words while semantic aspect specifies structures for interpretation and such provides the meaning. We propose different set of words and rules for different level of abstraction, such as using joint angles at the low level and using "walk" and "jump" at the high level of abstraction. The interpretation and the meaning are based on our framework that includes egocentric reference system, progressive quantized refinement, and automatic constraint satisfaction.

Our language and our framework (Choi & Chen, 2002) are unique in many ways comparing to other related research (Kanehiro et al. 2004). Our reference system simplifies specification of locomotion and allows motions to be described by uniform and deterministic expressions. Our concept of Progressive Quantized Refinement allows a humanoid robot to interact with its environments using different level of granularity. Our Automatic Constraint Satisfaction system reduces the complexity of specifying humanoid motions. Moreover, our underlining model using non-deterministic finite state machines allows humanoid robots to learn new motor skills.

4. Specifying Humanoid Motions

The proposed language and framework for specifying humanoid motions includes the following attributes: motion description layers, egocentric reference system, progressive quantized refinement, and automatic constraint satisfaction, each of which is described as follows.

4.1 Motion Description Layers

Table 1 outlines the concept of motion description layers. Each description layer is a level of abstraction. Joint Angle layer describes a motion in terms of changes in the joint angles, such as knee joint rotate to 30 degree or elbow joint rotate to 45 degree. This layer provides detail and precise information that can readily be used to control various actuators of a robot. Path layer describes a motion in terms of a connected line that is specified by points. A simple path can be specified using two points, such as Hand (v1) that moves hand from current position to point v1. More points provide more detail specification of the path; for example, Foot (v1, v2) denoted that foot moves from current position through v1 to v2.

Motion primitive (Nakaoka et al., 2004) layer describes a motion in terms of a given set of essential motions that can be combined to form more complex motions. The set of essential motions must first be identified. It must be complete so that we can describe all possible motions of a humanoid robot. We must also provide a set of rules for specifying how one motion primitive can be combined with another. In effect, we are creating a formal language and insuring that the language is both complete and consistent. This is an axiomatic approach to describe humanoid motions.

Motion sequence layer describes a sequence of motions in terms of motion blocks such as walk, run, jump, and turn. Using this high-level description, we can describe a complex task

Description Layer	Example
Motion Sequence	Walk, Run, Jump, Turn
Motion Primitive	Raise, Lower, Forward, Backward
Path	Hand (v1), Foot (v1, v2)
Joint Angle	Knee Joint 30, Elbow Joint 45

Table 1. Motion Description Layers

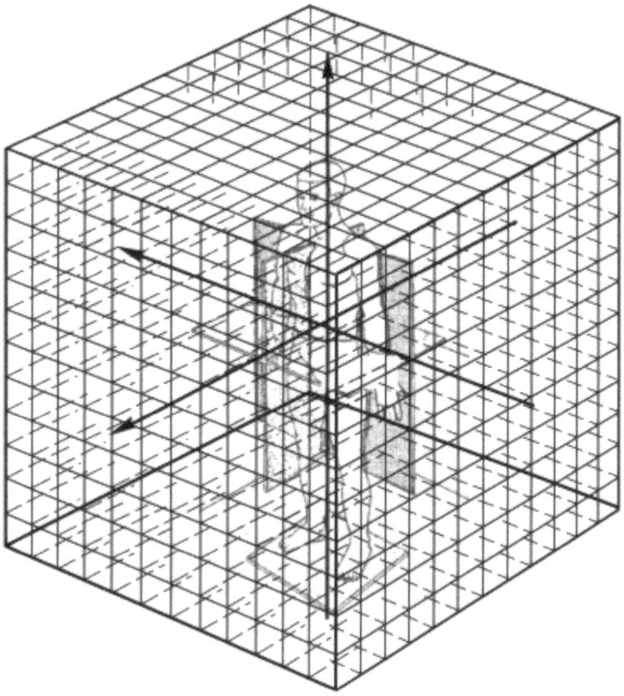

Fig. 1. Egocentric Space Reference System

with ease without having to specify the angle of each joint. However, this high-level description is not as precise as low-level description and thus leaves must room for interpretation that is addressed in this investigation by using Progress Quantized Refinement discuss in Section 4.3.

4.2 Egocentric Reference System

We proposed an egocentric reference system for specifying space-time in discrete finite four-dimensional hyperspace. Each point in our reference system is represented by a quintuple (x, y, z, t). Each of the variables, x, y, z, and t, is an integer ranging from -128 to +127. The origin of the reference system locates at (0, 0, 0, 0). In short, each point in our reference system can be stored using four bytes or 32 bits.

Our reference system is egocentric in that the origin of space is located at the center of the torso of a humanoid robot, as denoted in Figure 1. The origin of time is located at the beginning of a state transition.

In our system, a motion is defined by a sequence of state transitions. Each state transition begins at time 0 and must be completed in 127 time units or less. Negative time units represent the time units used during the last state transition. Each state transition begins with the origin of space located at the center of the torso. In short, a state transition begins at (0, 0, 0, 0). All changes during a state transition are specified within the egocentric reference system.

Translation between the egocentric reference system and its world reference system is done at the end of each state transition. For example, beginning at a standing position as shown in

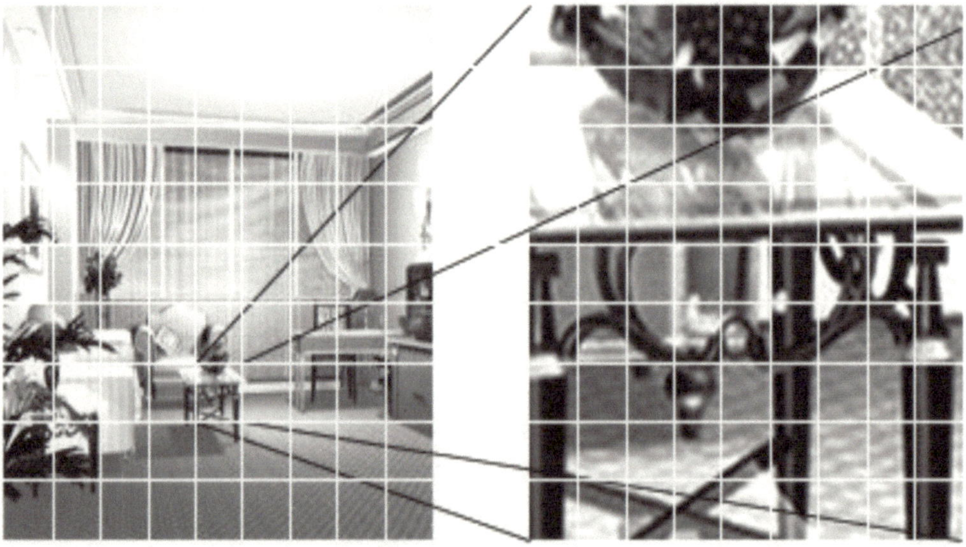

Fig. 2. Concept of Progressive Quantized Refinement

Figure 1, the robot moved three units forward in positive y-axis direction and completed at a standing position, and the whole transition takes 9 units of time. Now, the center of the torso is located at (0, 3, 0, 9). Assuming at the beginning of the transition R(0, 0, 0, 0) in the robot's egocentric reference system is located at W(3, 2, 4, 2) in its world reference system. Also assume that y-axes of both systems are parallel and have the same direction, and each unit in the egocentric reference system represents 2 units in the world reference system. To reset R(0, 3, 0, 9) back to R(0, 0, 0, 0), we makes R(0, 0, 0, 0) now to corresponding to W(3, 2+3*2, 4, 2+9*2).

4.3 Progressive Quantized Refinement

We proposed a concept called Progressive Quantized Refinement for a humanoid robot to interact with its environments using different level of granularity. Figure 2 illustrates the concept; on the left picture a 9x9 unit squares is used to display a room while on the right picture the same sized 9x9 unit squares is used to display part of a table. For a robot to put an object on the table, the robot can first use the left picture to move toward the table. Then, it can use the right picture to put the object on the table.

At different states a robot can change its unit scale factor as needed. For example, a unit length in the robot's egocentric space reference system can be scaled to 1 cm, 1 inch, or 1 meter in its world reference system. A unit time can be scaled, for example, to 1 second, 1 minute, or 5 minutes.

4.4 Automatic Constraint Satisfaction

We proposed to use Automatic Constraint Satisfaction to reduce the complexity of specifying humanoid motions. There are many implicit requirements for locomotion, such as maintaining balance and structural integrity. Automatic constraint satisfaction system will provide additional changes to meet the implicit requirements.

A system for providing automatic constraint satisfaction for locomotion is very complex and much research is being done on areas such as motion planning and constraint satisfaction. For example, we can simply specify that the robot must move its right hand from current position (3, 6, 2, 0) to new position (3, 50, 2, 6). The simpler the specification, in most cases, requires the more complex constraint satisfaction. In our example, the hand must reach the new position using 6 units of time, so that speeds for various actuators must be adjusted to meet this requirement. If the hand cannot reach the new position by simply raising it and reaching out, then the robot must move the whole body toward the new position.

5. Acquiring New Motor Skills

The ability for acquiring new motor skills is essential for mental developments. The trivial approach is to simply program a humanoid robot for new required motor skills (Ude et al., 2004), which can easily be done by an experienced programmer using our proposed language and framework. Thus, in the following we will focus on strategies for acquiring motor skills through learning from trial and error and learning by macro approach.

Learning motor skills has not yet been a central focus of Machine Learning researchers. Thus, much research remains to be done on automatic acquiring new motor skills. We briefly outline strategies for creating such a system, which in part is based on the author's work on automata for learning sequential tasks (Choi, 1998; 2003).

5.1 Learning from Trial and Error

One way for acquiring new motor skills is by trial and error. This approach requires first identifying an objective and then selecting actions or motions to achieve the objective. In particularly, using our proposed framework, identifying an objective can be described as identifying a goal state, while selecting actions or motions can be described as selecting a sequence of transitions from the current state to the goal state.

Using our proposed framework, the underlining model is a non-deterministic finite state machine (Choi 2002). From one state, there may be several transitions to follow to other states. Choosing one transitions or the other is why we call this a trial and error approach and is why it is non-deterministic. To achieve the objective is to find a sequence of transitions from the current state to the goal state. As soon as a sequence of transitions is found, it can be stored for future use.

5.2 Learning by Macro Approach

Macro approach can be used in a supervised learning environment (Bentivegna & Atkeson, 2001; Arsenic, 2004; Calinon et al. 2007). After a humanoid robot is repeatedly instructed to perform certain task, the robot can store the sequence of motions to associate with a name of the task. After which, the robot can be commanded to perform the sequence by simply specifying the name of the task. This is a simple record and play back approach.

A more sophisticated approach is provided by the author (Choi, 2002; 2003), in which a robot can build a non-deterministic finite state machine based on the repeated instructions to perform certain task. The resulting non-deterministic finite state machine can then be used for other purposes such as for learning from trial and error as discussed above.

6. Virtual Reality Simulation of Humanoid Robots

We developed a humanoid motion description language, called Cybele, based on the above proposed framework. Our development process in turn enhances the strength of our framework. We also developed a virtual reality system to simulate humanoid robots (Zhou & Choi, 2007). Virtual reality simulation of humanoids has been an important subject due to its wide range of applications such as motion picture production and robotics (Kanehiro et al., 2001). To achieve realistic motions, programmers currently must be highly experienced to define all control factors for the movements of individual body parts on each time instance. Since it is far too difficulty to control the detail moments of each individual body parts, currently motion capture systems are widely used to record motions enacted by a person and then to use the captured motions to animate humanoids (Riley & Atkeson 2000; Safonova et al., 2002; Nakaoka et al., 2003; Ude et al., 2004).
We attempted a high-level and goal-oriented approach. As shown in Figure 3, there are two extremes on the approaches to humanoid simulations, the goal-oriented and the motion-oriented. The goal-oriented approach specifies the objectives using high-level languages, that are easy to write, and that use Artificial Intelligent methods to generated detailed motions, but that are much more difficulty in terms of building the system. On the other hand, the motion-oriented approach can specify detail motion using degree of freedom (DOF) of each join. However, it is more complicated to write such specifications and programmers need to provide all the details, but such a system is earlier to be built.

In this chapter we present a new high-level goal-oriented language named Cybele. The language is designed for humanoid motion description (Choi & Chen, 2002) and is independent of virtual reality simulation systems. Our new language provides simple syntax for expressing parallel and sequential processing as well as the combination of these two. We also introduce syntax for expressing motions. In particular, to solve the motion mixing problem in parallel and complex blocks, we present an approach to synthesizing motions with combination of partial states. We define key poses as states or partial states and create primitives between two key states. We also extract a relatively small set of parameterized motions from infinite complex motion sequences and make them reusable as primitive motions. Our simulator interprets the programs, breaks down motions into primitives with time, scope, and priority. Final motion sequences are then generated using a synchronization approach.

To achieve constraints satisfaction, we define the scope and priority for each joint in the humanoid. This system checks constraints on the joints affected by the motion and determines which motion is in conflict with some others. In addition, to create a multi-platform system, we implement a prototype system with two parts following the same paradigm as the Java virtual machine. The first part interprets the program and generates motion sequences. The second part translates motion primitives to interface with systems such as virtual reality simulation systems, animation systems, or robotic systems (Kanayama & Wu, 2000; Hirai et al., 1998; Sony, 2008).

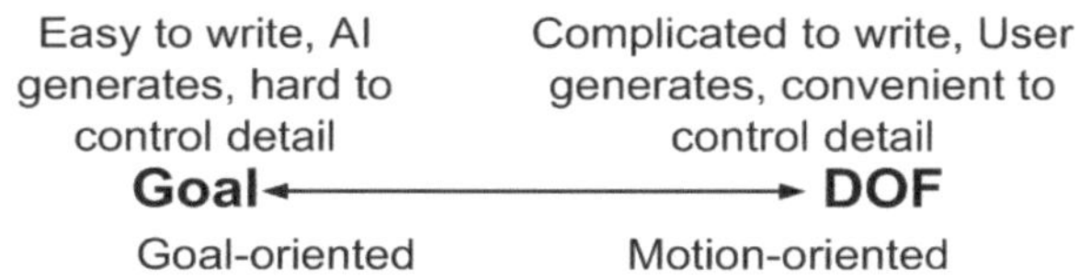

Fig. 3. Two extremes of motion languages

7. Humanoid Robotic Language Specification

Our new humanoid motion description language is called Cybele (Choi & Chen 2002). The features of Cybele language are object-oriented and scripting motion description language. Besides the conventional variable declaration, functions, and control flow, we create a new syntax for describing parallel actions. We also create a new complex motion statement and expand the use of the conventional motion control concepts using new parallel blocks.

7.1 Motion Statements

To describe humanoid motions, we abstracted a collection of predefined motions for the whole body and body parts. We decompose the humanoid in a standard way as specified in humanoid animation specification (Harris et al., 1999). Figure 4 shows the BNF of motion statements. In general, a motion statement is composed of a main part, parameter part, followed by a colon ':', a description part, and ends with a semicolon. A main part includes the object name and the motion name. An object name can be optional. If there is no object name, the system will take the current default object.

A parameter part takes motion parameters that provide additional detail to specify the motion. Each expression is separated by comma ','. The description part can be provided right after the motion function or at the end of a block. It provides a set of attributes to change the scale of the motion. We currently define two attributes namely starttime and speed. We can adapt approaches (Pollard, 1999) to retarget and change the scale of motion.

7.2 Specification of Complex Motions

Complex motions can be specified as combinations of sequential and/or parallel sub-motions. Sequential motions will execute one after another. Parallel motions will execute in parallel. The start time for parallel motions may not be the same and thus producing partially overlapping motions.

For ease of specification, we defined our language such that programmers can write overlapping sequential and/or parallel motions in compounded program blocks. Each block can be sub-block of other blocks. A compound block can be sequential, parallel, or mixed of the two.

```
<motion_statement>
        ::= <object_serial_option.><motion>'('<param_list>')'<description_part>';'
<object_serial_option>
        ::= <object> '.'| <object> '.' <object_serial_option> | NULL
<param_list>
        ::= <parameter>| <param_list> ',' <parameter>
```

```
<description_part>
        ::= ':' <description_list>|NULL
<description_list>
        ::= <attribute>|<description_list>','<attribute>
<attribute>
        ::= | speed '=' <float>| starttime '=' <float>        //unit second
```

Fig. 4. Motion statement BNF

```
[ // begin parallel block
  statement1;      // statement1 starts from time 0
  statement2;      // statement2 starts from time 0
  statement3 : starttime = t1; // statement3 starts from time t1
] // parallel block end
```

Fig. 5. Parallel block

```
{ // begin sequential block
  statement1; // statement1 starts from time t0 (duration=d1)
  statement2; // statement2 starts from (t0+d1) (duration=d2)
} // total duration is d1+d2
```

Fig. 6. Sequential block

We define the syntax of parallel block as statements enclosed within "**[ ]**". Figure 5 shows an example. All statements inside parallel block are considered being issued concurrently. They all start at the same time unless otherwise specified by the "starttime" parameter. In Figure 5 for example, the syntax "statement3 : starttime = t1" specified that statement 3 starts t1 units of time later than the other two statements.

We define the syntax of sequential block as statements enclosed within "**{ }**". Figure 6 shows an example, in which statement2 starts after statement1 completed. A delay can be used such that statement 2 does not need to starts right after the completion of statement 1.

Sequential and parallel blocks can be combined or embedded with each other to describe complex motion logic. Figure 7 shows an example of the structure. The code in Figure 7 illustrates the time specification as shown in Figure 8.

7.3 Sample Program

We show a sample Cybele program in Figure 9. In this program, Jack is the actor. He walks backward two steps. Then, he nods his head, while at the same time, he walks forward two steps. Then, he makes a right turn, a left turn, step right once, and step left once in a sequential way.

8. Humanoid Motion Framework in Cybele

To specify humanoid motions, we need to define humanoid framework. Meanwhile, complex motions require checking the constraints and the limitations of humanoid robots. We use H-Anim (H-Anim 2008), an international standard, as our humanoid model. We adopt the hierarchy tree structure of humanoids (Lee & Shin 1999), use degree of freedom (DOF) (Badler et al., 1993) to describe each joint, define states of humanoid, assign priorities to motions, and check constraints and limitations.

```
[
  { statement1; statement2; }
   // a sequence begin from time 0
  statement3;   // statement3 begins at time 0
  [ { statement4; statement5; }
     // a sequence begin from time 1
    statement6;   // statement6 begins at time 1
  ] : starttime = 1
]
```

Fig. 7. Complex block

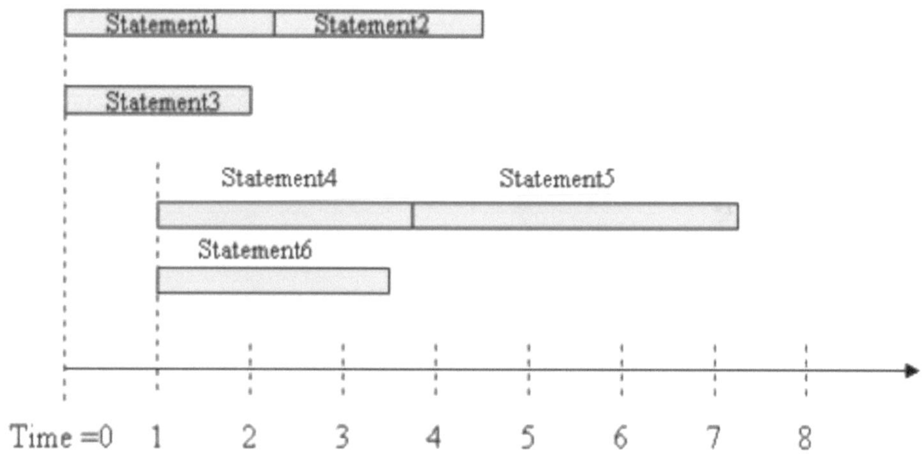

Fig. 8. Complex block time slot

8.1 Degree of Freedom and Joints

In humanoid robots, joints connect segments through attachment frames that are called sites. Each segment can have several sites. Joints connect sites on different segments within the same figure. Each joint can have at most six degree of freedoms (DOF), including three for rotational and three for translational. In our system, figures have only three rotational DOF, which is defined by the corresponding angle around the axis. Some constraints include rotation axis, joint angle, and upper and lower limits of the joint angle.

```
{
    Jack.backwalk(2) : speed = 1;
    [
      Jack.nod(1);
      Jack.walk(2) : speed= 1;
    ]
    Jack.turnright();
    Jack.turnleft();
    Jack.sidewalkright(1);
    Jack.sidewalkleft(1);
}
```

Fig. 9. Sample program

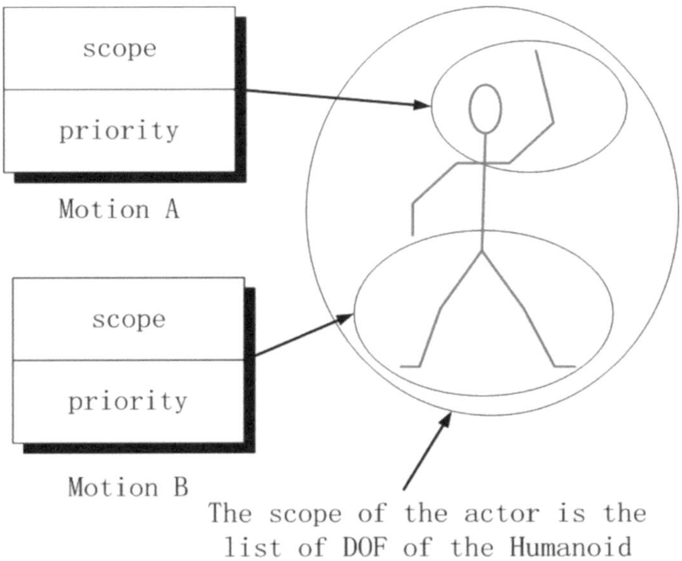

Fig. 10. Scope and priority

8.2 States of Humanoid

States are very important aspects of the humanoid model. Static and recognizable states such as stand, sit, crawl, prone, or supine is defined by the relative positioning of each body part, joint and DOF in our system. Programmers can create their customized states by creating new states, which can be a set of DOF for each affected joint. The same state can be applied on different humanoid models, which makes different humanoids move into certain posture.

We describe a partial state of a humanoid as a body part or a set of body parts associated with certain predefined DOF. Partial states are assigned scope and priority (Figure 10). Scope defines the subset of joints that the motion affected. Priority defines the strength and relative importance of sub-motion with respect to all other motions that are in current or different blocks. The scope and priority allows us to combine partial states to form whole body states or postures, and allows new motions to be generated and synchronized.

8.3 State Transition Approach

After we defined states of humanoid, we can define transitions between two states. Transitions between states (Badler et al, 1994) imply movement from one posture to a new posture. Some transitions are possible and some are not, based on the limitation of humanoid robots. Certain transitions are selected and defined as primitive motions. Some of the motion primitives we defined are walk, run, jump, and turn. Our system takes the motion primitives and automatically generates the complete motion sequence from the initial state to the final state.

Weight	Descriptions
$\omega = 1$	This weighted motion take highest priority over other motions within the same scope. All other existent motions within the scope are transitioned to finish then this weighted motion starts.
$0 < \omega < 1$	This weighted motion is combined with other weighted motions within the same scope, based on their relative weights.
$\omega = 0$	This weighted motion cannot be combined with any other motion within the same scope. The motion is executed after existing motions finished.

Table 2. Combination of Motions Based on Weights

8.4 Motion Combination with Constraints Checking

When two motions intersect with each other in terms of time and/or scope, we need to check the constraints imposed by the limitations of humanoid robots. For example, we cannot require an arm to move forward at the same time require the arm to move backward. When motions are organized in a sequential way, one after anther, and we do not need to check this type of conflicts.

We automated the process of combining parallel motions and resolving possible conflicts between motions. The process uses the predefined scope and priority of each motion. Table 2 shows some examples of priorities that is described here as weights. The weight ranges from 0 to 1 and 1 being the highest. For example, a motion with weight 0 cannot be combined with any other motion within the same scope, while a motion with weight 1 will be executed prior to any other motion within its scope. If two motions have the same predefined weight, then the relative position in the program block is used to determine their priorities. In this case, the first statement has higher priority than the second one. To reduce the complexity of assigned detailed weight, we predefined weight categories and grouped motions into the categories.

8.5 Three Dimensions of Complex Humanoid Motions

We defined three dimensions for complex humanoid motions. There are time, scope, and level. Figure 11 shows an example of their relationships. The time axis shows the beginning and ending time for each motion. The scope axis indicates which joints or body parts are affected by the motion. The level axis shows the hierarchical structure among various motions nested within complex sequential and parallel blocks. The level is used during motion combination process, in which lowest level motions are created first then process to the highest level.

9. Cybele Humanoid Simulation System

We developed a virtual reality simulator for humanoid robots based on our new motion description language and humanoid framework. The simulation system interprets the motions specified by the language and checks the constraints based on the humanoid framework. The system is designed to interact with user inputs and control humanoid robots in real time.

```
{   //Block 1: level 1, attribute 0
··· //statements: BlockID 1, level 1, attribute 0
    [   //Block 2: level 2, attribute 1
    ··· //statements: BlockID 2, level 2, attribute 1
    ]   //Block 2: end
··· //statements: BlockID 1, level 1, attribute 0
    [   //Block 3: level 2, attribute 1
    ··· //statements: BlockID 3, level 2, attribute 1
        {   //Block 4: level 3, attribute 0
        ··· //statements: BlockID 4, level 3, attribute 0
            [   //Block 5: level 4, attribute 1
            ··· //statements: BlockID 5, level 4, attribute 1
            ]   //Block 5: end
        }   //Block 4: end
    ··· //statements: BlockID 3, level 2, attribute 1
        [   //Block 6: level 3, attribute 1
        ··· //statements: BlockID 6, level 3, attribute 1
        ]   //Block 6: end
    ··· //statements: BlockID 3, level 2, attribute 1
    ]   //Block 3: end
... //statement: BlockID 1, level 1, attribute 0
}   //Block 1: end
```

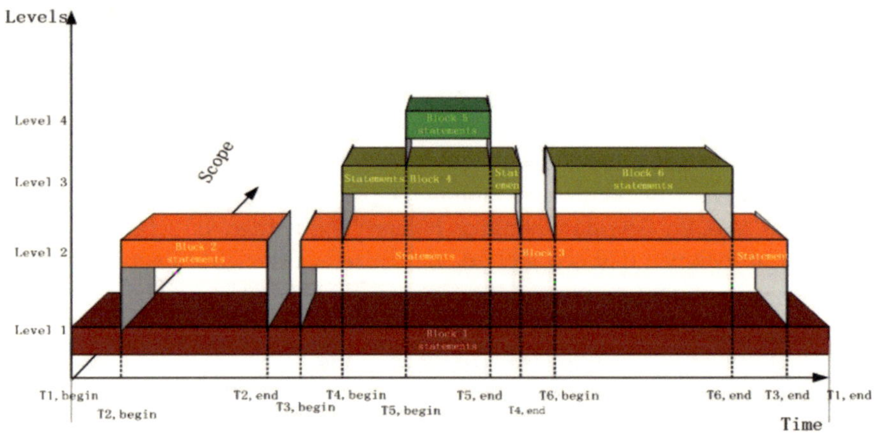

Sequential Block: Attribute = 0
Parallel Block: Attribute = 1

Fig. 11. Three Dimensions of Humanoid Motions

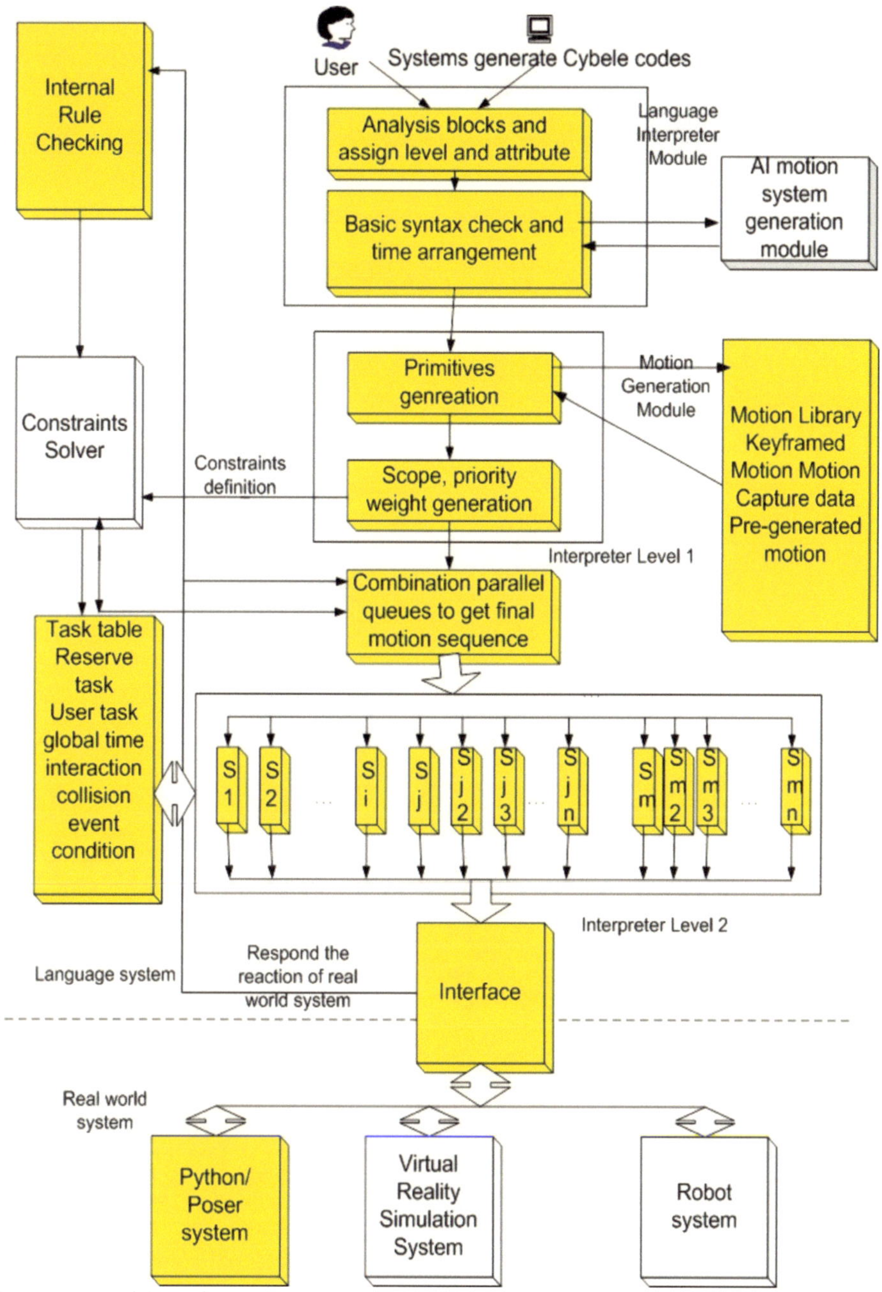

Fig. 12. Humanoid Simulation System Overview

Figure 12 shows an overview of our humanoid simulation system. The system first gets the Cybele program from some consoles or from other interfacing systems. It interprets the program by analyzing various syntax and structures. Then the language interpreter module passes the results to the motion generation module, which creates motion sequences. In this module, all instructions will be broken down into primitives and tagged with level and block information. Every primitive is generated based on parameters and attributes such as duration, speed, and so on. It performs the following functions:

1. Sub-motion generation. The interpreter checks the library to find the proper primitive sequences.
2. Time schedule. A time scheduler schedules the duration for each sub-motion.

Fig. 13. Humanoid Dancing in Virtual Environment

After all the primitives are generated, the motion sequences (labelled S's in the figure) are transferred to the interface. The interface has two functions. The first function is to transmit the elementary motion sequences to various real world systems, which consist of Python/Poser system, Virtual Reality Simulation system, and Robot system (as shown in Figure 12).

The second function of the interface module is to collect the feedbacks and other system parameters from the real world system. These feedbacks are used be the language system to adapt to the current constraints imposed be the environment.

There are two types of constraints in our system. One is an internal rules constraint which is defined by the system and cannot be changed after the environment is created. The other is a rules constraint created by programmers after the environment is created and that can be changed by programmers. All these constraints are resolved by the Constraints Solver module (shown on the right side of Figure 12).

As can be seen from the system overview (Figure 12), this system is quite complex. Currently, we have completed the implementation and testing of the highlighted modules. Our test results show that such a system is viable. One of our tests was to describe dance movements using Cybele. Figure 13 shows one of our simulations of a humanoid dancing in a virtual environment.

10. Conclusion and Future Research

In this chapter we described a description language and framework for specifying humanoid motions and for allowing humanoid robots to learn motor skills through interacting with the environments. The language and framework are unique and encompassing many areas of research interesting to researchers in epigenetic robots. The framework can also serve as an outline of strategies for future research programs in humanoid motion description and motor skill acquisition.

Based on our motion description framework, we developed a new humanoid motion description language called Cybele. New features of the language include complex parallel and sequential blocks, new syntax for motion statements, and built-in constraints for motion combination. The motion description links tightly to the underlying models. We defined a humanoid motion framework to make specification of humanoid motion possible. Then we created a simulation system to implement the framework and to execute the motion descriptions. For our system, we also created a solution to generate complex parallel and sequential motion sequences based on the limitations and constraints of the humanoid body. To solve the motion mixing problem in complex blocks, we presented an approach for synthesizing motions with partial state combinations. In the solution, scope and priority are defined, and primitives are combined to produce final motions with constraints checking. Test results show that Cybele is an effective motion description language, and also show that such virtual reality simulation of humanoid robots is viable.

The proposed simulation system is quite complex, which includes using artificial intelligent techniques to automatically generate detailed motion sequences based on given goal statements. Further work is required to develop such AI motion generation system and to further develop AI methods for solving various constraints imposed by the limitation of the humanoid robots and by the environments. Although our system is designed to interact directly with humanoid robots, we have yet to test the system using real robot. Although our virtual simulation shows such system is feasible, a real humanoid robot will encounter much more constraints when interacting directly with real human environments including interacting with human. Thus, future research would also focus on multi-agent systems in which robots, human, and their environments interacting with each others in real time.

For humanoid robots to be able to effectively function in real environments and interacting with people, they must be able to adapt and able to learn. Most researchers realize this requirement and many are working on various learning methods for humanoids. However, much research remains to be done for humanoid robots to learn motor skills. Although this chapter outlined some strategies including, creating a motion description language, using non-deterministic finite state machine to model sequence of motion states, and using automatic constraint satisfaction to generation motions plausible for the underlying humanoid bodies, much work remains to be done in this exciting area of research.

11. References

Adamson A. (1987). "Calaban", Demonstration at Birmingham University, England, June 1987, http://www.bham.ac.uk/calaban/.

Arsenic, A.M. (2004). "Developmental learning on a humanoid robot," *2004 IEEE International Joint Conference on Neural Networks*, vol.4, pp. 3167- 3172, 25-29 July 2004.

Arsenio, A.M. (2004). *Cognitive-Developmental Learning for a Humanoid Robot: A Caregiver's Gift*, Doctoral thesis, Massachusetts Inst. of Tech., Cambridge, Dept. of Electrical Engineering and Computer Science.

Badler, N. I., & Smoliar, S. W. (1979). "Digital Representation of Human Movement", *Computer Surveys*, Vol. 11, No 1, March 1979.

Badler, N. I., Bindiganavale, R., Granieri, J. P., Wei, S., and Zhao, X. (1994). "Posture Interpolation with Collision Avoidance," *In Proc. Computer Animation*, pp.13-20.

Badler, N., Phillips, C., and Webber, B. (1993). *Simulating Humans: Computer Graphics, Animation, and Control*, Oxford University Press.

Bentivegna, D.C. & Atkeson, C.G. (2001). "Learning from observation using primitives," *IEEE International Conference on Robotics and Automation*, vol.2, pp. 1988-1993.

Brooks, R.A. (2002). "Humanoid robots," *Communications of the ACM*, Special Issue: Robots: intelligence, versatility, adaptivity, Vol. 45, Issue 3, pp. 33-38, March 2002.

Brooks, R.A. (1996). "Prospects for human level intelligence for humanoid robots," Proceedings of the First International Symposium on Humanoid Robots (HURO-96), pp. 17-24.

Brooks, R.A., Breazeal, C.; Marjanovic, M.; Scassellati, B. & Williamson, M.M. (1998). "The Cog Project: Building a Humanoid Robot," *IARP First International Workshop on Humanoid and Human Friendly Robotics*, (Tsukuba, Japan), pp. I-1, October 26-27.

Burghart, C., Mikut, R., Stiefelhagen, R., Asfour, T., Holzapfel, H., Steinhaus, P., & Dillmann, R., (2005). "A cognitive architecture for a humanoid robot: a first approach," *2005 5th IEEE-RAS International Conference on Humanoid Robots*, pp. 357- 362, 5-7 Dec. 2005.

Calinon, S., Guenter, F., & Billard, A. (2007). "On Learning, Representing and Generalizing a Task in a Humanoid Robot," *IEEE transactions on systems, man and cybernetics*, Part B. Special issue on robot learning by observation, demonstration and imitation, vol. 37, num. 2, 2007, p. 286-298.

Calvert, T.W., Bruderlin, A., Mah, S., Schiphorst, T., & Welman, C. (1993). "The Evolution of an Interface for Choreographers", *Interchi*, pp. 24-29.

Causley, M. (1980). *An introduction to Benesh Movement Notation*, ISBN: 0836992806.

Choi, B. (1998). "Automata for Learning Sequential Tasks," *New Generation Computing: Computing Paradigms and Computational Intelligence*, Vol. 16, No. 1, pp. 23-54.

Choi, B. (2002). "Applying Learning by Example for Digital Design Automation," *Applied Intelligence*, Vol. 16, No. 3, pp. 205-221.

Choi, B. (2003). "Inductive Inference by Using Information Compression," *Computational Intelligence* 19 (2), 164-185.

Choi, B., and Chen, Y. (2002). "Humanoid Motion Description Language," *Second International Workshop on Epigenetic Robotics*, pp. 21-24.

Conway, M.J. (1997). *Alice: Easy-to-Learn 3D scripting language for novices*, Ph. D thesis, University of Virginia.

Eshkol-Wachman (2008). Eshkol-Wachman Movement Notation, http://www.movementnotation.com/

H-anim (2008). Humanoid animation work group. http://www.h-anim.org.

Harris, J.A., Pittman, A.M., Waller, M.S., and Dark, C.L. (1999). *Dance a while Handbook for Folk, Square, Contra and Social Dance*, Benjamin Cummings.

Hirai, K, Hirose, M. Haikawa, Y. Takenaka, T. (1998). "The Development of Honda Humanoid Robot", *Proceeding of IEEE Intl. Conference on Robotics and Automation (ICRA)*, pp. 1321-1326.

Honda (2008). Honda Humanoid robot: http://world.honda.com/ASIMO/

Huang, Z., Eliëns, A., and Visser, C. (2002). "STEP: a Scripting Language for Embodied Agents," *Proceedings of the Workshop of Lifelike Animated Agents*, Tokyo.

Hutchinson A. & Balanchine, G. (1987). *Labanotation: The System of Analyzing and Recording Movement*, ISBN: 0878305270.

Kanayama, Y., and Wu, C. (2000). "It's Time to Make Mobile Robots Programmable," *Proceedings of Intl. Conference on Robotic and Automation (ICRA)*, San Francisco.

Kanehiro, F., Hirukawa, H. & Kajita, S. (2004). "OpenHRP: Open Architecture Humanoid Robotics Platform," *The International Journal of Robotics Research*, Vol. 23, No. 2, 155-165.

Kanehiro, F., Miyata, N., Kajita, S., Fujiwara, K., Hirukawa, H., Nakamura, Y., Yamane, K., Kohara, I., Kawamura, Y., & Sankai, Y. "Virtual Humanoid Robot Platform to Develop Controllers of Real Humanoid Robots without Porting," *Proceedings of the 2001 IEEE/RSJ International Conference on Intelligent Robots and Systems*, Hawaii, Oct. 29 - Nov. 03, 2001, pp. 1093-1099.

Lee. J and Shin. S. (1999). "A hierarchical approach to interactive motion editing for human-like figures," *Proceedings of SIGGRAPH 1999*, pages 39-48.

Nakaoka, S., Nakazawa, A., Yokoi, K., & Ikeuchi, K. (2004). "Leg motion primitives for a dancing humanoid robot," *2004 IEEE International Conference on Robotics and Automation*, Vol.1, pp. 610- 615, 26 April-1 May 2004.

Nakaoka, S., Nakazawa, A., Yokoi, K., Hirukawa, H., & Ikeuchi, K. (2003). "Generating whole body motions for a biped humanoid robot from captured human dances," *IEEE International Conference on Robotics and Automation, 2003. Proceedings. ICRA '03*, vol.3, pp. 3905- 3910, 14-19 Sept. 2003.

Nishimura, Y., Kushida, K., Dohi, H., Ishizuka, M., Takeuchi, J., & Tsujino, H. (2005). "Development and psychological evaluation of multimodal presentation markup language for humanoid robots," *2005 5th IEEE-RAS International Conference on Humanoid Robots*, pp. 393- 398, 5-7 Dec. 2005.

Nozawa, Y., Dohi, H., Iba, H., & Ishizuka, M. (2004). "Humanoid robot presentation controlled by multimodal presentation markup language MPML," *13th IEEE*

International Workshop on Robot and Human Interactive Communication, pp. 153- 158, 20-22 Sept. 2004.

Oh, J.H., Hanson, D., Kim, W.S., Han, Y., Kim, J.Y., & Park, I.W. (2006). "Design of Android type Humanoid Robot Albert HUBO," *2006 IEEE/RSJ International Conference on Intelligent Robots and Systems*, pp. 1428-1433, Beijing, Oct. 2006.

Perlin, K, Gikdberg, A. (1996). "Improv: A System for Scripting Interactive Actors in Virtual Worlds", *Computer Graphics Proceeding*, pp.205-216.

Perlin, K., and Goldberg, A. (1996). "Improv: A System for Scripting Interactive Actors in Virtual Worlds," *Computer Graphics*; Vol. 29 No. 3.

Pollard, N. S. (1999). "Simple Machines for Scaling Human Motion," *Eurographics Workshop on Animation and Simulation*, Milan, Italy.

Python (2008). Python. http://www.python.org

Riley, M. & Atkeson, C.G. (2000). "Methods for motion generation and interaction with a humanoid robot: Case studies of dancing and catching," *Proc. 2000 Workshop on Interactive Robotics and Entertainment*, pp. 35-42, Robotics Inst., Carnegie Mellon Univ.

Ryman, R., Singh, B., Beatty, J., & Booth, K. (1984). "A Computerized Editor of Benesh Movement Notation," *Dance Research Journal*, 16(1): 27-34.

Safonova A., Nancy S.P., & Hodgins, J.K. (2002). "Adapting human motion for the control of a humanoid robot," *Proceedings of International Conference on Robotics and Automation*, pp. 1390-1397.

Scassellati, B.M. (2001). *Foundations for a Theory of Mind for a Humanoid Robot*, Doctoral thesis, Massachusetts Inst. of Tech., Cambridge, Dept. of Electrical Engineering and Computer Science.

Schiphorst, T. (1992). "LifeForms: Design Tools for Choreography", *Dance and Technology I: Moving Toward the Future*, pp. 46-52.

Schrand, R. (2001). *Poser 4 Pro Pack f/x & Design*, Coriolis Group.

Sony (2008). Sony QRIO humanoid robot, http://www.sony.net/SonyInfo/CorporateInfo/History/sonyhistory-j.html.

Ude, A., Atkesona, C.G., & Rileyd, M. (2004). "Programming full-body movements for humanoid robots by observation," *Robotics and Autonomous Systems*, Vol. 47, Issues 2-3, 30 June 2004, pp. 93-108.

Yokoi, K. (2007). "Humanoid robotics," *International Conference on Control, Automation and Systems*, pp. lxxiv-lxxix, Seoul, 17-20 Oct. 2007.

Zhou, Y. & Choi, B. (2007). "Virtual Reality Simulation of Humanoid Robots," *IEEE Industrial Electronics Society (IECON) 33rd Annual Conference*, pp. 2772-2777.

2

Emotion Mimicry in Humanoid robots using Computational Theory of Perception

Mohsen Davoudi[1], Mehdi Davoudi[2] and Nima Seif Naraghi[2]
1. *Politecnico di Milano, Milan, Italy*
2. *Eastern Mediterranean University, Gazimagusa, Cyprus*

1. Introduction

In robotics research one of the major issues is focused on the humanoids interaction modalities with humans. Humans have developed advanced skills in interpreting his emotions to the bodily expressions. If similar skills can be acquired by robots, it would allow them to generate behaviors that are familiar to us and thus increase their chances of being accepted as partners in our live.

It has been known that humans can express a lot of information and the emotional state of the performer using body movements. Human has some perception based behavior seen in his bodily expressions. Movements of hands, legs, head, etc show the inner emotion of the person in specific situations subconsciously. Reflecting the bounded ability of the human brain to resolve details of motions, perceptions are intrinsically imprecise.

The main goal of this Chapter is to introduce a novel approach for modeling an emotion-based motion generator for humanoids in order to present emotionally expressive styles of movements and emotion mimicry in different situations using Computational Theory of Perception (CTP).

Emotion mimicry is such an area, where expression of the proper emotion is based on a combination of various quantitative as well as qualitative measurements. In this perspective, a major shortcoming of existing approaches is that, as they are based on bivalent, they don't provide tools for dealing with perception-based information.

Humans have capability to perform a wide variety of physical motions without any measurement and computation. In performing such movements, humans employ perceptions of time, situation, possibility, inner emotions and other attributes of physical and mental objects. Motion characteristics in humans consist of measurement-based information and perception-based emotions. A big percent of human knowledge is perception-based.

The main goal of this Chapter is to introduce a novel approach for modeling an emotion-based motion generator for humanoids in order to present emotionally expressive styles of movements and emotion mimicry in different situations using CTP which contains a special capability to compute and reason with perception-based information.

A fuzzy logic-based analyzer that interprets the linguistic emotions that are common among people into what is called the Generalized Constraint Language (GCL) is introduced in this

chapter. Fuzzy Logic-Based probability theory has the fundamental ability to operate on perception-based information, which bivalent logic-based probability theory does not posse.
A particular GCL [1] is introduced for humanoid robots to categorize data needed for trajectory tracking for each joint of the robot which lead to emotion mimicry. An Artificial Neural Network generates velocity and acceleration for each joint of humanoid using GCL values as well. A concept that plays a key role in emotion mimicry with CTP is fuzzy logic based GCL.

An idea which underlies the approach described in this work is that an emotion may be viewed as a proposition that fuzzy analyzer approximates by means of the intensity and type of emotion. A proposition plays the role of a carrier of information. In our design fuzzy analyzer gets two types of linguistic proposition that are detected from a sentence:

1) The type of emotion such as happiness, anger, kindness, stress, surprise and sadness,
2) The intensity of emotion (very, more or less; extremely,...).

For example I say "move your hands very angry", two words are detected: angry and very.

In the next step, an Artificial Neural Network (ANN) generates velocity and acceleration for each joint of humanoid using GCL variables. GCL is an interface between two portions of this analysis: a) emotion interpreter (the fuzzy analyzer), b) a data generator for physical motions of the humanoid joints. The overall structure of proposed robot motion system is shown in figure1.

The computational theory of perceptions enhances the ability of Artificial Intelligence (AI) to deal with real-world problems such as humanoid interactions in which decision relevant information is a mixture of measurements and perceptions. Motions in an intelligent humanoid are compromise between perceptions of emotions in common sense and measured data for trajectory tracking in humanoid joints.

In order to deal with perception-based analysis of the motion, an array of tools centering on Computing with Words and Perceptions (CWP), Computational Theory of Perceptions (CTP) and Precisiated Natural Language (PNL), are required[3].

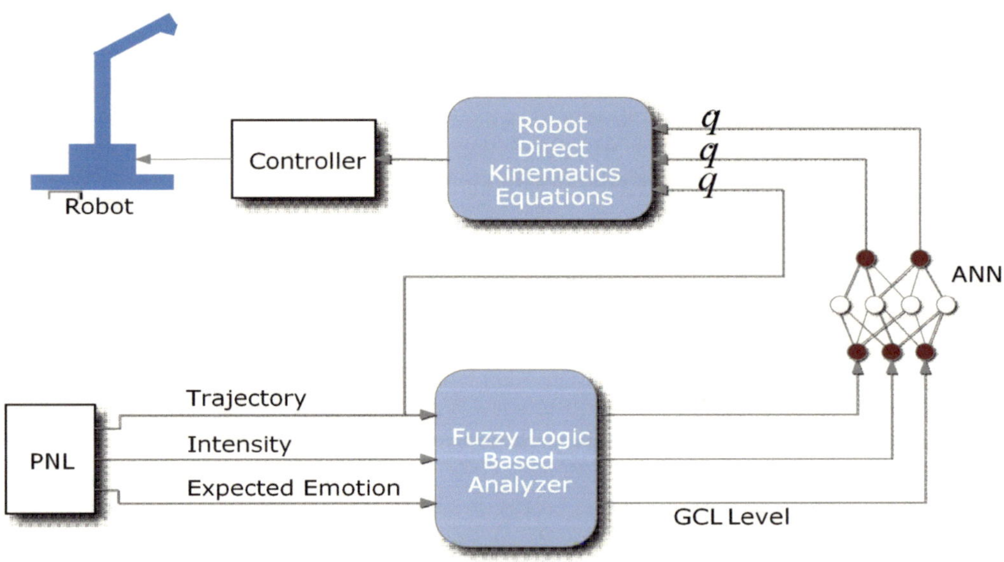

Fig. 1. The overall structure of intelligent robot motion system

2. Generalized Constrants for Intelligent Robot Motion

CW (Computing with Words) serves two major purposes in intelligent robot motion generation: a) provides a machinery for dealing with emotions in which precise information is not available, b) provides a machinery for dealing with emotions in which precise information is available, but there is a tolerance for imprecision which can be exploited to achieve tractability, robustness, simplicity and low solution cost.
Precisiated Natural Language (PNL) is a sublanguage of precisiable propositions in Natural Language (NL) which is equipped with two dictionaries: 1) NL to GCL, 2) GCL to PFL (Protoform Language) and a set of rules of deduction (rules of generalized constrained propagation) expressed in PFL.

Perceptions in Emotion mimicry are f-granular in the sense that a) the boundaries of perceived classes are fuzzy and b) the values of perceived attributes are granular, with a granule being a clump of values drawn together by indistinguishability, similarity, proximity or functionality of emotions.

$$p = \text{proposition in a natural language} \tag{1}$$

A proposition is a perception if it contains a) fuzzy quantifiers such as many, most, few, etc., b) fuzzy qualifiers such as usually, probably, possibly, typically, generally,... c) fuzzy modifiers such as very, more or less, extremely,… and d) fuzzy nouns, adjective or adverbs such as emotion nouns that is prevalent among people in a culture. We considered a number of inner emotions in this paper: *Angriness, kindness, stress, happiness.*
Generally, the mathematics of perception-based information has a higher level of generality than the mathematics of measurement-based information. The constrained variable, X, may assume a variety of forms. In particular,
X is an n-array variable, $X=(X_1, \ldots, X_n)$
X is a proposition, e.g., $X=$ *very angry*
X is a function
X is a function of another variable, $X=f(Y)$
X is conditioned on another variable, X/Y
X has a structure, e.g., $X=Emotion(Stress)$
X is a group variable. In this case, there is a group, $G[A]$; with each member of the group, $Name_i$, $i=1, \ldots, n$, associated with an attribute-value, A_i. A_i may be vector-valued. Symbolically

$$G[A]: Name_1/A_1+\ldots+Name_n/A_n. \tag{2}$$

$G[A]$ is a *relation* (r) and X is a generalized constraint [4].

<u>Possibilistic</u> of emotion mimicry (r=blank) is written such as equation 3.

$$X \text{ is } R \tag{3}$$

which R playing the role of the possibility distribution of X.

For example:

$$X \text{ is } [a, b] \tag{4}$$

means that *[a, b]* is the set of possible values of X. Another example:

$$X \text{ is ``}happy\text{''} \tag{5}$$

In this case, the fuzzy set labeled *happy* is the possibility distribution of X. If μ_{small} is the membership function of *happy*, then the semantics of *X is "happy"* is defined by

$$Poss\{X=happy\} = \mu_{small}(happy) \tag{6}$$

where *u* is a generic value of X [6]. Probabilistic in emotion mimicry (*r=p*)

$$X \text{ isp } R, \tag{7}$$

which *R* playing the role of the probability distribution of X. For example:

$$X \text{ isp } N(m, \sigma^2) \tag{8}$$

Means that X is a normally distributed random variable with mean m and variance σ^2.
For example, if X is a random variable which takes values in a finite set of emotions $\{e_1, \ldots, e_n\}$ with respective probabilities $p_1, \ldots, p_n$, then X may be expressed symbolically as equation 9:

$$X \text{ isp } (p_1 \backslash e_1 + \ldots + p_n \backslash e_n), \tag{9}$$

with the semantics

$$prob(X = e_i) = p_i, \quad i=1, \ldots, n. \tag{10}$$

For e_i= *happy* probability of *X* is given in equation 11:

$$prob(X \text{ is } happy) = \int \mu_{small}(u) f(u) du \tag{11}$$

Hence, test-score of this constraint on *f* is given by:

$$ts(f) = \mu_{likely}(\int f(u) \mu_{happy}(u)) \tag{12}$$

As other consideration, Bimodal ($r = bm$) is the bimodal constraint is symboled in equation 13,

$$X \; isbm \; R, \tag{13}$$

R is a bimodal distribution of the form:

$$R: \;\; \Sigma_i \; k_i \backslash A_i \qquad , i=1, \ldots, n \tag{14}$$

this means that

$$Prob(X \; is \; A_i) \; is \; k_i \tag{15}$$

Example:

$$prob(Emotion(happy) \; is \; High) \tag{16}$$

is likely

$$\int g(e_4)\mu_{high}(e_4)de_4 \tag{17}$$

or, in annotated form,

$$GC(g) = X / \int g(e)\mu_{high}(e)de \;\; \text{is } R/likely. \tag{18}$$

the test-score of this constraint on f and g is given by

$$ts(f) = \mu_{likely}(\int f(u)\mu_{high}(u)du) \tag{19}$$

$$ts(g) = \mu_{likely}(\int g(u)\mu_{high}(u)du) \tag{20}$$

Basically, we use two forms of X:

1) proposition form expressed as p,

2) function form (measure-based information) expressed as f.

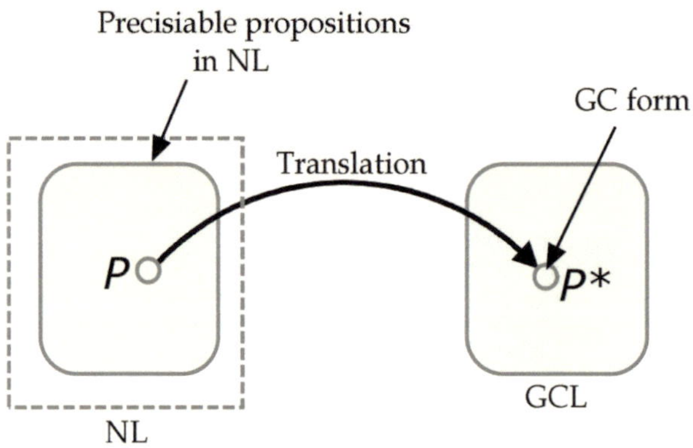

Fig. 2. GCL form of proposition expressed as *p*

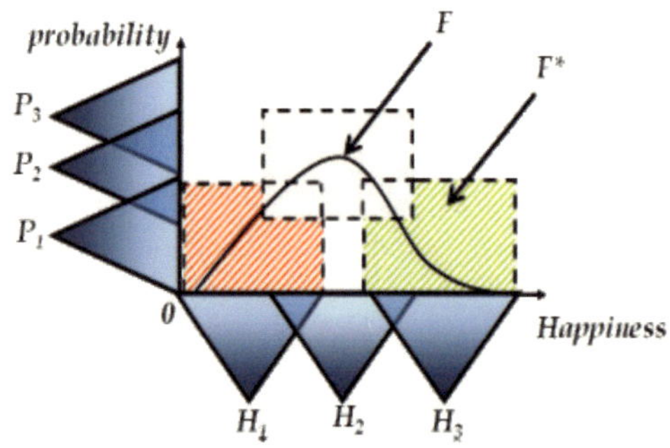

Fig. 3. Fuzzy-graph constraint. *f** is a fuzzy graph which is an approximate representation of function form (measure-based information) expressed as *f*

Considering figure3 we can express the probability of *happiness* as given in equation 16:

$$P\ (happiness) = P1 \backslash H1 + P2 \backslash H2 + P3 \backslash H3 \tag{21}$$

3. Fuzzy Analyzer

We supposed four emotions that are common among people:

$$U=\{e_1,e_2,e_3,e_4\}$$

e_1: *Kindness (k)*
e_2: *Stress (s)*
e_3: *Angriness (a)*
e_4: *Happiness (h)*

$$Decision=\begin{cases} Offensiveness = f\,(e_i), \quad i=1\ldots4 \\ Intensity = g\,(e_i), \quad i=1\ldots4 \end{cases} \tag{22}$$

Membership functions of *offensiveness*:

$$\mu_o(i) = \sup_e(\mu_o(\int_U \mu_{ei}(u) f(u) du)) \tag{23}$$

Membership functions of *intensity*:

$$\mu_i(i) = \sup_e(\mu_o(\int_U \mu_{ei}(u) g(u) du)) \tag{24}$$

An Example: *John is so much happy today. He wants to play guitar.*
Combined propositions are: (*happy + act of playing guitar*). Deduction with perceptions involves the use of protoformal rules of generalized constraint propagation. Playing guitar is an action that is done using organs specially hands. Therefore considering his inner emotion, his movement is so quick and we can guess his inner emotion (*happiness*) from the motions of his hands. As shown in table 1 and table 2, a phrase is converted to prototype form.

proposition in NL	precisiation
p	*p** (GC-form)
John is very happy	ΣCount (*intensity. happiness /happiness*) is much

Table 1. Proposition interpretation

precisiation	protoform
*p** (GC-form)	*PF(p*)*
ΣCount *(intensity. happiness /happiness)* is much	Q *A′*s are *B′*s

Table 2. Precisiation interpretation

After parsing *p* is converted to semantic parse that may be one of these forms:

1) Logical form
2) Semantic network
3) Conceptual graph
4) Canonical form

Semantic parse can be abstracted to more than one protoforms. In order to computation of GCL form for emotion sets we use fuzzy analyzer as inference motor.

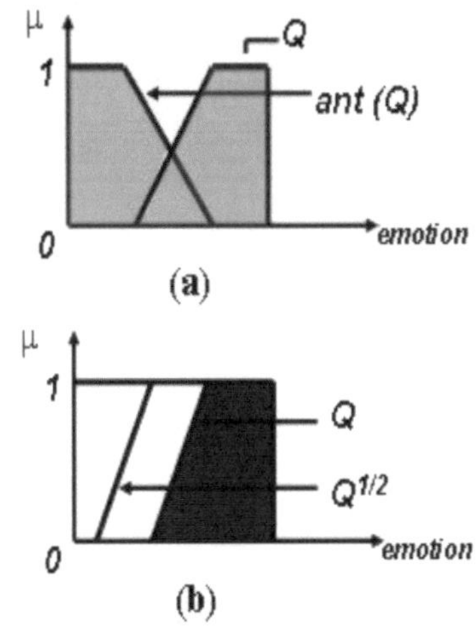

Fig. 4. Emotion definition, A simple fuzzification for happiness (a), culture variation is modeled as shift in the membership function (b).

A simple fuzzification for *happiness* is shown in figure 4 (a). μ is the membership function of the *intensity*. Naturally, Q is *happiness* and *ant(Q)* is *sadness*. The definition of *happiness* may be changed by cultures which is modeled as a shift in the membership functions (see figure4 (b)). The overall scheme of the fuzzy analyzer is demonstrated in figure 5.

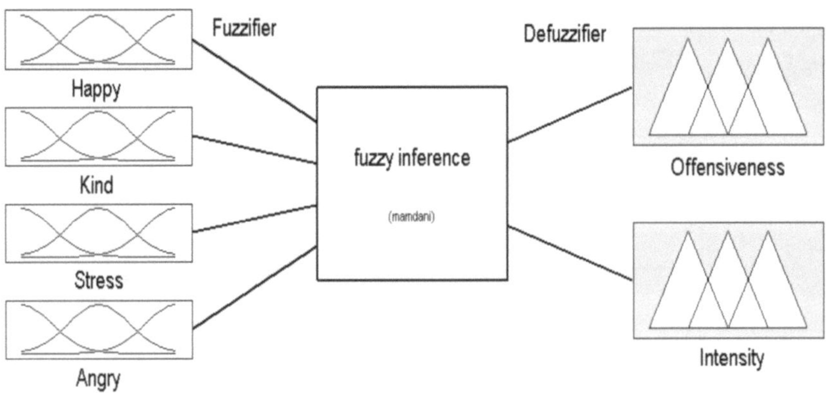

Fig. 5. Fuzzy analyzer

A standard GCL for humanoid robots is defined to categorize data needed for trajectory tracking for each joint which leads to emotion mimicry and intelligent robot motion. As discussed in section2, motion characteristics in humans consist of measurement-based information and perception-based emotions and a big percent of human knowledge is perception-based. Expected mimicry or right motion expression in each situation is the goal that we consider to design humanoid robots. So defining some emotion based rules for fuzzy analysis and creation of deduction database for emotion definition is a base for designing. (see figure 6)

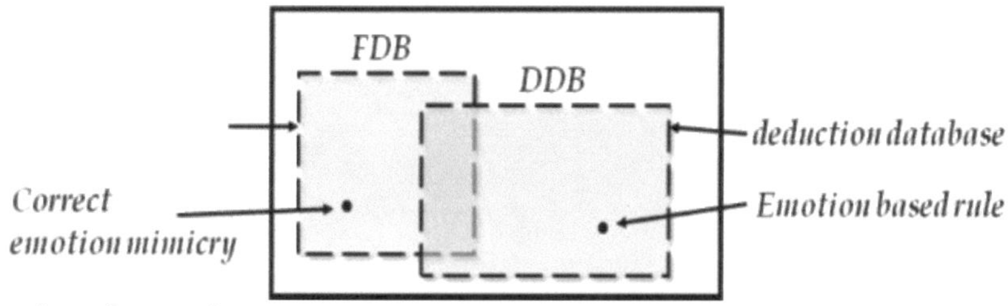

Fig. 6. Database interaction

Information about *trajectory* of each joints in a humanoid, the *time* that motions accrue and the supposed *age* for the humanoid are measurement-based information that can be retrieved from sensors and the motion generator algorithm in a humanoid. They are going to be given to an artificial neural networks (ANN) for motion generation.
The *offensiveness* and *intensity* variables which are GCL form of perception-based information are the other inputs of ANN as well.
The last step for mimicry is the generation of the movements from the GCL variables. Using an ANN with less than 30 neurons GCL form is translated to $\dot{q}, \ddot{q}$ and delay time value for each joint of humanoid. Figure 7 shows the structure of artificial neural networks designed for motion generation.

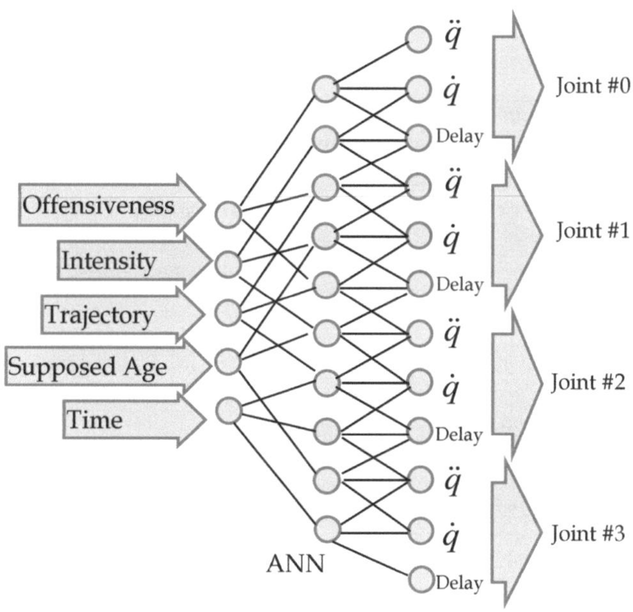

Fig. 7. Artificial Neural Networks

Multi Layer Perceptron (MLP) , RBF and Recurrent Neural Networks are candidate networks for this purpose. For simplicity, a MLP is used as the core of ANN. (See figure 8)

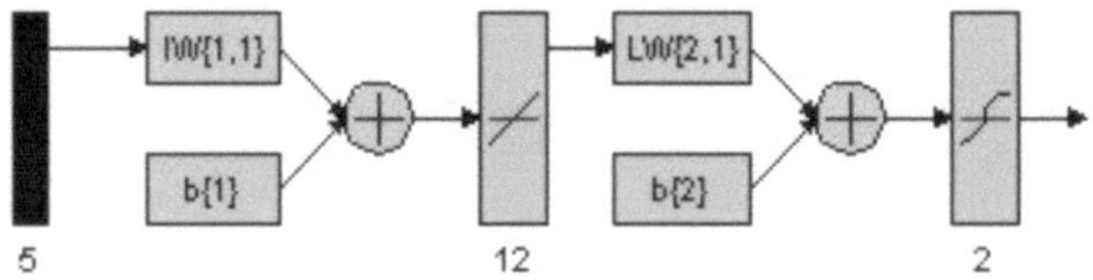

Fig. 8. Multi Layer Perceptron (MLP) as the core of ANN

4. Simulation Results

In this section a computer based simulation for control of a joint in a humanoid robot is demonstrated. Figure 9 shows the state in which the robot asked to move his arm kindly. Three curves with different amplitude and phase shows the intensity of the kindness.

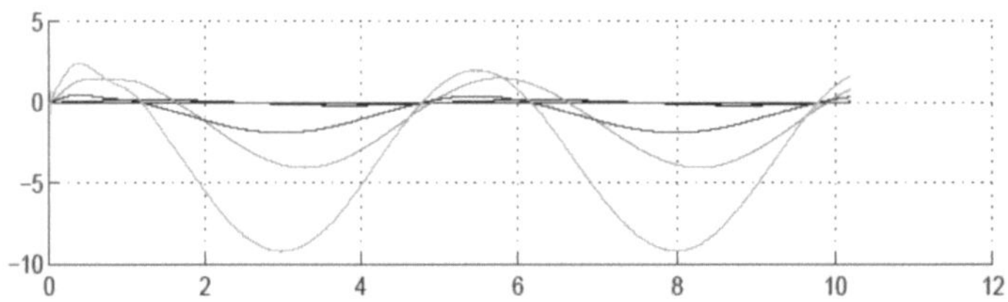

Fig. 9. Joint position control in kindness state

Another emotion that has been modeled in the simulation is *angriness* which is shown in figure10. In this figure the generated motions of an angry person is presented as sharp curves in control signal. Roughly, the *intensity* parameter effects on the control signal as increment of amplitude of control signal (speed of the motion).

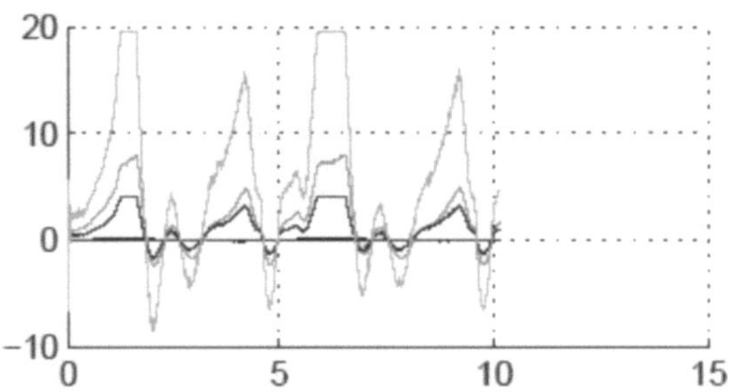

Fig. 10. Joint position control in angriness state

Since the mimicry of *kindness* is logically smoother than mimicry of *angriness* in the joint motions, comparing curves of figure 9 and figure 10 visually we can infer that the joint motions in *kind* state are more smoother.

figure 11 shows the practical result of the joint response to the position control signal tracking the desired position curve. The physical delay of trajectory tracking in humanoid robot's joints may cause some problems in emotion mimicry which can be modeled in the emotion mimicry system as well. Neglecting the differences in desired value and joint response in the joints and delays, the whole concept of the emotion mimicry is protected.

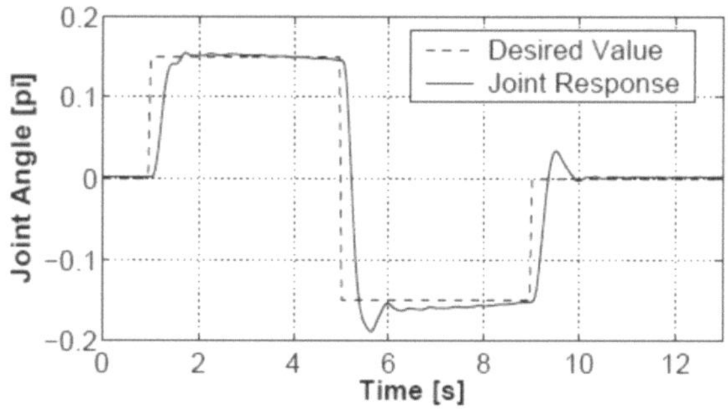

Fig. 11. Joint response to the position control signal

5. Conclusion

Emotion mimicry is such an area, where presentation of the proper emotion is based on a combination of various quantitative as well as qualitative measurements. In this perspective, a major shortcoming of existing approaches is that, based as they are on bivalent, they don't provide tools for dealing with perception-based information. It is not possible to construct a computational theory of perceptions within the conceptual structure of bivalent logic and probability theory in this area. With CTP and GCL using fuzzy logic we solved problems of existing approaches which neural networks and/or fuzzy logic can't solve lonely.

6. References

A. Bardossy, L. Duckstein(1995), Fuzzy Rule-based Modelling with Application to Geophysical, Biological and Engineering Systems, CRC Press

D. Dubois, H. Prade(1992), Gradual inference rules in approximate reasoning, Information Sciences: An International Journal 61, 103-122

D. Dubois, H. Fargier, H. Prade(1993), The calculus of fuzzy restrictions as a basis for flexible constraint satisfaction, in: Proc. 2nd IEEE Int. Conf. on Fuzzy Systems, San Francisco, CA, pp. 1131-1136

D. Dubois, H. Prade(1996), Fuzzy Information Engineering: a Guided Tour of Applications, John Wiley and Sons

D. Filev, R.R. Yager(1994), Essentials of Fuzzy Modeling and Control, Wiley-Interscience

M. Higashi, G.J. Klir(1993), Measures of Uncertainty and Information Based on Possibility Distributions, Fuzzy Sets for Intelligent Systems, Morgan Kaufmann Publishers, San Mateo, CA, pp. 217-232

M. Jamshidi, A. Titli, L.A. Zadeh, S. Boverie, (1997), Applications of Fuzzy Logic towards High Machine Intelligence Quotient Systems, Environmental and Intelligent Manufacturing Systems Series, vol. 9, Prentice Hall, Upper Saddle River, NJ

A. Kaufmann, M.M. Gupta(1985), Introduction to Fuzzy Arithmetic: Theory and Applications, Von Nostrand, New York

E.H. Mamdani, S. Assilian(1975), An experiment in linguistic synthesis with a fuzzy logic controller, International Journal of Man-Machine Studies,1-13.

V. Novak (1999), I. Perfilieva, J. Mockor, Mathematical Principles of Fuzzy Logic, Kluwer, Boston/ Dordrecht

H.T. Nguyen(1993), On Modeling of Linguistic Information Using Random Sets, Fuzzy Sets for Intelligent Systems, Morgan Kaufmann Publishers, San Mateo, CA, pp. 242-246

T.J. Ross(2004), Fuzzy Logic with Engineering Applications, second ed., Wiley

J. Yen, R. Langari(1998), Fuzzy Logic: Intelligence, Control and Information, first ed., Prentice Hall, Berlin

L.A. Zadeh(1975), The concept of a linguistic variable and its application to approximate reasoning, Part I: Information Sciences, 199-249

L.A. Zadeh (1983), A computational approach to fuzzy quantifiers in natural languages, Computers and Mathematics,149-184

L.A. Zadeh(2002), toward a perception-based theory of probabilistic reasoning with imprecise probabilities, Journal of Statistical Planning and Inference 105

3

A biologically founded design and control of a humanoid biped

Giuseppina C. Gini, Michele Folgheraiter, Umberto Scarfogliero and Federico Moro
DEI, Politecnico di Milano
Italy

1. Introduction

During the last decade many advancements in the fields of industrial and service robotics have produced robots that are well integrated inside the industry; they can operate faster and with higher precision in comparison to human beings. Nevertheless if we take a look at the kinematic structure of these systems, it is clear that the actual machines are limited in the mobility and in the number of tasks that they can perform.

This is more evident if we intend to apply those robots in an unstructured environment like home. First at all the robot should be able to move around avoiding obstacles, climbing stairs, opening doors. These movements should also be performed with a certain level of compliance for the safety of the human beings that are in the environment. Secondly, the robots should be able to use tools and other machines designed for human use, and based of the human manipulation and kinematic abilities.

A possible solution for mobility, that is well applied in mobile robotics, is the choice of a wheeled traction system. This usually is a simple manner to move on flat floors, and is efficient from the energetic point of view (during the movement the center of mass acts on a straight line). However it presents important limitations, for example it is not possible for such a robot to overcome obstacles bigger than the wheels dimensions.

Those limitations can be overcome if the robot is equipped with legs, that normally act by increasing the robot's DOF(Degrees of Freedom). Many studies were conducted on legged robot in order to improve the efficiency and stability during walking.

A pioneering contribution was done (Takanishi et al, 2004) at the laboratories of Waseda University (Tokyo). Several other modern robots are designed to walk and behave like humans (Hashimoto et al, 2002)(**3**) but until now the efficiency of the human gait is still far from being reached. In this sense, the work of McGeer (McGeer, 1990) can be considered exemplar. His passive dynamic walker made a stable gait without close position control, considering the walking motion as a natural oscillation of a double pendulum; and this is actually how humans seem to walk (Gottlieb et al, 1996) (Kiriazov, 1991). His results inspired many other works, such as the stability analysis (Garcia et al, 1998) and the physical implementation (Wisse et al, 2001) (Kuo, 1999)(Collins et al, 2001) of several prototypes.

In this paper we present LARP (*Light Adaptive-Reactive biPed*), our humanoid legged system, with the aim to explain how the mechanical design makes the robot able to adapt to the real operating environment. Our aim was to create a system that could represent a good model of human lower limbs, in order to understand how the natural walking motion is achieved and how it can be implemented in a humanoid robot. For this reason, we adopted anthropomorphic feet, knees and a mass-distribution similar to the human limbs.
According to (McGeer, 1990)} we designed an actuation system that can take advantage of the natural dynamics of the link. In addition, studying the results we got from our controller we found several similarities with the assumptions of the Equilibrium Point Theory. This is a widely debated theory, formulated in 1965 by A. Feldman (Asatryan, 1965) (Asatryan and Feldman, 1966 a) (Asatryan and Feldman, 1966 b) and still in evolution. This theory proposes that the segmental reflexes, together with the muscle-skeletal system, behave like a spring. Movement is achieved just by moving the equilibrium position of that spring (Latash and Gottlieb, 1991) (McIntyre and Bizzi, 1993) (Gottlieb et al, 1989) and this is actually how our actuator, provided with visco-elastic elements (Scarfogliero et al, 2004 a) performs the movement.
In the following Sections we concentrate our attention on the robot design and architecture, with particular emphasis on the knee, which presents several similarities to the human articulation, and the foot, developed with two passive degrees of freedom. Then we illustrate the spring-damper actuators and joint controller. We present the robot simulator and the results obtained in computing the energy consumption of LARP. Finally we shortly describe the controller for the gait. The last section outlines the conclusions we can draw from our work and proposes future developments.

2. The mechanical architecture

The LARP project started in 2003 in the Artificial Intelligence and Robotics Lab of DEI, at Politecnico di Milano. It soon provided a 3D model developed in ADAMS and a prototype, as shown in Figure 1. LARP has twelve active degrees of freedom plus two passive degrees for each leg; it is ninety cm tall and weights less than five kg. It is entirely made by pieces cut out from a polycarbonate sheet with the laser cutting technology; this material has a good strength-weigh ratio, and can be widely deformed before breaking.

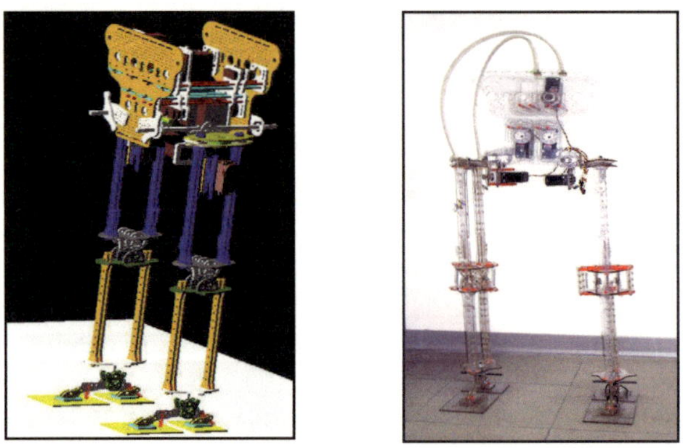

Fig. 1. The LARP robot. (a) the ADAMS model, (b) the mechanical prototype with some actuators installed.

Figure 2 shows the organization of the twelve degrees of freedom in the robot. The range of motion of each joint is similar to that of humans during normal walking.

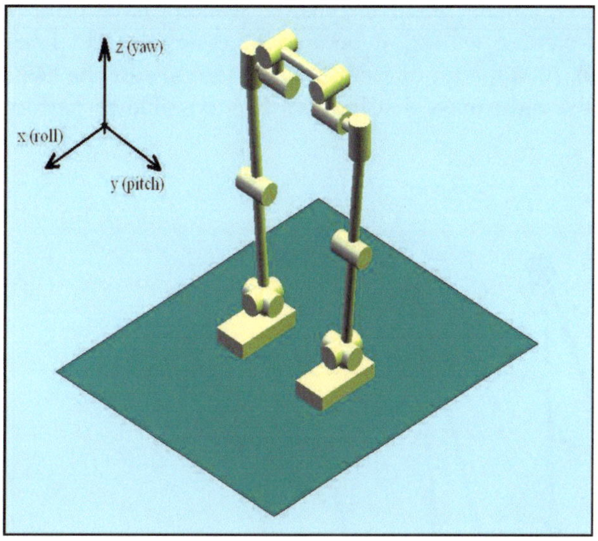

Fig. 2. The organization of the twelve degrees of freedom in LARP.

Each foot has two passive degrees of freedom to ensure a reliable base during the whole stance phase. Joint torques are provided by servo motors located in the upper part of the robot. The transmission is performed by a simple system of cables and levers. The servo motors are equipped with a spring and a damper to permit the control of joint stiffness.

The mechanical design of LARP focused on role and function of knee and foot.

Regarding the knee functions, the most obvious is lifting the shank for the foot clearance. In practice, if that was the only purpose of that joint, an hip articulation could make the job. Using stiff legs could actually simplify the motion and the robot structure (examples of this kind of robots go back to the simple Fallis's toy (Fallis, 1888) to the various 3D biped robots of MIT LegLab. In practice, however, the knee has several important functions in the walking dynamics, that suggest the introduction of the knee articulation.

Also the foot has important consequences on the dynamics of the step, and deserves more investigations. Usually legged robots have simple flat feet, or no feet al all; the foot in biped instead is flexible and accounts for important roles.

3. The design of an anthropomorphic knee

3.1 The importance of the knee in bipeds

It is well known that it is possible to walk with stiff legs, however it is also known that knee articulation is fundamental for the gait stability and energy efficiency. The story of artificial bipeds is plenty of simple prototypes walking without anthropomorphic knees. The importance of knee actuation became more clear when considering the gait efficiency. In 1990 Tad McGeer published "Passive Dynamic Walking" (PDW), where he demonstrated how it is possible to exploit the mass distribution of the robot to make it walk on a shallow slope without actuation (McGeer, 1990). The prototype was exploiting the gravity force to swing the leg forward, exactly as a double pendulum would do. The only power needed was the one necessary to shorten the leg in order to create foot clearance during the swinging motion. Today, several passive dynamic walkers have been developed, but in order to have a fully-passive walker, it became necessary to add knee joints (McGeer, 1990)(Wisse et al, 2002). As a matter of facts, this joint empowers the swinging motion due to gravity, and with the right mass distribution, it is possible to perform a fully-passive pace.

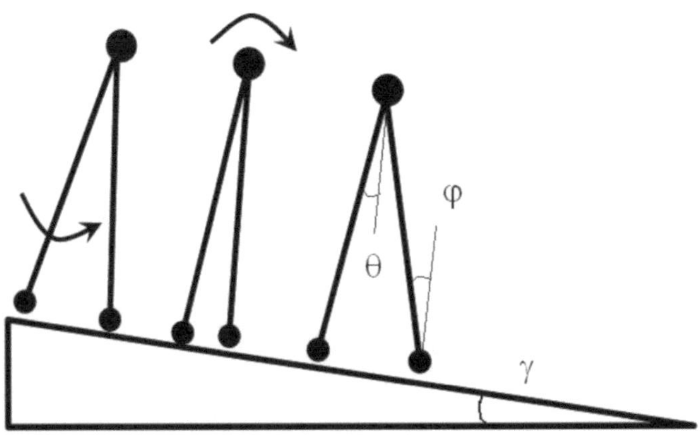

Fig. 3. With the right mass distribution of the robot it is possible to make it walk on a shallow slope without actuation, as demonstrated by Passive Dynamic Walkers

Apart from PDW, the knee is fundamental to ensure energetic efficiency. Let's consider a robot with straight legs. In this case the foot clearance would have to be created by an additional pelvic tilt. This means a reduced step length and a bigger energy consumption, as the pelvis is the heaviest part of the body while knee stretching just lift the foot. This has a big influence on walking efficiency (Koopman et al, 2001) Another effect of straight legs would be that, during the step, the double support time is decreased, on behalf of the single support time. As the former is the most stable position during the step, the straight leg walking is more critical from the stability point of view. So knee-walking needs less energy to ensure gait stability.

The knee is important also during the stance phase, while the supporting leg remains straight. In this case, while the swinging leg moves forward, the knee of the stance leg has to counteract the inertial load generated by gait motion, as shown in Figure 4. In this case the knee can be exploited to store energy, acting as a spring. This energy can be used to pull the hip forward, increasing the step length and the foot clearance. Using a force control to actuate the knee, as we discussed in (Scarfogliero et al, 2004 b) it is possible to store energy, exploiting in this way the natural dynamics of the walking motion. The same happens in humans: during the stance phase, the knee bends a bit, storing energy as a spring would do. This energy is then released to empower the hip forward motion, with a relevant increase in the step length and foot clearance. The result is that the gait performs, with the same energy consumption, a more stable walk. This behavior was also underlined by simulations on a PDW, the robot Mike, of the Delft University of Technology (Wisse et al, 2002).

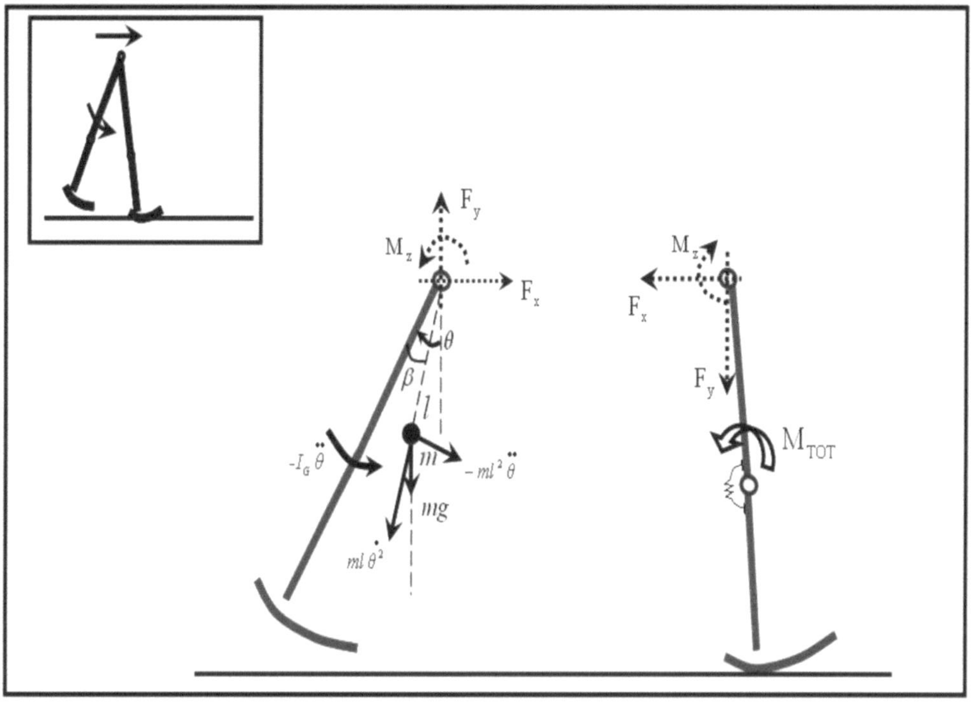

Fig. 4. During walking, the knee of the stance leg has to counteract the inertial loads due to the swinging motion.

3.2 The design of the knee joint

Regarding the knee structure the most obvious and more adopted solution in robotics is a simple pin joint. Usually the motor is applied directly to the joint, but there are also some examples where, for mass-distribution reasons, the motor is placed in the upper part of the robot (Pratt et al, 2001).

Looking at the prosthesis field, we find a completely different approach. Here the knee reveals its crucial importance, not only related to the prosthetic issues, but also for the walking motion. Passive prosthesis have to perform the knee bending using inertial torque generated by the forward acceleration of the thigh, in a similar manner as in passive dynamic walking. In addition, for obvious safety reasons, during the stance phase the knee has to be locked.

Today, prosthetic knees are build using multi-axial mechanisms. The characteristic of these mechanisms is that during the motion, the center of rotation *cr* is not fixed, as in a pin joint, but moves along a trajectory that depends on the mechanism structure. As the stability of the knee during the stance phase strongly depends on the position of *cr*, varying the mechanism proportions, it is possible to have different *cr* trajectories with different stability properties, as illustrated in Figure 5.

Fig. 5. Scheme of a four-axial mechanism for a prosthetic knee. The dotted line represents the path followed by the centre of rotation cr. The big dot represents the cr when the knee is extended. Placing this point more backward, results in a more stable stance phase

For LARP we designed a special joint based on the human articulation. If we sketch the human knee from an engineering point of view, we would define it as an hybrid juncture, similar to a compliant joint but with rolling surfaces. This structure is very close to the compliant rolling-contact joint designed by (Jeanneau et al, 2004) and illustrated in Fig. 6.

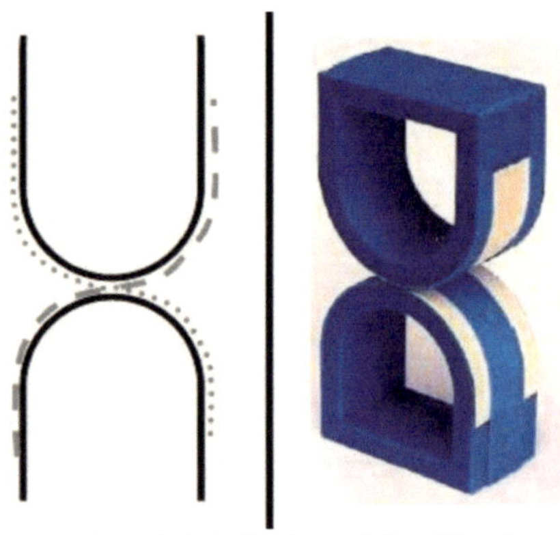

Fig. 6. The compliant rolling-contact joint designed by Herder. Reprinted with permission from J. Herder (Jeanneau et al, 2004)

Our joint is composed by two circular surfaces rolling on each other. Flexible bands constrain the joint, leaving only one degree of freedom, i.e. the rotation along the joint axis. During the motion, the tendons wrap on a surface or on the other, letting the joint rotate without scratch. This significantly reduces friction.
Critical in this kind of joint are the torsional stiffness and the rigidity against external forces. This issue is fundamental for the biped robots, where the knee is subject to high torsional and flexional disturbances. To solve this aspect, we strengthened the joint, designing the articulation shown in Fig. 7. Instead of flexible bands, we used three Coramide strings that can all be independently fastened. This makes the articulation more firm as well as allows a

fine joint calibration. In addition, we added two springs, which increase the contact force between the circular surfaces. Connecting the spring ends at the center of curvature of the two profiles, results in a constant spring-length - equal to two times the radius of the profile. In this case no torque interferes with the joint rotation. Anyhow, it is possible to arrange the springs in a way that they force the joint to rotate to a particular equilibrium position.

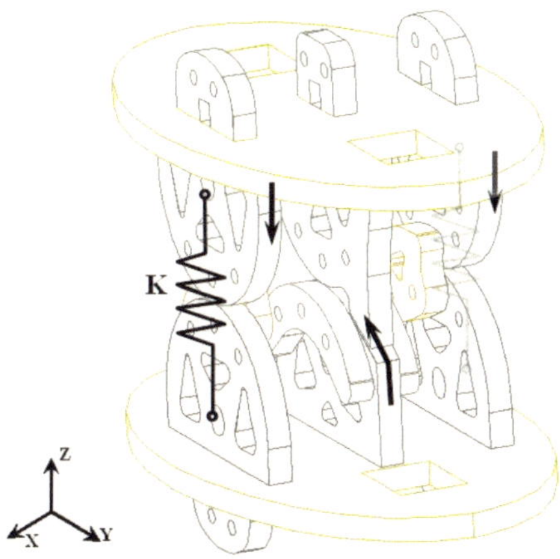

Fig. 7. The knee joint designed for our robot. The arrows show the way the tendons are wrapped.

In the knee we arranged the springs as shown in Fig. 8, so that during the rotation, the spring length has a maximum for a given θ (where θ is the angle that the shank forms with the vertical). In Fig. 8, γ represents the angle between the shank axis and the force direction. When γ is zero, the springs torque is zero too. This permits to find one equilibrium position, in particular, an instable equilibrium position, as the spring is at its maximum extension. Attaching the spring forward, with ψ <0, we can both help knee bending during leg swinging and knee stretching at the end of the step.

$$\gamma = \frac{\theta}{2} \qquad \gamma = \frac{\theta - \psi}{2}$$

$$\bar{\theta} = 0 \qquad \bar{\theta} = -\psi$$

Fig. 8. We can exploit the action of the springs to impose a suited torque on the joint. In particular, it is possible to generate the two positions of instable equilibrium to favour knee bending and knee stretching.

Adopting this kind of joint for the knee articulation has several advantages with respect to a pin joint. The first consideration is about energy efficiency. A reduced friction at the knee not only reduces knee actuation, but can influence the whole gait. As pointed out in the previous paragraph, using elastic actuators or even a passive knee, the leg can be bent exploiting inertial forces due to hip actuation. In this sense, efficient knee joint is fundamental not to demand higher hip torque. Another aspect that strongly characterizes this compliant joint is that the center of rotation *cr* is not fixed, as in a pin joint, but moves upward and backward during the rotation, as illustrated in Fig. 9.

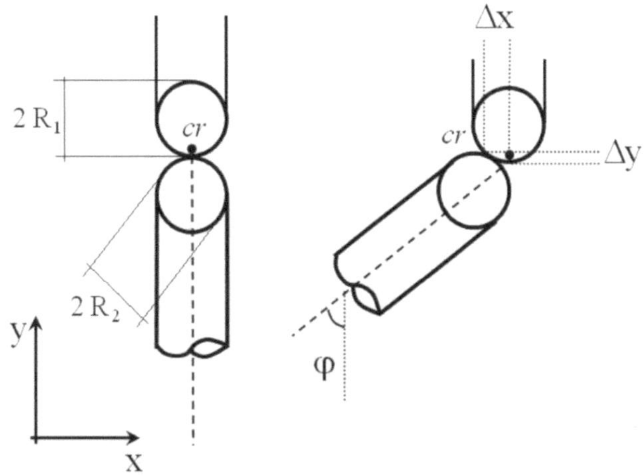

Fig. 9. When the leg is bending, the center of rotation moves upward and backward, according to the ratio between R1 and R2, the radius of the rolling surfaces.

This motion increases the foot clearance necessary to swing the leg, and the shank active rotation can thus be reduced. The effect is not only on energy consumption - i.e. the knee could be passive in some robots - but also on the gait stability. Actually the inertial load of knee-bending and knee-stretching is one of the most important in the dynamics of walking. This is also the reason why the foot must be designed as light as possible, as described in the next paragraph.
Regarding the radius of curvature of the two surfaces, we can look for an optimal design to maximize the foot clearance during the rotation. We can consider that if one contact surface (for example the upper one) has radius infinite or zero, the upward translation is null during the rotation. This means that there must be a finite rate value of the two radius that maximizes the upward motion. Considering that the two surfaces are in contact without slipping, the upward motion Δy, as in Fig. 10, is so computed:

$$\alpha R1 = \theta R2 \tag{1}$$

$$\Delta y = R1(1 - \cos(\alpha) = R1(1 - \cos(\theta\ R2/R1) \tag{2}$$

If R1 is fixed and we vary R2, the maximum can be found quite straightforward given θs:

$$R2 = (\pi / \theta s)R1 \quad (3)$$

θs is the angle of the bent knee in the instant the foot is closer to the ground.
According to this simple analysis, the radius R2 should be longer than R1; the bigger the ratio between the two radius, the smaller the shank rotation, till having a flat surface rolling on the upper one.

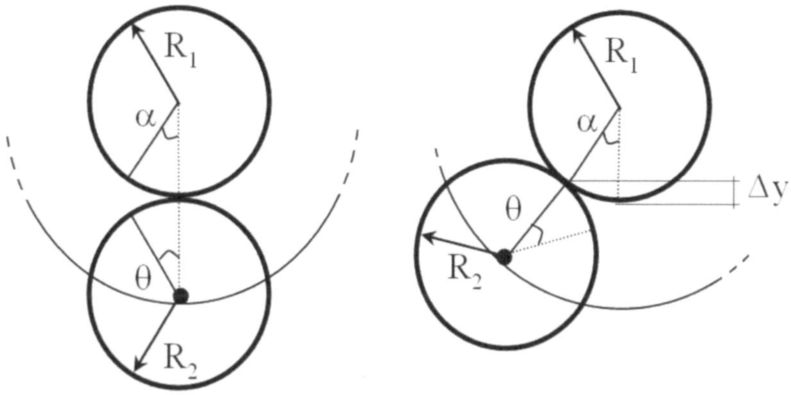

Fig. 10. The upward movement can be expressed as a function of R1, R2, and θ due to the constrain of rolling without slipping.

4. The role of the foot

4.1 The importance of foot in stability and efficiency of walking

The foot is probably the most challenging part for a biped robot to be anthropomorphic. Not only from the sensory point of view, but for the unique combination of mobility and lightness.
First of all, the foot inertia must be negligible with respect to the leg inertia. There are several evidences for this. One reason is the energy efficiency. To understand that, try to run or kick wearing heavy boots. During the swing-phase, the torque needed to move the leg forward is mainly due to inertial loads, that highly depend on the foot weight. In addition, these loads would act on the hip and the stance leg; their impact on the stability is more critical when the ratio body weight/foot weight is low.
But the foot does not only affect the dynamic balance. It is fundamental for stability to keep the center of mass as high as possible. At the beginning of the stance phase, the biped robot can be considered as an inverted pendulum, both in the fore-aft plane and in the frontal plane. Placing the center of mass higher increases the inertia of the pendulum respect to the hinge. It is well known that this implies slower changes respect to the initial position and thus a wider stability respect to external disturbances (Sardain, 1999). As already pointed out about the knee, stability and energy efficiency are strictly related. A more stable gait requires less motor action to counteract disturbances (Maloiy et al, 1986).
Another aspect that characterizes the human foot is its mobility and elasticity. (Ker et al. 1987) found that the foot behaves like an elastic body, returning about 78% of the energy in its elastic recoil. During running, the arc of the foot stores and returns 17% of the energy the

body loses and regains at each footfall, while till the 35% of this energy is stored and returned by Achilles tendon.
The foot mobility of course has a big influence on the whole kinematics and dynamics of the motion, especially on the ankle. In particular, during the stance phase, the contact point moves from the heel to the toe, and the foot is rotated before the toe-off. The position of the contact force plays a very important role in determining the joint torques, thus the energy consumption. As in normal walking the ground reaction is much higher than inertial forces, in first approximation, we can consider only this force acting on the stance leg (Vaughan, 1996). From this point of view, the bigger the arm between the joint and the contact force, the bigger would be the torque needed. In order to minimize energy consumption, while walking we naturally pose the leg joints close to the line of action of the contact force (Saunders et al, 1953) (Alexander, 1992). For this reason it is important to have a foot that let adapt the position of the ankle, and thus the other joints, without losing grip.
This aspect is particularly relevant at toe-off, when only a small region of the foot is in contact. Also here the mobility and elasticity of the foot plays a very important role (Doneland et al, 2002) (Kuo, 1998).
Fig. 11 shows a simple biped model at hell-strike: the rear leg is in the stance phase, and the fore leg is about at foot-fall. The energy loss at the impact depends on the vertical velocity of the center of mass *cm*. The ideal situation is when the *cm* velocity is parallel to the ground, and the legs simulate the wheel (McGeer, 1990). In normal walking, without toe-off the motion of the *cm* is rotational along the contact point of the stance leg. This means that at foot fall there is a component of *cm* vertical velocity that causes impact loss. Using toe-off, this component can be significantly reduced, resulting in a more efficient and smooth gait. (Kuo, 1998) figured out that providing all the energy necessary for walking by the toe-off muscle instead of the hip reduces the energy cost by a factor of 4.

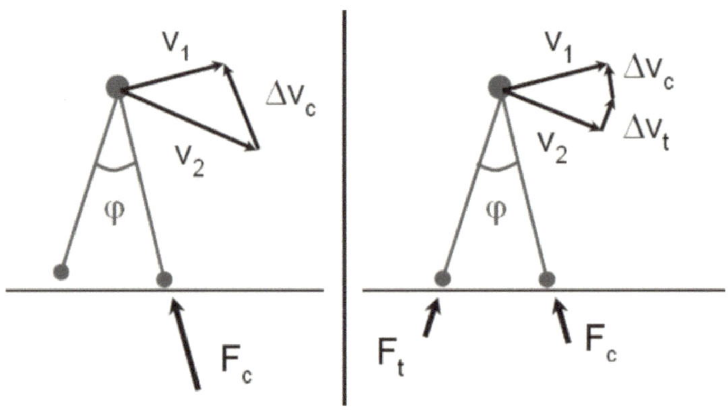

Fig. 11. Consider v1 and v2 as the velocities of the *cm*, respectively before and after heel-strike, while Fc and Ft are the ground reaction forces. With toe off (on the right) the *cm* vertical velocity is reduced, and the gait is smoother and more efficient.

4.2 The anthropomorphic foot

Nowadays, almost all the biped robots adopt a flat foot, with relatively heavy dampers to smooth the heel-strike; a noticeable exception is the design of toe joints proposed in (Sellaouti et al, 2006). The mobility of human foot is very difficult to reproduce, also because, for walking stability, it is fundamental to keep the foot light. In the previous paragraph, we underlined that the key issues for the foot design are:

- The possibility to change ankle position without losing grip. This is a key issue for energy efficiency (Alexander, 1992)
- A good elasticity to store and release part of the energy lost at footfall. Also a good damping is required to smooth the impact occurring at every step
- The capacity to adapt to different ground situations without losing grip in different step phases, as at toe-off

Using a flat foot implies that the ankle position is fixed during the whole stance phase and, at toe-off, the contact is reduced to the foot edge as in Fig. 12. On the other hand, a flat foot is probably the simplest design that can be conceived, and ensures a big base on which to lean during the stance phase.

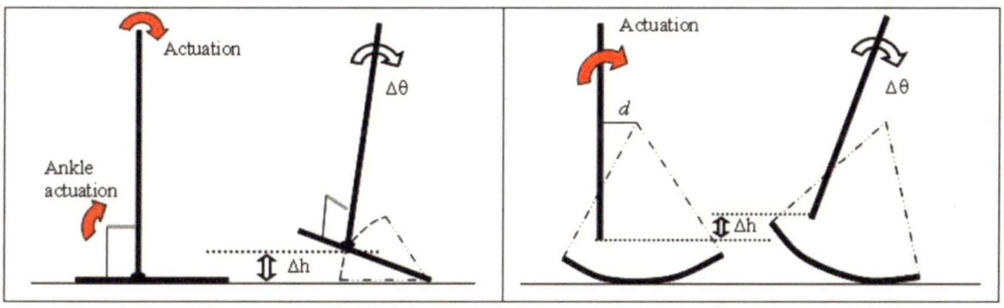

Fig. 12. A flat foot compared to a circular foot

Another type of simple foot profile, adopted mainly on passive dynamic walkers, is the round foot. The advantage of this kind of foot is that the ankle joint is moved forward during the rotation, minimizing in this way the torque needed at toe-off. The drawback of the round profile is that the contact surface is reduced to a thin area. That is way this kind of foot is mainly adopted on 2-D bipeds.

Thus, our goal was to develop a foot with the right trade-off between mobility and stability, keeping at the same time the structure as light as possible. So we adopted performing materials, mainly polycarbonate and carbon. Then, we designed the human-foot structure with a two-dof device, shown in Fig. 13. The foot has one passive degree of freedom that represent the heel, an arc, and another passive dof for the toe. In addition, we inserted an artificial Achilles tendon between the heel and the arc.

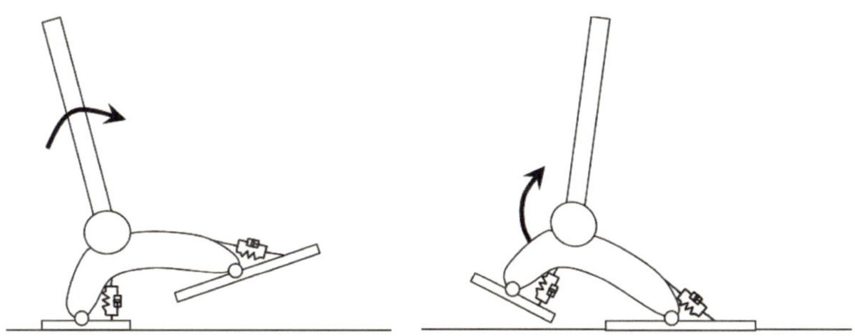

Fig. 13. The foot of LARP, developed to mimic the human one. It has two passive degrees of freedom, with a spring-damper system to smooth the heel-strike.

The articulations in the foot play an important role in determining the gait kinematics and dynamics. As shown in Fig. 13, at heel-strike and at toe-off, the ankle position is not constrained in one fixed position. This gives the ankle an addition degree of freedom, which makes it possible to minimize energy consumption as stated above.
Generally speaking, during the stance phase the contact position moves from the heel to the toe. With our foot, the center of rotation follows the same motion. This means that the lever arm of the ground reaction force is already reduced respect to a flat foot, where the ankle and the center of rotation are constrained in the same fixed point. Moreover, the foot keeps a firm base to lean even at toe-off, when the ankle is moved forward and upward for knee-bending. In this way the double support time - the time when both feet lean on the ground - can be increased, resulting in a more stable walk.
For simplicity, the foot proportions have been chosen similar to the human foot. Anyway, it is possible to optimize the arc proportions, which represent for the ankle the arm of the contact-force at heel-strike and toe-off, according to stability or efficiency criteria.

5. Actuators control

5.1 The spring-damper actuator

The twelve actuated degrees of freedom are actuated by an elastic actuator. The actuator is composed by a servo motor (a big servo with 24 kg cm torque), a torsional spring and a damper, as illustrated in Fig. 14. The resulting assembly is small, lightweight and simple. Using a spring between the motor and the joint let us have a precise force feedback simply measuring the deflection of the spring. The resulting actuator has a good shock tolerance; fundamental in walking, as impacts occur at every step. In addition, we can exploit the natural dynamic of the link storing energy in the spring.
Similar actuators, with a DC motor and a spring, have been successfully used in biped robotics by (Pratt et al. , 1995) (Takanishi et al, 1997)

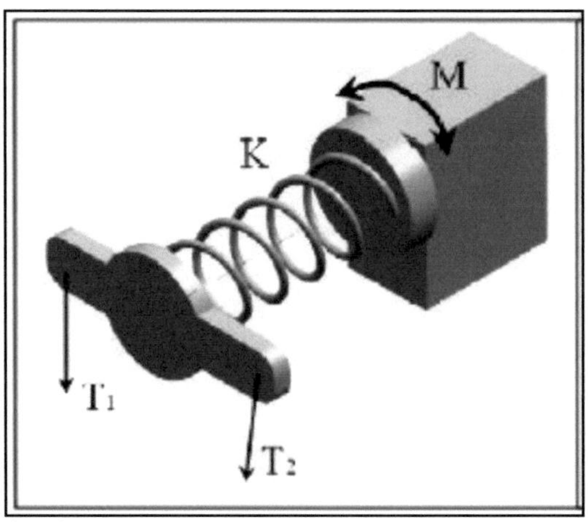

Fig. 14. The schema of the elastic actuator.

The choice of the servos and the materials was made basically on cheap and off-the-shelf components. The main characteristic of this actuator is that the joint stiffness is not infinite,

as it is in servo motors, and it can be changed in real time despite the constant stiffness of the spring. This has been achieved through a right choice of spring-damper characteristics and thanks to an intuitive control algorithm.
Let define the joint stiffness kg as:

$$kg = Me/\varepsilon \tag{4}$$

where Me is the external load and ε is the position error.
A first prototype of our actuator was composed by two motors and two springs, working as agonist and antagonist muscles in humans. This let us to vary the joint stiffness even when no external load is acting, pre-tensioning the joint. With only one motor and one spring, the initial stiffness of the joint is fixed by the spring constant, since the motor needs some time to tension the spring and counteract the external torque. Also, in this conditions, the presence of the damper in parallel to the spring permits to avoid high initial errors due to rapidly varying loads.

The damping factor can be chosen constant, at its critical value $\xi=1$.

$$wn = \sqrt{(kg/I)} \text{ and } d = 2\,\xi wnI \tag{5}$$

or can be varied during motion, in order to save motor torque and make the system faster. In the following paragraph we present the first option.

5.2 The control algorithm for a fixed damping factor

The spring-damper actuator can be used in a torque control loop: the high-level controller assigns the torque to be delivered and, measuring the spring deflection, the low-level regulator makes the actuator perform the task.
A way to assign joint torques is the Virtual Model Control (Pratt et al. 2001). In this approach, the controller sets the actuator torques using the simulation results of a virtual mechanical component. In such a manner the robot can benefits of the component behavior without having it really.
In other classical approaches (Kwek et al, 2003) the calculation of the joint torques is based instead on the dynamic model of the robot, usually complicated and imprecise. Indeed the biped robot can be formalized with a multi-input-multi-output (MIMO) non linear system, that sometime presents also time variant dynamical behavior. In these conditions a classical PID (Proportional Integral Derivative) controller is not suitable and more complex control strategies are needed. On the other hand, if we apply only a simple position controller we lack the control of the joint stiffness.
To solve these issues we developed a simple algorithm that can control the joint stiffness and position providing the worth torque without complex calculations. While a high-level controller assigns the trajectories, as in classical position control, the elastic low-level regulator varies the joint stiffness in real time and makes a smooth motion.
In addition, we developed a more articulated algorithm with acceleration and velocity feedback; it provides an estimation of the external torque acting on the link, and modifies

the joint stiffness accordingly. These algorithms are described in detail in the next two subsections.

5.2.1 The controller using position feedback

The basic control algorithm is simple and very closed to a classical model of the Equilibrium Point hypothesis. It takes in input the reference position ϕr and the joint stiffness kg and gives in output the motor position α_0 The only state information needed is the actual joint position, that must be measured and feed-backed to the regulator. We may remind that the difference between the actual position and the motor one is covered by the spring deflection. The control law is expressed by equation 6:

$$\alpha_0 = (kg/k)\,(\phi r - \phi) + \phi \tag{6}$$

where k represent the spring stiffness, ϕr and ϕ the target and actual angular position respectively. The result is that a virtual spring with kg stiffness is acting between the reference angle and the actual position. For kg =k, $\alpha 0 = \phi r$, as the spring and joint stiffness coincide. If kg<k the motor rotation will be lower than the reference, as the spring stiffness is higher than the one required for the joint. Dually, if kg>k the motor has to rotate more to generate higher torques. Thus, the choice of kg and k can depend on the motor characteristics: kg>k attenuates the effects of a motor position error, while kg<k is suited when the motor limit is the speed.

For the other input, the reference position, to avoid high initial acceleration ϕr should be defined with second order functions with suited time constants. The finite joint stiffness betokens the presence of an error and one may define the time when the desired position must be reached, accordingly with the joint stiffness.

If stiffness is very high, the error will be small, and the actual trajectory very close to the assigned one; this means that in presence of a step in ϕr high acceleration peaks can be generated.

If the joint stiffness is small, one may expect relevant differences between the reference and actual trajectories, as the inertia and the damping oppose to fast movements. The static error ε depends anyway on the external load T_{ext} as in equation 7:

$$\varepsilon = T_{ext}/kg \tag{7}$$

Equation 7 represents also a way to determine the joint stiffness, deciding the maximum error tolerance and estimating the external maximum load. Note that kg can be changed in real time, according to the precision needed in critical phases of the motion. To define the reference trajectory we used a step function filtered by a second order filter defined by a suited time constant T. In this way we can characterize the reference pattern with a single parameter. For simplicity the damping factor is set to a constant value that keep the system at the critical damping, as in equation 5

We simulated the control of a simple 1-dof pendulum to confirm the theoretical approach. In the simulation, gravity and external loads were included. Also friction was included to test the robustness of the algorithm.

We set the system parameters as: m=1.2 kg; l=0.3 m; Ig=7.35*10-2 kgm2; k=6 Nm/rad; kg=10 Nm/rad (where l is the distance between the center of mass and the joint axis).
Fig. 15 (a) shows the joint angles and motor positions of the system for a commanded movement from 0 to 0.3 rad at 0.1 sec, and from 0.3 rad to -0.3 rad at 1.2 sec with a constant time T=0.08 s. Here, only gravity is acting, but tests were made including variable external disturbances, which could mimic the inertia load of other moving links. With "static angle", we denote the position the joint would have if the link inertia was zero and the damper was not present. The chosen stiffness is quite weak, and the error is about 0.1 rad only due to gravity. Looking at the motor position, we can notice that it is always opposite to the angle respect to the reference since the spring stiffness is chosen lower than the joint stiffness. In this way the motor has to rotate more, but the system is less sensitive to motor position error. At about 1.4 sec., the motor rotation changes velocity due to servo maximum torque limit. In the simulation also servo speed limitation was included.
About the resulting rotational acceleration, we can notice in Fig. 14 (b) only two peaks, acceleration and deceleration, with no oscillation. This pattern, typical of damped systems, is useful when it is needed to exploit the natural dynamics of multi-link systems. For instance, when starting a step, the acceleration of the thigh can be used to bend the knee, as in passive dynamic walkers (McGeer, 1990)(Collins et al, 2001) or, before foot-fall, the deceleration of the swing motion can be exploited to straight the leg, as in passive lower-limb prosthesis.

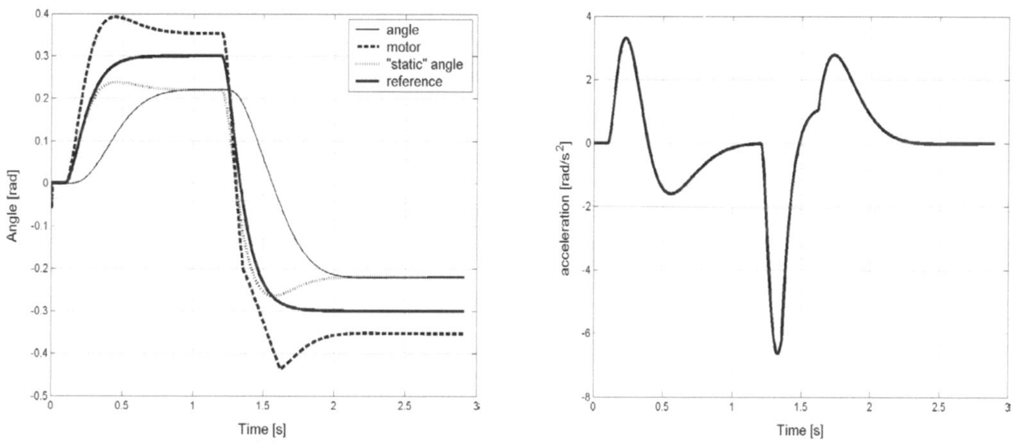

(a) (b)

Fig. 15. (a) The link rotation and the motor position referred to the commanded angle. The actual angle approaches the reference accordingly to the stiffness and external load ("static" angle). (b) The acceleration pattern presents two peaks, characteristic of damped systems. The change at about t=1.5 s is due to the limit on servo maximum torque.

To figure out the influence of rapidly external loads, we studied a positioning task under step-varying external torque. Here the stiffness was set high, since a keep-position task was to be performed: k=10 Nm/rad; kg=50 Nm/rad. Figure 16 shows the result of the system under the action of an external load composed by a sinusoidal and constant action; at 0.1s there is a positive step; at 1s a negative one.

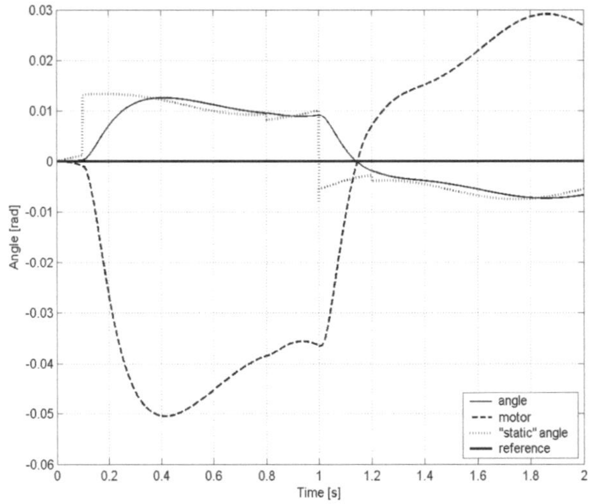

Fig. 16. The system behavior under rapidly-varying external torques. These can be seen in the "static angle" changing accordingly to the sinusoidal and step components of the load.

Using this simple control law, we do not need to solve any inverse dynamic problem, but just decide the joint stiffness, using for example equation (7), and define the suited reference pattern. Different is the case when, given a reference trajectory, we want to follow it controlling the motor torque; in this case, the external load plays a very important role, while, with the elastic control, we just need a rough estimate of it when the joint stiffness is fixed.

The following subsection describes a more complete algorithm that can automatically adapt joint stiffness to the external load, given system inertia, or its average value for a multi-link system.

5.2.2 Force estimation through acceleration feedback

In trajectory planning, not only the position is constrained, but also the velocity and acceleration must respect some limitations. This is especially important when we want to exploit the natural dynamic of the multi-body system; the acceleration of the thigh can be used to bend the knee when starting the step (McGeer, 1990) or to straight it before the foot-fall, as in passive leg prosthesis. Also velocity and acceleration limitations are needed where inertial loads, due to the movement of one part, can interfere with the motion of the rest of the robot; this is particularly relevant in bipedal walking.

To consider acceleration constrains, we included in our controller a sort of impedance control. By this term, we refer to the fact that the algorithm tracks the delivered torque and studies the resulting acceleration, creating a function relating these two quantities. In this way, we can create a simple dynamic model of a multi-body system without solving any inverse dynamic problem. The model can also get a good estimate of the external load acting on the joint, including gravity and the interaction force with other links.

This can be obtained using, in the control loop, the equations (8):

$$T^{i-1}_{ext} = -k \cdot (\alpha_0^{i-1} - \varphi^{i-1}) + I \cdot \ddot{\varphi}^{i-1} + d \cdot \dot{\varphi}^{i-1} \tag{8}$$

where d is the damping factor (as in equation 5), α_0 is obtained from equation (6), I is the inertia and k an elastic constant. We can assume that between the instants i-1 and i of the control loop the external load remains constant, so $T^{i-1}_{ext} = T^{i}_{ext.}$
Given the values of k, d, I, the position of the motor a_0 and the estimation of $T_{ext.}$, the acceleration can be estimated from equation (9):

$$Ai = (k \cdot (\alpha_0^{i} - \varphi^{i}) + T^{i-1}_{ext} - d \cdot \dot{\varphi}^{i}) / I \tag{9}$$

In this way we implement a kind of impedance control: if the acceleration (system output) in the next step is different from the foreseen one, given the calculated α_0 (system input), system infers that a different load is acting (system model has changed) and thus the motor position α_0 is corrected accordingly. In some way this is also how we sample the properties of objects in real word. For instance, to check whether a bin is empty or not we lift it, and according to the resulting motion we estimate the mass. In a positioning task, we make this sample-evaluation-correction every instant.
The simulations on a single joint, with parameters: m=1.2 kg; l=0.3 m; Ig =7.3510^{-2} kg m^2, k=10 Nm/rad; kg=50 Nm/rad, are discussed, evaluating position, acceleration, and external load. In Fig. 17 we illustrate the results on the angle, with and without limiting the motor torque and using as external load only the gravitational one.

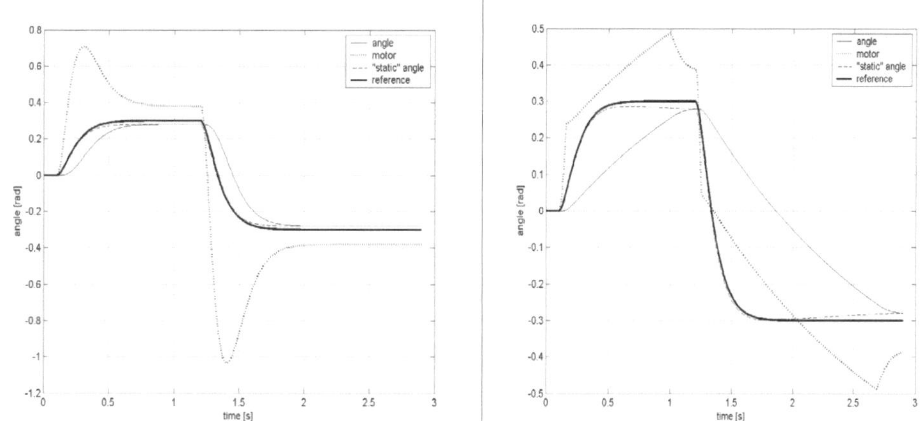

Fig.17. the angles with and without torque limitation

We can notice in Fig 18 the effect of limiting motor torque on the acceleration pattern. The characteristic is similar to the human electro-myographic activity, composed by there phases: acceleration-pause-deceleration (Kiriazov, 1991) (Gotlieb et al, 1989) and suitable for exploiting the natural dynamic of the links, i.e. in leg swinging as pointed out before. We can also notice that the system perform a pretty good estimation of the external load acting on the link.

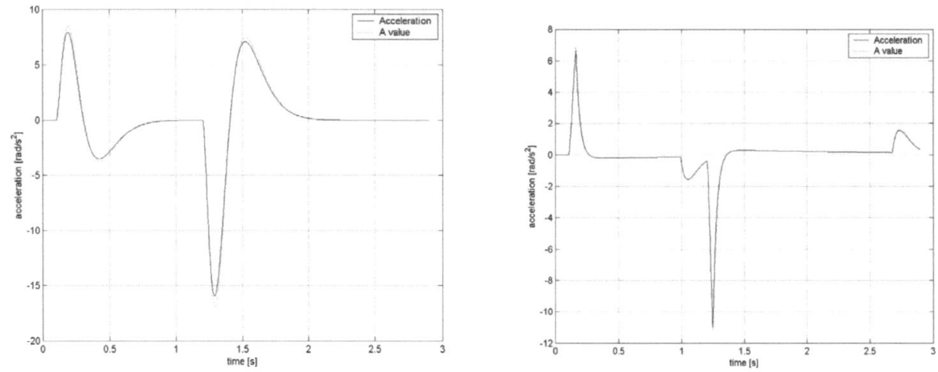

Fig. 18. the acceleration with and without motor torque limitation is considered.

If we impose a joint stiffness too high for the load applied, or if the reference angle changes too quickly, the controller decreases joint stiffness during the motion to prevent too high accelerations. This is performed using the calculated acceleration for the incoming iteration (equation 9) If, with the imposed stiffness, the acceleration A^i is too high, the low-level controller modifies kg (given by the high-level algorithm), to respect acceleration limits. In this way the real value of the acceleration is kept below its maximum value, despite wrong high-level commands.

Setting joint stiffness can be done with equation 7, or with a trial-and-error procedure. For example, a learning algorithm could be used to kg and the time constant of the reference trajectory. The choice of these two parameters as inputs for the low-level regulator is relevant since they can greatly influence the joint behavior, without hampering the final positioning.

The only information the controller needs about the system is its inertia. In multi-link systems it can be approximated with a constant average value computed on all the links, or it can be calculated during the motion. In any case, the controller seems to be quite robust respect to inertia uncertainties, showing no relevant changes even for errors of about 30% (see figure 19). The difference in inertia load is considered by the controller as an additional external torque. The damping, equation (5) can be rewritten as(10):

$$d = 2\xi\sqrt{kgI_i} \tag{10}$$

This means that the damping factor is proportional to the square root of inertia errors: while a too high inertia makes the system over-damped, an underestimation can let the system oscillate. Anyway, the error in the inertia must be very high (such as 50%) to see noticeable effects on the damping.

In the external torque estimation (Fig. 19) we can notice the effect of wrong inertia input in the controller: for instance, if the real inertia value is higher, the controller acts as an additional external load is braking rotation during positive accelerations, as the real inertia is higher than what expected. In this way, the system is "automatically compensated".

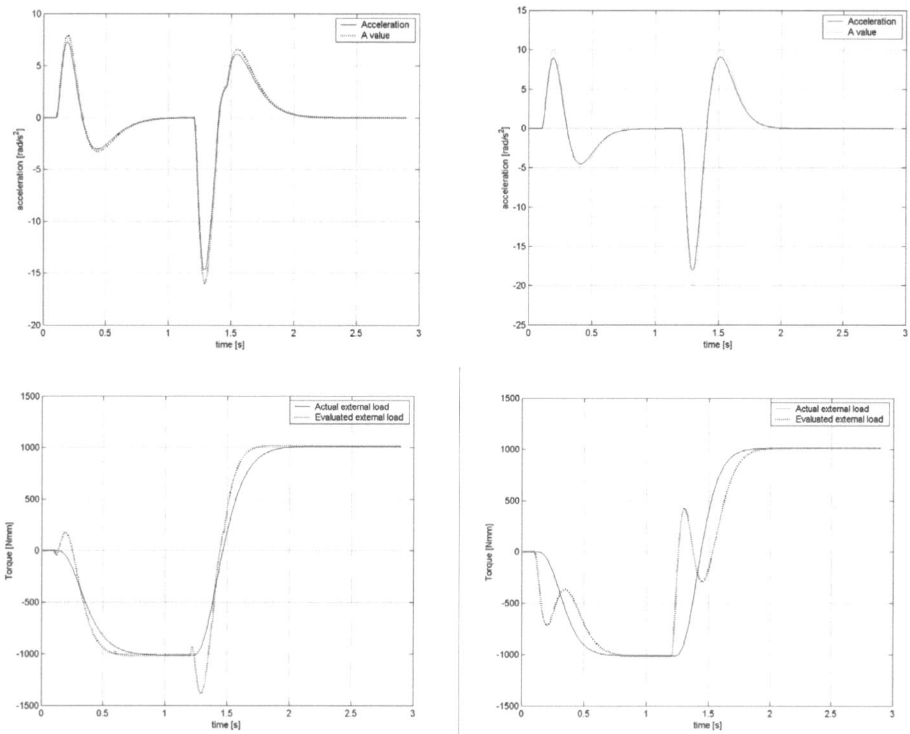

Fig. 19. Overestimated, on the left, and underestimated inertia on the right. Top row shows acceleration, bottom row torque.

From Fig 19 we may observe that an error of 30% in inertia does not compromise the positioning. If the computed inertia is lower than the real one, for example when the system is accelerating, the algorithm interprets the too small acceleration as an external load. When the computed inertia is higher than the real one, the system is over-accelerated, and a virtual additional positive torque is considered acting.

5.3 Results on LARP

The spring-reactive control has been implemented in a computer simulation on the simple robot model of Fig. 2, and after on the real prototype. In the first test, the robot had to preserve the equilibrium despite external disturbances. To run this test we implemented also a simplified physical prototype of LARP, with two dof in the ankle (pitch and roll) and one in the hip (yaw) for each leg.

Figure 20 shows the external disturbances applied on the robot. The joint stiffness is set according to equation 7, where ε is the maximum error and T_{ext} is the corresponding gravitational load. The value of inertia is calculated focusing on the resulting damping more than on the real value, that should be computed along the closed kinematic chain formed by the biped. Thus, for the ankle, we figure out the inertia of the robot considering the two feet coincident. Given this inertia value, we evaluate the needed total damping factor d. As in the feet two dampers in parallel are present, we split the inertia so that the sum of the two dampers equals the total damping needed. Regarding the hip, we proceed in the same way, neglecting the leg beneath the joint for computing inertia.

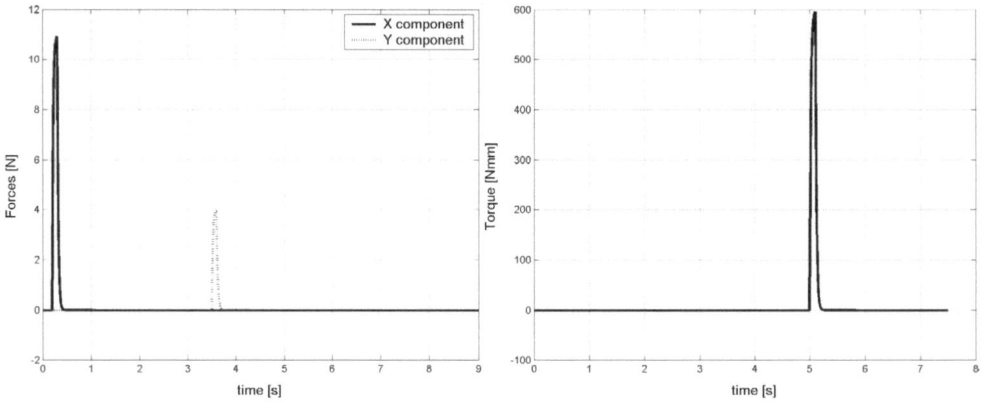

Fig. 20. The external disturbances applied to the robot, forces and torque.}

The results are shown in figure 21, where we can notice that a position error appears when the disturbance is applied, as the actual angle differs from the reference position zero. The dotted line shows the motor rotation, that counteracts the disturbance and brings the joint back to the reference. In this way the robot is able to "react" to external loads, admitting a positioning error in order to preserve the whole balance.

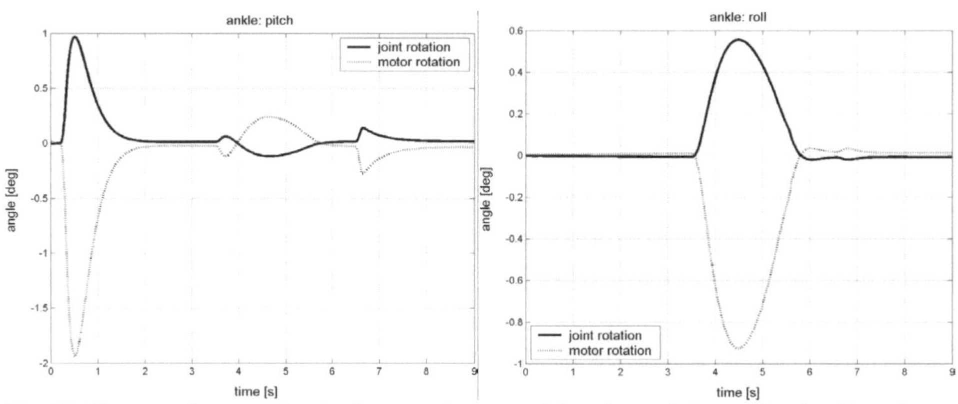

Fig. 21. The angular position in the two degrees of freedom of the ankle: the disturbances are adsorbed and the robot returns in its initial position.

6. Simulation of a static gait and energy consumption

In this section we present some results we have obtained using the direct/inverse kinematic model of our biped. We do not enter in details of this models, but we prefer concentrate our attention on energetic considerations. We also want to stress here that simulations are very important for complex projects like our robot, and this is more important if the hardware is not yet completed. The movement of the biped robot was performed using elastic actuators

and a controller based on the Equilibrium Point Hypothesis (Scarfogliero et al, (a and b), 2004), as illustrated in the previous Section.
Using the inverse kinematic solution, we can set a reference trajectory for the foot and calculate the relative joints positions. Each gait is characterized by

- the step length and height,
- the minimum height that the pelvis is allowed to reach during the motion, and
- the maximum lateral movement admissible (oscillations in frontal plane).

During all the motion the robot assumes only stable configurations, this means that if we arrest the movement the robot will keep the balance. A sufficient condition for the static stability for the robot is that the projection of its center of mass falls inside the convex area that cover the contact surface of the two feet. In our simulation the static stability is always guarantee by a software module that adjusts the pelvis position when the stability condition is not verified.
To evaluate the energy required to complete a step we made the following assumptions and approximations:

- Each robot link is modelled by a mass located in its center of mass.
- The center of mass for the entire robot is calculated by a weighted average of each link's center of masses.
- The robot moves very slowly, therefore inertia forces are neglected.
- We do not consider friction forces present in the joints.
- We consider that kinetic energy during the falling phase (the lifted foot approaches the ground) is completely lost during the movement.

The energy to lift each single link was therefore calculated with equation 11, where m is the mass of the link-i, g the gravity constant and Δhi the excursion along the z-axis for the center of mass of link-i.

$$W_i = m_i \, g \, \Delta h_i \tag{11}$$

Clearly we have approximated the real energy required for the movement, but our aim is to distinguish between efficient and non efficient gaits.
In the simulation shown in Fig. 22, we settled the step length at 0.5m, the minimum height for the pelvis at 0.68m and the maximum lateral excursion at 0.09m (oscillation in frontal plane).

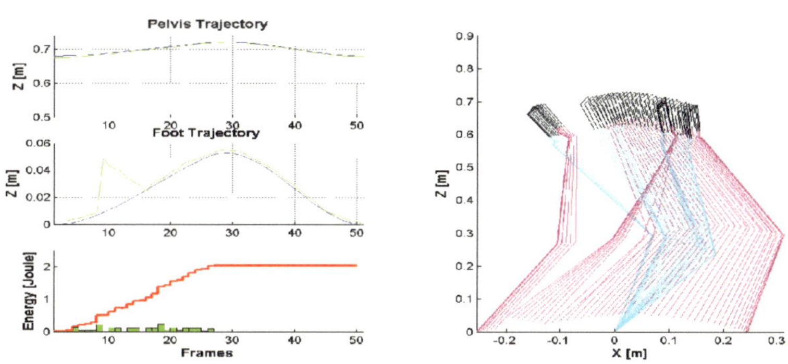

Fig. 22. Energy consumption with minimum height for pelvis at 0.68m}

The first two graphs in the left part of fig. 21 represent the Z coordinates for the pelvis and the foot during the motion (in blue the reference trajectory, in green the real trajectory performed by the robot). From the third graph in the left side we see that the total energy consumed to perform the gait is about 2 joules. It is possible to note that at the eighth frame the real foot trajectory deviates significantly from the reference, as a result of the stability algorithm that tries to maintain the balance. Finally the graph in the right site shows a stick model for the robot in the lateral plane and gives us an overview of the all gait's phases.
In the simulation shown in Fig. 23, we set the minimum height allowed for the pelvis at 0.55m. This means that the robot, when following the reference trajectory, is allowed to lower more its *cm*. In this case the reference trajectory for the foot is well followed, nevertheless the energy required for this gait is increased to about 6 joules. This is due to the fact that the links of the robot have a greater excursion along the z axis.

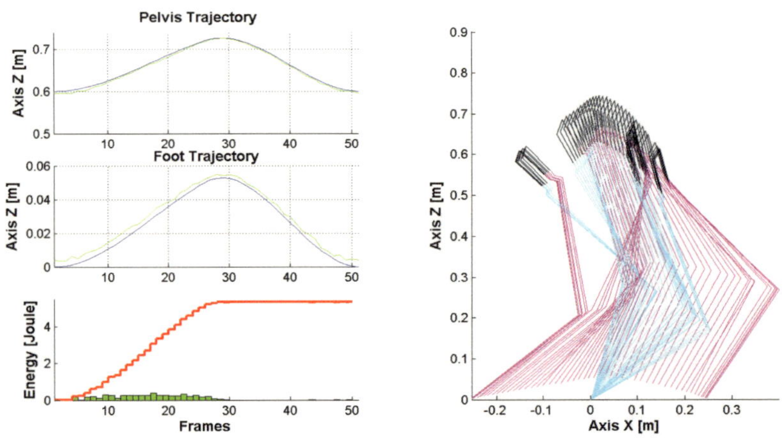

Fig. 23. Energy consumption with minimum height for pelvis at 0.60m

These preliminary results confirm that the position of the *cm* of the robot has a great impact on energy consumption. If the *cm* is maintained low the stability algorithm does not interfere with the trajectory following for the foot in static condition. Nevertheless these kind of postures require more energy, while limiting the vertical movement of the pelvis can save energy during the gait. We can also assert that the knee joint covers a very important role during the gait, indeed without this degree of freedom it is difficult move down the robot's *cm* and therefore stabilize the posture.
The strategy to decrease the height of the *cm* is advantageous to control the robot stability if we are in static conditions (at low acceleration and velocities the inertia forces can be neglected), nevertheless with this kind of posture the robot is not able to perform fast walking, and also the energy required for the movement is high. The human walking, on the contrary, can be assumed as dynamical; indeed in each instant the body is not in a stable position. Furthermore the pelvis is maintained high and as fixed as possible to reduce the energy required for the movement, as indicated in our results.
To be energetically efficient our robot should be tested also in dynamical conditions, to take advantage of the knee and foot design that were thought to store the inertia and impact forces.

7. Simulation of the robot's dynamic behaviour

To better account for the many elements that act in the real situations we developed a MSC Adams model of the robot, to compute in real time the forces through a Simulink application, whose general architecture is illustrated in Fig. 24. Three are the main modules.

- The force generation module is a closet loop system that generates the correcting forces to apply to the links, proportional to the difference between the target position and the sensed position.
- The ZMP controller is a closed loop controller that generates a correction force, proportional to the difference between the ideal and the real position of the ZMP.
- The 3D simulator animates the ADAMS model, as illustrated in Fig. 24.

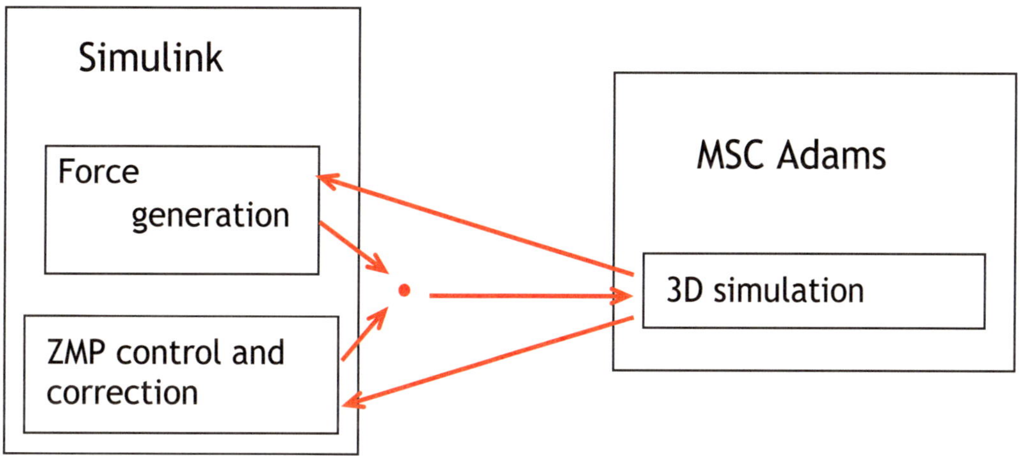

Fig. 24. The architecture of the LARP simulator.

To simulate a step using our dynamic simulator we start from recorded joint positions, that are used to generated both the static and dynamic robot gaits. Those positions are called *frames* and are obtained from measures from human beings.
The gate is composed of 3 phases, each represented by different frames:

- start - contains 11 frames and moves the robot from the initial idle configuration to the move phase;
- move - contains 44 frames; the robot makes 2 steps for each move phase, and returns to the initial configuration. We can repeat this phase as many times as we want;
- stop - contains 11 frames and is the reverse of the start phase; the robot goes from the move phase to the initial idle position.

Starting from the joint positions, and using the Newton-Eulero equations, in theory we can compute from equation 12 forces and moments that act on the center of mass of each link and send the so obtained values, with a given frequency, to the actuators to obtain the trajectory.

$$F_i = m\dot{v}c_i$$
$$N_i = {}^{Ci}I\dot{\omega}_i + {}^{Ci}I\omega_i \tag{12}$$

To obtain dynamic stability we use the Zero Moment Point (ZMP) criterion. Using the data from the sensors on the links we compute the x and y coordinates of the Zero Moment Point according to equation 13:

$$x_{ZMP} = \frac{\sum_{i=1}^{n} m_i(\ddot{z}_i + g)x_i - \sum_{i=1}^{n} m_i\ddot{x}_i z_i - \sum_{i=1}^{n} T_{yi}}{\sum_{i=1}^{n} m_i(\ddot{z}_i + g)}$$
$$y_{ZMP} = \frac{\sum_{i=1}^{n} m_i(\ddot{z}_i + g)y_i - \sum_{i=1}^{n} m_i\ddot{y}_i z_i - \sum_{i=1}^{n} T_{xi}}{\sum_{i=1}^{n} m_i(\ddot{z}_i + g)} \tag{13}$$

where:

i indicates the link number
T is the torque applied by the motor
m is the mass
g is the gravity acceleration
(x,y,z) is the center of mass of the link
$(\ddot{x}, \ddot{y}, \ddot{z})$ is the link acceleration

We developed an algorithm for the dynamic stability, where the real ZMP is compared with the ideal ZMP. Since the most instable phase is the one of single support, we correct the errors in this phase. The parameters of the algorithm are:

- *nframes*, number of frames in the single support phase;
- *lfoot* and *wfoot*, length and width of the foot;

For each value of the independent variable *actualframe* we obtain the ideal x and y coordinates of the ZMP, x_{ID} and y_{ID}, using the equation 14:

$$x_{ID}(actualframe) = \frac{lfoot}{nframes} * actualframe$$
$$y_{ID}(actualframe) = \frac{3}{4}lfoot - \left(\left(actualframee - \frac{nframes}{2} \right)^2 * \frac{\frac{wfoot}{2}}{\left(\frac{nframes}{2}\right)^2} \right) \tag{14}$$

The coordinates are relative to a reference system in the foot, with origin in the intersection point between the back and the inner sides of the foot, with the x-axis in the direction of the foot point and y-axis that points to the external side of the foot.

Suppose the ideal trajectory of the ZMP point linearly moves along the x-axis and quadratically moves along the y-axis from $\frac{1}{4} wfoot$ to $\frac{3}{4} wfoot$ in $\frac{1}{2} nframes$, *and* simmetrically, from $\frac{3}{4} wfoot$ to $\frac{1}{4} wfoot$ in $\frac{1}{2} nframes$.

The obtained trajectory is parabolic, as in humans, and is illustrated in Figure25.

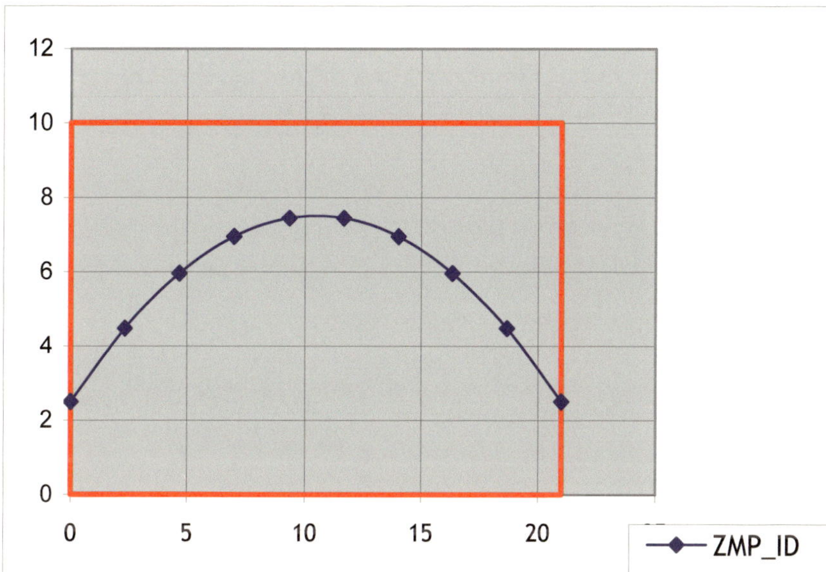

Fig. 25. Trajectory of the ideal ZMP for the left foot of LARP (in red the area of the foot)

Comparing x_{ID}, y_{ID} and x_{ZMP}, y_{ZMP} we compute the errors with equation 15:

$$\begin{aligned} \varepsilon_x(actualframe) &= x_{ID}(actualframe) - x_{ZMP}(actualframe) \\ \varepsilon_y(actualframe) &= y_{ID}(actualframe) - y_{ZMP}(actualframe) \end{aligned} \quad (15)$$

We compute the corrective force F_{CORR} to apply to the hip joints as in equation 16:

$$F_{CORR}(actualframe) = \alpha * \varepsilon(actualframe) + \beta * (\varepsilon(actualframe-1) - \varepsilon(actualframe)) \quad (16)$$

with α and β constants. The correction is proportional to the error and to its trend. The actuators in the hip apply the total force F_{TOT} (equation 17):

$$F_{TOT}(actualframe+1) = F_T(actualframe+1) + F_{CORR}(actualframe) \quad (17)$$

where F_T is the offline computer force necessary to obtain the wanted trajectories.
The controller schema in Simulink is illustrated in Appendix A.

7.1 Results in simulation

We have experimented on LARP two situations, namely maintaining the standing position, and walking.

7.1.1 Maintaining the standing position

To keep the robot in standing position we assign zero values to the angles of movement, and detach the subsystem devoted to the stability control of walking. In the following figures 26, 27, and 28 we see the positions (in meters) and the torques (in Nm) applied to the left hip, left knee, and left ankle respectively, all around the y axis.

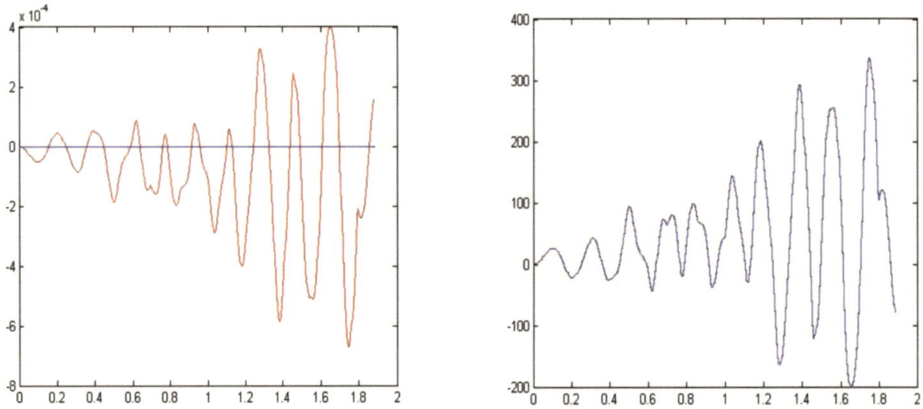

Fig. 26 Sensed position and moment applied to the left hip around the y axis.

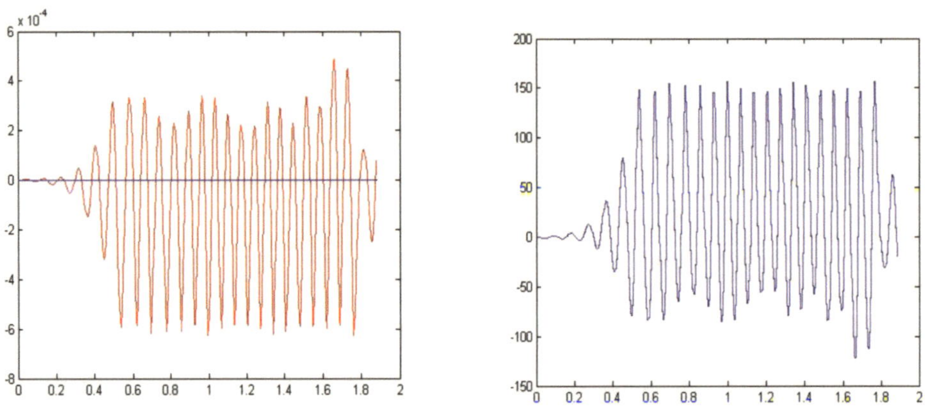

Fig. 27. Sensed position and moment applied to the left knee around the y axis.

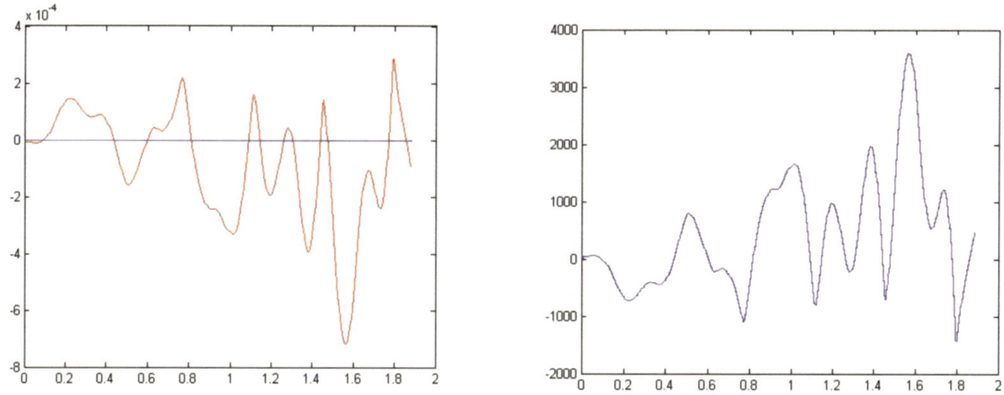

Fig. 28. Sensed position and moment applied to the left ankle around the y axis.

The three joints have a similar behavior; after few oscillation it stabilizes. Observe that the maximum error is really small, about 0,5 mm.

7.1.2 Walking

In simulating the dynamic walking new problems emerged.

First, since walking for bipeds is a really dynamic situation, more and different parameters are involved.

The most important problem in LARP is the position of the center of mass. Until now LARP has no torso, so the maximum height for the center of mass is the pelvis. The inertia acting on the pelvis cannot be compensated, as in humans, from the arm movements, so it generates oscillations that in some cases cannot be compensated.

Another problem is the foot. The real foot has a flat surface, but in simulation a flat surface requires to consider too many points, and is impractical. To speed up the simulation we model the flat surface as eight small spheres, with only eight point contacts. This solution reduces the friction and the robot can slip. For those reason the simulation of the gait should be stopped after a few steps. The addition of the torso will be done for future simulations.

8. Conclusions

LARP is a project about developing a biped robot in parallel with the study of human walking. Some images of the mechanical construction are in Appendix 2.

Today several humanoids robots are able to walk and perform human-like movements. Anyhow, the structure of such robots significantly differs from the human's one. This causes the robots to be energetically inefficient, as they are unable to exploit the natural dynamics of the links, and very poorly adaptable to unstructured terrains.

Studying the human knee and foot we found several advantages in adopting human-oriented design for these parts. In particular, a compliant knee was developed, having two circular contact surfaces and five tendons. This articulation is highly efficient and permits to increase the foot clearance during the swing phase. Regarding the foot, two passive joints were introduced to mimic the high mobility of the human foot. To ensure stability both at

heel-strike and toe-off we used two planar surfaces connected to the arc of the foot by two passive degrees of freedom.
Further work has to be done for the complete design of a human-like robot, starting from a new design of hip and ankle articulation. In this case we should investigate the role of the third dof in the human ankle (torsion of the foot along the leg axis), which is omitted in most of the modern humanoid robots. Also, it remains to fully test our model in dynamical condition, in order to find the more efficient and efficacious gait.
The similarity between the behavior of our robot and of human walking can be exploited to promote a further research comparing the biped behavior with human theories assumptions.

9. References

R. Alexander, *The Human Machine*. Columbia University Press. New York, 1992.

F. A. Asatryan, D.G., "Functional tuning of the nervous system with control of movement or maintenance of a steady posture - mechanographic analysis of the work of the joint or execution of postural task," *Biofizika*, vol. 10, pp. 837–846, 1965.

D.G. Asatryan, A.G. Feldman.(a), "Functional tuning of the nervous system with control of movement or maintenance of a steady posture - ii controllable parameters of the muscle," *Biofizika*, vol. 11, pp. 498–508, 1966.

D.G. Asatryan, A.G. Feldman (b), "Functional tuning of the nervous system with control o movement or maintenance of a steady posture - iii mechanographic analysis of the work of the joint or execution of a postural task, *Biofizika*, vol. 11, pp. 667–675, 1966.

S. Collins, M. Wisse, and A. Ruina, "A three dimensional passivedynami walking robot with two legs and knees," *The Internationa Journal of Robotics Research*, vol. 20, no. 7, pp. 607-615, 2001.

J. Doneland, R. Kram, and D. A. Kuo, "Simultaneous positive and negative external mechanical work in human walking," *Journal o Biomechanics*, vol. 35, pp. 117–124, 2002.

G. Fallis, "Walking toy," *U.S. Patent No.376588*, 1888.

M. Garcia, A. Chatterje, A. Ruina, M.J. Coleman, "The simplest walking model Stability, complexity and scaling," *ASME Journal of Biomechanica Engineering Vol. 120 p.281-288*, vol. 120, pp. 281–288, 1998.

G. Gottlieb, Q. Song, D. Hong, G. Almeida, D. Corcos., "Co-ordinating movement a two joints: a principle of linear covariance," *Neurophysiology*, vol. 75 no. 5, pp. 1760–1764, 1996.

G. L. Gottlieb, D.M. Corcos, G.C. Agarwal., "Strategies for the control of single mechanical degree of freedom voluntary movements," *Behavioral an Brain Sciences*, vol. 12, no. 2, pp. 189–210, 1989.

S. Hashimoto at al., "Humanoid robots in waseda university hadaly 2 an wabian," *Autonomous Robots*, vol. 12, pp. 25–38, 2002.

K. Hirai, M. Hirose, Y. Haikawa, T. Takenaka., "The development of honda humanoid robot," *IEEE International Conference on Robotics and Automation* pp. 1321–1326, 1998.

A. Jeanneau, J. Herder, T. Lalibert´e, and C. Gosselin, "A compliant rolling contact joint and its application in a 3-dof planar parallel mechanism with kinematic analysis," *In Proc. DETC 2004 Salt Lake City, Utah*, 2004.

R. Ker, M. Bennett, S. Bibby, R. Kerster, and R. M. Alexander, "The spring in the arc of the human foot," *Nature*, vol. 325, pp. 147–149 1987.

P. Kiriazov., "Humanoid robots: How to achieve human-like motion," *Journal of Biomechanics*, no. 24, pp. 21-35, 1991.

B. Koopman, F. van der Helm, and P. Veltink, *Human motion control*. University of Twente, Enschede; University of Technology, Delft, 2001.

A. D. Kuo., "Stabilization of lateral motion in passive dynamic walking, *The International Journal of Robotics Research*, vol. 18, no. 9, pp 917–930, 1999.

A. D. Kuo, "Energetics of actively powered locomotion using the simples walking model," *ASME Journal of Biomechanical Engineering* vol. 124, pp. 281–288, 1998.

L. C. Kwek, E. K. Wong, C. K. Loo, M. V. C. Rao, "Application of active force control and iterative learning in a 5-link biped robot," *Journal of Intelligent and Robotic Systems*, vol. 37, no. 2, pp. 143–162, 2003.

M.L. Latash., G.L. Gottlieb, "An equilibrium-point model for fast single-join movement. similarity of single-joint isometric and isotonic descendin commands," *Journal of Motor Behavior*, vol. 23, pp. 163–191, 1991.

H.-O. Lim, S. A. Setiawan, and A. Takanishi, "Position-based impedance control of a biped humanoid robot," *Advanced Robotics* vol. 18, no. 4, pp. 415–435, 2004.

G. Maloiy, N. Heglund, L. Prager, G. Cavagna, and C. Taylor "Energetic costs of carrying loads: have african women discovere an economic way?" *Nature*, vol. 319, pp. 668–669, 1986.

T. McGeer, (a) "Passive dynamic walking," *The International Journal o Robotics Research*, vol. 9, no. 2, pp. 62–82, 1990.

T. McGeer, (b)"Passive walking with knees," *IEEE International Conferenc on Robotics and Automation*, vol. 2, pp. 1640–1645, 1990.

J. McIntyre and E. Bizzi, "Servo hypotheses for biological control o movement," *Journal of Motor Behavior*, vol. 25, no. 3, pp. 193–202 1993.

J. Pratt, C. Chew, A. Torres, P. Dilworth, and G. Pratt, "Virtual model control: An intuitive approach for bipedal locomotion," *The International Journal of Robotics Research*, vol. 20, no. 2, pp. 129 143, 2001.

G.A. Pratt , M.M. Williamson, "Series elastic actuators," *IEEE International Conferences on Intelligent Robots and Systems*, no. 1, pp. 399–406 1995.

P. Sardain, M. Rostami, E. Thomas, and G. Bessonnet, "Biped robots Correlations between technological design and dynamic behaviour, *Control Engineering Practice*, vol. 7, pp. 401–411, 1999.

J. Saunders, V. Inman, and H. Eberhart, "The major determinants in normal and pathological gait," *Journal of Bone and Joint Surgery*, vol 35-A, pp. 543–558, 1953.

U. Scarfogliero, M. Folgheraiter, G. Gini, (a) "Advanced steps in biped robotics: Innovative design and intuitive control through springdampe actuator," *In Proc. IEEE Humanoids 2004, Los Angeles, USA* 2004.

U. Scarfogliero, M. Folgheraiter, G. Gini (b) "Larp: Biped robotic conceived as human modelling," *In Proc. Neurobotics KI 2004, Ulm Germany*, 2004.

R. Sallaouti, O. Stasse, S. Kajita, K. Yokoi, A. Kheddar, "Faster and smoother walking of Humanoid HRP-2 with passive toe joints", Proc IEEE IROS 2006,.

C. Vaughan, "Are joint torques the holy grail of human gait analysis? *Human Movement Science*, vol. 15, pp. 423–443, 1996.

T. A. Yamaguchi J., "Design of biped walking robot having antagonistic driven joint using nonlinear spring mechanism," *IROS 97*, pp 251-259, 1997.
M. Wisse and J. Frankenhuyzen, "Design and construction of mike; a 2d autonomous biped based on passive dynamic walking." *Delft University of technology*, 2002.
M. Wisse, A. L. Schwab, R. Q. vd. Linde., "A 3d passive dynamic biped wit yaw and roll compensation," *Robotica*, no. 19, pp. 275-284, 2001.

APPENDIX 1 - The Simulink schema of the controller for the dynamic stability of the robot

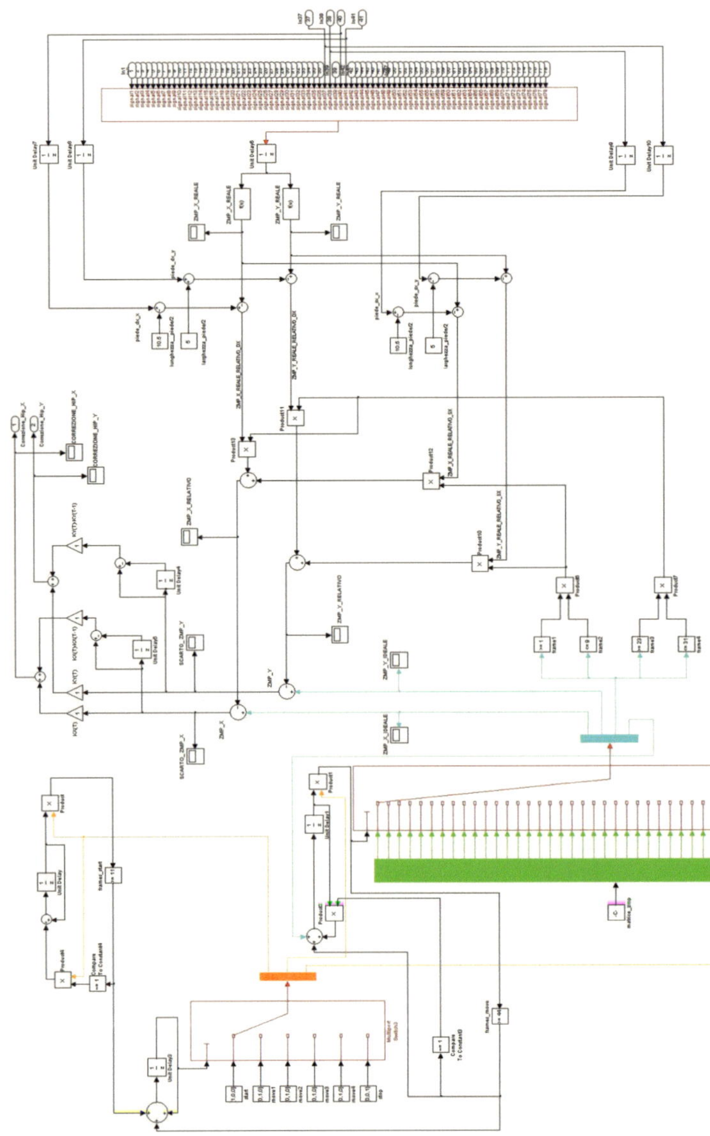

APPENDIX 2 – Images of LARP

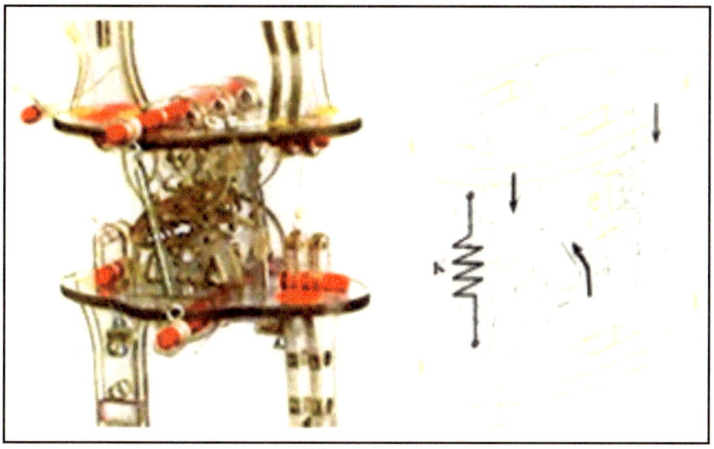

The knee

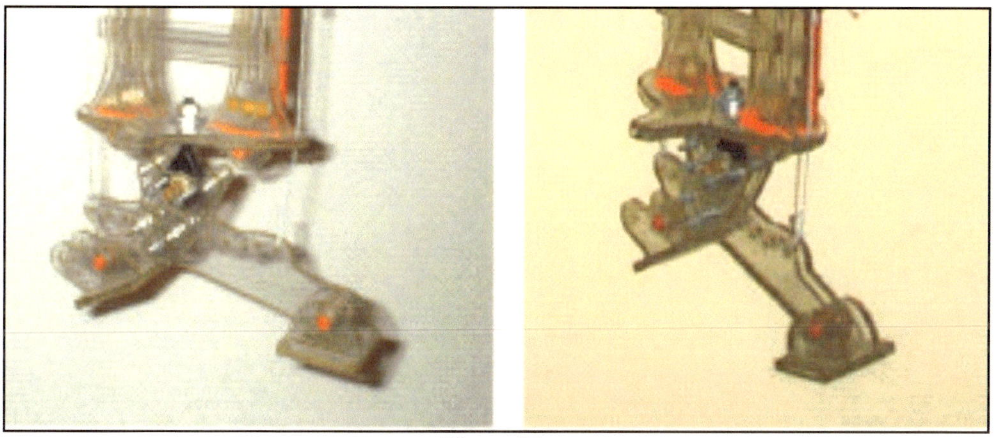

The ankle

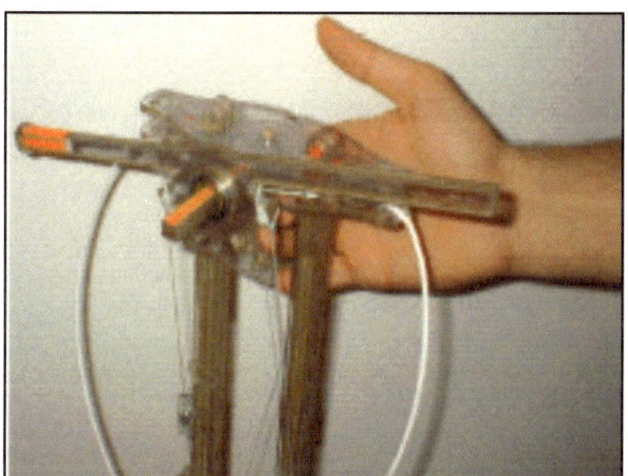

The hip

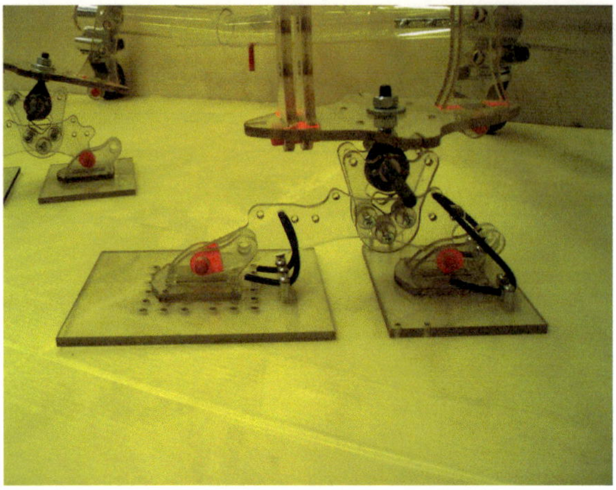

The foot

The actuators area over the hip

4

Connectives Acquisition in a Humanoid Robot Based on an Inductive Learning Language Acquisition Model

Dai Hasegawa, Rafal Rzepka & Kenji Araki
Hokkaido University
Japan

1. Introduction

1.1 Background

Humanoid robots, like the Honda's ASIMO[1], have been developed and such machines will be expected to help people to do a variety of tasks in every day life. Then robots that exist in the same dynamic environment as humans should be able to interact with humans using natural language.

However, when it comes to developing robots that understand language and work in dynamic environment, there are two problems to be solved in usual methods of the natural language processing.

One problem is that of the language grounding which was also pointed out by Roy (Roy, 2003). In natural language processing systems, meanings of words are defined by other words circularly. That is, the words are not connected to objects, movements, colors or other physical features in the real world. However, connecting words to the real world is needed when a robot is to perform a concrete action following a human's utterance. For example, when a person orders a robot to "kick the red ball", if the word "red" is not connected to a specific color representation in accordance with the real world then the robot cannot realize the order.

Another problem is that the top-down approach, which is very common in natural language processing systems, cannot deal with a dynamic environment. For example, SHRDLU (Winograd, 1972) is a system that connects words to a virtual world where some blocks exist. This system can understand utterances that refer to the virtual world such as "put the red cylinder on the green box" and can perform such actions when told, because the designer describes knowledge about the limited and static virtual world to the system in advance. On the other hand, SHRDLU cannot reply to utterances that refer to things out side of the virtual world such as "is the weather fine today?". Hence, the top-down approach is not appropriate to design a robot that has to understand language and work in an open and dynamic environment. Therefore, the bottom-up approach is needed. In the bottom-up approach, a robot dynamically learns the connection between words and the real world and

---

[1] ASIMO, http:/www.honda.co.jp/ASIMO

can update this knowledge through interaction with humans. Recently, a field called Cognitive Developmental Robotics (Asada et al., 2001) has been proposed, paving the way for this type of research. Cognitive Developmental Robotics aims to construct an intelligent robot which can behave adaptively by learning through interaction with the environment. The behaviorism approach "Embodied Intelligence", advocated by Brooks (Brooks, 1991), is another bottom-up approach.

For above mentioned reasons, a framework in which the robot makes connections between words and the real world by itself is an efficient way to develop machines that can perform universal tasks and understand natural language in dynamic environments. Thus approaches where robots acquire language developmentally like humans have attracted attention as one possible method to design the robot. Here we would like to suggest such a language acquisition mechanism for a humanoid robot. Our target language is Japanese, and we will use italic when giving Japanese examples.

1.2 Related Work

There are several research groups working on language acquisition for embodied systems. Iwahashi et al. proposed a mechanism whereby a robot arm acquires nouns, verbs, word order and concepts of movements from pairs of movement video and an audio explanation (Iwahashi et al., 2003). In this research, movement was modeled using a hidden Markov model (HMM). The system acquires nouns from audio using statistical learning. Verbs are also acquired by statistical learning and an embedded mechanism that can represent the trajectory of moving object.

Tani et al. described a system using a recurrent neural network (RNN), where a movable arm robot acquires nouns and verbs from pairs of an action pattern and a two word phrase (Tani et al., 2005).

Ogura et al. developed a mechanism where a humanoid robot acquires nouns, verbs, and word order using Self Organizing Incremental Neural Network (SOINN), and understands three words utterances (Ogura et al., 2006).

In the above research, utterances lack the naturalness of natural Japanese language we use, because they do not use particles at all. Fukui et al. experimented with machine language understanding of more natural utterances using an AIBO[2] (Fukui et al., 2004). In their research, the action representation is a row of numbers which indicates an action, such as kicking, heading or making steps, etc. and the AIBO learns words or phrases from pairs of a simple sentence and a pattern of movement.

However, the above mentioned systems do not deal with the understanding of compound sentences containing more than one verb. It is rather obvious that it is much more natural to give orders to robots by using compound sentences such as "go to the next room and bring me a book". Earlier research has concentrated only on spatial concepts such as colors, shapes movements or actions, ignoring the necessity to acquire other concepts, such as time or negation, for understanding compound sentences.

1.3 Our language Acquisition Model

We will develop a language acquisition mechanism that can deal with compound sentences.

---

[2] AIBO, http://www.sony.jp/products/Consumer/aibo/

In a compound sentence, connectives (in Japanese for instance indicating conjunction or negation) join main and subordinate clauses. We try to acquire concepts of time and then ground connectives to them.

Below are explanations of four levels of our language acquisition model based on research on human language acquisition.

(a) First, infants acquire the phonology of their mother tongue. In this acquisition step, infants use categorical perception (Eimas & Miller, 1980), though we are not concerned with phonology acquisition. Assuming that phonology is already acquired and phonologies are connected to characters, we use textual inputs for our system.

(b) Second, the morphology is acquired. Infants learn segmentation of words from continuous speech gradually, using prosody and statistical information (Jusczyk, et al., 1999). Such statistical segmentation has already been developed as a morphological analyzer for textual Japanese such as MeCab[3]. Therefore, we use such an analyzer to overcome the problem of word segmentation in our system. Utilizing this analyzer, our system has knowledge of word segmentation from the beginning.

(c) Third, the words which are segmented are grounded to appropriate meanings in the real world. Nouns are acquired by fast mapping from very few language inputs, because infants ground nouns to their meanings using some innate constraints of cognition (Markman, 1989). Verbs and adjectives have not been studied that thoroughly. Our system does not abstract and acquire nouns, verbs, or adjectives yet.

(d) Finally, we know that infants also infer things inductively (Heit, 2000). Therefore, we believe that infants ground abstract words to meaning, such as connectives that include partial time concept, by inductive inference from actual examples.

Our system acquires connectives by abstracting from pairs of a sentence and an action using the "Three Examples Based Inductive Learning" method which is based on Inductive Learning (Araki & Tochinai, 2000). In this algorithm, we use our system's innate learning ability and innate cognitive ability. The former is the ability to compare if one string contains the other string, and then to parameterize the common part; the latter is the ability to recognize movements based on the final posture of the robot. This learning algorithm is one of original points of this paper, and the acquired connectives have compositionality and can create various new meanings by combining known words or phrases. Moreover, the humanoid robot is taught actions by users using the direct physical feedback. Direct physical feedback means a teaching method where the user teaches the robot by moving its arms, legs, or head directly. Thus teaching method is also original.

2. Suggested Method

2.1 System Overview

Our system is shown in Fig. 1 and it is outlined below.

(a) A user inputs a command in natural language (Japanese) from a keyboard for the humanoid robot. The user can input both a command representing a simple action (see 2.2) and a command representing a compound action (see 2.3).

[3] MeCab: Yet Another Part-of-Speech and Morphological Analyzer, http://mecab.sourceforge.jp/

(b) The robot applies rules (see 2.8) or examples (see 2.7) that have been previously acquired, and performs an action. If there is no rule or example which should be applied, the robot does not perform any action.
(c) When the robot performs an action, the user makes the judgment whether the performance is correct or not, and if it is not, the user teaches it the right action (feedback process). After that, the system adds a pair of input command and taught action to the example database.
(d) If the humanoid robot does not perform any action, the user teaches it a proper action by direct physical feedback (see 2.4). Then the system adds a pair of command and action to the example database.
(e) Finally, the system generates rules which represent meanings of connectives from the example database by Three Examples Based Inductive Learning process (see 2.8).

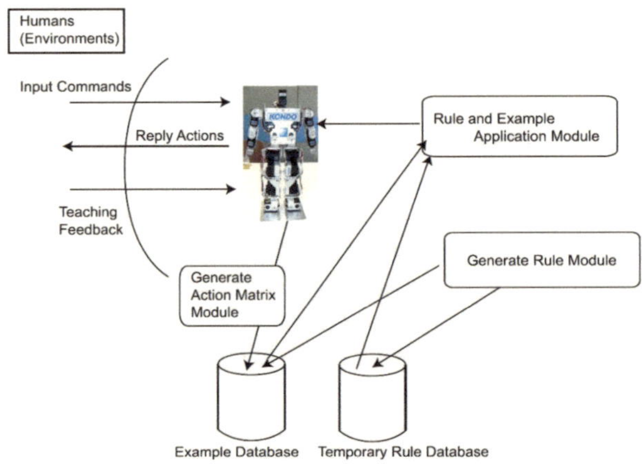

Fig. 1. System overview

2.2 Element of Action

In their work on understanding of order expressions for actions, Shinyama et al. discussed about "the vagueness in instructions and the mistiness in spatial points" (Shinyama et al., 2001). Regarding the vagueness in instructions, robots have to determine whether a user's utterance instructs to perform something or not. Although, for example, "can you raise your hand?" has a form of a question, it may instruct the robot to raise its hand. Concerning the vagueness of spatial points, robots have to determine where the point of the user's instruction is. For example, "raise your hand" has vagueness of height which the hand should be raised to.f
In our previous work (Hasegawa et al., 2007), to make the acquisition of connective simpler, we defined an action as a body movement trajectory of the shortest distance between starting and final positions. The final position which the robot should reach is obtained from user's input. Furthermore, we resolve the vagueness by teaching and averaging. The robot determines a final position of an action as an average point from all inputs taught by all users who input the same command. Within this definition, the system can not deal with actions where the final position changes continuously and intermittently, for example, "to wave a right hand". However, the action is represented as a combination of two actions

which are "to move the right hand to the right side" and "to move the right hand to the left side".
Then, it is necessary to define a minimum unit of action for segmentation of complex movements. We try to perform segmentation based on language inputs from users. Thus, we define that a minimum unit of action is an action which is represented by one sentence containing only one verb. We call it Element of Action (hereafter abbreviated as EoA). We also call a command representing EoA an EoA command. It can be summarized as follows.
(a) EoA shows final physical position of action.
(b) EoA command contains only one verb.
For example, "*migi-te wo agete* (raise your right hand)" or "*migi wo muite* (look to the right)" are EoA commands. This definition of actions has one important problem. The problem is an ignorance of a spatial context dependence of actions. For example, in the input "put your right hand on your left hand", the final position of the right hand depends on the spatial position of the left hand. By our definition of actions, the system can not learn the correct action. However, we regard this as the verb acquisition or the verb understanding problem, because the system has to understand that the verb "put" requires two objects usually. We plan to handle it in our future work. Therefore, in our current system verbs are not analyzed. In the connective acquisition, our system deals with the verbs which require only one object in a limited way.

2.3 Connectives

A connective is defined as follows.
(a) A word which is a conjunction or a conjunction particle.
(b) A segment connecting two sentences, which contains a conjunction or a conjunction particle.
Connectives connect two EoA commands. We also call a command that contains a connective a connective command. For example, "*migi-te wo agete kara migi wo muite* (face to the right after raising your right hand)" is a connective command, and "*kara* (after)" is a connective. According to our definition of connectives, they can contain more than two verbs to connect two EoAs. We define an action which contains more than one verb as a compound action to distinguish it from a single EoA. Compound actions are performed by inputting connective commands. There are not so many types of connectives, but their usage slightly depends on individual interpretations regarding timing, speed or quantity of action. Thus we consider that robots should learn the meanings of connectives from various users. In this case, such knowledge is being acquired like common sense. We use average of all inputs and feedbacks to simulate this process.

2.4 Teaching Method

A humanoid robot learns EoAs and compound actions by being taught by human how to move its body.
There are some methods already developed where a human supervisor teaches actions to humanoid robots. In one of them, a user shows actions to a humanoid robot equipped with vision and the robot imitates "seen" actions (Mataric, 2000) (Schaal, 1999). Another is where a user makes a humanoid robot learn actions from human's motion capture (Nakaoka et al., 2003). However, these methods can teach actions beyond an allowance of a robot's body

structure while our method also helps the user and robot itself to know the functional limits, because its body is not exactly the same as a human's. The robot can not perform an action immediately if it exceeds its physical capabilities. Though there is a way where a user teaches combinations of primitive movements implemented to a robot beforehand, it is difficult to teach new primitive movements and perform corrections.
Therefore, we decided to implement a direct physical feedback method where humans teach actions to a robot by actually moving its body parts. We claim it is a universal and natural method which allows teaching within the limits of any humanoid robot's body structure. This approach also allows robots to reproduce movements with very high certainty. Our direct physical feedback has the merits, but on the other hand the method to teach very complicated actions where many joints must be moved simultaneously and one user is not enough to perform the feedback. Therefore, we do not consider our teaching method as a general teaching method but as an additional method which complements other methods.
The direct physical feedback has an advantage in teaching the haptic actions. For example, in the teaching method through a vision or a motion capture, it is difficult to teach an action where a robot pushes a button or a robot touches something. Because the strength and quantity of such actions is very delicate, our method should be faster and easier to use.

2.5 Humanoid Robot

For our experiments, we used a humanoid robot (KHR2-HV[4]) shown in Fig. 2. The robot is equipped with 17 motors, no sensors and it sends signals describing its motors' state only. We use all 7 motors which are placed in the upper half of the robot's body. The particular motors are abbreviated as follows. [H]:Head, [LS]:Left Shoulder, [LA]:Left Arm, [LH]:Left Hand, [RS]:Right Shoulder, [RA]:Right Arm, [RH]:Right Hand. All motors' 180 degrees movements were divided into 10 ranks, 18 degrees each.

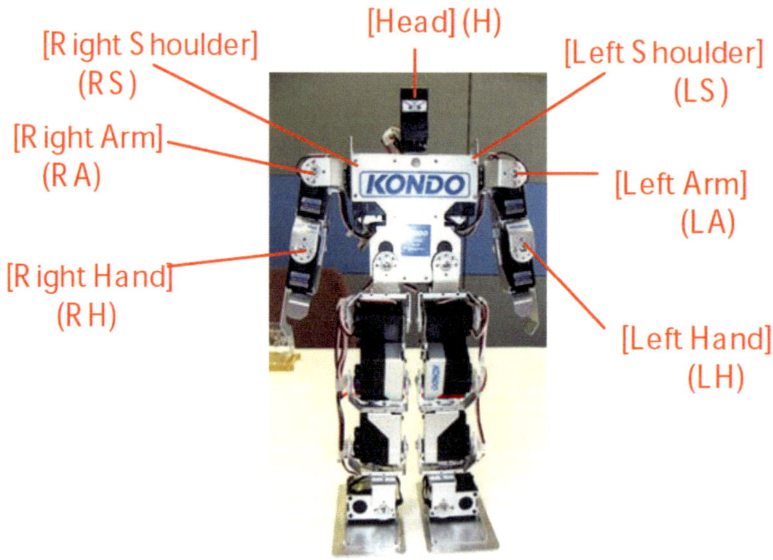

Fig. 2. KHR2-HV

[4] Kondo Kagaku Co. Ltd, http://www.kondo-robot.com/

2.6 Representation of Action

An EoA or compound action is represented as a matrix shown in Fig. 3. We call the matrix as an action matrix. If a user moves the robot's body, then it measures all the angles of motors in degrees and digitalizes them every 1 second. A row of an action matrix corresponds to 1 second, and each row corresponds to every motor's state in this particular second. If a motor does not move, degree of its angle is described as "x". An action matrix has 28 rows and 8 columns, but the number of lines depends on hardware's restrictions. To make the acquisition of connectives simpler we do not use motor speed information yet but we plan to use it when acquiring adverbs. Therefore the row of speed is still not used and the matrix can not represent EoAs containing changing speed, for example "*migi-te wo hayaku agete* (raise your hand quickly)".

Fig. 3. Action matrix

2.7 Example

The system makes examples from actual inputs from users. An example is a pair of a natural language part and an action matrix part. A natural language part is an EoA command or connective command. EoA commands and connective commands input by users are segmented into morphological elements first using MeCab. Then the system changes ending of verbs (which vary depending on conjugation) to original forms by a morphological analyzer. We perform morphological analysis to make learning more effective. It means that analyzing morphological elements and absorbing a variety of changing verb endings will be accomplished only by increasing the number of inputs. Therefore, we use a morphological analyzer instead of a vast amount of inputs. An action matrix is generated automatically based on its correspondence to the natural language part. The correspondence between those parts is being learned from users' input and feedback.

3. Generation of Rule by Three Examples Based Inductive Learning

3.1 Distinction Examples of EoAs and Compound Actions

All inputs from users are stored in an example database. Therefore, the system needs to distinguish examples of EoAs and compound actions (Fig. 4). Firstly, the system brings out any three examples from the example database. In natural language part, if one example's string includes the other two example's string, then the system distinguishes such example as an example of compound action, and the other two examples as an example of EoAs. Next, the system extracts EoA example's final state of motors from action matrix part (Fig.

5). Because there are some examples which are the same language part and different final state of motors, their final states are averaged.

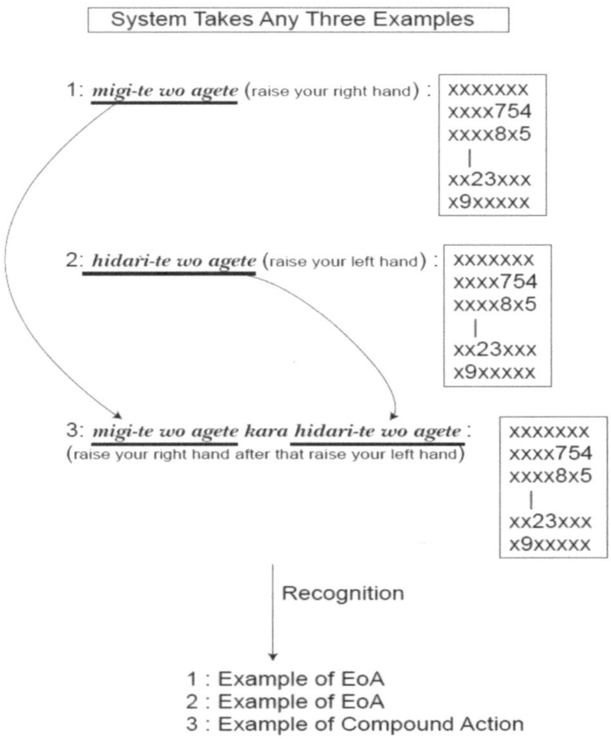

Fig. 4. Distinction of EoA and compound action

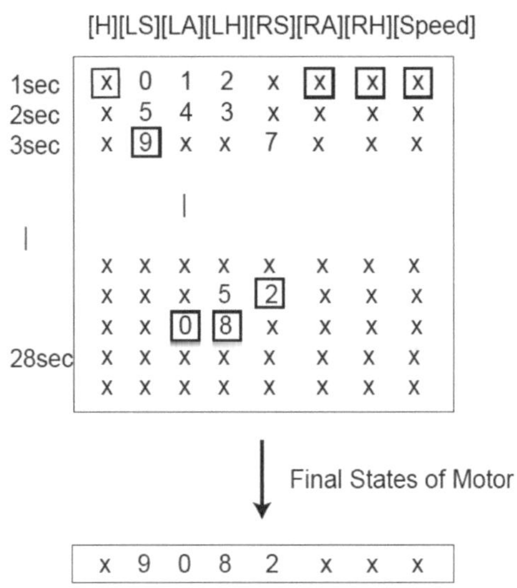

Fig. 5. Final states of motor

3.2 Abstraction of Meanings of Connectives by Three Examples Based Inductive Learning

Meanings of connectives are abstracted from examples of compound actions which contain connectives (Fig. 6). In this paper, as our targets are compound sentences which have one connective, we do not implement recursiveness of Inductive Learning in our method. Therefore, the abstraction of the method is performed one time.

(a) The system brings out any example of compound action containing two EoAs.

(b) In natural language part, the system parameterizes substrings of EoA commands of the connective command string as @1 and @2. Then, the remaining part of the string is assumed to be a connective.

(c) Next, in the action matrix part of compound action, the system finds rows which accomplished final positions of EoAs, and parameterizes the rows as @1 and @2. Then, the structure of the remaining matrix represents the meaning of the connective.

(d) Lastly, a rule becomes a pair of a parameterized natural language part and a parameterized action matrix part.

In a parameterized action matrix, the structure, for example the sequence of parameters or the time delay from the first parameter to the next parameter of action matrix, represents meaning of connective. Originally Inductive Learning makes abstraction reflexively by abstraction among rules. However, in our system, Three Examples Based Inductive Learning makes one abstraction for the first step. Therefore, we can deal only with two EoAs combinations.

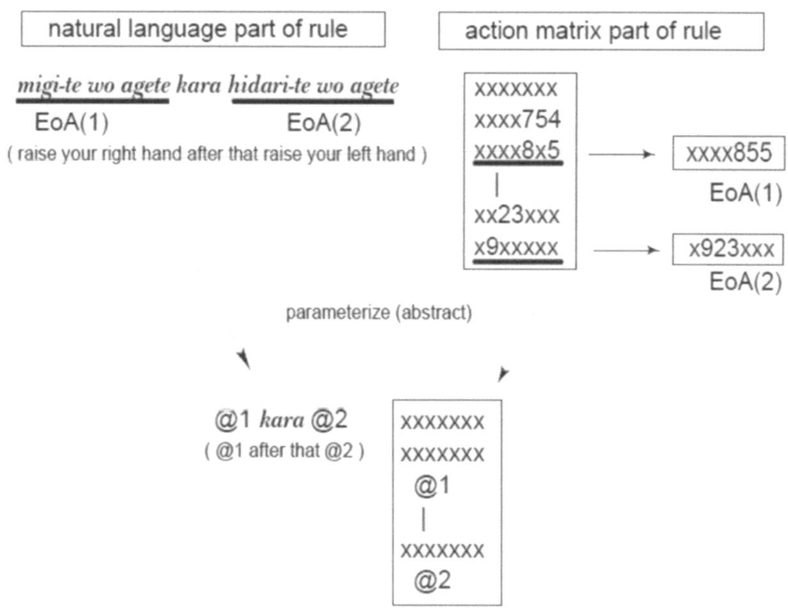

Fig. 6. Gengeration of rules by Three Examples Based Inductive Learning

3.3 Averaging of Rules

Rules are classified into 6 structures as follows.

(a) There is no parameter.
(b) There is only parameter @1.
(c) There is only parameter @2.
(d) There are @1 and @2 in normal ascending order.
(e) There are @1 and @2 in reverse ascending order.
(f) There are @1 and @2 in the same raw.

There are rules about a connective having different structure of action matrix, because the users' feedbacks slightly differ for actions with the same connectives. Therefore, the system has to average the examples contained in each structure. Firstly, the system distinguishes each rule about one connective as 6 structures. Then, in each structure, positions of parameter rows are averaged.

3.4 Adaptation Value of Rules

When there are averaged rules which have different structure, the system needs a criterion to select the best one. For this reason, the system calculates adaptation values of rules. The system counts the number of rules which have identical structure. Then the system regards the number as the adaptation value of the rule. Because the more number of teachings of the same rule's structure increases the higher the credibility of the rule becomes, a rule which has the largest adaptation value is referred if there are more rules which can apply.

3.5 Application of Examples and Rules

When a user makes a command in natural language, the system tries to perform the given action by applying previously taught examples and abstracted rules. The process is outlined below.

(a) Firstly, the system tries to apply rules. If input commands contain a connective string inside the rules database, then a rule which have the highest adaptive value is chosen.

(b) Then, the system chooses two EoA examples for parameters in natural language part from the rules database. Following the structure of the action matrix of the rule, final states of motors of the two EoA examples are inserted in parameters and the action matrix is performed by the robot.

(c) Secondly, if the system can not apply any rules, the system tries to perform the input command by applying actions from previous examples. Then, an example which has the same string as the input command is chosen and the robot performs an action based on the example's action matrix corresponding to its equivalent in language part.

(d) Finally, the robot performs no action if there is no rule or example which can be applied to the input command. Because the meanings of connectives are abstracted, the system can deal with unknown combinations of EoAs. There is an uncountable number of EoAs. Their combinations are a product of the EoAs number and EoAs number minus one, however it is not necessary to teach a robot this amount of combinations when using our method.

4. Experiment

We implemented the system and experimented on learning connectives. Learning experiment and evaluation experiment were conducted.

4.1 Learning Experiment

In this experiment, we determine if the learning system works and the learning process converges. We made four participants (age 20-30, all male students of graduate school majoring in science) input commands and taught actions to the robot. The system accumulates previously input knowledge without deleting it when users change. The flow of the learning experiment is shown in Fig. 7.

1. A participant inputs a command in natural language (Japanese) from a keyboard for the humanoid robot. A participant can input one of ten designated EoA commands or one connective command containing two of those ten EoAs and any connective he wants. Then all connectives are free to choose by participants to make this connective learning experiment fair.

2. The robot applies rules or examples that have been previously acquired, and performs an action. If there is no rule and no example which could be applied, the robot does not perform any action.

3. In the end the user evaluates the robot's performance by using marks shown below and teaches the robot the action when needed.

(a) Correct Response: The system's reply is regarded by participant as a correct movement.

(b) Semi-correct Response: The system's reply is close to the correct response but is not perfect in the participant's opinion. We define correct response and semi-correct response as proper responses.

(c) Erroneous Response: Participant regards the system's reply as incorrect and teaches the correct action to the robot.

(d) No Response: There is no response from the robot due to the lack of particular connective rule(s) and previous teaching of the EoA. The participant has to teach the robot a correct action.

4. Repeat them.

The system has no knowledge in the beginning, therefore all EoAs are taught by participants. In this situation, thing we have to pay attention to is the variety of language. For example, "raise your right hand" and "rise up your right hand" mean the same action. Because there are many ways how to represent the same action, users have to teach a vast number of EoAs to the robot. This problem can probably be solved by paraphrasing or other techniques; however we regard the problem as one of the future works. Therefore, for this stage we restricted the number of EoAs to 10 in order to make learning of connectives more effective. We prepared in advance a questionnaire which was answered by 10 participants. The questionnaire asked what action the participant would like this robot to perform. From this questionnaire results we collected 100 actions, and we chose randomly 10 EoAs which our robot can perform with its upper body. We used the 10 following EoAs.

(a) *migi-wo muite* (look to the right)

(b) *hidari-wo muite* (look to the left)

(c) *migite-wo atama-ni oite* (put the right hand on your head)

(d) *hidarite-wo koshi-ni oite* (put the left hand on your waist)

(e) *Ain shite* (perform the Ain3)

(f) *komanechi shite* (perform the Komanecchi4)

(g) *keirei shite* (salute [me])

(h) *akushu shite* (shake hands [with me])

(i) *gattsu pozu shite* (clench [hold up] fists in triumph)

(j) *hidarite-wo mae-ni dashite* (hold out your left arm)

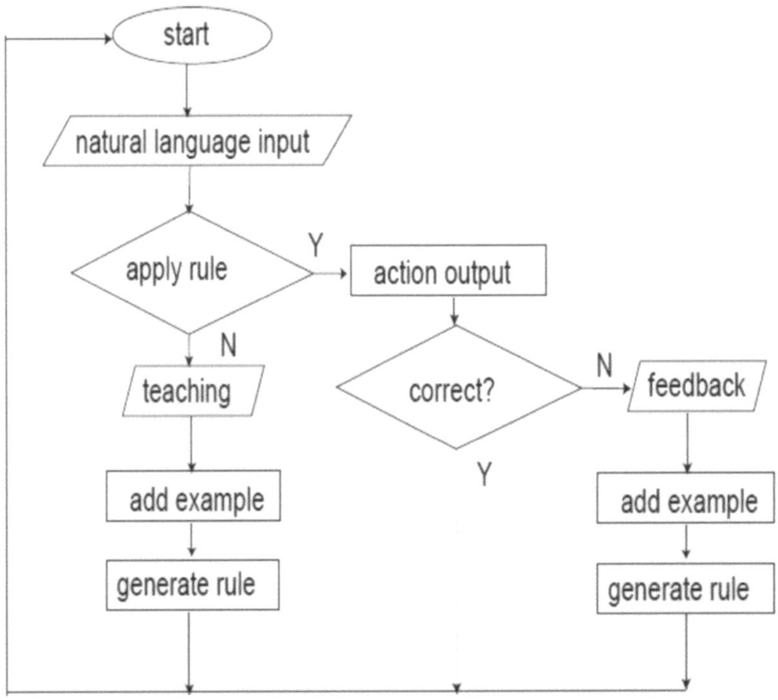

Fig. 7. System flowchart

4.2 Result of Learning Experiment

The participants input a total number of 210 commands which contained 157 connective commands (all 157 connective commands differed). The number of connectives actually used by participants was 25 (see Table 1). The process of learning experiment is shown in Fig. 8 and 9. Fig. 8 shows accumulative shifts of proper response ratio, error response ratio and no response ratio of all responses. The proper response line shows gradual increase, while the no response line shows a radical decrease in the beginning and a gradual decrease later. This proves that our system kept learning EoAs and connectives from the state of no knowledge, therefore it gradually made the robot perform proper movements. We can also observe that the system can correct wrong rules, as the erroneous response line shows a slight, gradual decrease. It proves that the system could adapt to changing participants (four of them in this case) and could change rules correctly due to the feedback process. Fig. 9 shows a shift per 10 commands of proper response ratio to 157 connectives commands excluding EoA commands. The proper response line comparatively gets steady after inputting about 80 connective commands. It means, though the system had no enough knowledge to reply with proper responses in the beginning, the robot has corrected its knowledge enough to perform proper movements later.

-kara (and then) **
nagara (and at the same time) **
mae-ni (before) **
ato (after) **
-naide (without) **
– (-) **
-to doji-ni (while) **
-kara yukkuri (and after that slowly) *
sugu (do just after) *
-to misekakete (pretend to do and after that) *
ato-de (after that)
sukoshi[Kanji] matte-kara (after waiting a while)
ato-de takusan matsu (wait a few minutes and after that)
tsutsu (and)
shibaraku shite-kara (after some time)
sono mama (keep on)
ato-de (Kanji) (after that)
soshite (after)
soshite takusan matte-kara (after waiting couple of minutes ...)
toki (when ...)
kara sugu (soon after)
sukoshi [Kana] matte-kara (after waiting a second)
takusan matte-kara (after waiting several seconds)
-to omowasete (after making one think that you do...)
... *furi-shite* (pretend to ... and after that)

Table 1. Connectives from participants

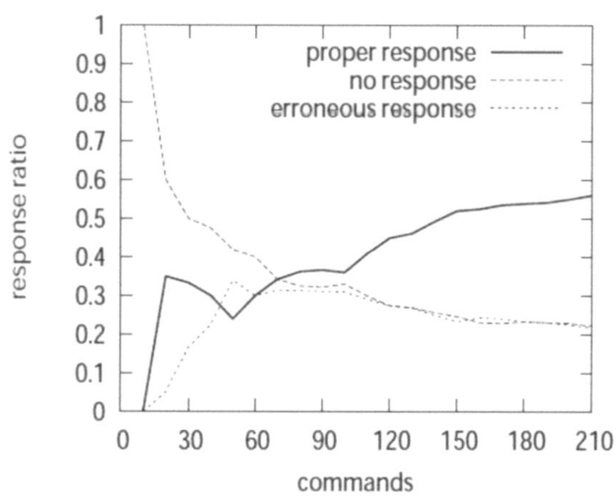

Fig. 8. Shift of the system response

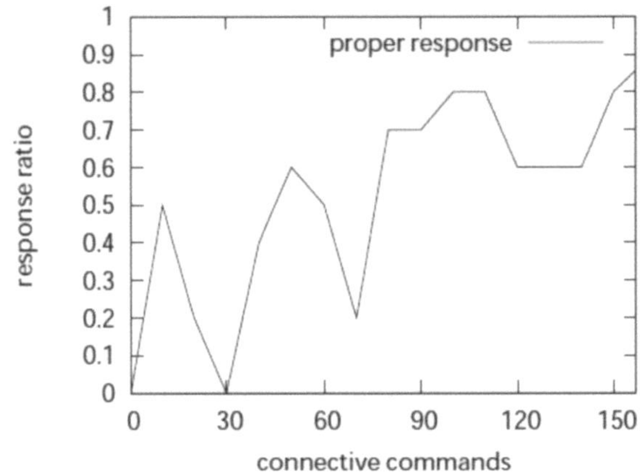

Fig. 9. Shift of the system response (connective commands)

4.3 Evaluation Experiment

In this experiment, we determine if rules acquired in the learning experiment are correct. We asked four participants to evaluate the rules. The participants (age 20-30, all male and majoring in science) were different from the learning experiment participants. 25 connectives were proposed by participants during the learning experiment. However, there were cases where participants made mistakes while teaching. We took them into consideration and chose connectives input by participants more than three times as target rules of this evaluation. There were ten of such rules obtained (rules marked "**" or "*" in Table 1).

By adding two EoAs to each of ten target connectives, we made ten connective commands. Then the robot performed those connective commands, and the four participants were shown the commands and the actions. The participants evaluated actions with the following evaluation values.

(a) Success: Score=3

(b) Almost Success: Score=2

(c) Failure: Score=1

4.4 Result of Evaluation Experiment

We define a "correct rule" as a rule which was evaluated as "Success: Score=3" or "Almost Success: Score=2" by all participants. The overall precision was calculated by Formula (1), and the precision of connectives acquisition was 0.7.

From the evaluation results we can find that the system can acquire connectives which signify simple relation of order conjunction (for instance, "*nagara* [and at the same time]" or "*mae-ni* [before]"). We partially acquired a concept of time using the embodied system. Furthermore the system acquired a connective that represents meaning of logical negation. That is "*naide* (without)". The system could not acquire three connectives. First of them, "-*kara yukkuri* (and after that slowly)" was scored 1 by only one participant. This connective contains an adverb. However, other participant scored 3 for the same performance. It shows that the evaluation of performance requested with the same command depends on participants. In the second case, "*sugu* (just before)" was scored only 2 or 1. This connective also contains an adverb. The last one, "*misekakete* (pretend to do and after that)" is scored 1 by all participants. It contains a verb and the system still can not understand verbs correctly. The robot still can not represent verb meaning. Therefore, the connectives including verbs are not acquired. We plan to work on a mechanism to acquire verbs in the near future.

$$\text{Precision} = \frac{\text{the number of correct rules}}{\text{the number of rules applied three or more times}} \quad (1)$$

Rule	Evaluation (participants A, B, C, D)	sum
@1 *kara* @2 (@1 and then @2)	3, 3, 2, 2	10
@1 *nagara* @2 (@1 and at the same time @2)	3, 2, 3, 3	11
@1 *mae-ni* @2 (@2 before @12)	3, 3, 2, 3	11
@1 *ato* @2 (@2 after @1)	3, 3, 2, 3	11
@1 *naide* @2 (@2 without @1)	3, 3, 3, 3	12
@1 @2 (@1 @2)	3, 3, 3, 3	12
@1 *-to douji-ni* @2 (@1 while @2)	3, 2, 2, 3	10
@1 *-kara yukkuri* @2 (@1 and after that slowly @2)	1, 2, 2, 3	8
@1 *sugu* @2 (@1 just before @2)	2, 2, 1, 2	7
@1 *-to misekakete* @2 (pretend to @1 and after that @2)	1, 1, 1, 1	4

Table 2. Evaluation

5. Discussion

5.1 Compositionality of Rules

In Fig. 9, the system responded to unknown connective commands with an accuracy of about 0.7 after 90 commands. These commands are new combinations of EoAs. The number of possible commands that can be made by such combinations is 2,500, because the number of EoAs is ten and the number of connectives that were input by users is 25. Still the system could respond with high accuracy by learning only a small number of commands. Thus, we can conclude that the acquired connectives that are abstract rules have compositionality and the rules can handle many new meanings by combination of simple sentences. Such compositionality is one important aspect of natural language. The system is able to produce different (and not previously input) results by varying EoAs. This is one advantage of our learning algorithm.

5.2 Learning Algorithm

In the learning algorithm which we suggested, the system can only connect two simple

sentences, though humans can create complex orders with many clauses following each other, for example "@A and after that @B and then @C and after that ...". This is because the process of comparing and parameterization is not recursive. We have to extend the learning algorithm to compare not only examples, but also rules recursively.
Furthermore, we utilized an innate learning ability and an innate cognitive ability as a basis for the learning algorithm. The cognitive ability is the ability to recognize movements based on the final posture of the robot. However, this is too simplistic to be able to recognize many types of movements. With this limited basic ability, the system cannot correctly recognize for example waving movements or movements in which the final posture depends to the context. Therefore, we have to extend the cognition ability.

5.3 The Number of EoAs

In this paper, we do not abstract the EoAs themselves. Therefore, the number of EoAs grows large. If one user inputs a command expressed a little bit differently than the other users' commands, then the system recognizes it as a completely different input and does not perform any action. The users may have to teach EoAs endlessly. For example, we have to teach "raise your right hand" and "raise your left hand" separately, because the commands mean different actions. However, if the system acquired the meaning of the verb "raise" after being taught "raise" a few times, then the system could autonomously create the unknown action represented as "raise your left hand" by combining it with meaning of "left hand". To solve this problem, we need acquisition of verbs, adverbs and nouns. To acquire the meanings of verbs, we need a representation system for them. Such system has to be able to handle both abstraction and symbol manipulation.

5.4 Two Types of Ambiguity of Connectives

We believe that there are two types of ambiguity in connectives. One is ambiguity depending on a user's peculiarities and the other is ambiguity depending on other context factors where the connective usage patterns may change for the same user. The former ambiguity can be resolved by distinguishing users, though we do not currently implement that. However, the latter ambiguity remains a problem to be solved. For example, "raise your right hand before raising your left hand" is ambiguous because the robot cannot decide if the right hand should be lowered before raising left hand or the right hand should remain raised.

6. Conclusions and Future Work

We proposed a language acquisition model based on an Inductive Learning Language Acquisition Model. Then we focused on connectives acquisition under the assumption that our system already acquired phonology and morphology. The proposed connectives acquisition algorithm is that the system inductively abstracts meanings of connectives from pairs of connectives and motor patterns. As a result of two experiments, we confirmed that the system learns connectives with an accuracy of 0.7 and the acquired connectives have compositionality.
In our future work, we will concentrate on improving the recursive learning algorithm. After that, we will work on a verb acquisition algorithm for humanoid robots.

7. References

Araki, K. & Tochinai, K. (2001). Effectiveness of Natural Language Processing Method Using Inductive Learning, *Proceedings of the IASTED International Conference on Artificial Intelligence and Soft Computing*, 0-88986-264-8, Cancun , pp.295-300

Asada, M.; MacDorman, K.; Ishiguro, H. & Kuniyoshi, Y. (2001). Cognitive developmental robotics as a new paradigm for the design of humanoid robots, *Journal of Robotics and Autonomous System*, Vol.37, pp.185-193, 0262-8856

Brooks, R. A. (1991). Intelligence without representation, *Artificial Intelligence*, Vol.47, pp.139-159, 0004-3702

Eimas, P. D. & Miller, J. L. (1980). Contextual effects in infant speech perception, *Science*, 209, pp.159-207, 0036-8075

Fukui, H.; Araki, K. & Tochinai, K. (2004). The Method of Acquisition of Corresponding Relation between Language and Act Using Inductive Learning with Genetic Algorithm (in Japanease), *Journal of Infomation Processing Society of Japan*, Vol.45, No.10, pp.2428-2437, 0387-5806

Hasegawa, D.; Rzepka, R. & Araki, K. (2007). Evaluation of Connectives Acquisition in a Humanoid Robot Using Direct Physical Feedback, *Advances in Artificial Intelligence*, Vol.4830, pp.664-668, 978-3-540-76926-2, December 2007, Springer, Surfers Paradise Australia

Heit, E. (2000). Properties of Inductive Reasoning, *Psychonomic Bulletin & Review*, Vol.7, pp.569-592, 1069-9384

Iwahashi, N. (2003). Language acquisition by robots? towards a new paradigm of language processing, *Journal of Japanese Society for Artificial Intelligence*, Vol.48, No.1, pp.49-58, 1346-0714

Jusczyk, P. W.; Houston, D. M. & Newsome, M. (1999). The beginnings of word segmentation in English-learning infants, *Cognitive Psychology*, 39, pp.159-207, 0010-0285

Markman, E. (1989). Categorization and naming in children: Problems of induction, Cambridge, Mass.: The MIT Press, 0-26263-136-9

Mataric, M. J. (2000) Getting Humanoids to Move and Imitate, *IEEE Intelligent Systems*, v.15 n.4, p.18-24, 0-7803-9731-7

Nakaoka, S.; Nakazawa, A.; Yokoi, K. & Ikeuchi, K. (2003). Symbolic Motion Description for a Dancing Humanoid Robot (in Japanease), *Technical Report of IEICE, PRMU2003-137*, 0913-5685

Ogura, T.; He, X.; Satou, A. & Hasegawa, O. (2006). Developmental Word Acquisition And Grammer Learning by Humanoid Robots through SOINN (in Japanease), *IEICE Technical Report PRMU2006-100*, pp.51-56

Roy, D. (2005). Semiotic schemas: A framework for grounding language in action and perception, *Artificial Intelligence*, Vol.167, pp.170-205, 0004-3702

Schaal, S. (1999). Is Imitation Learning the Route to Humanoid Robots?, *Trends in Cognitive Science*, Vol.3, No.6, pp.233-242, 1364-6613

Shinyama, Y.; Tokunaga, T. & Tanaka, H. (2001). "kairai" - Software Robots Understanding Natural Language (in Japanease), *Journal of Infomation Processing Society of Japan*, Vol.42, No.6, pp.1359-1367, 0387-5806

Sugita, Y. & Tani, J. (2005). Learning samentic combinatoriality from the interaction between linguistic and behavioral processes, *Adaptive Behavior*, Vol.13, No.1, pp.33-52, 1059-7123

Winograd, T. (1972). Understanding Natural Language, *Academic Press*, 0127597506, NY

Performance Assessment of a 3 DOF Differential Based Waist joint for the "iCub" Baby Humanoid Robot

W. M. Hinojosa, N. G. Tsagarakis and Darwin. G. Caldwell
The University of Salford
United Kingdom

1.Introduction

Anthropomorphic type robots combine many desirable features such as natural human like locomotion and human friendly design and behavior. As a result of this multi degree of freedom human like robots have become more and more common and many humanoid robots have recently been designed and fabricated.

The first biped humanoid robot was WABOT developed at Waseda University in 1973. This biped which was able to perform simple static walking was followed by the development of WABIAN I and II. WABIN-RII (Yamaguchi et al., 1999). Following these first prototypes a number of other human like robots were developed including the H6, H7 at the University of Tokyo (Nishiwaki et al., 2000), the impressive humanoid robots P2, P3 and ASIMO developed by HONDA (Hirai et al., 1998) (Hirose et al., 2001) and some more recent prototypes such as the JOHNNIE the anthropomorphic autonomous biped robot (Gienger et al., 2001) developed at University of Munich and the HRP, HRP-2 developed by METI in Japan (Kaneko et al., 2002) (Kaneko et al., 2004) (Kanehira et al., 2002).

Other less well know medium and small size humanoids include SAIKA (Shirata et al., 2004) and KENTA (Mizuuchi et al., 2002), the MK.5 a compact size humanoid robot constructed by Aoyma Gakuin University (Furuta et al., 2000), the PINO platform constructed by ERATO (Yamasaki et al., 2000) and the SDR-3X (Sony Dream Robot-3X) and SDR-4X developed mainly for entertainment (Kuroki et al., 2001) (Fujita et al., 2003).

In the above examples the waist joint is usually implemented using a simple serial mechanism with 2 DOF. This waist configuration is adequate for these robots since the tasks that are usually performed are limited to entertainment and amusement applications and in demonstrating walking capabilities.

The concept behind the development of iCub is to provide the cognition research community with an open architecture human like hardware/software platform for understanding of cognitive systems through the study of cognitive development. The iCUB will replicate a 2 and a half year old human baby acting in cognitive scenarios, performing tasks useful to learning and interacting with the environment and humans. In the early definition of the project two main tasks were considered from which the design requirements for the waist mechanism were derived. These are crawling and manipulation

(Metta et al.). Based on the requirements implied by these two tasks the design of the waist mechanism of the iCub was realized. A 3 DOF differential based mechanism was employed to provide not only increased stiffness but also increased range and flexibility of motion for the upper body.

This paper presents the design and control of this differential based waist mechanism of the iCub. The paper is organized as follows: Section II gives the general specifications of the waist joint in terms of DOF, range of motions and torque requirements. The following section describes the mechanical design adopted for the waist mechanism and highlights the advantages of this approach. Section IV introduces the system model used for the design of the control scheme while section V presents the control system design. Estimated performance measures of the control scheme are presented in sections VI and VII, by means of simulation and experimental result. Finally, section VIII introduces the conclusions of this work.

2. Waist Specifications

The kinematics specifications of the waist joint of the iCub include the definition of the number of D.O.F required and their actual location as well as the range of motions. These were defined with attention given to address the requirement for crawling and manipulation and in general to imitate the human baby form. As has been mentioned the iCub will have the approximate size of a two and a half year old child (Metta et al.). The D.O.F required for the waist was determined by considering both crawling and manipulation scenarios. Crawling simulation analysis showed that for effective crawling a 3 D.O.F waist is essential, Table I.

	Degrees of Freedom (°)	
Joint	Human	iCub
Waist	3 Roll Pitch Yaw	3 Roll Pitch Yaw
		=3DOF

Table 1. Waist mechanism number of D.O.F.s

An additional advantage that a 3 D.O.F waist will offer is the increased range and flexibility of motion for the upper body. This increased flexibility results in an amplified workspace of the iCub when performing manipulation tasks using its hands when in a sitting position. As manipulation is directly related to learning which is an essential task for the iCub the 3 D.O.F waist will provide significant benefits. Based on the above the iCub waist needs to provide 3 D.O.F enabling pitch, roll and yaw of the upper body.

As the iCub is a human-like robot and will perform tasks similar to those performed by a human, the range of motions of a standard human were used as a starting point for the selection of the movable ranges for the waist joints of the iCub.

Table II below introduces the range of motion specifications for the joints of the waist mechanism in comparison with the corresponding ranges found in the human.

Waist	Range of motion (°)	
Joint	Human	iCub
Waist roll	-35, +35	-90,+90
Waist pitch	-30, +70	-10,+90
Waist yaw	-40, +40	-60,+60

Table 2. Range of motion of the waist joint

Looking at Table II it can be observed that the range of motion in some joints has been extended or modified. In particular, the range of the waist yaw and roll has been increased while the range of the pitch motion was modified to increase the upper body forward angle to improve the front workspace of the robot. This extends the area where the iCub can reach and manipulate objects.

The last specifications to be discussed are the required torques for the waist mechanism. This forms the starting point for the selection of the actuation groups. To optimize the selection of actuators and reduction ratios, iterations of the mechanical design of the waist and simulated analysis of the system were carried out.

The selection of the type of actuator to power the waist of the iCub involved simulations of the robot while performing crawling motions with different speeds and transitions from a sitting to crawling pose and vice versa. From these simulations, the peak torque requirements of each joint of the waist mechanism were identified as presented in Table III.

Waist joint	Torque Required(Nm)
Roll	30.1
Pitch	45.8
Yaw	27.2

Table 3. Torque required for the waist mechanism

3. Waist mechanical design

The CAD model and the first prototype of the waist mechanism of the iCub baby humanoid robot are shown in Figure 1 and 2. The iCub waist was designed using the 3D CAD software *Pro Engineer Wildfire 2 from PTC.*

In the previous section, the role of the waist joint in the flexibility of motion of the upper body was highlighted. Such flexibility must be accompanied by high positional accuracy of the upper body as is required during manipulation.

For the pitch motion of the waist, the two actuator groups that power the pitch and yaw motion apply a synchronous motion to the two differential input wheels using the torque of both motors. For the yaw motion, the motors turn in opposing directions. This causes a yaw motion on the upper body, again using the torque of both motors.

This differential mechanism has several advantages when compared with traditional serial mechanisms used in humanoid robots. These are:

i) Increased stiffness and accuracy.
ii) The sum of the torque generated by the two actuators that power the differential joint can be distributed in both joints.

iii) As a result, smaller actuators can be used to achieve the maximum output torques required for the pitch and yaw motions.

The roll motion is implemented with the incorporation of a pulley shaft that is directly connected to the upper body frame. Torque is conveyed through a cable transmission system that provides also additional gearing to meet the torque requirements of the roll joint, Table III. For the first prototype DC motor actuators were employed to power the waist joints.

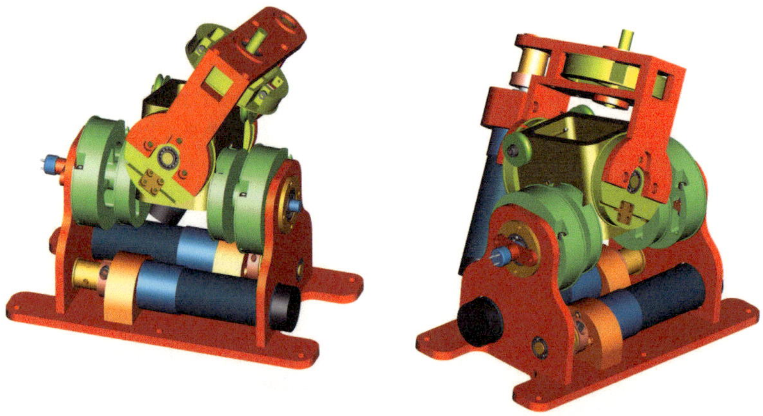

Fig. 1. CAD captures of the differential drive to be used as the robot waist.

Fig. 2. Prototype of the differential based waist mechanism.

4. System model

In order to achieve effective control of waist motions an accurate model of the system has been developed and extensive simulations performed.

3.1 Motor actuator model

For the development of this system model, a model of the motor actuator used to power the waist joints was required. Equations (1) and (2) characterize a general DC motor.

$$J\frac{d^2\theta}{dt^2} = T - B\frac{d\theta}{dt} \Rightarrow \frac{d^2\theta}{dt^2} = \frac{1}{J}\left(K_t i - B\frac{d\theta}{dt}\right) \tag{1}$$

$$L\frac{di}{dt} = V - R\times i - e \Rightarrow \frac{di}{dt} = \frac{1}{L}\left(V - R\times i - K_e\frac{d\theta}{dt}\right) \tag{2}$$

This mathematical model was implemented using MATLAB Simulink, Figure 3, and the results were referenced with the real system.

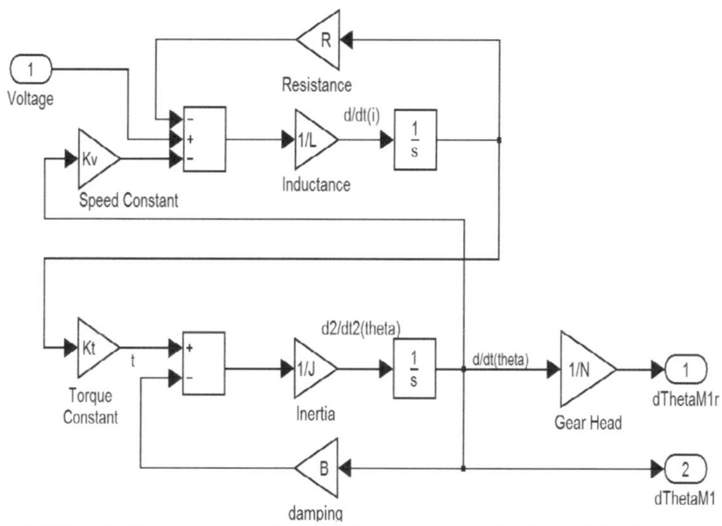

Fig. 3. MATLAB Simulink representation of the mathematical model of the two DC motors used to drive the differential waist mechanism.

3.2 Differential mechanism model

There are 2 motors controlling the pitch and roll motion of the robot waist, both located in parallel one in front of the other, but with its shafts pointing in opposite directions. From the kinematics analysis of the joints, the differential equations for the differential driver are:

$$R\times\left(\theta_{M1} + \theta_{M2}\right) = \theta_{roll} \tag{3}$$

$$R\times\left(\theta_{M1} - \theta_{M2}\right) = \theta_{pitch} \tag{4}$$

Where R is the gear head reduction ratio for motor 1 and 2, θ_{M1} and θ_{M2} are the rotor angles for motor 1 and 2 respectively and θ_{roll} and θ_{pitch} are the waist roll and pitch rotation angles.

3.3 Waist dynamics

In order to have an accurate mathematical description of the system the dynamics of the system must be included in the simulation; this includes the weights of the mechanical assembly, friction of the contact joints and motors and the inertia of the system. In Figure 4 the block named "spine dynamics" includes the waist kinematics (combines the motion of the two DC motors in differential mode) and the dynamics of a generic limb with a variable weight used to perform a wide range of tests under different sets of conditions.

3.4 Complete system modelling

In the simulations performed, the position feedback is calculated by integrating the speed output from the DC motor model, in the real system the speed is calculated using the derivative of the position output acquired from incremental encoder readings. Figure 4 shows the final control scheme including the differential equations block and the system dynamics for both motors, for a 2 DOF motion.

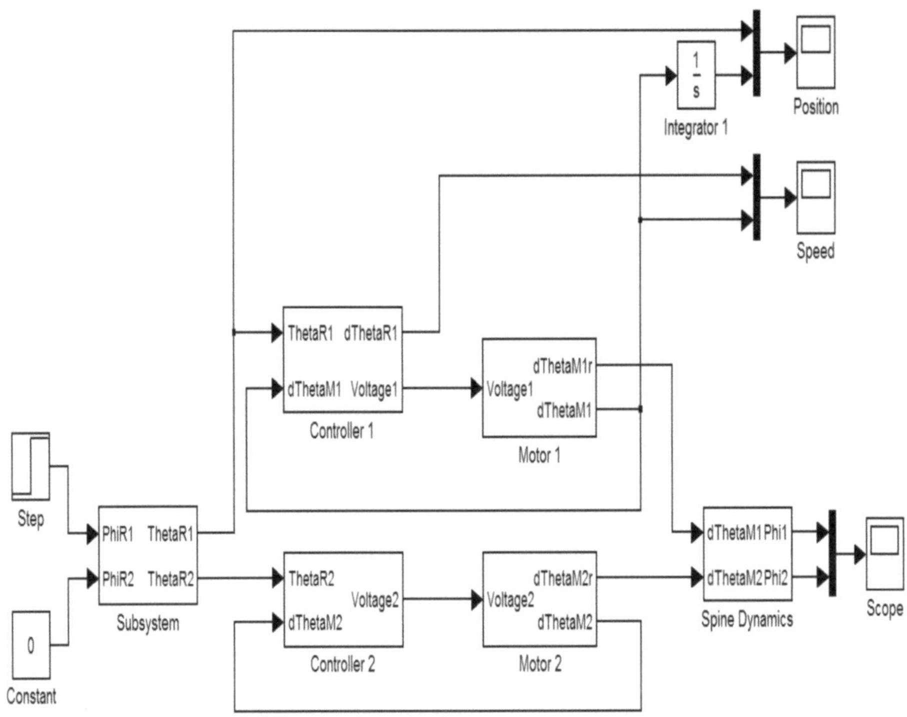

Fig. 4. Complete system for the 2 DOF differential joint. Speed and position control of 2 DC motors configured as a differential drive.

5. Control system

Many different approaches exist for control, each one with advantages and disadvantages, but the classic PID control scheme is still implemented in about 90% of real systems (Åström, 2005). In this section, the design, model and simulation of a dual loop PID control system with a dynamic anti-windup scheme will be described.

A traditional PID controller uses the derivative and integral of the error signal; but when the reference input changes, the tracking error changes rapidly. The derivative signal of the tracking error can generate shock on the system (Åström, 2005). To avoid this, the controller proposed uses the derivative of the output signal as shown in Figure 5.

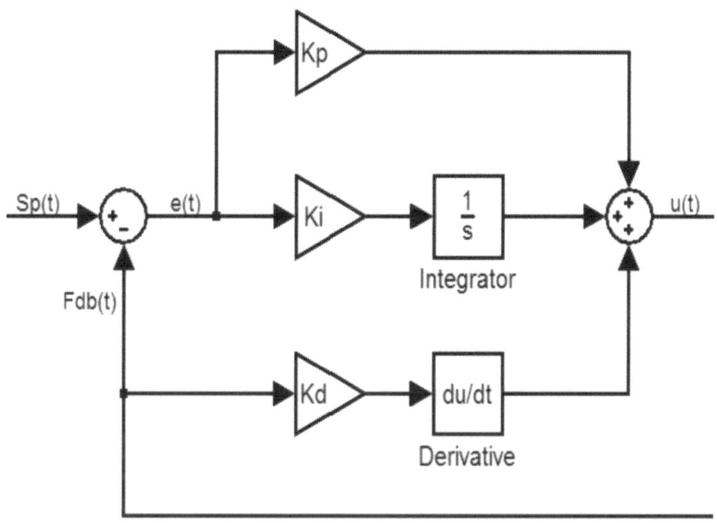

Fig. 5. Block Diagram of the classic three-term control scheme with derivative term using the feedback signal.

In order to perform position and speed control, a cascade controller was implemented. As observed in Figure 6 a first inner loop is used to control speed, this loop uses the position controller (outer loop) output as a reference signal. The speed is then established by setting the maximum and minimum values of the output limiter of the position controller. The fact that the speed is to be adjusted by setting a maximum and a minimum value to the limiter implies that the system will saturate. A simpler limiter will limit the position control output to a maximum value, which represents the desired maximum rotor speed. If this speed is low enough, the rotor can take a considerable amount of time to reach its final position, which in turn will produce a high integral term value.

In order to avoid this, a dynamic anti-windup scheme (Åström, 2005) was implemented; this scheme, Figure 7, takes the output of the system and compares it with a pre-saturated output to obtain a saturated error value, this value is then multiplied by a previously set gain and then introduced to the integrator, reducing its value. Using this scheme it is possible to ensure that the integral term will not have a high value so when the signal reaches its set point it will effectively reduce the steady state error, further more reducing the overshoot. Figure 6 shows the mentioned scheme.

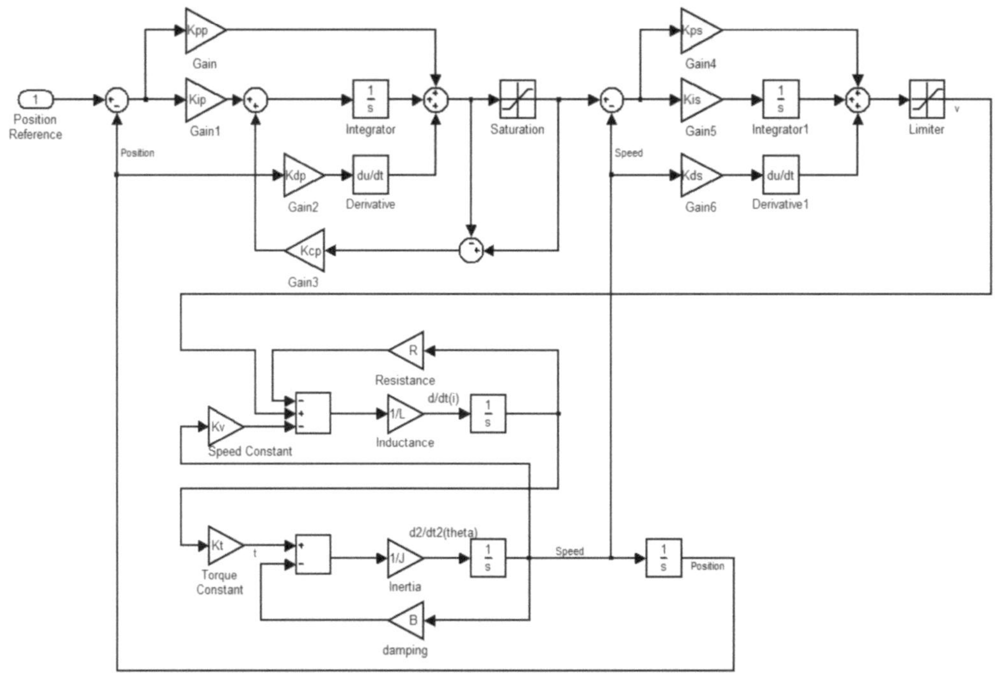

Fig. 6. Cascade control scheme for position and speed control, including the DC motor model.

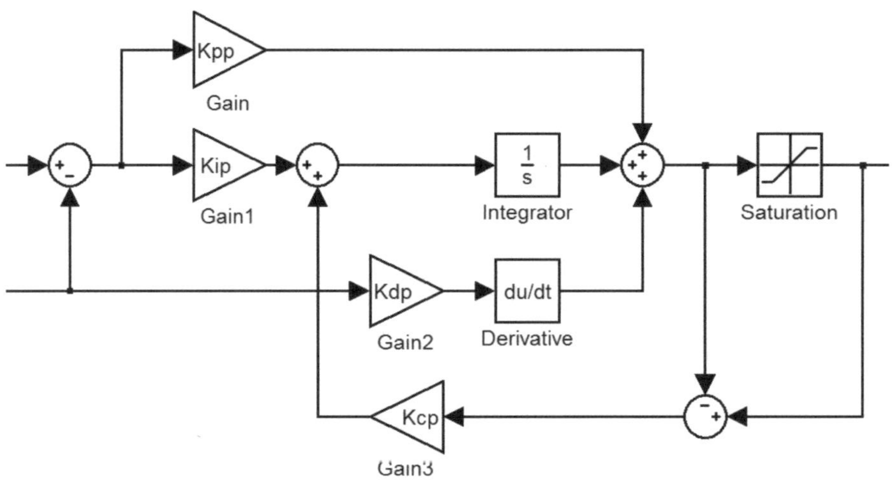

Fig. 7. Controller with dynamic anti-windup scheme, the integral term is effectively reduced, preventing overshoot and the saturation of the integral term.

6. Simulation results

Testing of the control schemes were carefully evaluated and rigorously analyzed through simulations. For the following simulations, 2 types of reference signal where used: a step input signal and a sinusoidal signal at a range of different frequencies. In all cases, the solid lines represent the reference signals and the dashed lines the rotor position. In the following graphs, the only load considered was the weight of the limbs; the frictions in the joints were not considered.

6.1 Speed control

The balance of the body of the robot is greatly affected by the speed of motion due to inertial forces generated by the body weight. The following graphs show the simulated results of the speed controller, with a forced reference input signal (the actual speed reference is generated by the position controller). These simulations were performed using MATLAB Simulink. The actuator groups used for this test consist of two MAXON DC motors and gearboxes with a maximum output speed after the reduction of 440 deg/s.

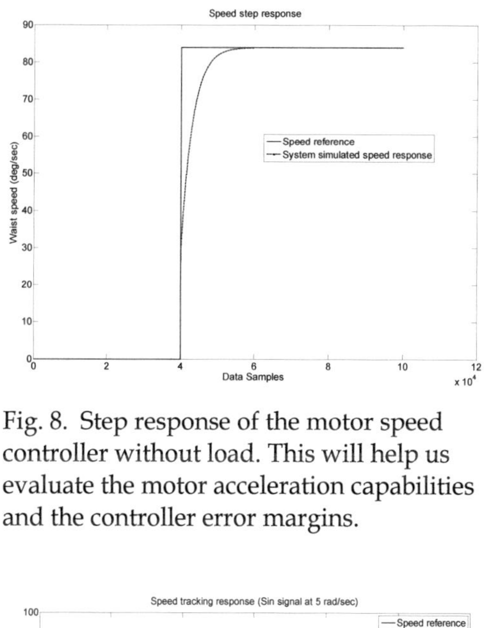

Fig. 8. Step response of the motor speed controller without load. This will help us evaluate the motor acceleration capabilities and the controller error margins.

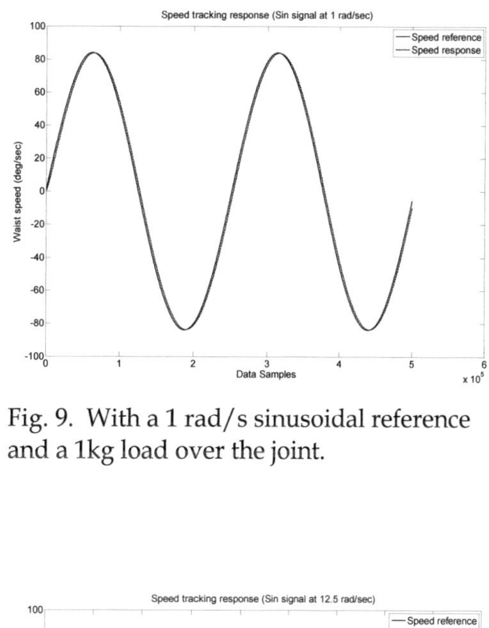

Fig. 9. With a 1 rad/s sinusoidal reference and a 1kg load over the joint.

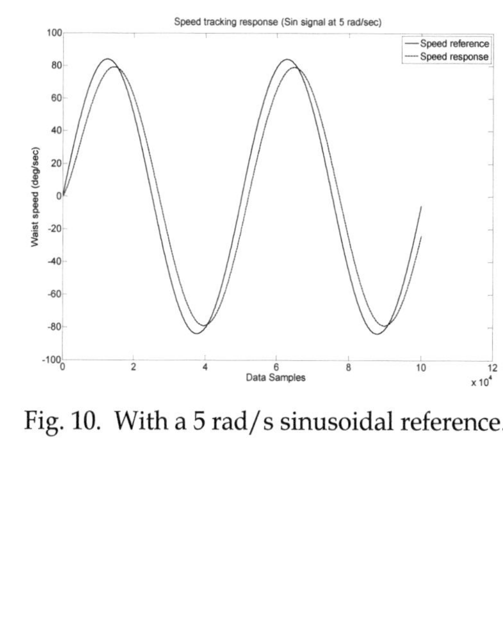

Fig. 10. With a 5 rad/s sinusoidal reference.

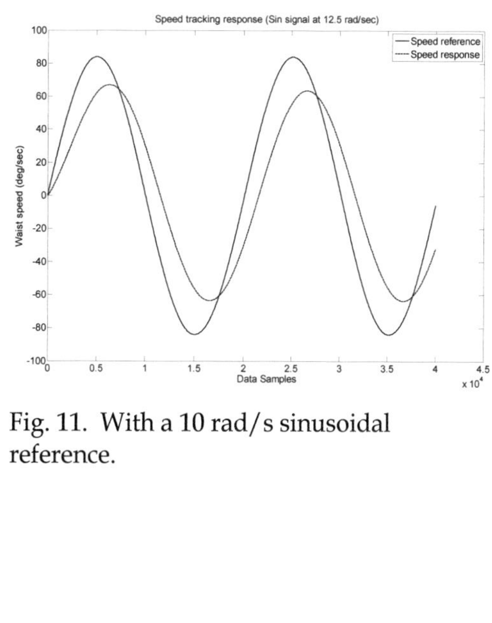

Fig. 11. With a 10 rad/s sinusoidal reference.

For the simulations results presented, reference signals of 84 deg/s (1.46 rad/s) were used and these represent the rotor speed for one of the motors.
This allowed the evaluation of the motor speed control capabilities under loaded and unloaded conditions and different motion directions (the load presented to the motor changes as the position of the centre of gravity of the robot's body moves)
The simulation results shows a good speed response time capable of moving the body of the robot at acceptable speeds.

6.2 Position control

The following graphs show the simulation results of the position using the complete model (position and speed controller cascaded).

The effect of a 1kg load in the tracking capabilities of the position controller can be seen; the load slows down the motion of the joint, even with a speed set near the motor maximum speed, though the position controller was able to position the waist joints with errors around 0.03deg in steady state.

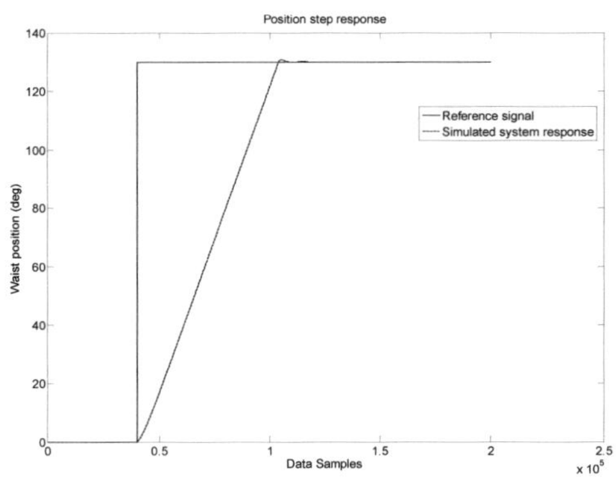

Fig. 12. Response for a square signal reference.

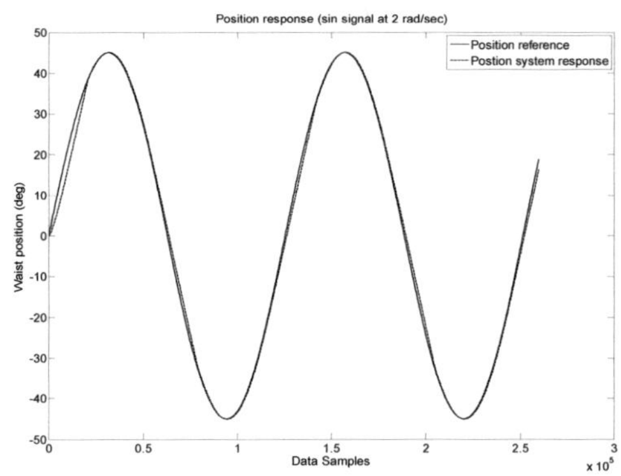

Fig. 13. Position tracking response with a 2 rad/s sinusoidal reference with a 90deg. of motion range.

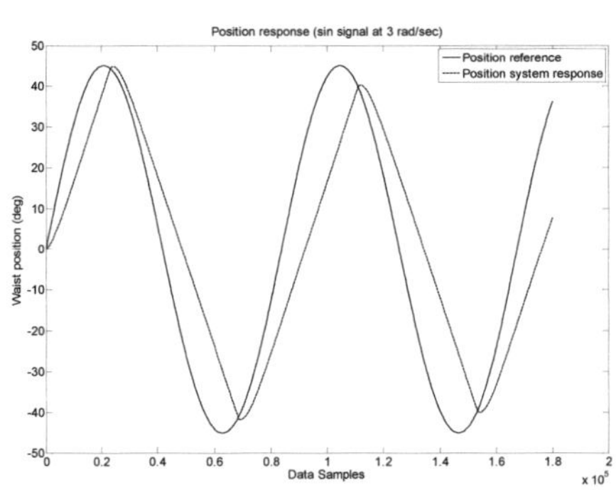

Fig. 14. Position tracking response with a 3 rad/s sinusoidal signal and 90deg of motion range.

Due to mechanical limitations, it can be observed from Figure 14 that the robot waist was not able to follow the reference signal accurately at rates of change higher than 2 rad/s even when setting the speed limit at its maximum.

7. System experimental results

The following results were obtained by collecting measurements from the real system through the microcontroller (TMS320F2810, DSP from Texas Instruments) on which the control system was implemented, through its JTAG interface to a PC. Figure 15-17, show the results for the position control, when a step input is presented, with a set speed of 84 deg/s in the joint. The joint is set to move forward and backwards between 22 and 65 deg.
After evaluating the results from the real system, different factors not considered in the simulation model, like friction in the waist joints, were observed to have little effect on the results.

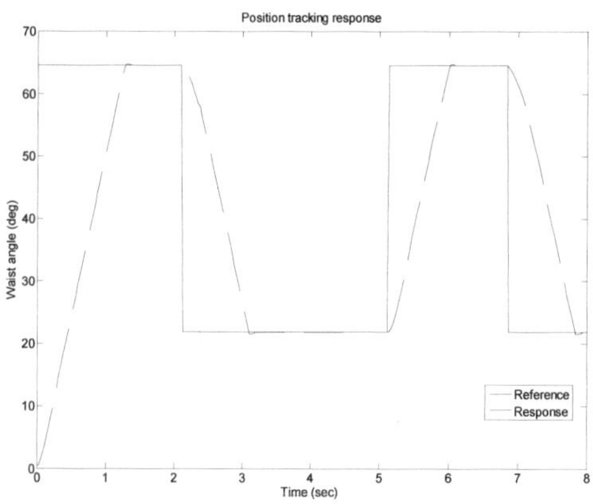

Fig. 15. Position Tracking Response of the real system.

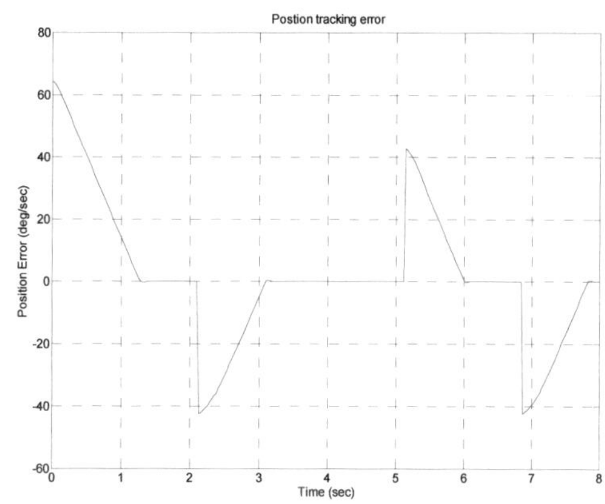

Fig. 16. Position tracking error of the real system.

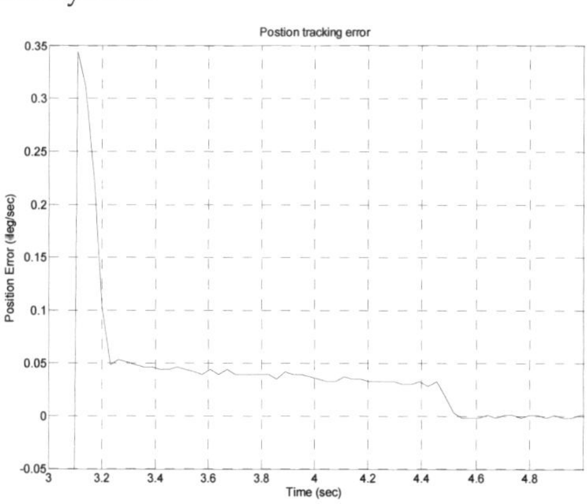

Fig. 17. Position tracking error in steady state. Actual results show errors under 0.05deg in steady state with 0.35deg overshoot.

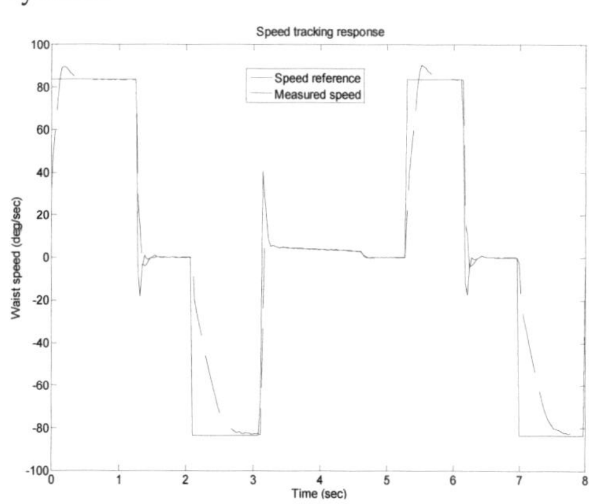

Fig. 18. Speed tracking response of the real system. Reference signal in solid line, actual waist speed in dashed line.

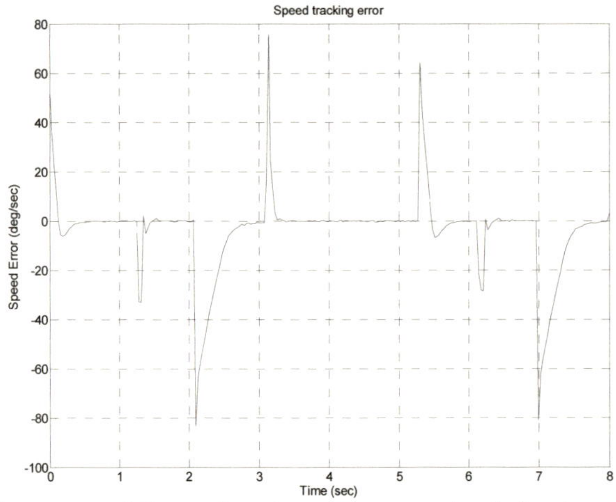

Fig. 19. Speed tracking error of the real system. Errors under 1 deg/s can be observed.

8. Conclusions and future work

This work presented the design of a differential based mechanism developed to form the waist joint of a baby humanoid robot. A cascade PID based position and speed controller was developed and its characteristics, such as overshoot, settling time and steady state error, have been evaluated through both experimentation and simulation. A control system consisting on a PID controller was established to achieve accurate position control of the joints.

It has been demonstrated through experimental implementation that the proposed control system can achieve control accuracy of 0.05 deg in step responses. In addition, a favorable speed control for sinusoidal and step trajectories was achieved. The control results presented in this study demonstrate that the proposed mechanism and control system can offer the desired motion range with high positional accuracy.

Future work will include a performance evaluation of the system using variable length and weight bodies to evaluate the effect of inertia on the system as well as mechanical (fatigue, maximum torques, etc.), electronic (current consumption, noise, etc.) and thermal effects.

9. References

Åström, K. J. (2005). *Advanced PID Control,* ISA - The Instrumentation, Systems, and Automation Society, ISBN-10: 1556179421

Fujita, M.; Kuroki, Y.; Ishida, T. and Doi, T. Autonomous Behavior Control Architecture of Entertainment Humanoid Robot SDR-4X. *Proceedings of IEEE/RSJ International Conference on Intelligent Robots and Systems (IROS 2003),* Vol. 1, pp. 960–967.

Furuta, T.; Tawara, T.; Okomura, Y. and Tomiyama, K. (2001). Design and Construction of a Series of Compact Humanoid Robots and Development of Biped Walk Control Strategies, *Robotics and Autonomous Systems,* Vol. 37, November 2001, pp. 81-100.

Gienger, M.; Löffler, K. and Pfeiffer, F. (2001). Towards the Design of Biped Jogging Robot, *Proceedings of the IEEE International Conference on Robotics and Automation*, pp. 4140-4145.

Hirai, K.; Hirose, M.; Haikawa, Y. and Takenaka, T. (1998). The Development of Honda Humanoid Robot, *Proceedings of the IEEE International Conference on Robotics and Automation*, Vol. 2, pp. 1321-1326, ISBN: 0-7803-4300-X, May 1998, Leuven, Belgium.

Hirose, M.; Haikawa, Y.; Takenaka, T. and Hirai, K. (2001). Development of Humanoid Robot ASIMO, *Proceedings of the IEEE/RSJ International Conference on Intelligent Robots and Systems*, Maui, October 2001.

Kanehira, N.; Kawasaki, T.; Ota, S.; Akachi, K.; Isozumi, T.; Kanehiro, F.; Kaneko, K. and Kajita, S. Design and Experiment of Advanced Leg Module (HRP-2L) for Humanoid Robot (HRP-2) Development, *Proceedings of the 2002 IEEE/RSJ International Conference on Intelligent Robots and System*, Vol. 3, pp. 2455-2460, ISBN 0-7803-7398-7.

Kaneko, K.; Kanehiro, F.; Kajita, S.; Yokoyama, K.; Akachi, K.; Kawasaki, T. ; Ota, S. and Isozumi, T. (2002). Design of Prototype Humanoid Robotics Platform for HRP, *Proceedings of the 2002 IEEE/RSJ International Conference on Intelligent Robots and Systems*, Vol. 3, pp. 2431-2436, ISBN: 0-7803-7398-7.

Kaneko, K.; Kanehiro, F.; Kajita, S.; Hirukawa, H.; Kawasaki, T.; Hirata, M.; Akachi, K. and Isozumi, T. (2004). Humanoid Robot HRP-2, *Proceedings of the 2004 IEEE International Conference on Robotics and Automation*, Vol. 2, pp. 1083-1090, ISBN: 0-7803-8232-3, New Orleans, LA, USA, April 2004.

Kuroki, Y.; Ishida, T.; Yamaguchi, J.; Fujita, M. and Doi, T. (2001). A Small Biped Entertainment Robot, *Proceedings of IEEE-RAS International Conference on Humanoid Robots*, pp. 181-186, Maui, Hawaii.

Metta, G.; Vernon, D. and Sandini, G. *D8.1 Initial Specification of the CUB Open System*, http://www.robotcub.org

Mizuuchi, I.; Tajima, R.; Yoshikai, T.; Sato, D.; Nagashima, K.; Inaba, M.; Kuniyoshi, Y. and Inoue, H. (2002). The Design and Control of the Flexible Spine of a Fully Tendon-Driven Humanoid "Kenta", *Proceedings of the 2002 IEEE/RSJ International Conference on Intelligent Robots and Systems*, Vol. 3, pp. 2527-2532, ISBN: 0-7803-7398-7, Lausanne, Swizerland, September 2002.

Nishiwaki, K.; Sugihara, T.; Kagami, S.; Kanehiro, F.; Inaba, M. and Inoue, H. (2000). Design and Development of Research Platform for Perception-Action Integration in Humanoid Robot: H6, *Proceedings of the International Conference on Intelligent Robots and Systems*, vol. 3, pp. 1559-1564, Takamatsu, Japan.

Shirata, S.; Konno, A. and Uchiyama, M. (2004). Design and Development of a Light-Weight Biped Humanoid Robot Saika-4, *Proceedings of the 2004 IEEE/RSJ International Conference on Intelligent Robots and System*, Vol. 1, pp. 148-153, ISBN 0-7803-8463-6, Sendai, Japan, 28 September - 2 October, 2004.

Yamaguchi, J.; Soga, E.; Inoue, S. and Takanishi, A. (1999). Development of a Bipedal Humanoid Robot - Control Method of Whole Body Cooperative Dynamic Biped Walking, *Proceedings of the IEEE International Conference on Robotics and Automation*, vol. 1, pp. 368-374, Detroit, MI, USA.

Yamasaki, F.; Matsui, T.; Miyashita, T. and Kitano, H. (2000). PINO The Humanoid that Walks, Proceedings of the IEEE-RAS International Conference on Humanoid Robots, Cambridge, MA.

6

Emotion-based Architecture for Social Interactive Robots

Jochen Hirth and Karsten Berns
Robotics Research Lab, Department of Computer Science, University of Kaiserslautern
D-67653 Kaiserslautern, Germany

1. Introduction

Applications of robots as assistance systems in household, health- or elderly care lead to new challenges for roboticists. These robots need to fulfill complex tasks and navigate in complex human "real world" environments. One of the most difficult challenges in developing assistance robots for everyday life is the realization of an adequate human-robot interface. Traditional interfaces like mouse, keyboard, or joystick appear inappropriate because even untrained non-robot or computer specialists should be able to control the robot.

From our point of view, the best way to achieve this goal is, to enable the robot to "natural" human interaction. Since approx. 65% of the information in human-human interaction is conducted non-verbally (Birdwhistell, 1970), the robots need the abilities to recognize and express these non-verbal signals.

The now upcoming question, whether humans would accept and use these interaction abilities of a robot at all, can be answered by investigating two studies.

Firstly, in (Reeves & Nass, 1996) it is proved that humans tend to use their experiences and customs from human-human interaction in human-machine interaction. They treat their computer as if it has a mind, they talk to it, they have some special emotional connection to it, and they even give non-verbal signals that are used in human-human interaction to it although they know that it has no possibility to recognize or even understand them. These results suggest that humans would use the humanoid interaction abilities of a robot. In (Breazeal, 2002) it is shown that the usage of emotional expressions of a robot enhance the human-robot interaction. So if a robot would have the abilities for humanoid interaction humans would use these features.

Secondly, in (Watzlawick, 2000) it is pointed out, that social interaction takes place in every situation in which two or more persons are facing each other. So it is clear, that if a robot should act as partner of a human, it needs the abilities for social interaction.

In the following, psychological fundamentals that are necessary for the realization of social interactive robots are explained. Afterwards, the related work on social interactive robots is summarized and discussed. Then the emotion-based control architecture of the robot ROMAN of the University of Kaiserslautern is introduced and the results of some tests to evaluate this architecture are presented. Finally at the end, a summary of this chapter as well as an evaluation of the developed architecture and an outlook for future work are given.

2. Psychological Insights

To enable a robot to interact socially, it is suggestive to investigate how social interaction between humans is realized. Psychological theories make statements about the involved components and their meaning to humans. In first place these components are motives and emotions. To have a social interactive robot it seems appropriate to transfer these components to the technical system.
In the following the definition of "social interaction" is discussed. Afterwards the involved psychological components, emotion and motives, are investigated. The function of emotions for the social human-human interaction as well as the importance of a motivation component is illustrated. Finally the relation between these psychological approaches for the social human-human interaction and the emotion-based architecture of ROMAN are pointed out.

2.1 Social Interaction

Psychologists differ between the two terms: "social interaction" and "social communication". Both terms are central for the human life in social groups.
Social interaction describes the alternating sequence of actions between humans that react on each others actions (Ulich, 2000).
Social communication describes the percipience and the exchange of information between two or more humans (Hobmair et al., 2003).
These definitions forebode that social interaction and social communication are closely connected to each other. It is not possible to have social interaction without social communication and vice versa. If people are influencing each other by their actions they are also transmitting information. If someone gives information to a person, this person is automatically influenced in some way. Because of this, in the following, the term "social interaction" will be used for both, interaction and communication.
As mentioned in (Canamero, 1997; Canamero & van de Velde, 1997) emotions play a major role in guiding and motivating actions in social interaction situations. The emotional state reflects the level of success of the interaction; it is conducted to the interaction partner by the emotional expressions. This information enables one interaction partner to react more precisely to the other one. That way misunderstanding can be prevented.

2.2 Emotion

In psychology there exist several theories about the generation of emotions (Hobmair et al., 2003). With the help of these theories the meaning of emotions to social interaction can be worked out. An overview of emotion theories can be found in (Ulich & Mayring, 1992). Two of the theories mentioned there are discussed in the following. Darwin postulated the evolutionary biological theory stating that the correlation between feeling and the expression are the same all over the world and they are inborn. That means that these expressions can be used all over the world for communication what means an enormous benefit unlike to speech. Functionalist oriented component-stress-models like postulated by

Lazarus or Leventhal, mention that emotions are the results of the evaluation of specific goals. This already happened in the beginning of mankind. That means that emotions are the result of the adaption of humans to a highly complex environment. In summary, emotions are correlated to specific expressions and they are necessary for the interaction with a complex environment.
Emotions include changes in physical and psychical state. These changes are perceived by humans and are rated as comfortable or uncomfortable. Therefore emotion mainly influences the behavior of humans.

In (Hobmair et al., 2003) several functions of emotions are pointed out:

- **Selective function**: Emotions influence the perception of the environment as well as the perception of internal stimuli.
- **Expressive function**: This is the most important function for social interaction. Facial expressions, gesture, body posture, and the tone of the voice are changed by emotions. These changes are used to conduct non-verbal information.
- **Motivational function**: Emotions activate and control the behavior of humans. Humans try to experience comfortable rated emotions again and try to avoid uncomfortable emotions.

This motivational function of emotions is the main reason for humans to interact. They want to reach a certain goal and as long this goal is not reached negative emotions are experienced. To explain this motivating and goal generating processes the theoretical construct of motives and motivation was invented (Hobmair et al., 2003). Therefore it is also necessary to investigate motives in order to explain the human interaction.

2.3 Motives and Motivation

Motivation is, unrecognizable from outside, a reason that activates and controls human acting in order to achieve a specific target. These reasons are named motives (Mietzel, 2000). When motives activates, goal directed behaviors are stimulated. This process is called motivation. So motivation is a state of activity that is controlled by motives. This state of activity lasts until the specific goal is reached or a motive of a higher priority gets active (Hobmaier, 2003).

According to (Hobmair et al., 2003) a motive is defined by five criterions:

- **Activation**: Specific behaviors are activated
- **Direction**: The activity is directed to a specific goal and lasts until this goal is reached or a motive of a higher priority level gets active
- **Intensity**: The intensity of the activity of a motive can vary
- **Duration**: In most cases the activity is maintained until the specific goal is reached.
- **Motivation**: A theoretical construct to explain the reasons for the observed behavior

The intensity of a motive does not only depend on the strength of the releasing stimulus. As mentioned in (Toman, 1973), motives have a time depending **cyclical character**. That means the more time passed since the last stimulation, much higher the intensity of the motive will

be. On the other hand, if a motive was active, just a few seconds ago, the same stimulus will only cause a little or even no intensity.
The activation of motives is explained using the control-loop-theory (Hobmair et al., 2003). The difference between the actual-state and the target-state leads to activity of the motives.

In summary, to realize social interaction an emotional state is necessary, because it is responsible for the human adaption to complex situations - like an interaction. This emotional state influences the perception system. It is also responsible for the expressions during the interaction. These expressions are of enormous importance for the interaction since they can be understood all over the world. Finally emotions are necessary to generate goals. The satisfaction of these goals is the reason why humans interact at all. So, the model to describe the human interactive system consists of an **emotional state**, a **perception system**, an **expressive system**, and a **motivational system**.

3. Social Interactive Robots and Agents

A number of research projects focus on human-robot interaction. In general the generated systems can be divided in embodied (robot or avatar) and disembodied systems. Since disembodied systems have the disadvantage that they can not use non-verbal signals for their interaction they are not taken into account in this chapter, as non-verbal signals are essential for social human interaction. In the following some representative projects for the generation of embodied systems for social human-robot interaction are presented.
There are a lot of systems that have the abilities to express emotions but they are lacking an architecture consisting of an internal emotion or motivation system, but as mentioned in section 2 this is essential for realizing adapted behavior in interaction situations. The different expressions are triggered by some specific events, e.g. the recognition of a special word. One of these systems is the humanoid robot Barthoc (Hackel et al., 2006). This robot has a huge number of degrees of freedom; therefore it is able to use gestures, facial expressions, body posture, and speech to interact with humans. It also has the abilities to recognize speech and to detect humans by its camera system. Another project in this area is Fritz (Bennewitz et al., 2007). Fritz is a communication robot that is able to walk - a biped robot - to interact by speech, facial expressions and gestures. A very interesting aspect in this project is the emotion expression by the tone of the voice. Therefore, the emotional state is mapped to speed, frequency, and volume of the voice.
A very complex and advance emotion architecture for social interaction is realized in the virtual agent Max (Becker-Asano et al., 2008). The architecture consists of a cognition module and an emotion module. The cognition module generates goals, depending on the achievement of these goals the emotion module is changes. The emotion module influences the cognition module by directly triggering behaviors or by defining goals, like experiencing positive emotions. That way, Max is able to generate dynamic sequences of emotional expressions in interaction situations. Although, as mentioned in (Bartneck et al., 2004), virtual agents have the disadvantage that they are not physically present and can not manipulate the real world, move, react on or use touching, this approach provides a lot of interesting aspects. Especially the experience in the combination of cognition and emotion in one control architecture is useful for future work on social interacting robots.

Certainly one of the best known projects in the area of social interactive robots is Kismet (Breazeal C. & Brooks R., 2005). Kismet was the first robot that is able to generate sequences of behavior and emotional expressions in an interaction situation. Besides the facial expressions Kismet is also able to generate sounds. These sounds are also adapted to the actual emotional state. The control architecture of Kismet consists of an emotional state, a perception system, a drive system for intrinsic motivation, and an expression system for generating emotional expressions and behavior. The emotional state also selects the group of possible behaviors. For behavior activation a drive system is used. The drives calculate their satisfaction depending on sensor data and activate different behaviors of the predefined group in order to reach a satisfied state again. Although the emotion representation and - depending on this - the adaption to changing situations are limited to the predefined set of emotions and the corresponding behaviors, Kismet proved that social human-robot interaction is possible. The psychologically grounded architecture of Kismet can be used as starting point for further development of social interactive robots.
A project that is closely geared to Kismet is Mexi (Esau et al., 2007). Mexi is able to generate facial expression depending on its internal state. According to (Breazeal C. & Brooks R., 2005) this state depends on emotions and drives, however the underlying architecture is less complex than the architecture of Kismet. Because of this it is more limited in its interaction abilities. Furthermore Mexi uses a chatbot to communicate verbally. But the chatbot has no direct influence on the emotional state of the robot. So the emotional state can not be changed directly depending on the verbal interaction.
One of the most advanced projects in the area of social human-robot interaction is Qrio (Aoyama & Shimomura, 2005). Qrio is a complex biped humanoid robot that is able to navigate in human environment. Qrio was developed as entertainment robot. It has the abilities to interact with humans using speech, body posture, and gestures. In context of social interaction Qrio is missing the most important medium, the facial expressions. As already mentioned before, most information in an interaction is conducted non-verbally and the facial expressions are the most important non-verbal signals. For realizing the emotional influence on Qrio's behavior the emotion grounded architecture was developed. This control architecture of Qrio consists of a perception part, an emotion representation, predefined behaviors, and a memory part. A very interesting aspect of this architecture is the memorization of the experienced emotion in correlation to the perceived object. They call this method emotionally grounded symbol acquisition, which seems to be very useful for the development of social interactive robots. That way the emotions experienced in the interaction with a special person can be remembered if this person is met again.

4. The Emotion-based Architecture of the University of Kaiserslautern

The emotion-based control architecture supporting human-robot interaction is tested on the humanoid robot ROMAN (figure 1). For social interaction, ROMAN is equipped with 24 degrees of freedom (11 in the face, 3 in each eye, 4 in the neck and 3 in the upper body) to realize non-verbal signals. To detect its interaction partner as well as its operational environment, two Point Grey Dragonfly cameras are installed in the eyes, six microphones and four infra red distance sensors mounted in the upper body. More information on the mechatronics system can be found in (Hirth & Berns, 2007). Besides this, the dialog system –

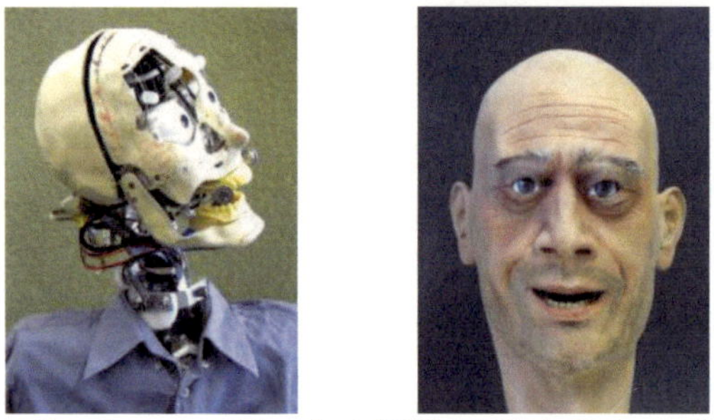

Fig. 1. The social interactive humanoid robot ROMAN, left: bare head, right: silicon skin clued on the head

based on speech synthesis and recognition - explained in (Koch et al., 2007) - is integrated to communicate using verbal speech.

For the realization of the emotion-based architecture, the integrated Behavior-based Control (iB2C) (Proetzsch et al., 2007) is used. According to section 2 the architecture consists of an emotional state, a perception system called percepts of interaction, the habits of interaction as an expressive system, and motives as motivational part (see figure 2). To achieve "natural" social interaction the goal of the implementation is to realize the functions of emotions: selective function (here: percepts of interaction), expressive function (here: habits of interaction), and motivational function (here: motives) besides this the characteristics of motives should be considered: activation, direction, intensity, duration, motivation, cyclical. The percepts of interaction perceive information of the environment. Depending on this information, direct responses - reflexes - performed by the habits of interaction are activated and the motives calculate their satisfaction. This satisfaction changes the actual emotional state of the robot. Besides this the motives activate several habits of interaction to change the robots behavior in order to reach a satisfied state. The actual emotional state influences the percepts of interaction, the motives, and the habits of interaction. In different

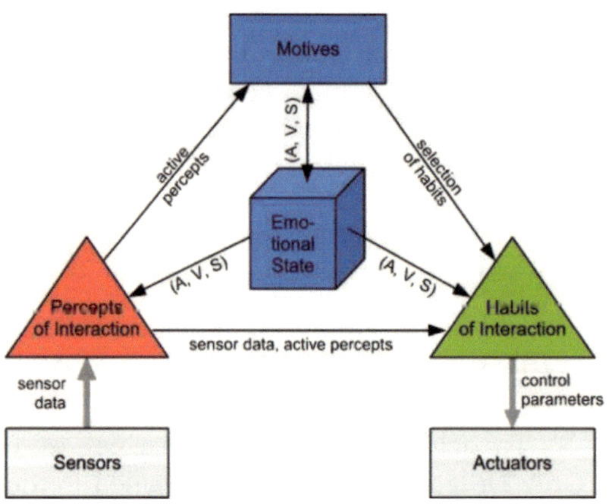

Fig. 2. The emotion-based control architecture, consisting of the four main groups, motives, emotional state, habits of interaction, and percepts of interaction; the emotional state of the system is represented by the 3 dimensions arousal (A), valence (V), and stance (S)

emotional states, the robot will interpret its environmental information in different ways and it will also act in a different way. For example if it is very nervous, its movements will be executed much faster and also its speech will be conducted much faster, than otherwise. Besides this, the actual emotion is displayed by facial expressions and body postures.

4.1 Emotional State

When creating emotion-based control architecture, the first question is how to describe an emotional state for using it in a humanoid. Looking to psychology leads to the so-called affective or emotion spaces. Several psychologists and cognitive scientists have designed different emotion representations: Like discrete affective vector representations (Ekman, 1992; Ortony et al., 1988), two dimensional representation (Plutchik, 1980; Russell 1980), and three dimensional representations (Broekens & de Groot, 2004; Mehrabian, 1996). A disadvantage of the discrete affective vector representation is that it lacks precision, because no continuous changes of the strength of a certain emotion are possible. In the two dimensional and especially in the three dimensional representations, the strength emotions can be described more precise and smooth continuous transitions between the several states can be realized. In two dimensional emotion representations the state is described by some kind of intensity value and a positive/negative rating. This does not suffice for the description of the complex state of an interaction. Therefor a third dimension representing the willing to interact is used. Because of this a three dimensional representation provides much more possibilities to specify the actual state. According to the studies of (Breazeal & Brooks, 2005; Lee et al., 2006), the emotional space shown in figure 3 was selected. The three axes of the emotional space are arousal (A), valence (V), and stance (S). Arousal specifies how thrilling a stimulus is to the robot. Valence describes how favorable or unfavorable a stimulus is to the system. High valence means the robot is very content with the situation and low valence means the robot is discontent. Stance specifies the attitude of the robot to the environment in a certain situation. High stance means the robot is willing to get new stimuli; low stance means the robot is not willing. In most emotional representations for every emotion a subarea in the emotional space is reserved. If the actual emotional state is

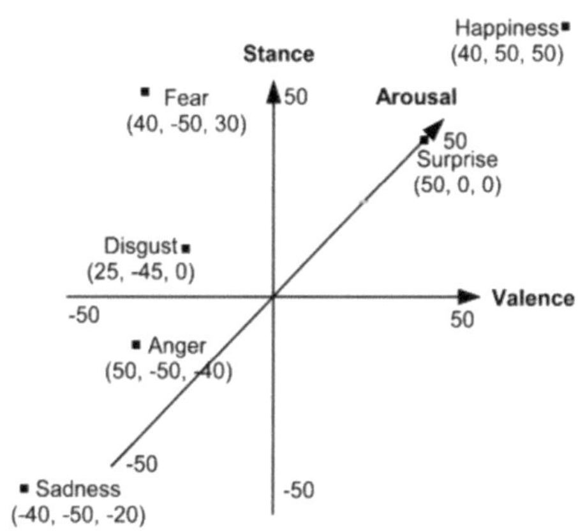

Fig. 3. The emotion space of ROMAN, this representation is used to define the actual emotional state of the system; every emotion can be described by the three parameters arousal, valence, and stance

within this area, the corresponding emotion is activated. That means every emotion that should be used has to be predefined. In psychology six basic emotions are distinguished: anger, disgust, fear, happiness, sadness, and surprise (Ekman & Friesen 1978) and all other emotions are mixtures of them. Because of this, the architecture does not work with a predefined number of emotions, but with the 3D-coordinates of an emotional state. This brings the advantage that every basic emotion, as well as mixtures of them, can be represented in the emotion space. An advantage in comparison to discrete emotion representations depends on the fact, that two emotional states that are located in the same subarea of the emotion space can be differentiated easily by using the coordinates.

According to the expressive and selective function of emotions, the whole robot behavior is influenced by the emotional state. If the robot is in a depressive state, for example, its movements and its speech will be slower and its voice will be deeper. By contrast, if the robot is very nervous, its movements will be very fast, the eyes and the eyelids will move with high frequency, and also its speech will be faster and the voice will be much higher.

The perception of the environment is also influenced. A certain situation will be interpreted in different ways depending on whether the emotional state of the robot is in a positive range or in a negative range.

4.2 Percepts and Habits of Interaction

The percepts of interaction are responsible for the detection of the environment. The sensor data is interpreted and combined to more complex information. That way a detailed description, e.g. of the interaction partner could be generated. Every environmental stimulus is represented by one percept. If the stimulus is detected, the corresponding percept is activated. Depending on the active percepts the motives determine their satisfaction with regard to the active percepts and the emotional state. Depending on the satisfaction of the motives, the emotional state is changed and the habits of interaction are stimulated (Strupp et al., 2008).

The habits of interaction describe the expression mechanism of the robot. As basic modules for the habits of interaction (Hirth & Berns, 2007) the behavior modules of iB2C, figure 4, are used.

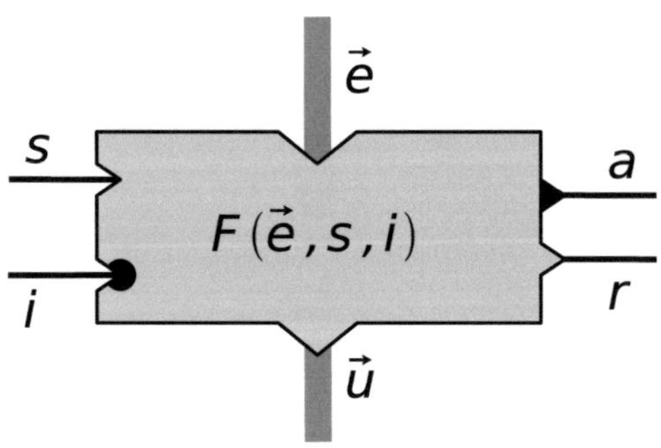

Fig. 4. The iB2C module: all habits of interaction are built out of these modules

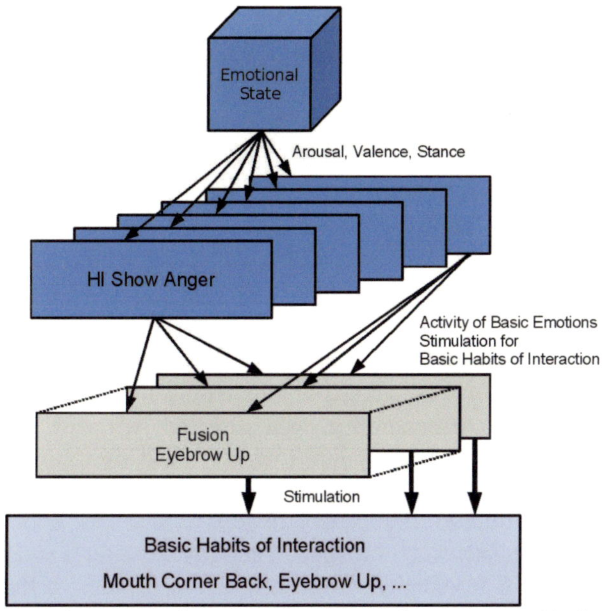

Fig. 5. The generation of emotional expressions using the concept of habits of interaction

These modules have three inputs: stimulation s, inhibition i and data input $\vec{e}$, and three outputs: activity a, target rating r and data output $\vec{u}$. In addition these modules have a function for
the calculation of the output data depending on the input: $F(\vec{e},s,i)=\vec{u}$ this function is called transfer function. The activation ι of a behavior is calculated depending on the stimulation and the inhibition: $\iota = s\cdot(1-i)$. More complex habits of interaction, e.g. emotional expressions, can be generated as a combination of basic habits of interaction. Habits of interaction concerning the same or similar body parts can be grouped. Based on the hierarchical ordering of the habits of interaction, different complexity levels can be generated. That way a complex behavior network for the generation of social behavior in an interaction situation is generated.

The most important function of the habits of interaction is the expression of ROMAN's emotional state. Therefore six habits of interaction representing the six basic emotional expressions (anger, disgust, fear, happiness, sadness, and surprise) are generated (figure 5). The (A,V,S)-vector representing the actual emotional state is the input of these modules. Depending on this vector every module calculates its activity. The activity of emotion$_i$, a_i where $i \in \{anger, disgust, fear, happiness, sadness, surpise\}$, is calculated in equation 1, where d denotes the diagonal of the emotion space (figure 3), P_i represents emotion$_i$ and $\vec{e}$ the input vector (A,V,S).

$$a_i = \frac{d - |P_i - \vec{e}|}{d} \tag{1}$$

For every basic emotion the specific expression, including facial and body movements, is defined depending on (Ekman & Friesen, 2002). To achieve more than just the basic emotional expressions the output is merged by a weighted fusion depending on their activity. That way it is possible to display a huge number of emotions instead of just the six basic ones.

4.3 Motives

To have a robot that uses an emotional state and that has the ability to express these emotions, leads to the question: how to activate these components. As explained in the model of (Mietzel, 2000) in section 2, a key factor for the social human-human interaction is to reach a certain goal. Without these goals humans would not interact with each other. As already mentioned, motives are generating goal directed behavior sequences. A certain motive is defined by its activation, direction, intensity, and duration. They represent the combination of emotions, needs, and their satisfaction.

The motives in the emotion-based architecture of ROMAN (Hirth & Berns, 2008) have an internal goal function that needs to be satisfied. They calculate their satisfaction depending on the activated percepts of interaction. This satisfaction influences the actual emotional state. The motives also try to change the robot's behavior in a way to reach a maximum of satisfaction. The output of different motives, that means the change of the emotional state and the activation of habits of interaction, is merged depending on the satisfaction of the motives. The lower the satisfaction of a motive, the higher is its influence on the robot behavior. In a robot system motives can be used to define some basic goals of the robot. For the use in a system for the interaction with humans some kind of communication motive seems to be appropriate. Also some kinds of safety motives are useful.

The motives are built using behavior modules of the iB2C-Architecture of the Robotics Research Lab (Proetzsch et al., 2007), see section 4.1. A motive (see figure 6) consists of three inputs (stimulation s sensor data $\vec{e}$ and inhibition i) as well as three outputs (control data $\vec{u}$, activity a, and target rating (satisfaction) r). In addition, it has four internal functions.

The first function $r(\vec{p},\vec{w},d,t)$ (see equation 2), calculates the target rating (satisfaction) of the motive by comparing the goal of the motive and the active percepts.

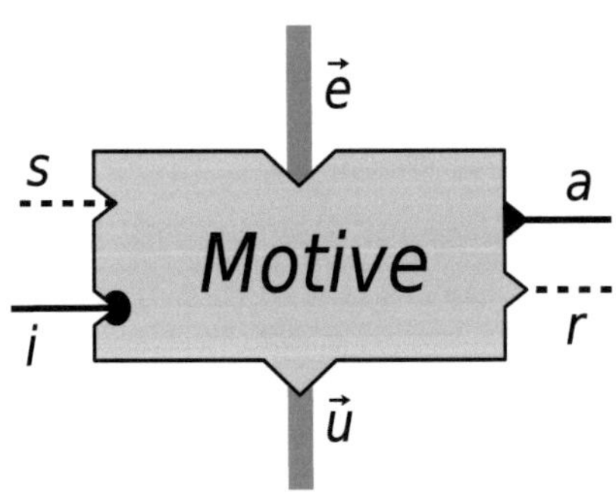

Fig. 6. A single motive module, of the emotion-based architecture of the humanoid ROMAN

$$r(\vec{p},\vec{w},d,t)=sigmoid\left(\left(\sum_{i=0}^{N}\frac{w_i}{\sum_{j=0,j\neq i}^{N}w_j}\cdot p_i\right)+t\cdot d\right) \quad (2)$$

Where N denotes the number of percepts, every percept $\vec{p}$ ($\vec{p}$ is part of the sensor data $\vec{e}$) represents one complex environmental information, e.g. the position of the interaction partner. The strength of the corresponding stimulus, e.g. the probability for the detection of a face, is represented by the activity of the percept. The weight vector $\vec{w}$ contains one weight for every percept; the weight represents the importance of the specific percept to the current motive. If the information, represented by the percept, is of maximum importance the weight value is 1 if it is completely unimportant the weight value is 0. As mentioned in section 2 every motive has a time depending characteristics, this is represented by $d\in[0,\infty]$. The time since the last stimulation of the motive is denoted by t, the strength of the influence of this time is specified by d. The sigmoid function is shown in equation 3.

$$sigmoid(x)=\frac{1}{2}+\frac{1}{2}\cdot\sin\left(\left(\frac{x-t_0}{t_1-t_0}\cdot\pi-\frac{1}{2}\right)\cdot\pi\right) \quad (3)$$

The second function calculates the activation $\iota(s,i)=s\cdot(1-i)$, where $i\in[0,1]$ denotes the inhibition input.

Depending on the activation and the target rating, the third function of a motive calculates the activity $a(r(\vec{e},\vec{w},d,t),\iota)$ (see equation 4). The activity-function of a motive is a piecewise defined function in which the interval $[0,t_0]$ denotes the inactive area of the motive, $[t_0,t_1]$ the area in which the activity is calculated based on $r(\vec{e},\vec{w},d,t)$ and $[t_1,1]$ means the satisfaction area. The codomain for the satisfaction function as well as for the activity function is $[0,1]$, corresponding to the control-loop-theory explained in section 2.

$$a(r(\vec{e},\vec{w},d,t),\iota)=\tilde{a}(r(\vec{e},\vec{w},d,t))\cdot\iota \quad (4)$$

$$\tilde{a}(r(\vec{e},\vec{w},d,t))=\begin{cases}0, \text{if } r(\vec{e},\vec{w},d,t)<t_0\\ 1, \text{if } r(\vec{e},\vec{w},d,t)>t_1\\ r(\vec{e},\vec{w},d,t), \text{else}\end{cases}$$

Finally the fourth function of a motive determines the control data $\vec{u}=F(\vec{e})$ that change the emotional state and contain stimulation and control values for the Habits of Interaction.

That way the characteristics of a motive mentioned in section 2 are realized:

- Activation, by $\iota(s,i)$
- Direction, by the internal goal function
- Intensity, by $a(r(\vec{e},\vec{w},d,t),\iota)$
- Duration and Cyclical character, by $r(\vec{e},\vec{w},d,t)$

- Motivation, by $\vec{u}$

As described above a motive gets active if the discontent value exceeds t_0 . The motive than calculates $\vec{u}$ for changing the robot's behavior in order to reach a saturated state. If the saturation of the motive is getting higher, the activity of the motive is getting lower. In addition every motive consists of different states. In every state the weight vector $\vec{w}$ as well as the dependence of time d may change. That way the generated motives realize the above mentioned request that they act in a completely different manner depending on their progress and on the environmental situation.

Motives that realize similar behaviors are grouped together. Within these groups the different motives can be realized on different priority levels or also on the same priority level. Motives on higher priority levels inhibit motives on lower levels, depending on their activity. For generating the output of a motive group, the output of a single motive is merged by a weighted fusion. The different motive groups are also realized on different priority levels. Groups on higher priority levels inhibit groups on lower levels. The activity of a motive group is represented by the activity of the corresponding fusion module and the inhibition of a motive group is realized by inhibiting this fusion module. Because of this modular set up the motive system remains maintainable and it can easily be exchanged or extended. The motives that are implemented at the moment and their functions are listed in table 1.

Motive	**Function**
Obey humans	If a human gives the robot the order to do something it will stop its actual work and obey this order.
Self-protection	Generate an evasive movement, if a close object is detected.
Energy consumption	If the robots energy is low it will try to get new energy, by telling humans in its surrounding.
Avoid fatigue	If the load is to high this motive avoids the starting of new processes.
Communication	If a person is detected this motive tries to start a conversation and takes care that the person is focused.
Exploration	If the robot is getting bored because of the absence of stimuli, this motive starts the exploration of the robots surrounding.
Entertainment	If ROMAN is exploring its environment but does not get any interesting stimulus, this motive gets active. ROMAN starts to sing a song especially to attract the interest of humans in its surrounding.

Table 1. All motives of interaction implemented at the moment in the control architecture of ROMAN

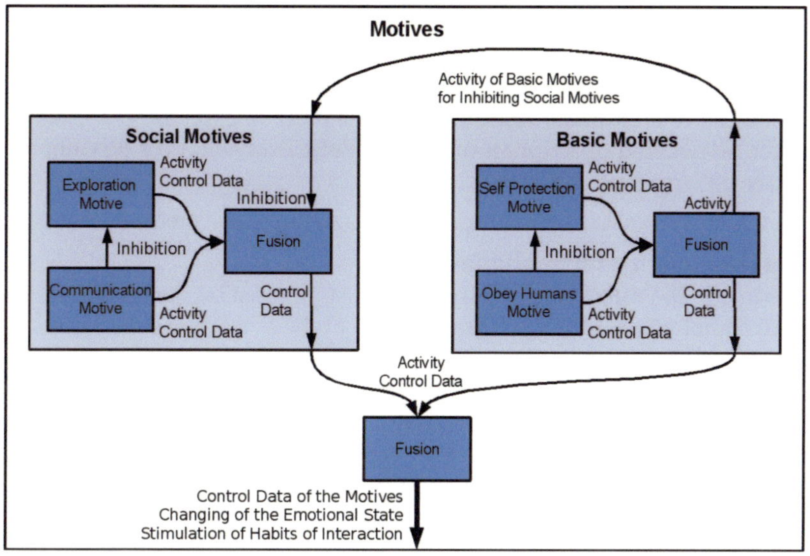

Fig. 7. An extract of the realized motive system; basic motives have a higher priority level than social motives

As an example an extract of the realized motive network is explained (see figure 7). In this example two motive groups are illustrated. The first group is the "social motives" that contains the "exploration motive" and the "communication motive" and the second group are the "basic motives" that contains the "self protection motive" and the "obey humans motive". Within the "social motives" the "communication motive" is on a higher priority level than the "exploration motive", therefore the "exploration motive" is inhibited by the "communication motive". The control data output of both motives is merged by a weighted fusion depending on the activity of the different motives. In the "basic motives" group, the "obey humans motive" has a higher priority than the "self protection motive". That way multiple groups of motives can be generated. These groups are also considered as motives. They are included in a larger group called "motives". In this example the "basic motives" are on a higher priority level than the "social motives", so the "social motives" are inhibited by the "basic motives". The output of the single motive groups is also merged by a weighted fusion depending on the activity of the different groups. This activity represents the activity of the single motives within a group.

5. Evaluation and Results

To test and verify the emotion-based architecture different experiments have been developed and conducted. In these experiments the single parts of the emotion-based architecture - percepts of interaction, habits of interaction, emotional state, and motives - were tested.

5.1 Percepts of Interaction

The percepts of interaction were used to realize a camera based emotion detection system. First experiments of this system show correct detection rates of about 60% for the emotions happiness, sadness, and surprise. Since the percepts of interaction are not in the main focus of this chapter no detailed description of these experiments will take place here. For more information see (Strupp et al., 2008).

5.2 Habits of Interaction and Emotional State

In this experiment the subsystem of the habits of interaction for generating emotional expressions in combination with the representation of the emotional state were tested.

Experimental Setup: Nine images (figure 8) and nine videos showing the facial expressions of ROMAN were presented to 32 persons (13 women and 19 men) of the age ranging from 21 to 61 years. Every person has to rate the correlation between presented expression and the six basic facial expressions with grades from 1 to 5 (1 means a weak and 5 a strong correlation). The videos were chosen instead of the robot to achieve the same presentation of the expression for every subject. This is important because the position of the robot and the subject may influence the rating of the expression strength. To avoid these problems and in order to get comparable results the video presentation was preferred. The program used for the analysis of the evaluation was the SPSS[1] (Statistical Package for the Social Scientist). The results of the evaluation are shown in table 2 and in table 3. The left column contains the shown facial expression. The right column contains the average values of the detected correlation between the six basic facial expressions and the current picture or video.

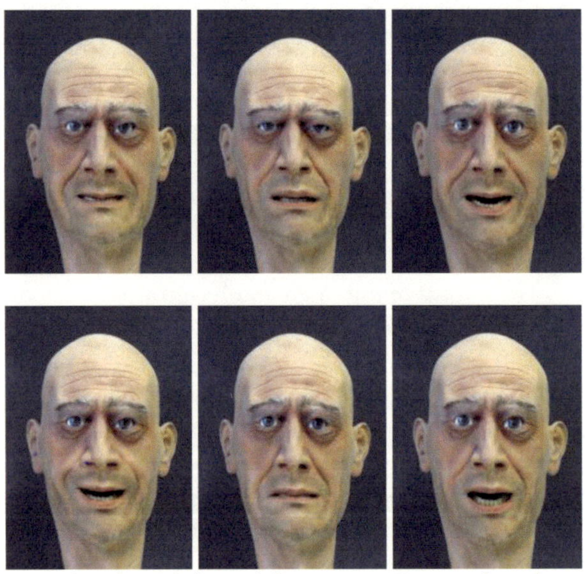

Fig. 8. Basic facial expression from left to right: anger, disgust, fear, happiness, sadness, and surprise

---

[1] http://www.spss.com/de/

Presented emotion	Detected strength
Anger	Anger: 4.5, Disgust: 1.8, Fear: 1.5, Happiness: 1.0, Sadness: 1.4, Surprise: 1.2
Disgust	Anger: 1.7, Disgust: 2.6, Fear: 1.0, Happiness: 2.6, Sadness: 1.2, Surprise: 3.7
Fear	Anger: 1.4, Disgust: 1.8, Fear: 3.6, Happiness: 1.5, Sadness: 1.8, Surprise: 3.8
Happiness	Anger: 1.1, Disgust: 1.0, Fear: 1.2, Happiness: 4.3, Sadness: 1.0, Surprise: 2.3
Sadness	Anger: 2.2, Disgust: 1.3, Fear: 2.8, Happiness: 1.0, Sadness: 3.9, Surprise: 1.3
Surprise	Anger: 1.3, Disgust: 1.3, Fear: 2.7, Happiness: 1.4, Sadness: 1.6, Surprise: 4.2
50% Fear	Anger: 1.5, Disgust: 1.5, Fear: 3.0, Happiness: 1.6, Sadness: 1.8, Surprise: 2.8
Anger, Fear, and Disgust	Anger: 3.0, Disgust: 2.4, Fear: 2.0, Happiness: 1.0, Sadness: 2.5, Surprise: 1.4
50% Sadness	Anger: 2.2, Disgust: 2.4, Fear: 2.8, Happiness: 1.0, Sadness: 3.7, Surprise: 1.3

Table 2. The results of the experimental evaluation of the images

Experiment Results: The results of the evaluation show that the correct recognition of the facial expressions anger, happiness, sadness and surprise is significant (significance $\alpha < 5\%$). But the facial expressions fear and disgust are not identified that good. Furthermore the results show that in most cases there are no significant differences between the evaluation of the pictures and the videos (significance $\alpha < 5\%$). Due to continuous human machine interaction the result of the video experiment is more important. The analysis of the understated emotions (fear 50% activation and sadness 50% activation) show, the subjects recognize that the facial expression is not that strong than in the case of 100% activation. The evaluation of the anger, fear and disgust mixture shows that in this case no facial expression is identified. But the selected facial expressions are named in the most cases (similar to the interpretation of comparable human faces expressions). Compared to psychological experiments for the recognition of human facial expressions there was no significant difference in the evaluation of the basic emotions anger, happiness, sadness and surprise.

5.3 Motives

To check the functionality of the motive system another experiment was conducted. In this experiment, the activation, activity and target rating of the involved motives and some other, interesting for the actual scenario, are monitored. These values are displayed in diagrams where the y-axis represents the monitored value and the x-axis represents the time in seconds. In the following we present the results of two representative runs.

Presented emotion	Detected strength
Anger	Anger: 3.7, Disgust: 2.1, Fear: 2.7, Happiness: 1.4, Sadness: 1.7, Surprise: 1.7
Disgust	Anger: 3.0, Disgust: 2.6, Fear: 1.9, Happiness: 1.0, Sadness: 2.3, Surprise: 1.5
Fear	Anger: 1.9, Disgust: 1.5, Fear: 3.3, Happiness: 2.5, Sadness: 1.5, Surprise: 3.3
Happiness	Anger: 1.1, Disgust: 1.0, Fear: 1.1, Happiness: 4.3, Sadness: 1.1, Surprise: 2.5
Sadness	Anger: 1.6, Disgust: 1.5, Fear: 1.9, Happiness: 1.1, Sadness: 3.6, Surprise: 1.5
Surprise	Anger: 1.6, Disgust: 1.6, Fear: 3.5, Happiness: 1.3, Sadness: 1.3, Surprise: 4.6
50% Fear	Anger: 2.0, Disgust: 1.8, Fear: 3.2, Happiness: 1.4, Sadness: 1.8, Surprise: 3.5
Anger, Fear, and Disgust	Anger: 1.9, Disgust: 1.4, Fear: 2.4, Happiness: 1.2, Sadness: 3.4, Surprise: 1.5
50% Sadness	Anger: 1.6, Disgust: 1.4, Fear: 2.0, Happiness: 1.1, Sadness: 3.5, Surprise: 1.5

Table 3. The results of the experimental evaluation of the videos

Experimental Setup: For the first experiment, the exploration-motive, communication motive and the self-protection motive are activated. A person stands at one side of ROMAN. That means the robot needs to turn its head to see this person. The activation, the activity, and the target rating of the different motives are recorded.

Expected Behavior: It is expected that the exploration motive gets discontent, which will be displayed by a rising target rating value. This will initiate a turning of the head. If the robot recognizes the person; the communication motive should get discontent because it wants to start a dialog with the person. If the target rating-value of the communication motive reaches the maximum, this motive will get active. It should inhibit the exploration motive and start a dialog. During the dialog the person should step closer towards the robot. This should cause an activation of the self-protection motive. The robot should move backwards and asks the person to step back. Afterwards, when the person has stepped back, the dialog should be resumed to completion.

Experiment Results: The results of this experiment are shown in figure 9. The exploration motive gets discontent and active, turns the robot's head and detects the person. After approx. $5s$ the communication motive gets active and the exploration motive is inhibited. After approx. $45s$ the target rating of the communication motive increases as the probability rate of the human detection decreases. Approx. at $50s$ the target rating of the communication reaches 1 (maximum) and the exploration gets active again. But a few

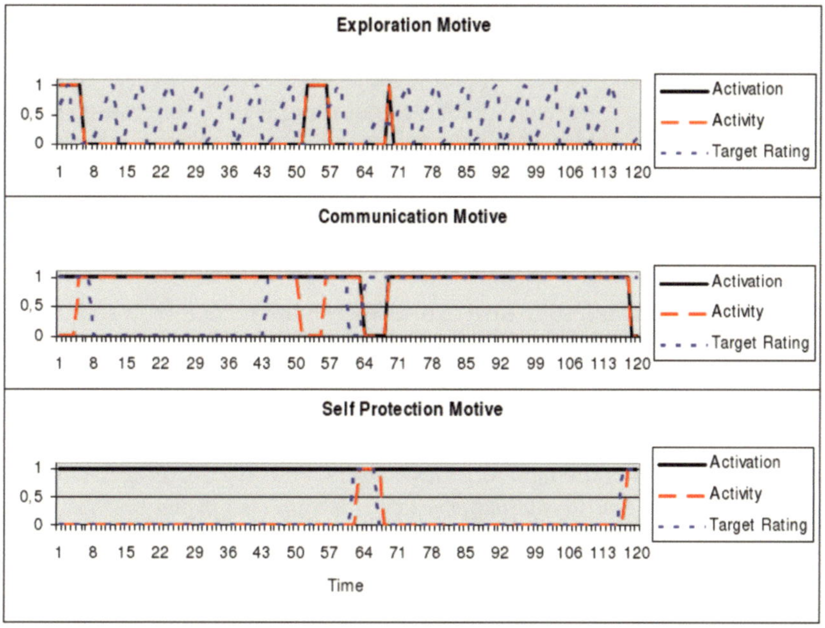

Fig. 9. The results of the experiment: the activities and discontents of the exploration motive, communication motive, and follow object behavior during a communication are measured

seconds later the subject is detected again and the communication continues. After approx. $58s$ the target rating of the self protection motive increases as the subject is too close to ROMAN. The activated motive inhibits all other motives, activates an evasion behavior and tells the human to step back. After approx. $68s$ the target rating and the activity of the self protection motive reaches 0 again and the communication is restarted.

6. Conclusion

This chapter explained the benefit of social interaction for humanoid robots by presenting psychological background of social interaction and ways to realize social interaction on a robot. The state-of-the-art in developing social robots was discussed in detail. Afterwards the humanoid robot ROMAN was introduced and the developed emotion-based control architecture that enables ROMAN to interact socially was explained in detail. Finally the results of some tests to evaluate this architecture were presented.

Although the structure of the emotion-based architecture will not be changed, the several subsystems need to be extended and the communication between these systems needs to be improved. For future, the amount of observable behaviors by the percepts of interaction will be increased. The habits of interaction will be extended by the integration of arms and hands in order to use gestures for the communication. That will increase the expressiveness of ROMAN enormously. In order to realize situation adapted behavior in interaction situations it is necessary to investigate, which motives are necessary. Therefore an interaction situation needs to be developed and studied in cooperation with psychologists. Furthermore

psychological experiments need to be conducted where the impressions of ROMAN to humans are evaluated and also the quality of the interaction has to be evaluated.
As final test-scenario, an interaction situation could be mentioned where the robot should assist a human in solving a problem. This problem might be a tangram-game which the robot knows how to solve. By investigating its human partner, the robot should then adapt its expression and helps to the situation.

7. References

Aoyama K. & Shimomura H. (2005). Real World Speech Interaction with a Humanoid Robot on a Layered Robot Behavior Control Architecture, *Proceedings of the IEEE International Conference on Robotics and Automation (ICRA)*, pp. 3825-3830, 2005 April 18-22, Barcelona, Spain

Bartneck C.; Reichenbach J. & van Breemen A. (2004). In Your Face, Robot! The Influence of a Character's Embodiement on How Users Perceive Its Emotional Expressions, *Proceedings of the Fourth International Conference on Design and Emotion*, 2004 July 12-14, Ankara, Turkey

Becker-Asano C.; Kopp S.; Pfeiffer-Leßmann N. & Wachsmuth I. (2008). Virtual Humans Growing up: Prom Primary Toward Secondary *Emotions, Künstliche Intelligenz (KI)*, No. 1, (2008 February), pp. 23-27

Bennewitz M.; Faber F. Joho D. & Behnke S. (2007). Fritz - A Humanoid Communication Robot, *Proceedings of IEEE International Symposium on Robot and Human Interactive Communication (RO-MAN)*, pp. 1072-1077, 2007 August, Jeju Island, Korea

Birdwhistell R. L. (1970). *Kinesics and Context: Essays in Body Motion Communication*, University of Pennsylvania Press, Philadelphia, USA

Breazeal C. (2002). Regulation and Entrainment in Human-Robot Interaction, *International Journal of Experimental Robotics*, Vol. 21, No. 10-11, (2002) , pp. 883-902

Breazeal C. & Brooks R. (2005). Robot Emotion: A Functional Perspective, In: *Who Needs Emotions? The Brain Meets the Robot*, Fellous J.M. and Arbib M.A. (Eds.), Oxford University Press

Broekens J. & de Groot D. (2004). Scalable and Flexible Appraisal Models for Virtual Agents, *Proceedings of the International Conference on Computer Games, Artificial Intelligence, Design and Education (CGAIDE)*, pp. 208-215, 2004, Reading, UK

Canamero D. (1997). Modeling motivations and emotions as a basis for intelligent behavior, *Proceedings of the International Conference on Autonomous Agents (AA)*, pp. 148-155, 1999 February, Marina del Rey, USA

Canamero D. & van de Velde W. (1997). Socially Emotional: Using Emotions to Ground Social Interaction, *Proceedings of the AAAI Fall Symposium on Socially Intelligent Agents*, pp. 10-15, 1997 November 8-10, Cambridge, USA

Ekman P. (1992). An Argument for Basic Emotions, *Cognition and Emotion*, Vol. 6, (1992), pp. 169-200

Ekman P. & Friesen W. (1978). *Facial Action Coding System*, Consulting psychologist Press, Inc

Esau N.; Kleinjohann L. & Kleinjohann B. (2007). Integration of Emotional Reactions on Human Facial Expressions into the Robot Head MEXI, *Proceedings of the IEEE/RSJ*

International Conference on Intelligent Robots and Systems (IROS), 2007 October 29–November 2, San Diego, USA
Hackel M.; Schwope S.; Fritsch J. & Wrede B. (2006). Designing a Sociable Humanoid Robot for Interdisciplinary Research, *Advanced Robotics*, Vol. 20, No. 11, (2006), pp. 1219–1235
Heckhausen, H. (1980). *Motivation und Handeln*, Springer Verlag
Hirth, J. & Berns K. (2007). Concept for Behavior Generation for the Humanoid Robot Head ROMAN based on Habits of Interaction, *Proceedings of the IEEE-RAS International Conference on Humanoid Robots (Humanoids)*, 2007 November 29–December 1, Pittsburgh, USA
Hirth, J. & Berns K. (2008). Motives as Intrinsic Activation for Human-Robot Interaction, *Proceedings of IEEE/RSJ International Conference on Intelligent Robots and Systems (IROS)*, 2008 September 22–26, Nice, France
Koch, J.; Wettach, J.; Jung, H. & Berns, K. (2007). Dynamic Speech Interaction for Robotic Agents, *Proceedings of the International Conference on Advanced Robotics (ICAR07)*, pp. 665–670, 2007 August 21–24, Jeju, Korea
Hobmair, H; Altenhan, S.; Betcher-Ott S.; Dirrigl W.; Gotthardt, W. & Ott W. (2003). *Psychologie*, Bildungsverlag EINS, Troisdorf, Germany
Kelley, A.E. (2005). Neurochemical Networks Encoding Emotion and Motivation - An Evolutionay Perspective, In: *Who Needs Emotions? The Brain Meets the Robot*, Fellous J.M. and Arbib M.A. (Eds.), Oxford University Press
Lee, H.S., Park, J.W. & Chung M.J. (2006). An Affect-Expression Space Model of the Face in a Mascot-Type Robot, *Proceedings of the 2006 6th IEEE-RAS International Conference on Humanoid Robots (Humanoids)*, pp. 412–417, 2006 December 4–6, Genoa, Italy
Martin, L.L. & Clore, G.L. (2001). *Theories of Mood and Cognition*, Lawrence ErlbaumAssociates, Inc.
Mehrabian, A. (1996). Pleasure-arousal-dominance: A general framework for describing and measuring individual differences in temperament, *Current Psychology: Developmental, Learning, Personality, Social*, Vol. 14 (1996)
Mietzel, G. (2000). *Wege in die Entwicklungspsychologie*, Stuttgart
Ortony A., Clore, G.L. & Collins, A. (1988). *The cognitive structure of emotion*, Cambridge University Press
Proetzsch M., Luksch T. & Berns K. (2007). The Behaviour-Based Control Architecture iB2C for Complex Robotic Systems, *Proceedings of the German Conference on Artificial Intelligence (KI)*, pp. 494–497, 2007 September 10–13, Osnabrück, Germany
Plutchik, R. (1980). A General Psychoevolutionary Theory of Emotions, *Emotion: Theory research and experience*, Vol. 1 (1980), pp. 3–33
Reeves, B. & Nass, C. (1996). *The Media Equation: How People Treat Computers, Television, and New Media Like Real People and Places*, CSLI Publications, Stanford, USA
Russel, J.A. (1980). The circumplex model of affect, *Journal of Personality and Social Psychology*, Vol. 39 (1980), pp. 1161–1178
Rolls, E.T. (1999). *The Brain and Emotion*, Oxford University Press
Strupp, S.; Schmitz, N. & Berns, K. (2008). Visual-based emotion detection for natural man-machine interaction, *Proceedings of the German Conference on Artificial Intelligence (KI)*, 2008, Kaiserslautern, Germany

Toman W. (1973). *Einführung in die Allgemeine Psychologie -- Band II Affektivität, Motivation, Persönlichkeit Soziale Kontexte*, Verlag Rombach, Freiburg, Germany

Ulich, D. (2000). *Einführung in die Psychologie*, Stuttgart

Watzlawick, P.; Beavin J.H. & Jackson D.D. (2000). *Menschliche Kommunikation*, Huber, Bern

7

A Walking Pattern Generation Method for Humanoid robots using Least square method and Quartic polynomial

Seokmin Hong*, †, Yonghwan Oh†
Young-Hwan Chang† and Bum-Jae You†
** University of Science and Technology(UST), Korea*
† Korea Institute of Science and Technology(KIST), Korea

1. Introduction

Humanoid robot has been developed for human's convenience in the human environment. Humanoid robots like ASIMO, HRP, WABIAN and Johnnie were successively developed (Hirai et al., 1998; Yamaguchi et al., 1999; Kajita et al., 2003; Löffler et al, 2003). Researches on humanoid have been done about balancing, walking pattern generation, motion generation, whole body cooperation and so on. In particular, many walking pattern generation methods have been developed to prevent a robot from tipping over. There are generally two groups for the walking pattern generation (Kajita et al., 2003; Hirukawa, 2006). The first group uses forward dynamics with multiple rigid-body (Hirai et al., 1998; Yamaguchi et al., 1999). This group demands precise information of the robot such as mass, inertia, and center of mass(CoM) of each link, and so on. In this group, the footprints of biped robot are changed for keeping the planned configuration of the robot (Hirukawa, 2006). On the other hand, the second group (Huang et al., 2001; Kajita et al., 2003; Löffler et al., 2003; Harada et al., 2004; Zhu et al., 2004; Oh et al., 2006) utilizes limited knowledge such as the total center of mass and total angular momentum. And this group makes use of the inverted pendulum model for the walking pattern generation and changes the configuration of the robot for keeping the planned footprints (Hirukawa, 2006). Our walking pattern method belongs to the second group. The inverted pendulum model is transferred the complex dynamic equation of humanoid robot into second order differential equation with some assumptions. Most researches on walking pattern generation dealt with periodic walking. However it is difficult to implement various gaits by periodic walking generator. Harada et al. introduced the analytical walking pattern method on real-time generation for coping with change of gait. Kajita et al. proposed the omni-direction walking pattern generation method using the preview control. Zhu et al. proposed walking pattern generation method with fixed ZMP and variable ZMP in order to make the biped walking pattern more human-like and more agile. This paper describes the omni-directional walking pattern method for the humanoid robots using the least square method with the quartic polynomials. And we design that the ZMP trajectory has the slope in single support phase. The CoM trajectory becomes effective in regard of velocity compared to the ZMP trajectory without the slope. We utilize a simple inverted pendulum model to represent the complex dynamics of the humanoid robot. This paper is organized as follows: It shows ZMP equation as the simplified model for the biped robot and arranges the relation of the ZMP and CoM in section 2. Three step modules are introduced for generating the ZMP and CoM trajectory in section 3 and we explain the procedure of generating the trajectory from footprints in section 4. Section 5,6 shows the results of simulations and experiment, and we conclude this paper in section 7.

2. Linear Inverted Pendulum Model

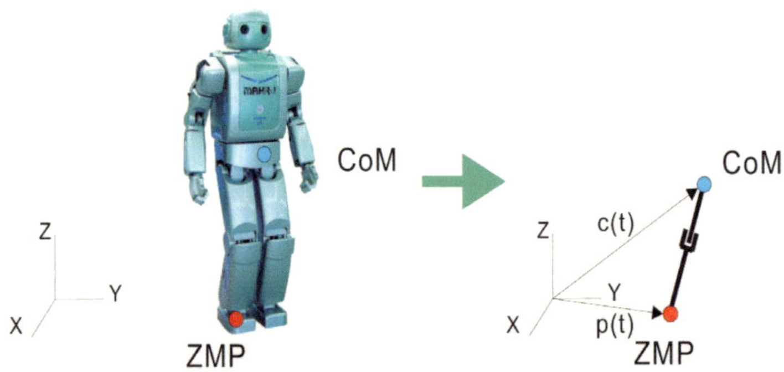

Fig. 1. Linear Inverted pendulum model

The humanoid robot is designed like human for doing various behaviors and adapting to human's environment. For this purpose, this robot consists of many links and joints. If the walking pattern took all dynamic properties of a number of links and joints into consideration, the waking pattern could make good performance. However, since it is difficult and complicated to calculate the walking pattern including all dynamic properties, a simplified model is required to control the biped robot. Fig. 1 shows the inverted pendulum model as the simplified biped robot. In Fig. 1, $p(t)$ and $c(t)$ denote the ZMP and CoM, respectively. The motion of the robot is designed by CoM trajectory instead of the full dynamics.

For simplicity, we use following assumptions:

1. The time derivative of the angular momentum about the CoM is zero.
2. The difference of CoM and ZMP at z-axis is constant.
3. The acceleration of ZMP at z-axis is zero.

Under these assumptions, the relation between ZMP and CoM can be represented as follows.

$$p(t) = c(t) - \frac{1}{\omega^2}\ddot{c}(t) \tag{1}$$

where ω is $\sqrt{g/(Z_{CoM} - Z_{ZMP})}$. Z_{CoM}, Z_{ZMP} and g are the values of the CoM and the ZMP at the z-axis and the gravity separately. The walking pattern of the ZMP and the CoM is made with Eq. (1). As mentioned above, the CoM is related to robot motion. Generally the robot motion has to keep the continuous jerk of the CoM for preventing the robot's motion from imposing a heavy burden to the robot. For the continuous jerk of CoM, more than cubic polynomials of the ZMP are sufficient to generate the ZMP and CoM trajectory. In case of the periodic walking pattern, it is proper to use cubic polynomial of the ZMP. But in case of various step length walking, these might take place large fluctuation of the ZMP trajectory. Since cubic polynomials constrain the shape of the ZMP, the ZMP trajectory makes the large fluctuation occur in order to satisfy Eq. (1) according to various steps. Therefore we choose quartic polynomials of the ZMP and then need an optimal method to obtain the solution owing to more numbers of unknowns than equations. In next section, the least square method will be explained as an optimal method in detail. Let us reconsider that the initial conditions $c(0)$, $\dot{c}(0)$ are known and the ZMP trajectory is given by quartic polynomials in Eq. (1). The general form of the ZMP and CoM could be found as follows.

$$p(t) = b_4t^4 + b_3t^3 + b_2t^2 + b_1t + b_0 \tag{2}$$

$$\begin{aligned} c(t) &= (c(0) - b_0 - \frac{2}{\omega^2}b_2 - \frac{24}{\omega^4}b_4)\cosh(\omega t) \\ &+ \frac{1}{\omega}(\dot{c}(0) - b_1 - \frac{6}{\omega^2}b_3)\sinh(\omega t) \\ &+ b_4t^4 + b_3t^3 + (b_2 + \frac{12}{\omega^2}b_4)t^2 + (b_1 + \frac{6}{\omega^2}b_3)t \\ &+ (b_0 + \frac{2}{\omega^2}b_2 + \frac{24}{\omega^4}b_4) \end{aligned} \tag{3}$$

$$\begin{aligned} \dot{c}(t) &= \omega(c(0) - b_0 - \frac{2}{\omega^2}b_2 - \frac{24}{\omega^4}b_4)\sinh(\omega t) \\ &+ (\dot{c}(0) - b_1 - \frac{6}{\omega^2}b_3)\cosh(\omega t) \\ &+ 4b_4t^3 + 3b_3t^2 + 2(b_2 + \frac{12}{\omega^2}b_4)t + (b_1 + \frac{6}{\omega^2}b_3) \end{aligned} \tag{4}$$

Equations (2), (3), and (4) present the fact that the coefficients of the CoM are presented by c(0), $\dot{c}(0)$ and those of the ZMP. Since c(0), $\dot{c}(0)$ are the provided values as the initial parameters, we focus on obtaining the ZMP coefficient. At the time interval of i-th support phase, Eq. (2), (3) and (4) are arranged by a matrix form as follows

$$\mathbf{X}_i = \mathbf{A}_i\mathbf{X}_{i-1} + \mathbf{B}_i\boldsymbol{\beta}_i \tag{5}$$

where

$$\mathbf{X}_i = \begin{Bmatrix} \omega(c(T_i) - p(T_i)) \\ \dot{c}(T_i) \end{Bmatrix}$$

$$\mathbf{A}_i = \begin{bmatrix} CT_i & ST_i \\ ST_i & CT_i \end{bmatrix}$$

$$\mathbf{B}_i = \begin{bmatrix} -\frac{24}{\omega^3}CT_i + \frac{12}{\omega}T_i^2 + \frac{24}{\omega^3} & -\frac{24}{\omega^3}ST_i + 4T_i^3 + \frac{24}{\omega^2}T_i \\ -\frac{6}{\omega^2}ST_i + \frac{6}{\omega}T_i & -\frac{6}{\omega^2}CT_i + 3T_i^2 + \frac{6}{\omega^2} \\ -\frac{2}{\omega}CT_i + \frac{2}{\omega} & -\frac{2}{\omega}ST_i + 2T_i \\ -ST_i & -CT_i + 1 \end{bmatrix}^T$$

$$\boldsymbol{\beta}_i = \begin{pmatrix} b_{i4} & b_{i3} & b_{i2} & b_{i1} \end{pmatrix}^T$$

CTi and *STi* denote cosh(ωt) and sinh(ωt), respectively. b_{in} (n = 1, . . . , 4) denotes the n-th order coefficient of the ZMP during the i-th support phase. Equation(5) will be used a basic equation of the ZMP and the CoM .

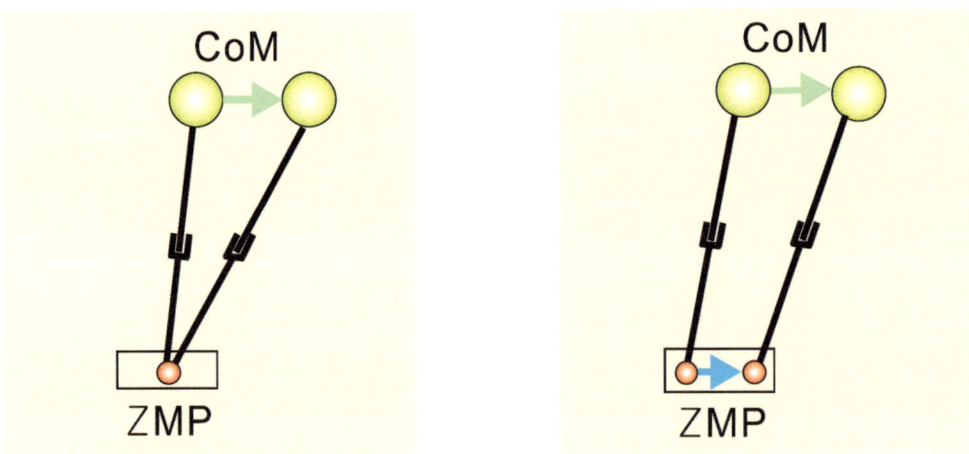

a) whitout the slope b) with the slope

Fig. 2. Inverted pendulum model without and with the slope

3. Step Modules

This section introduces three step modules which are proposed for making walking trajectory of the ZMP and the CoM: periodic step module, absorption step module, F-step module. Each step module can have the slope in single support phase. As shown in Fig. 2, the ZMP movement can be designed like the human walking in which the ZMP moves from heel to toe in single support phase. The CoM movement with a slope of the ZMP in single support phase becomes smaller than that with the flat line because the ZMP movement from heel to toe makes the CoM movement reduce. Section 4 shows the efficiency of the ZMP with the slope in single support phase through the simulation. The walking pattern method of the sagittal plane and the frontal plane is the same procedure. So we explain the walking pattern at the sagittal plane, hereafter.

3.1 Periodic Step Module

For generating the walking pattern, the initial values are needed. This step module provides the initial values as the seed of the walking pattern generation, because it makes the walking pattern independently with the position and velocity information of the ZMP at periodic steps. This consists of cubic polynomials (b_{i4} = 0) in double support phase and first order polynomials (b_{i4} = 0, b_{i3} = 0, b_{i2} = 0) in single support phase. Using the information of the ZMP, the solution can be obtained analytically. The obtained values by the periodic step module are used as the final values of position and velocity of the CoM at the first step. But the ZMP coefficients by the periodic step module are not used in making the trajectory because of not being proper to the arbitrary step. We only uses information of a particular position and velocity of the CoM obtained from this step module. It consists of 4 phases as shown in Fig. 3.

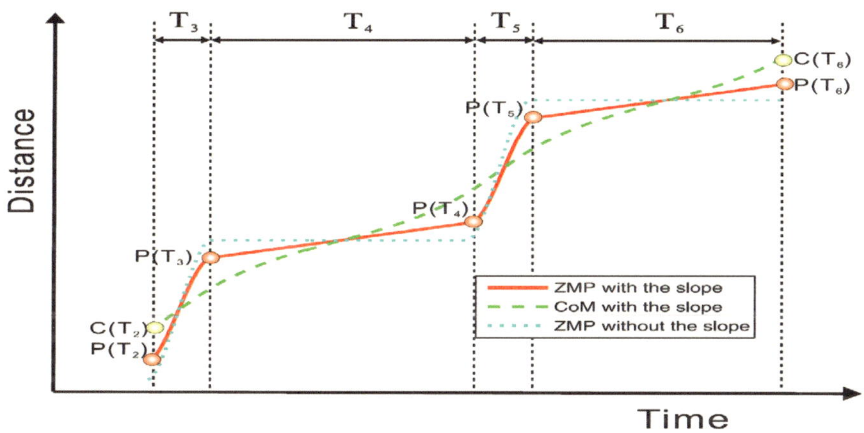

Fig. 3. Constitution of periodic step module

in Fig. 3., i.e. double support phase, single support phase, double support phase and single support phase. T_i means the interval time at the i-th support phase. The numbering is written down in Fig. 3 taking the first step into account. In Fig. 3, T_3 and T_5 are double support time, and T_4 and T_6 denote single support time. And it is known that the velocity of the ZMP is

$$b_{i1} = \begin{cases} \frac{p(T_6)-p(T_5)}{T_6} & \text{for } i = 3, 6 \\ \frac{p(T_4)-p(T_3)}{T_4} & \text{for } i = 4, 5 \end{cases}$$

at the meeting point of each phase. The values of $p(T_4)-p(T_3)$ and $p(T_6) - p(T_5)$ are the same value as the ZMP margin in single support phase. The value is selected less than a half of foot size considering keeping the stability margin. And as the term of the "periodic", it is assumed that this walking trajectory has the relation between initial and final condition of the ZMP and CoM as follows.

$$\omega(c(T_6) - p(T_6)) = \omega(c(T_2) - p(T_2)) \tag{6}$$

$$\dot{c}(T_6) = \dot{c}(T_2) \tag{7}$$

The relation of initial and final conditions as Eq. (6) and (7) causes the characteristics of periodicity. This relation also applies to the absorption step module as a constraint equation.The solutions of the periodic step module are as follows

$$\begin{Bmatrix} \mathbf{X}_2 \\ \boldsymbol{\beta}_3 \\ \boldsymbol{\beta}_5 \end{Bmatrix} = \mathbf{D}^{-1}\mathbf{E} \tag{8}$$

Where

$$\mathbf{D} = \begin{bmatrix} \mathbf{A}_6\mathbf{A}_5\mathbf{A}_4\mathbf{A}_3 - \mathbf{I} & \mathbf{A}_6\mathbf{A}_5\mathbf{A}_4\mathbf{B}_3^u & \mathbf{A}_6\mathbf{B}_5^u \\ \mathbf{0} & \mathbf{F}_3 & \mathbf{0} \\ \mathbf{0} & \mathbf{0} & \mathbf{F}_5 \\ \mathbf{0} & \mathbf{G}_3 & \mathbf{0} \\ \mathbf{0} & \mathbf{0} & \mathbf{G}_5 \end{bmatrix}$$

$$\mathbf{E} = \begin{Bmatrix} -\boldsymbol{\Gamma} \\ \alpha_3 T_3 - \alpha_6 T_3 \\ \alpha_5 T_5 - \alpha_4 T_5 \\ \alpha_4 - \alpha_6 \\ \alpha_6 - \alpha_4 \end{Bmatrix}, \quad \boldsymbol{\Gamma} = \begin{bmatrix} (\mathbf{A}_6\mathbf{A}_5\mathbf{A}_4\mathbf{B}_3^k)^T \\ (\mathbf{A}_6\mathbf{A}_5\mathbf{B}_4^k)^T \\ (\mathbf{A}_6\mathbf{B}_5^k)^T \\ (\mathbf{B}_6^k)^T \end{bmatrix}^T \begin{Bmatrix} b_{31} \\ b_{41} \\ b_{51} \\ b_{61} \end{Bmatrix}$$

$$\mathbf{F}_i = (T_i^3 \;\; T_i^2), \quad \mathbf{G}_i = (3T_i^2 \;\; 2T_i), \quad \boldsymbol{\beta}_i = (b_{i3} \;\; b_{i2})^T$$

$\mathbf{B}_i$ in Eq. (5) is divided into B_i^u and B_i^k by the coefficients of the periodic step module.

$$\mathbf{B}_i^u = \begin{bmatrix} -\frac{6}{\omega^2}ST_i + \frac{6}{\omega}T_i & -\frac{6}{\omega^2}CT_i + 3T_i^2 + \frac{6}{\omega^2} \\ -\frac{2}{\omega}CT_i + \frac{2}{\omega} & -\frac{2}{\omega}ST_i + 2T_i \end{bmatrix}^T$$

$$\mathbf{B}_i^k = \left(-ST_i \quad -CT_i + 1 \right)^T \tag{9}$$

and $\alpha_i = \frac{p(T_i) - p(T_{i-1})}{T_i}$ is the slope which is connected by two points $p(T_{i-1})$ and $p(T_i)$ in the i-th phase. This step module provides final values to F-step module for first step. The values are $c(T_3)$, $\dot{c}(T_3)$ obtained by $\mathbf{X}_3$ in order to increase the relation with first step.

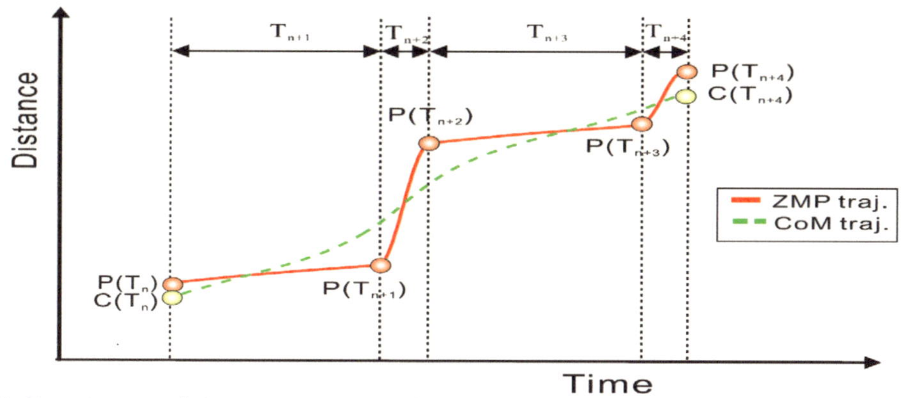

Fig. 4. Constitution of absorption step module

3.2 Absorption Step Module

The absorption step module generates the trajectory with positions of the ZMP from footprints and initial position and velocity of the ZMP and the CoM. This study defines a stride period as time consisted of 2 single support phases and 2 double support phases in Fig. 4. This step module makes the trajectory repetitively with the ZMP position information of the next stride period in every phase like Fig. 5.

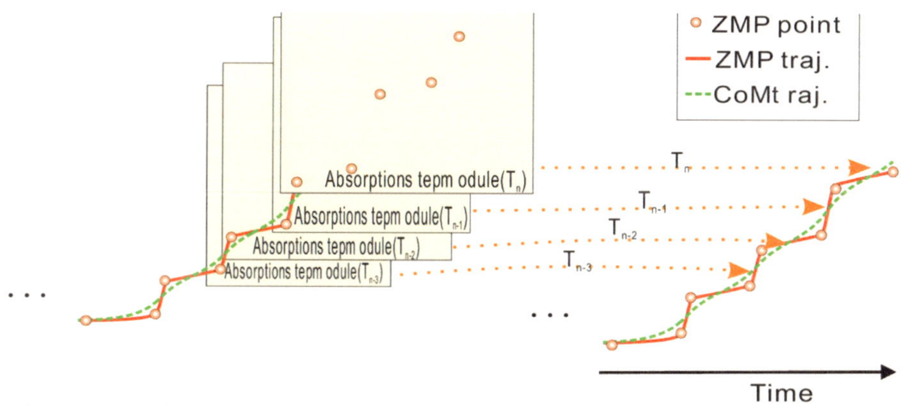

Fig. 5. Generation of ZMP trajectory using absorption step module

This is why it copes with variation of the ZMP trajectories when footprint is changed to walk with various steps and implements real-time walking pattern generation. This step module consists of 4 support phases: 2 single support phases and 2 double support phases. This uses the quartic polynomial of the ZMP trajectory. The numbers of unknown factors are more than the numbers of equations in this case. The redundant unknown factor plays an important role to reduce the fluctuation of the ZMP trajectory. Using Eq. (5), a matrix can be arranged from T_n to T_{n+4},

$$\boldsymbol{\kappa} = \boldsymbol{\gamma}^T \boldsymbol{\tau} \tag{10}$$

where

$$\boldsymbol{\kappa} = \mathbf{X}_{n+4} - \mathbf{A}_{n+4}\mathbf{A}_{n+3}\mathbf{A}_{n+2}\mathbf{A}_{n+1}\mathbf{X}_n$$

$$\boldsymbol{\gamma} = \begin{bmatrix} (\mathbf{A}_{n+4}\mathbf{A}_{n+3}\mathbf{A}_{n+2}\mathbf{B}_{n+1})^T \\ (\mathbf{A}_{n+4}\mathbf{A}_{n+3}\mathbf{B}_{n+2})^T \\ (\mathbf{A}_{n+4}\mathbf{B}_{n+3})^T \\ (\mathbf{B}_{n+4})^T \end{bmatrix}$$

$$\boldsymbol{\tau} = \begin{bmatrix} \boldsymbol{\beta}_{n+1}^T & \boldsymbol{\beta}_{n+2}^T & \boldsymbol{\beta}_{n+3}^T & \boldsymbol{\beta}_{n+4}^T \end{bmatrix}^T$$

$$\boldsymbol{\beta}_i = \begin{pmatrix} b_{i4} & b_{i3} & b_{i2} & b_{i1} \end{pmatrix}^T$$

β_i means the coefficient vector of the ZMP at the i-th phase. And there are 9 constraint equations for keeping the continuity of position and velocity of the ZMP trajectory on each phase. The equations are presented by a matrix as follows.

$$\boldsymbol{\varphi} = \boldsymbol{\Psi}\boldsymbol{\tau} \tag{11}$$

where

$$\boldsymbol{\varphi} = \begin{Bmatrix} p(T_{n+1}) - p(T_n) \\ 0 \\ dp(T_n) \\ p(T_{n+2}) - p(T_{n+1}) \\ 0 \\ p(T_{n+3}) - p(T_{n+2}) \\ 0 \\ p(T_{n+4}) - p(T_{n+3}) \\ dp(T_{n+4}) \end{Bmatrix}$$

$$\boldsymbol{\Psi} = \begin{bmatrix} \mathbf{F}_{n+1} & \mathbf{0} & \mathbf{0} & \mathbf{0} \\ \mathbf{G}_{n+1} & -\mathbf{H} & \mathbf{0} & \mathbf{0} \\ \mathbf{H} & \mathbf{0} & \mathbf{0} & \mathbf{0} \\ \mathbf{0} & \mathbf{F}_{n+2} & \mathbf{0} & \mathbf{0} \\ \mathbf{0} & \mathbf{G}_{n+2} & -\mathbf{H} & \mathbf{0} \\ \mathbf{0} & \mathbf{0} & \mathbf{F}_{n+3} & \mathbf{0} \\ \mathbf{0} & \mathbf{0} & \mathbf{G}_{n+3} & -\mathbf{H} \\ \mathbf{0} & \mathbf{0} & \mathbf{0} & \mathbf{F}_{n+4} \\ \mathbf{0} & \mathbf{0} & \mathbf{0} & \mathbf{G}_{n+4} \end{bmatrix}$$

$$\mathbf{F}_i = \begin{pmatrix} T_i^4 & T_i^3 & T_i^2 & T_i \end{pmatrix}$$
$$\mathbf{G}_i = \begin{pmatrix} 4T_i^3 & 3T_i^2 & 2T_i & 1 \end{pmatrix}$$
$$\mathbf{H} = \begin{pmatrix} 0 & 0 & 0 & 1 \end{pmatrix}$$

Equations (10) and (11) are changed into a matrix.

$$\boldsymbol{\xi} = \boldsymbol{\Phi}\boldsymbol{\tau} \tag{12}$$

where

$$\boldsymbol{\xi} = \begin{Bmatrix} \boldsymbol{\kappa} \\ \boldsymbol{\varphi} \end{Bmatrix}, \quad \boldsymbol{\Phi} = \begin{bmatrix} \boldsymbol{\gamma}^T \\ \boldsymbol{\Psi} \end{bmatrix}$$

The coefficients of each support phase were obtained with Eq. (12) and least square method. The least square method is used as the optimal method. The objective of using the method is that variation of the ZMP velocity is minimized. This prevents the ZMP trajectory from changing abruptly in order to satisfy Eq. (1) according to footprints. In order to do it, the cost function is defined as follows.

$$\begin{aligned} \mathcal{J} &= \frac{1}{2}\sum_{i=1}^{4}\int_0^{T_{n+i}} w_{n+i}[\dot{p}(t) - \alpha_{n+i}]^2 dt \\ &= \frac{1}{2}\boldsymbol{\tau}^T\mathbf{W}\boldsymbol{\tau} - \mathbf{U}\boldsymbol{\tau} + c \end{aligned} \tag{13}$$

where

$$\mathbf{W} = \begin{bmatrix} \mathbf{W}_{n+1} & \mathbf{0} & \mathbf{0} & \mathbf{0} \\ \mathbf{0} & \mathbf{W}_{n+2} & \mathbf{0} & \mathbf{0} \\ \mathbf{0} & \mathbf{0} & \mathbf{W}_{n+3} & \mathbf{0} \\ \mathbf{0} & \mathbf{0} & \mathbf{0} & \mathbf{W}_{n+4} \end{bmatrix}$$

$$\mathbf{W}_i = w_i \begin{bmatrix} \frac{16}{7}T_i^7 & 2T_i^6 & \frac{8}{5}T_i^5 & T_i^4 \\ 2T_i^6 & \frac{9}{5}T_i^5 & \frac{3}{2}T_i^4 & T_i^3 \\ \frac{8}{5}T_i^5 & \frac{3}{2}T_i^4 & \frac{4}{3}T_i^3 & T_i^2 \\ T_i^4 & T_i^3 & T_i^2 & T_i \end{bmatrix}$$

$$\mathbf{U} = \begin{Bmatrix} \alpha_{n+1} w_{n+1} \mathbf{F}_{n+1}^{T} \\ \alpha_{n+2} w_{n+2} \mathbf{F}_{n+2}^{T} \\ \alpha_{n+3} w_{n+3} \mathbf{F}_{n+3}^{T} \\ \alpha_{n+4} w_{n+4} \mathbf{F}_{n+4}^{T} \end{Bmatrix}$$

Using Lagrange multiplier,

$$\mathcal{L}(\boldsymbol{\tau}, \boldsymbol{\lambda}) = \frac{1}{2}\boldsymbol{\tau}^{T}\mathbf{W}\boldsymbol{\tau} - \mathbf{U}\boldsymbol{\tau} + c + \boldsymbol{\lambda}^{T}(\boldsymbol{\xi} - \boldsymbol{\Phi}\boldsymbol{\tau}) \tag{14}$$

The coefficients of each support phase are acquired.

$$\boldsymbol{\tau} = \boldsymbol{\Phi}_{W}^{+}\boldsymbol{\xi} - \boldsymbol{\Phi}_{W}^{+}(\boldsymbol{\Phi}\mathbf{W}^{-1}\mathbf{U}^{T}) + \mathbf{W}^{-1}\mathbf{U}^{T} \tag{15}$$

where

$$\boldsymbol{\Phi}_{W}^{+} = \mathbf{W}^{-1}\boldsymbol{\Phi}^{T}(\boldsymbol{\Phi}\mathbf{W}^{-1}\boldsymbol{\Phi}^{\mathbf{T}})^{-\mathbf{1}} \tag{16}$$

And w_i is the weighting factor. Generally, the stability of humanoid robot is whether the ZMP is located on the supporting zone which is shaped by supporting foot or not. The weighting factor suppresses the fluctuation of the ZMP trajectory for guaranteeing the stability of biped robot according to the various ZMP positions. The ZMP trajectory in single support phase is easy to diverge owing to small supporting zone. Therefore the weighting factor of single support phase is larger than that of double support phase. We find that the value of weighting factor w_n in single support phase is 100 times of the value in double support phase by simulation. The large fluctuation of the ZMP trajectory can occurs when more than the values of 3 ZMP positions are successively same at a stride period. That is why this phenomenon takes place due to Eq. (6) and (7). The ZMP trajectory moves abruptly to satisfy the constraint Eq. (6) and (7). There are three methods of absorbing the fluctuation of the ZMP trajectory. One is resetting the ZMP position, another is readjusting the weighting factor and other is readjusting the support phase time. In the first method, it is needed to set new extra ZMP positions to reduce the fluctuation irrespective of the footprints. Therefore, in this case, the double support time is increased and weighting factor is readjusted in a stride period with more than 3 same ZMP positions. The single support time don't need to change the value of phase time and weighting factor because the single support phase is already enough to be suppressed by weighting factor. The double support time is changed when the ZMP positions are more than 3 same values of the ZMP position in a stride period. The value of weighting factor at this double support phase is changed to the value of single support phase. But if the first phase time of a stride period was double support time, we did not change the support phase time and the weighting factor of the first double support phase for not having influence on total walking pattern time.

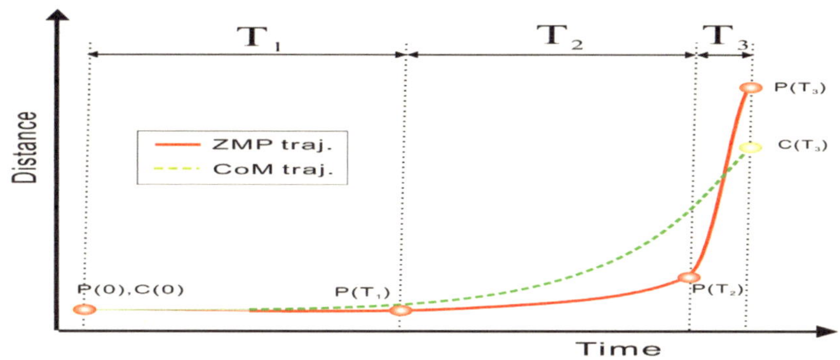

Fig. 6. Constitution of first step module

3.3 F-Step Module

This step module is for first step and final step. This module consists of 3 phases: double support phase, single support phase and double support phase. The difference of first step and final step is the procedure. The final step is in reverse order of the first step. The fluctuation of the ZMP trajectory might arise at the first phase time of the first step and the third phase time of the final step. That is why Eq. (1) has the non-minimum phase property. When the CoM moves forward at the starting point of walking, the ZMP moves naturally backward owing to the non-minimum phase property. We increase the phase time to suppress the non-minimum phase property at the first phase time of the first step and the third phase time of the final step. The changed phase time is selected within 2s.

F-step module is provided with the initial value of the ZMP and CoM which are set by the robot itself. Ideally the initial values of the ZMP and CoM are zero, but these values may not be zero because of the irregular ground condition and robot's kinematic error. It causes a problem that initial values are not zero when the trajectory is generated. Although this initial problem may be not critical effect on the robot, it could make the initial motion of the robot unnatural. We cope with the problem using least square method and the readjusted phase time in order that the more natural trajectory of the ZMP and CoM can be generated. The coefficients of each phases is decided with similar procedure which is introduced in the previous absorption step module section.

4. Implementation of Walking Pattern Generation

In this section, we show the procedure to generate the trajectory, which uses three step modules demonstrated in the previous section. The procedure is as follows.

Firstly, the walking parameters are selected. The parameters are step length, stride width, turning angle, single support time, double support time, user setting time, weighting factor values, initial value of the ZMP and CoM, and the ZMP positions according to footprints. Secondly, using periodic step module with the preset ZMP positions and velocities, final values of the first step are found. Thirdly, F-step module is used for the first step with initial values from user setting, final values from periodic step module and the preset ZMP positions according to the footprints in Fig.6. Fourthly, absorption step module is utilized in every phase. It makes the trajectory according to the ZMP positions of the next stride period.

And for the final step generation, absorption step module finally has to finish single, double, single and double support phase before the final step. And if more than 3 successive ZMP positions are the same value in generating the trajectory by the absorption step module, the abrupt fluctuation of the ZMP trajectory arises for satisfying the periodic characteristic. Double support phase time and weight factors are readjusted except that the first support phase of a stride period is double support phase. And finally, F-step module is used for the final step with initial values from the absorption step module and final values from user.

5. Simulation

We show the advantages of the proposed method in this section. The proposed method has the characteristic of being insensitive to initial values of the ZMP and CoM. As shown in Fig. 7 and Fig. 8, the proposed method makes the effect of initial values be reduced, although the robot has the initial values. The ZMP trajectory is changed automatically to reduce the effect of initial value of the ZMP and CoM in the first phase of first step. The fluctuations in Fig. 7 and Fig. 8 look large. But this variation do not arise the unstable condition because both feet of the robot stay on the ground.
And the proposed method has the slope in single support phase as shown in Fig. 9 and Fig. 10. The ZMP trajectory with the slope in single support phase has an advantage of reducing the CoM velocity because of the ZMP movement. Therefore it helps more efficient robot motion having slope in single support phase. In Fig.10, the robot does the same motion with less CoM velocity. It means the efficiency of the joint angular velocity enhances. Therefore the robot can implement longer stride with the same joint angular velocity.
And finally to demonstrate the usefulness of the proposed method, we generate 'U' walking pattern in Fig. 11 and Fig. 12. This walking pattern is mixed with straight waling and turning walking.

6. Experiment

We did an experiment with real humanoid robot MAHRU-R which is improved mechanical part design of a humanoid robot platform MAHRU at KIST.
Let us Mahru-R introduce briefly. Fig. 13 shows MAHRU-R which has 1350mm height, 50 kg weight including batteries. It has 12-DOF in two legs and 20- DOF in two arms including hands with 4-DOF grippers. Also, it is actuated by DC servo motors through harmonic drive reduction gears. The body is equipped with an IMU (Inertial Measurement Unit) sensor which consists of 3-axis gyroscope and 3-axis G-force sensors. Each ankle and wrist is equipped with a force/torque sensor. A distributed system was built for humanoid using

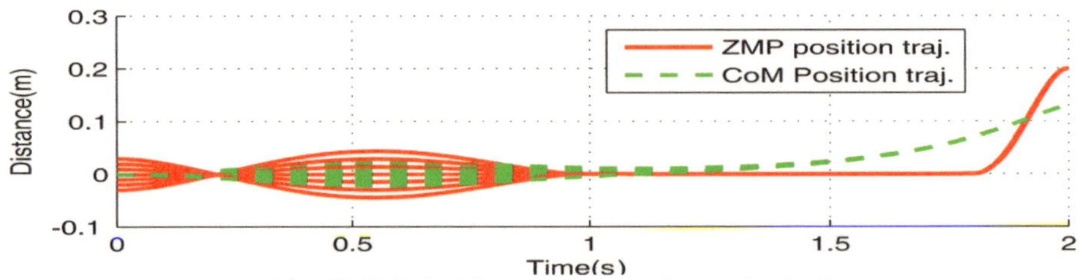

(a) ZMP& CoM trajectory at the sagittal plane

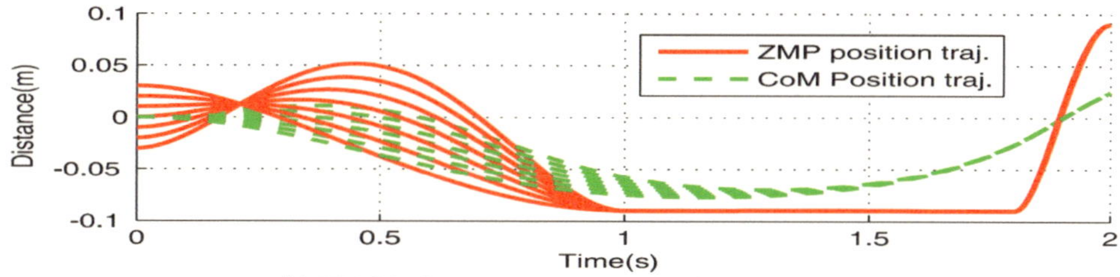

(b) ZMP& CoM trajectory at the lateral plane

Fig. 7. ZMP& CoM trajectory with initial ZMP position (from $-0.04m$ to $0.04m$), $T_1 = 1.0$s, $T_2 = 0.8$s, $T_3 = 0.2$s, step length = $0.2m$, stride width = $0.18m$

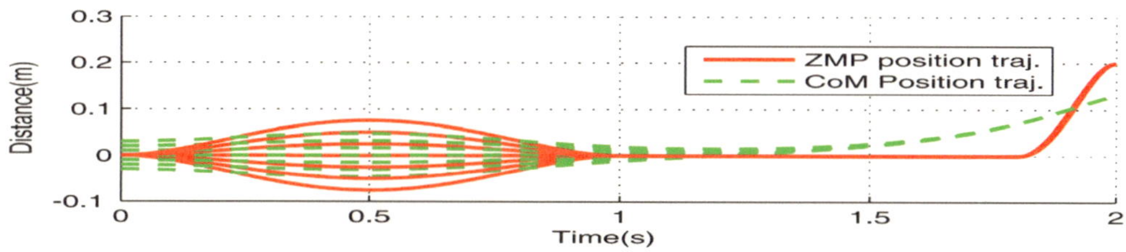

(a) ZMP& CoM trajectory at the sagittal plane

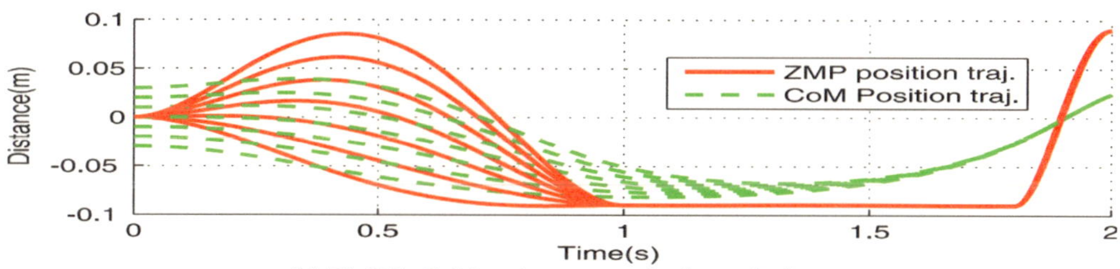

(b) ZMP& CoM trajectory at the lateral plane

Fig. 8. ZMP& CoM trajectory with initial CoM position (from $-0.04m$ to $0.04m$), $T_1 = 1.0$s, $T_2 = 0.8$s, $T_3 = 0.2$s, step length = $0.2m$, stride width = $0.18m$

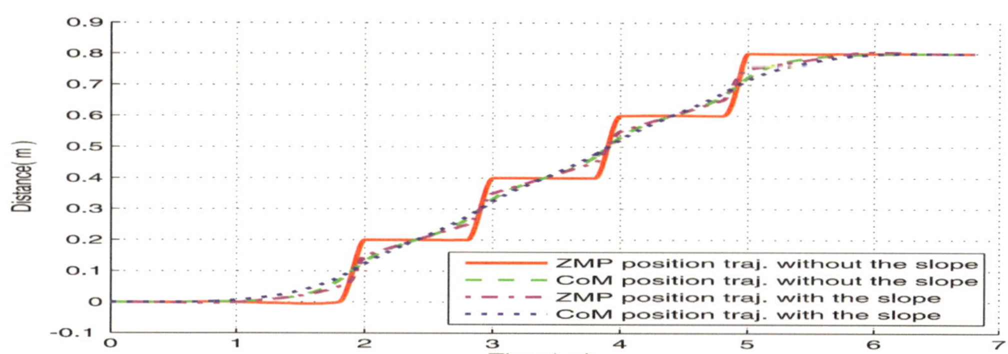

Fig. 9. ZMP& CoM position trajectory with $T_s = 0.8s$, $T_d = 0.2s$, step length=0.2m and slope = 0.25m/s

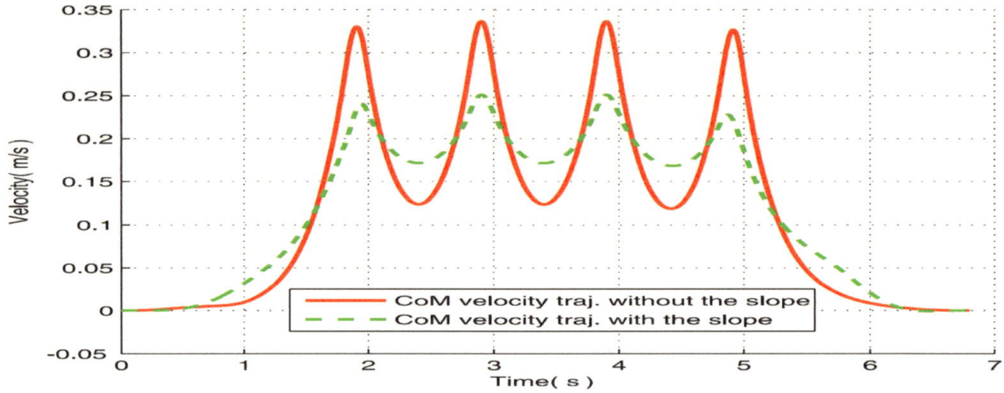

Fig. 10. ZMP& CoM velocity trajectory with T_s = 0.8s, T_d = 0.2s, step length = 0.2m and slope = 0.25 m/s

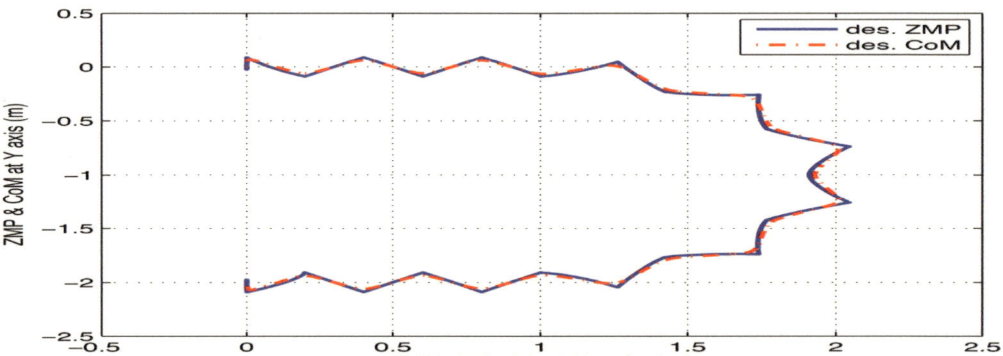

Fig. 11. 'U' walking pattern with T_s = 0.8s, T_d = 0.2s, step length = 0.2m, turn rate = 15 deg./stride and slope = 0

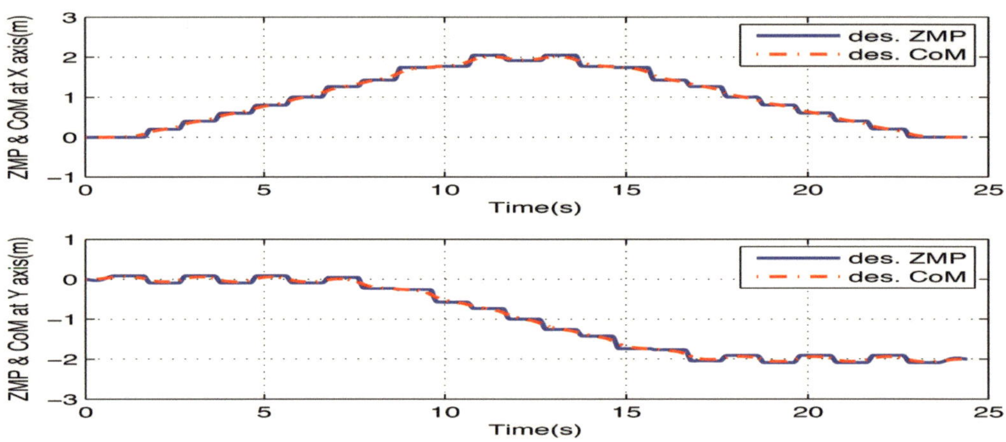

Fig. 12. 'U' walking pattern of ZMP and CoM at X and Y axis

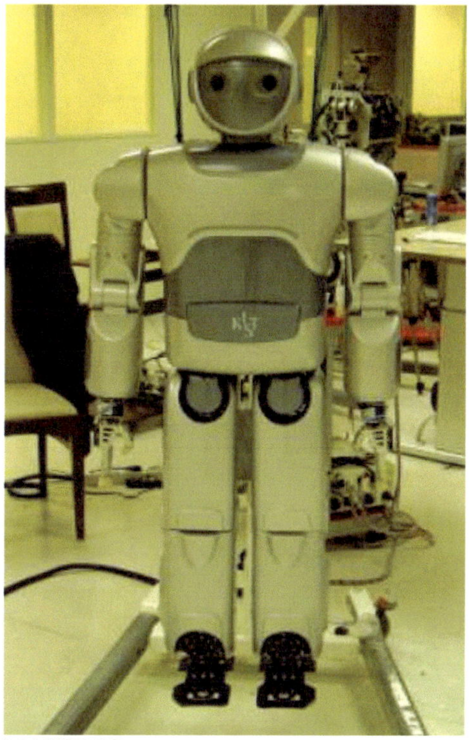

Fig. 13. Humanoid robot MAHRU-R

sub-controllers and IEEE 1394 protocol communication lines between the main controller and sub-controllers. The main real-time control algorithm runs on a micro-ATX CPU board in the backpack of MAHRU-R, whose operating system is real-time Linux (RTAI/Xenomai). It allows user to make timer interrupt service routine with the highest priority to control the robot in real-time. To confirm the effectiveness of the proposed method, the experiment was done using MAHRU-R. As shown in Fig. 14, the walking pattern was zigzag walking which consisted of the straight walking and turning. Fig. 15 and Fig. 16 show the desired ZMP and CoM are useful to the real humanoid robot walking pattern. When the humanoid robot started turning after 6 second, we increased the step width from 0.17m to 0.19m in order to avoid the collision of both feet. So the end point of Fig. 16 is different to the starting point at Y axis. Fig. 17 shows the snapshot of zigzag walking.

7. Conclusion

This paper proposed a new walking pattern method for the humanoid robot. The propose method was based on the linear inverted pendulum model as the simple model of humanoid robot. And this method consisted of three step modules for generating walking pattern. Using these step modules, this paper addressed robustness of the proposed method against initial values of the ZMP and CoM, and validated the efficiency with the slope at the single support phase through simulations. And finally we showed zigzag walking to verify the usefulness of the proposed method using real humanoid robot MAHRU-R.

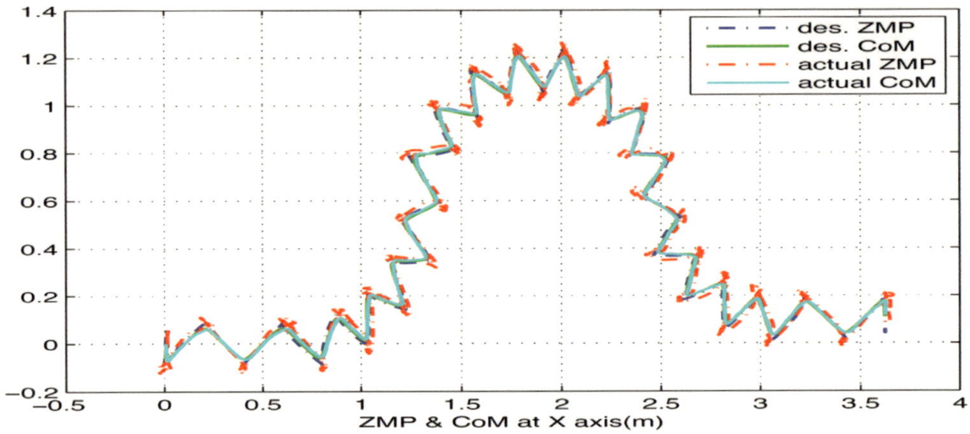

Fig. 14. Zigzag walking pattern with Ts = 0.95s, Td = 0.15s, step length = 0.2m, turn rate = 10 deg./stride and slope = 0

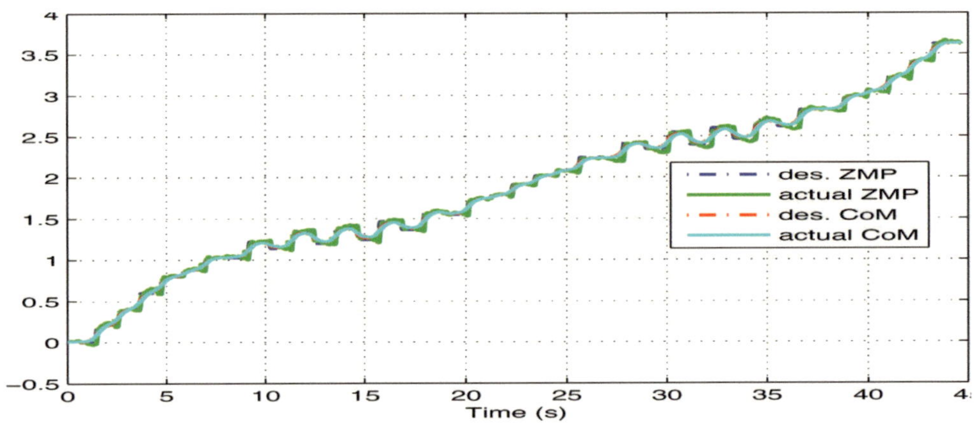

Fig. 15. ZMP and CoM of zig zag walking at X axis

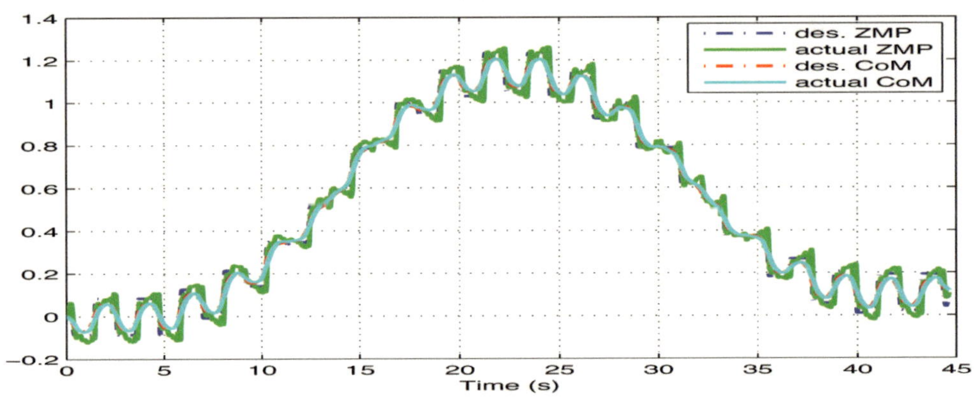

Fig. 16. ZMP and CoM of zig zag walking at Y axis

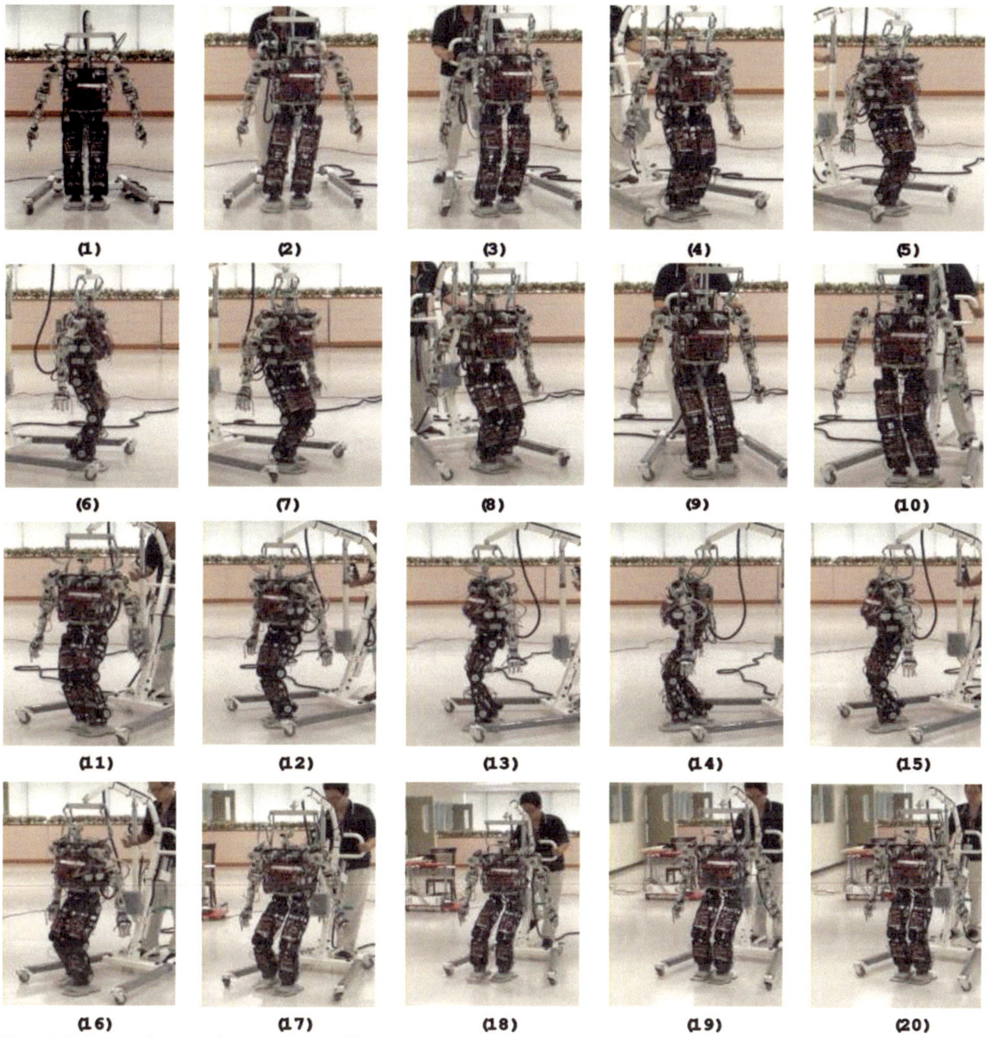

Fig. 17. Snapshots of zig zag walking

8. Acknowledgements

This work was supported in part by Ministry Knowledge Economy(MKE) & Institute for Information Technology Advancement(IITA) through IT Leading R&D support Project.

9. References

Harada, K.; Kajita, S.; Kaneko K. & Hirukawa, H. (2004). An Analytical Method on Real-time Gait Planning for a Humanoid Robot, *IEEE-RAS/RSJ Int. Conf. on Humanoid Robots*, Los Angeles, USA, pp. 640-655, November 2004.

Hirai, K.; Hirose, M.; Haikawa, Y. & Takenaka, T. (1998). Development of Honda Humanoid Robot, *Proc. of IEEE Int. Conf. on Robotics and Automation*, Leuven, Belgium, pp.

1321-1326, May 1998.

Hirukawa, H. (2006). On Humanoid Control, *Onassis Foundation Lecture Series 2006.*

Huang, Q.; Yokoi, K.; Kajita, S.; Kaneko, K.; Arai, H.; Koyachi N. & Tani, K. (2001) Panning Walking Patterns for a Biped Robot, *IEEE Trans. On Robotics and Automation*, pp. 280-289.

Kajita, S.; Kanehiro, F.; Kaneko, K.; Fujiwara, K.; Harada, K.; Yokoi K. & Hirukawa, H. (2003). Biped Walking Pattern Generation by using Preview Control of Zero-Moment Point, *Proc. of IEEE Int. Conf. on Robotics & Automation*, Taipei, Taiwan, pp. 1620-1626, September 2003.

Löffler, K.; Gienger, M.; & Pfeiffer, F. (2003). Sensor and Control Design of a Dynamically Stable Biped Robot, *Proc. of IEEE Int. Conf. on Robotics and Automation*, Taipei, Taiwan, pp.484-490, September 2003.

Oh, Y.; Ahn, K.; Kim, D. & Kim, C. (2006). An Analytical Method to Generate Walking Pattern of Humanoid Robot, *Pro. Of IEEE Int. Conf. On Industrial Electronics Society*, Paris, France, pp. 4159 - 4164, November 2006.

Yamaguchi, J.; Soga, E.; Inoue S. & Taknishi, A. (1999). Development of a Bipedal Humanoid Robot -control Method of Whole Body Cooperative Dynamic Biped Walking-, *Pro. of IEEE Int. Conf. on Robotics & Automation*, Detroit, USA, pp. 2299-2306, May 1999.

Zhu, C.; Tomizawa, Y.; Luo X. & Kawamura, A. (2004). Biped Walking with Variable ZMP, Frictional Constraint and Inverted Pendulum Model, *Pro. of IEEE Int. Conf on Robotics & Biomimetics*, Shenyang, China, pp. 425-430, August 2004.

8

Stable Walking Pattern Generation for a Biped Robot Using Reinforcement Learning

Jungho Lee
Department of Mechanical Engineering, KAIST,
335 Gwahangno Yuseong-gu, Daejeon, 305-701, Republic of Korea
Phone: +82-42-869-5223, Fax: +82-42-869-8900
E-mail: jungho77@kaist.ac.kr

Jun Ho Oh
Department of Mechanical Engineering, KAIST,
335 Gwahangno Yuseong-gu, Daejeon, 305-701, Republic of Korea

Abstract

In this research, a stable biped walking pattern is generated by using reinforcement learning. The biped walking pattern for forward direction is chosen as a simple third order polynomial and sinusoidal function is used for sideway direction. To complete the forward walking pattern, four boundary conditions are needed. In order to avoid jerk motion, initial position and velocity and final position and velocity of the joint are selected as boundary conditions. Also desired motion or posture can be achieved by using the initial and final position. The final velocity of the walking pattern is related to the stability but it is hard to choose proper value. So the final velocity of the walking pattern is chosen as a learning parameter. In order to find the proper boundary condition value, a reinforcement learning algorithm is used. For the sideway movement, a sway amount is selected as learning parameter and a reinforcement learning agent finds proper value for sideway movement. To test the algorithm, a three-dimensional simulator that takes into consideration the whole model of the robot and the environment is developed. The algorithm is verified through a simulation.

Keywords: Biped walking; Reinforcement learning; Robot learning; Humanoid robot

1. Introduction

In various research fields involving humanoid robots, this research centers on the mobility. For a humanoid robot, its method of locomotion is critical. Numerous movement methods have been considered, including the use of wheels and caterpillar-type motion, quadruped motion, and hexapod walking methods inspired from the motions of animals and insects. While these methods have respective strengths, it is preferable that humanoid robots to be used in human society be capable of biped motion using two legs, resembling a human, as our environment is geared to biped walking. By employing biped motion, the humanoid robot will be able to navigate stairs easily and walk along common walkways. In addition, it will be more familiar to humans.

The realization of biped walking is, however, relatively difficult because a biped walking robot is a highly complex system that is inherently unstable. The realization of biped walking started with the idea of static walking. The robot known as WABOT-1, which was developed by Waseda University in the 1970s, required 45 seconds per step but was the first biped walking robot that utilized the concept of static walking [23]. Static walking is characterized by slow biped movement, and the effect of linear or angular momentum of the robot is neglected.

Dynamic walking was considered after the general idea known as ZMP (Zero Moment Point) was introduced to a biped walking robot by Vukobratovic [1]. The ZMP is a point on the ground plane at which the total moments due to ground reaction force becomes zero. If the ZMP is located in a support region, the robot will never fall down when the robot is walking. The first robot to which the idea of ZMP was applied successfully was the WL-10 series from Waseda University. This robot can walk with 1.3 seconds between each step. After the appearance of the first dynamic biped walking robot, researchers developed a variety of working algorithms based on the ZMP.

Since the first successful presentation of biped walking at Waseda University in the 1970s and 1980s, there have been numerous trials to realize stable biped walking robustly and efficiently, with a number of notable results [2][3][4][5][6]. The humanoid robots developed by these research groups can walk steadily on flat or inclined ground and can even run [7][8][9]. Many notable algorithms developed for stable walking and the methods can be categorized into four paradigms [10]: (a) the use of passive walking as a starting point for the design of active walkers; (b) the use of the "zero moment point" control; (c) the use of fixed control architecture and application of a parameter search to find the parameter settings that yield successful walking gaits; and (d) the development of feedback laws based upon insights into balance and locomotion.

HUBO, the first humanoid robot in Korea, was developed by Oh et al. at KAIST in 2004 [2][11][12][13]. It is a child-sized (125 cm tall) biped walking robot with 41 DOF (Degree Of Freedom). This humanoid robot combines several biped walking methods for stable walking. For the walking strategy, a walking pattern specific to a given environment is initially designed and a ZMP (Zero Moment Point) feedback controller and other sub-controllers are then used to maintain stability for a dynamically changeable environment. Many researchers use only a ZMP feedback controller. While stable walking can be maintained in this manner, however, it is difficult to generate desired motions. Hence, HUBO uses the aforementioned (b) and (c) paradigms to overcome this problem.

But the key challenge with the existing method used by HUBO is the determination of the proper parameters for designing or generating a stable walking pattern. It is difficult to find proper parameters as they are influenced by many factors such as the posture of the robot and the ground conditions. The existing robot HUBO determines these parameters through many experiments along with an analysis of walking data using a real system. This process is, however, difficult and time-consuming. Furthermore, only an expert can tune these parameters because an unconfirmed walking pattern is tested using a real robot, there is an inherent risk of accidents. This is the starting point of the present research.

In order to overcome these problems, a HUBO simulator and an algorithm that automatically determines an appropriate walking pattern were developed. The HUBO simulator describes the dynamics of the entire system using a physics engine and includes interactions between the robot and its environment, such as reaction forces and collision analysis. The function of this simulator is to test walking patterns. Also reinforcement learning is used to find suitable walking pattern parameters automatically in order to ensure stable walking and tested on this simulator. Reinforcement learning is a learning method that mimics the human learning process (i.e., learning from experience). Furthermore, this control method is usable if the information or the model of the given system is unclear. With the exception of reinforcement learning, many other methods such as those utilizing a neural oscillator, neural network or fuzzy logic can be used to solve this problem. However, these methods are complex compared to reinforcement learning and require an expert or reliable data. Thus, reinforcement learning is used for the generation of stable walking patterns in this study.

Earlier research on the subject of biped walking using reinforcement learning focused primarily on stable walking. However, the posture of a robot is as important as stable walking. For example, posture is particularly important when the robot is climbing stairs or walking across stepping stones. In these cases, foot placement by the robot is very important. Each foot should be placed precisely or the robot can collapse. Thus, the main goal of this research is to determine a walking pattern that satisfies both stable walking and the required posture (foot placement) using reinforcement learning. Particularly, the Q-leaning algorithm is used as the learning method and CMAC (Cerebellar Model Articulation Controller) serves as the generalization method. The Q-learning algorithm is easy to implement and its convergence is not affected by the learning policy. Hence, it has been used in many applications.

2. Related work

Former studies concerning the realization of stable biped walking using reinforcement learning are categorized below.

(a) The use of reinforcement learning as a sub-controller to support the main controller
(b) The use of reinforcement learning as a main controller or a reference generator
In Case (a), reinforcement learning is normally used as a gain tuner of the main controller

or as a peripheral controller for stable walking. In Case (b), reinforcement learning is used directly to generate a stable walking pattern or as the main controller for stable walking.

Chew and Pratt [14][54] simulated their biped walking robot, Spring Flamingo (Fig. 2-1), in the planar plane (two-dimensional simulation). This seven-link planar bipedal robot weighed 12 kg, was 0.88m in height and had bird-like legs. A reinforcement leaning system was used as the main controller. The following states were chosen as follows: (a) velocity of the hip in the forward direction (x-coordinate); (b) the x-coordinate of an earlier swing ankle measured with reference to the hip; and (c) the step length. The goal was to enable the robot to walk with constant speed; thus, the learning system received '0' when the robot walked within the boundary speed or was given a negative value as a reward. The position of the swing foot was used as the action. Additionally, a torque controller in each ankle was used to control the ankle joint torque. The same type of torque controller was also used to maintain the velocity of the body. The ankle joint torque was limited to a certain stable value; hence, the robot could walk stably without considering the ZMP. However, because the goal was to realize walking with constant speed, the posture of the robot was not considered.

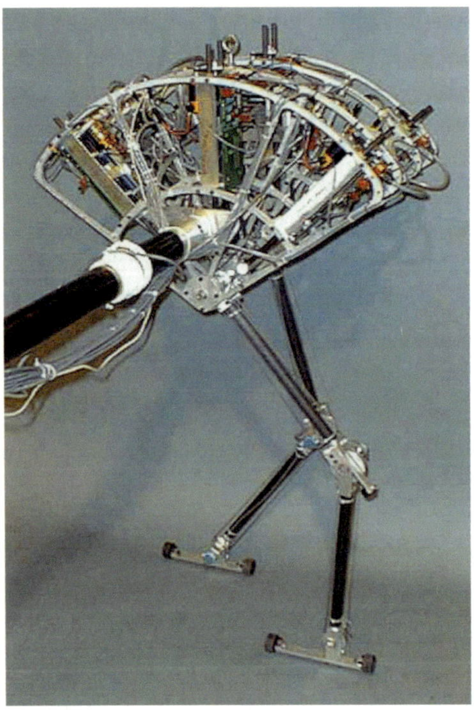

Fig. 1. Spring Flamingo

Benbrahim and Franklin [15] used a reinforcement learning system as both the main and sub-controllers. To achieve dynamic walking with their planar robot, central and other peripheral controllers were used. The central controller used the experience of the peripheral controllers to learn an average control policy. Using several peripheral controllers, it was possible to generate various stable walking patterns. The main controller activated specific peripheral controllers in an approach that was suitable for specific situations. However, the architecture of the controller was complex, and this approach required many learning trials and a lengthy convergence time.

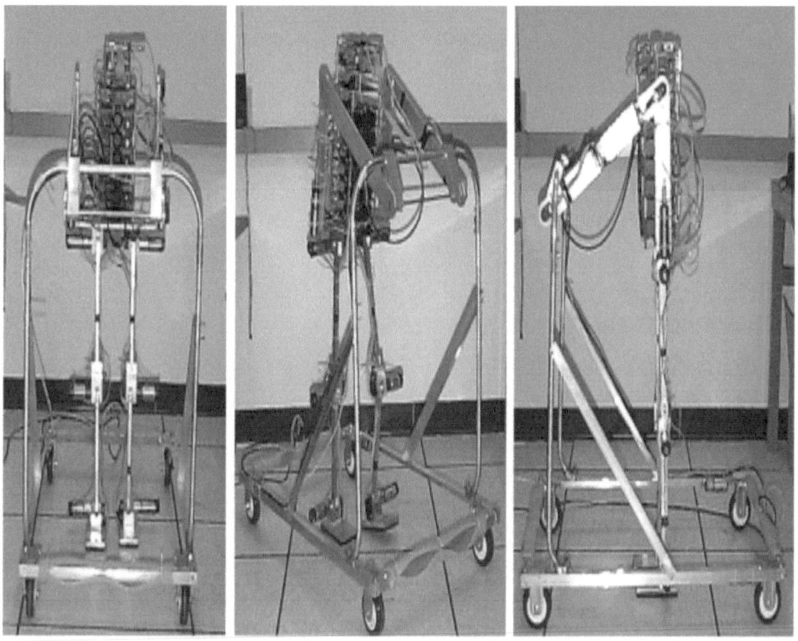

Fig. 2. Benbrahim's biped planar walking robot

Morimoto, Cheng, Atkeson, and Zeglin [16][60] (Fig. 2-3) used a simple five-link planar biped robot to test their reinforcement learning algorithm. The foot of each leg had a shape resembling a 'U', and no joints were used in the ankle. Consequently, it moved in the manner of a passive walker. The goal of the learning system was to walk with constant speed and the states were as follows: (a) velocity of the hip in the forward direction; and (b) forward direction distance between the hip and ankle. The reward was simply falling down or remaining upright, and the action was the angle of the knee joint. The hip joint trajectory was fixed but the step period could vary. If the swing leg touched the ground before the current step period, the next step period was decreased. In addition, if the swing leg touched the ground after the current step period, the step period was increased. This work concentrated only on stable walking; the posture of the robot was not considered.

Schuitema et al. [17] also used reinforcement learning to simulate their planar robot. Their robot, termed Meta, is a passive dynamic walking robot with two hip active joints. The goal of their learning system was to have the robot walk successfully for more than 16 steps.

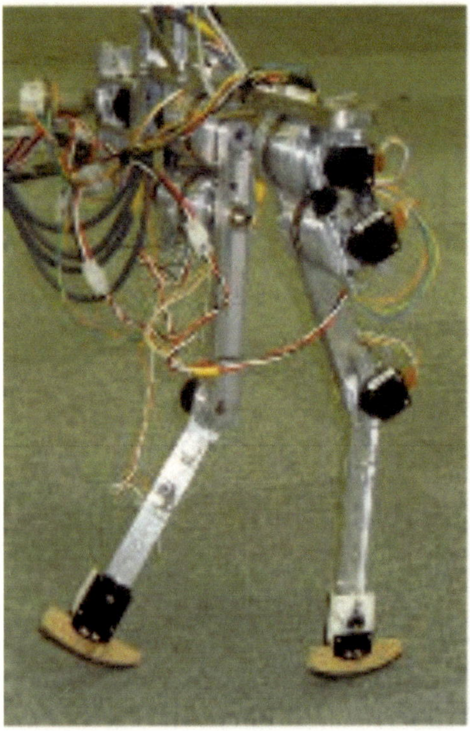

Fig. 3. Morimoto, Cheng, Atkeson, and Zeglin's five-link planar robot

The state space consisted of six dimensions: the angle and angular velocity of the upper stance leg, the upper swing leg, and the lower swing leg. To avoid conditions in which the system would not learn the same thing twice, symmetry between the left and right leg was implemented by mirroring left and right leg state information when the stance leg changed. There was one action dimension, the torque that was applied to the hip joint, which was given a range between -8 and 8 Nm. If the robot walked forward successfully, the learning system received a reward. Additionally, if the body of the robot moves backward, the learning system was penalized. Various experiments were simulated under various constraints and compared the results of each experiment.

Kim et al. [18] used a reinforcement learning system for ZMP compensation. In their research, two-mode Q-learning was used as ZMP compensation against the external distribution in a standing posture. The performance of the Q-learning system was improved using the failure experience of the learning system more effectively along with successful experiences. The roll angle and the roll angular velocity of the ankle were selected as the states. For the action, ankle rolling was given three discrete levels (±0.5°, 0°) during a period of 20 ms. If selecting an action in the opposite direction of the external force, the agent received a reward. If the angle and ZMP constraints were exceeded, the agent was penalized.

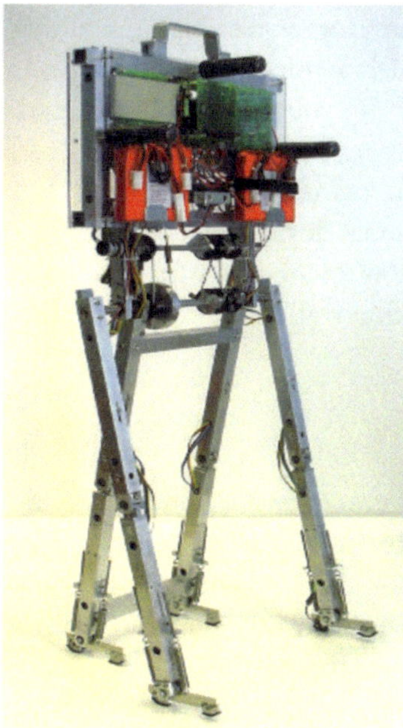

Fig. 4. Meta passive dynamic walking robot

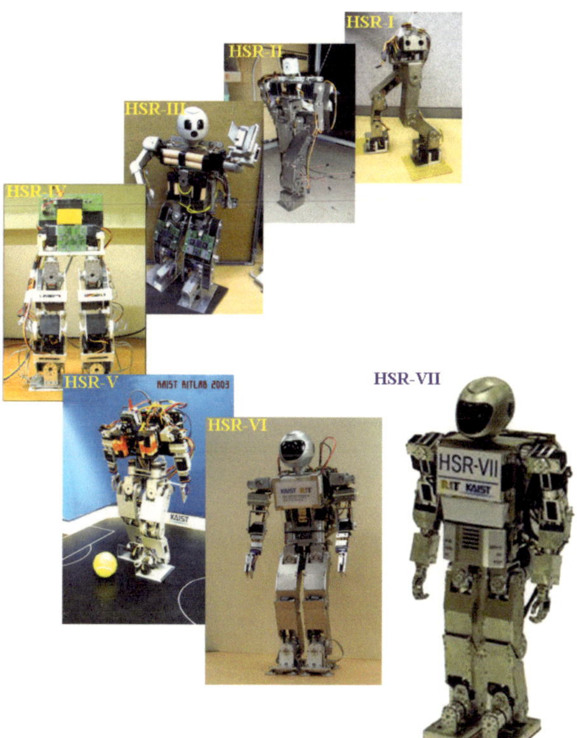

Fig. 5. HanSaRam Series

Other researchers have also proposed learning approaches for stable biped walking [19][20][21]. However, existing research in this area in which reinforcement learning is utilized concerns only stable walking while the posture of the robot is not considered. These researchers normally utilize the position of the swing leg for stable walking. It is, however, difficult to locate the foot in a desired position using only the swing leg. Moreover, it was necessary to use another controller or additional devices for the motion of the support leg. Therefore, this research focuses on the control of the support leg as opposed to control of the swing leg for stable and desired motion walking.

3. Walking pattern

3.1 Sagittal plane

There are several methods of designing a stable walking pattern. But recent researches can be categorized into two groups [22]. The first approach is the 'inverted pendulum model control method' [46][47]. In this method, a simple inverted pendulum model is used as a biped walking model. Based on this model, a proper ZMP reference is generated and a ZMP feedback controller is designed to follow this reference. As this method uses a simple inverted pendulum model, its control structure is very simple. Furthermore, because it follows the ZMP reference for stable walking, stability is always guaranteed. However, it requires a proper ZMP reference and it is difficult to define the relationship between the ZMP reference and the posture of the biped walking robot clearly and accurately. Therefore, it is difficult to select the proper ZMP reference if the posture of the biped walking robot and its walking stability is important. A pattern generator, which translates the ZMP reference to a walking pattern, is also required. Fig. 3-1 shows a block diagram of the 'inverted pendulum model control method'.

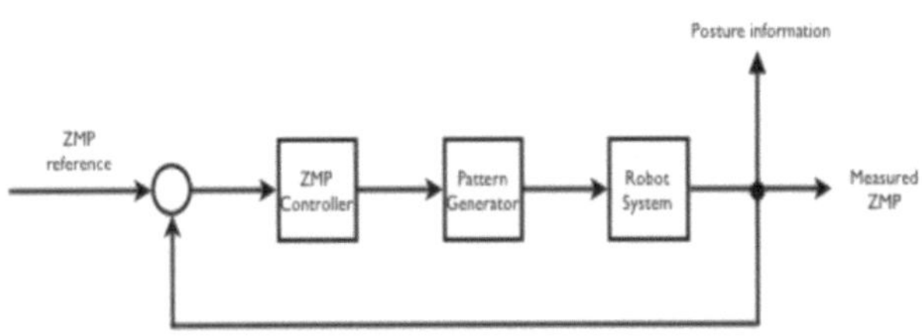

Fig. 6. Inverted pendulum model control method

A second method is known as the 'accuracy model method' [44][45][48][49]. This model requires an accurate model of the biped walking robot and its environment. In this method, a stable walking pattern is generated in advance based on the abovementioned accurate model and the biped walking robot follows this walking pattern without a ZMP feedback controller. One advantage of this method is that it allows control of the biped walking robot with a desired posture. Additionally, it does not require a ZMP controller. However, the generated walking pattern is not a generally functional walking pattern. For example, the walking pattern that is generated for flat ground is not suitable for inclined ground.

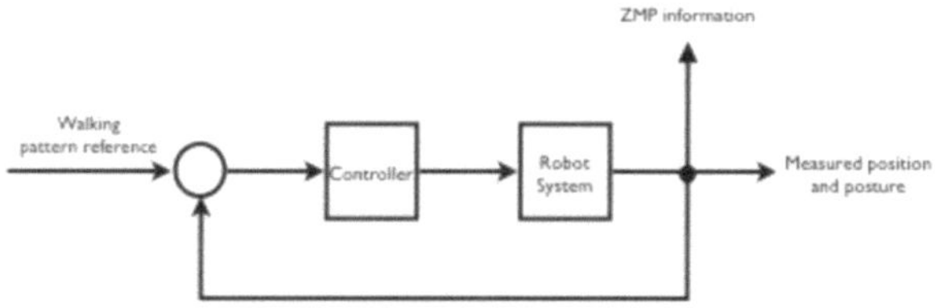

Fig. 7. Accuracy model method

Therefore, with different conditions (e.g., different ground conditions, step lengths, or step periods), new walking patterns should be generated. Fig. 3-2 shows the process of the 'accuracy model method'.

Compared to the 'inverted pendulum model control method', the 'accuracy model method' does not guarantee stability against disturbances; however, it has its own strengths. First, it is possible to control the motion of the biped robot using this method. The 'inverted pendulum model control method' only guarantees stability if the ZMP reference is correct. And it is not possible to control the motion. Second, this method is more intuitive compared to the 'inverted pendulum model control method'. Thus, it is easy to imply physical intuitions using this method. Third, a ZMP controller is not required. Hence, the overall control architecture is simpler with this method compared to the 'inverted pendulum model control method'.

However, an additional problem with the 'accuracy model method' involves difficulty in obtaining an accurate model of the robot and its environment, including such factors as the influence of the posture of the robot, the reaction force from the ground, and so on. Consequently, the generated walking pattern should be tuned by experiments. The generated walking pattern for a specific environment is sensitive to external forces, as this method does not include a ZMP controller. However, when the precise posture of the biped walking robot is required, for example, when moving upstairs or through a doorsill, the 'accuracy model method' is very powerful [9].

In an effort to address the aforementioned issues, the algorithm generating walking patterns based on the 'accuracy model method' was developed using reinforcement learning. To generate a walking pattern, initially, the structure of the walking pattern should be carefully selected. Selection of the type of structure is made based on such factors as polynomial equations and sine curves according to the requirements. The structure of the walking pattern is selected based on the following four considerations [24].

(a) The robot must be easy to operate. There should be minimal input from the operator in terms of the step time, stride, and mode (e.g. forward/backward, left/right) as well as commands such as start and stop.
(b) The walking patterns must have a simple form, must be smooth, and must have a continuum property. It is important that the walking patterns be clear and simple. The

trajectory of the walking patterns should have a simple analytic form and should be differentiable due to the velocity continuum. After the walking patterns are formulated, the parameters for every step are updated.
(c) The calculation must be easy to implement in an actual system. The calculation burden and memory usage should be small and the pattern modification process should be flexible.
(d) The number of factors and parameters that are to be tuned must be small. The complexity of the learning process for the walking patterns is increased exponentially as the number of factors and parameters is increased.

In this research, based on these considerations, a third-order polynomial pattern for the support leg was designed as the walking pattern. This pattern starts from the moment one foot touches the ground and ends the moment the other foot touches the ground (Fig. 3-3).

$$\begin{aligned} z(t) &= Z \\ x(t) &= at^3 + bt^2 + ct + d \end{aligned} \tag{1}$$

To create or complete the third-order forward walking pattern, as shown in Eq. 3-1, four boundary conditions are needed. These boundary conditions were chosen with a number of factors taken into account. First, to avoid jerking motions and formulate a smooth walking pattern, the walking pattern must be continuous. For this reason, the position and velocity of the hip at the moment of the beginning of the walking pattern for the support leg were chosen as the boundary conditions. Additionally, when the foot of the robot is to be placed in a specific location, for example traversing uneven terrain or walking across stepping stones, the final position of the walking pattern is important. This final position is related to the desired posture or step length, and this value is defined by the user. Hence, the final position of the hip can be an additional boundary condition. Lastly, the final velocity of the walking pattern is utilized as the boundary condition. Using this final velocity, it is possible to modify the walking pattern shape without changing the final position, enabling the stabilization of the walking pattern [24]. From these four boundary conditions, a third-order polynomial walking pattern can be generated.

Fig. 8. Sequence of walking

However, it is difficult to choose the correct final velocity of the pattern, as exact models include the biped robot, ground and other environmental factors, are unknown. The existing HUBO robot uses a trial-and-error method to determine the proper final velocity parameter, but numerous trials and experiments are required to tune the final velocity. Thus, in order to find a proper value for this parameter, a reinforcement leaning algorithm is used.

Table 3-1 summarizes the parameters for the sagittal plane motion. And to make problem simpler z-direction movement is fixed as Z (Eq. 3-1).

Boundary condition	Reason
Initial velocity	To avoid jerk motion
Initial position	To avoid jerk motion and continuous motion
Final position	To make wanted posture
Final velocity	To make the walking pattern stable (Unkown parameter)

Table 1. Boundary conditions for the walking pattern

3.2 Coronal plane

Coronal plane movements are periodic motions; it the overall movement range of these movements is smaller than the sagittal plane motion, a simple sine curve is used. If the movement of the z direction is constant, the coronal plane motion can be described by Eq. 3-2, where Y is the sway amount and w is the step period.

$$\begin{aligned} z(t) &= Z \\ y(t) &= Y\sin(wt) \end{aligned} \tag{2}$$

From the simple inverted pendulum model, the ZMP equation can be approximated using Eq. 3-3, where l denotes the length from the ankle joint of the support leg to the mass center of the robot.

$$ZMP(t) = y(t) - \frac{l}{g}\ddot{y}(t) \tag{3}$$

From Eq. 3-2 and Eq. 3-3, the ZMP can be expressed using Eq. 3-4

$$ZMP(t) = Y(1+\frac{l}{g}w^2)\sin(wt) \quad (4)$$

The length l and the step period w are given parameters and the acceleration of gravity g is known parameter. The only unknown parameter is the sway amount. The sway amount can be determined by considering the step period, the DSP (Double Support Phase) ratio and the support region. If the amplitude of the ZMP is located within the support region, the robot is stable. It is relatively easy to determine the unknown parameter (the sway amount) compared to the sagittal plane motion. However, it is unclear as to which parameter value is most suitable. The ZMP model is simplified and linearized, and no ZMP controller is used in this research. Thus, an incorrect parameter value may result from the analysis.

Therefore, using the reinforcement learning system, the optimal parameter value for stable walking using only low levels of energy can be determined.

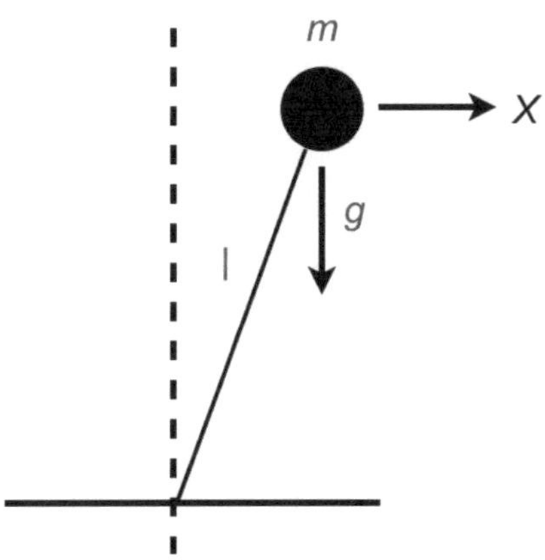

Fig. 9. Inverted pendulum model

4. Simulation

4.1 Simulator

4.1.1 Introduction

Reinforcement learning is based on trial-and-error methodology. It can be hazardous to apply a reinforcement learning system to an actual biped walking system before the learning system is trained sufficiently through many trials, as walking system likely has not been fully analyzed by the learning system. In particular, when such a system is inherently unstable, such as in the case of a biped walking robot, attention to detail is essential. Therefore, it is necessary to train a learning system sufficiently before applying it to a real system. For this reason, simulators are typically used.

A simulator can be used purposes other than for the training of a learning system. For example, simulators can be used for testing new control algorithms or new walking patterns. Various research groups investigating biped walking systems have developed simulators for their own purposes [32][33][34][35][36][37].

The HUBO simulator, which was developed for this study, is composed of a learning system that is in charge of all leaning processes, a physics engine that models a biped robot and its environment, and utility functions to validate the simulation results. Fig. 4-1 shows these modules and the relationships between them. As shown in the figure, learning contents or data obtained from the reinforcement learning module are stored through generalization process. In this study, the CMAC algorithm is used as the generalization method; however, other generalization methods can be easily adapted. The dynamics module, which contains a physics engine, informs the reinforcement learning module of the current states of HUBO. It also receives the action (final velocity of the walking pattern and the sway amount) from the reinforcement learning module, generates a walking pattern, and returns a reward. For the visualization of the movement of a biped walking robot, the OpenGL library is used. Because all components of the HUBO simulator are modularized, it is easy to use with new algorithms or components without modification.

The HUBO simulator contains all of the components necessary for simulating and testing biped walking systems and control algorithms. In addition, all modules are open and can be modified and distributed without limitation. The HUBO simulator follows the GPL (GNU General Public License) scheme.

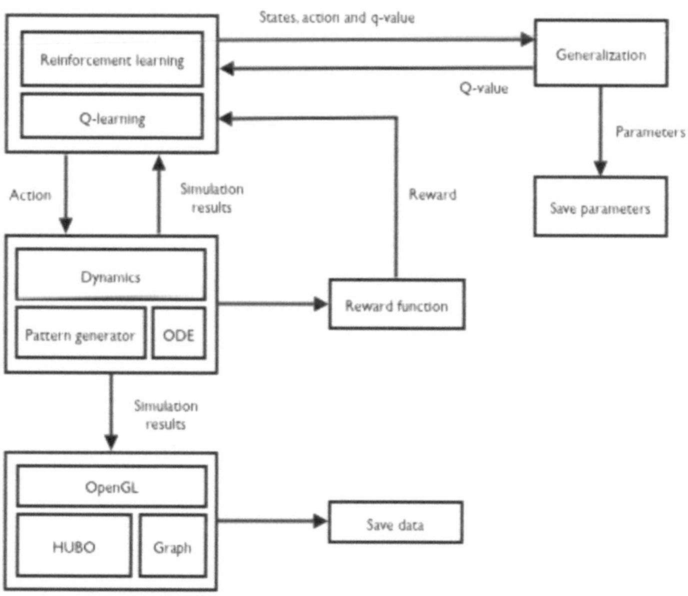

Fig. 10. Structure of the HUBO simulator

4.1.2 Physics Engine

To obtain viable simulation results, the dynamics model of a simulator is very important. If the dynamics model differs greatly from a real model, the result of the simulator is useless. Therefore, it is important to ensure that the simulation model resembles the actual model to the greatest extent possible. Essentially, the model of a biped walking system should contain a robot model as well as a model of the environment of the robot. Many researchers only consider the robot model itself and neglect a model of the environment, which is in actuality more important in a realistic simulation of a biped walking.

For this reason, a physics engine was used to build realistic dynamic model in this study. A physics engine is a tool or APIs (Application Program Interface) that is used for computer simulation programs. In this research, ODE (Open Dynamics Engine) [38] was used to develop the robot and environmental model in an effort to represent the actual condition of the robot accurately. ODE is a rigid body physics engine initially developed by Russell Smith. Its source code is open and is governed by the open source community. ODE provides libraries for dynamics analyses, including collision analyses. The performance of ODE has been validated by various research groups [37][39][40], and many commercial and engineering programs use ODE as a physics engine.

4.1.3 Learning System

The learning system of the HUBO simulator consists of a learning module and a generalization module. The reinforcement learning module uses the Q-learning algorithm, which uses the Q-value. To store the various Q-values that represent actual experience or trained data, generalization methods are needed. Various generalization methods can be used for this. In the present work, the CMAC (Cerebella Model Articulation Controller) algorithm is employed. This algorithm converges quickly and is readily applicable to real systems.

Setting up states and a reward function is the most important process in the efficient use of reinforcement learning. When setting up states, using physical meanings is optional; however, it is important that the most suitable states for achieving the goal are selected. Additionally, the reward function should describe the goal in order to ensure success. The reward function can represent the goal directly or indirectly. For example, if the goal for a biped walking robot is to walk stably, the learning agent receives the reward directly if the robot walks stably without falling down. Otherwise, it is penalized. In addition, the reward function describes the goal of stable walking indirectly, including such factors as the pitch or roll angle of the torso while walking and the walking speed. However, it is important that the reward should suitably describe the goal.

4.1.4 Layout

Fig. 4-2 shows the main window of the HUBO simulator. The motion of HUBO calculated using ODE is displayed in the center region of the HUBO simulator using OpenGL. Each step size or foot placement can be modified from the main window. Fig. 4-3 shows the learning information window. This window shows information such as the current states and the reward associated with the learning module. In addition, the learning rate and the update rate can be modified from this window. Fig. 4-4 shows the body data window. This window shows the current position and orientation of each body. As lower body data is important for the system, only the data of the lower body is represented. Fig. 4-5 shows the joint angle of the lower body. The data of the force and torque for each ankle joint is shown in the force-torque data window in Fig. 4-6.

The HUBO simulator was developed using the COCOA$^{\circ,R}$ library under a Mac OS X $^{\circ,R}$ environment. As COCOA is based on the Object-C language and all structures are modulated, it is easy to translate to other platforms such as Linux$^{\circ,R}$ and Windows$^{\circ,R}$.

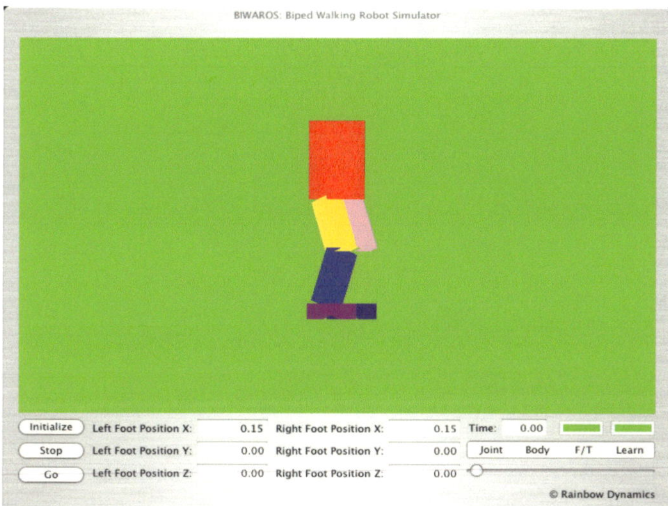

Fig. 11. Main window of the HUBO simulator

Fig. 12. Learning information window of the HUBO simulation

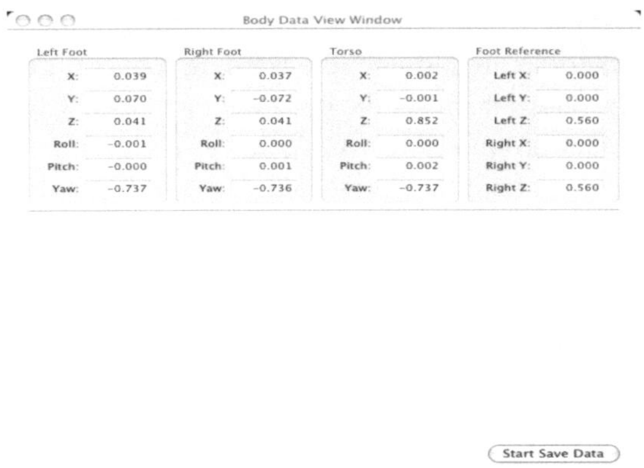

Fig. 13. Body data window of the HUBO simulator

Fig. 14. Joint data window of the HUBO simulator

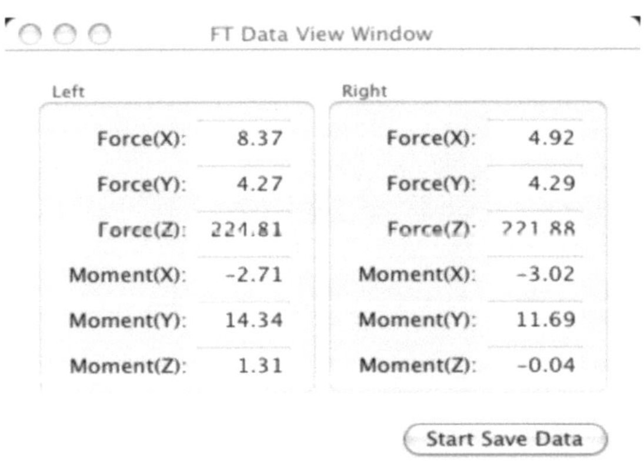

Fig. 15. Force-Torque data window of the HUBO simulator

4.2 States, action and reward

The biped walking pattern generation system can be viewed as a discrete system. Before a new walking step begins, the learning module receives the information of the current states and generates the walking pattern. The robot then follows the generated walking pattern. Following this, the walking pattern is finished and the process starts again. Therefore, this system can be viewed as a discrete system in which the time step is the walking pattern period or the step period. In this study, the walking pattern starts at the moment of the SSP (Single Support Phase) and ends at the moment of the next SSP. At the beginning of the SSP, the learning module receives the current states and calculates the action. Simultaneously, an evaluation of the former action is carried out by the learning module.

4.2.1 Sagittal plane

To set up proper states for the sagittal plane motion, a simple inverted model is used. From the linearized inverted pendulum model, the ZMP equation can be formulated, as shown in Eq. 4-1.

$$ZMP(t) = x(t) - \frac{l}{g}\ddot{x}(t) \tag{5}$$

From Eq. 4-1, the position and acceleration of the mass center is directly related to the ZMP. As the ZMP is related to the stability of the biped walking system, it is feasible to select the position and acceleration of the mass center as states. In addition, to walk stably with minimal energy consumption, the robot should preserve energy, implying that the robot should utilize its momentum (angular or linear). The momentum reflects current and future states; it is related to the velocity of the mass center. Therefore, the velocity of the mass center was chosen as the state in this study. Selected states and the reasons for their selection are summarized in Table 4-1.

All states are normalized to -1.0 ~ 1.0. However, the reinforcement learning agent has no data regarding the maximum values of the states. It receives this data during the training and updates it automatically. First, these maximum values are set to be sufficiently small; in this research, the value is 0.1. The reinforcement learning agent then updates the maximum value at every step if the current values are larger than the maximum values.

State	Reason
The position of the mass center with respect to the support foot	Relation between the position of the mass center and ZMP and the body posture
The velocity of the mass center	Angular or linear momentum
The acceleration of the mass center	Relation between the position of the mass center and ZMP

Table 2. States for the sagittal plane motion

The learning parameter learnt through reinforcement learning is the final velocity. It is an unknown parameter in the initial design of the walking pattern. The boundary conditions of the walking pattern were discussed in Chapter 3. Eq. 4-2 shows these conditions again.

$$
\begin{array}{ll}
X(t) = at^3 + bt^2 + ct + d & \text{Walking pattern} \\
\text{When } t = 0 & \\
X = current\ position & \text{Condition 1} \\
\dot{X} = current\ velocity & \text{Condition 2} \\
\text{When } t = T & \\
X = final\ position & \text{Condition 3} \\
\dot{X} = final\ velocity & \text{Unknown parameter}
\end{array}
\tag{6}
$$

From Eq. 6, Conditions 1 and 2 are determined from the former walking pattern and Condition 3 is the given parameter (the desired step size) from the user. However, only the final velocity is unknown, and it is difficult to determine this value without precise analysis. Hence, the action of the reinforcement learning system is this final velocity (Table 4-2).

Action	Reason
Final velocity of the walking pattern	Only the final velocity is unknown parameter and it is related to the stable walking

Table 3. Action for the sagittal plane motion

The reward function should be the correct criterion of the current action. It also represents the goal of the reinforcement learning agent. The reinforcement learning agent should learn to determine a viable parameter value for the generation of the walking pattern with the goal of stable walking by the robot. Accordingly, in this research, the reward is 'fall down or remain upright' and 'How good is it?' Many candidates exist for this purpose, but the body rotation angle (Fig. 4-7) was finally chosen based on trial and error. Table 4-3 shows the reward and associated reasons. If the robot falls down, the reinforcement learning agent then gives a high negative value as a reward; in other cases, the robot receives positive values according to the body rotation angle. The pitch angle of the torso represents the feasibility of the posture of the robot.

Reward	Reason
Fall down	This denotes the stability of the robot(or absence of stability)
Pitch angle of the torso	It represents how good it is for stable dynamic walking

Table 4. Reward for the sagittal plane motion

Fig. 16. Pitch angle of the torso (This angle is used as the reward in the sagittal motion)

Fig. 4-8 shows the overall structure of the generation of the walking pattern using reinforcement learning. The reinforcement learning system receives the current states, calculates the proper action, and the walking pattern generator generates the walking pattern based on this action. The reinforcement learning system learns the suitability of the action from its result, and this process is repeated until the reinforcement learning system shows reasonable performance.

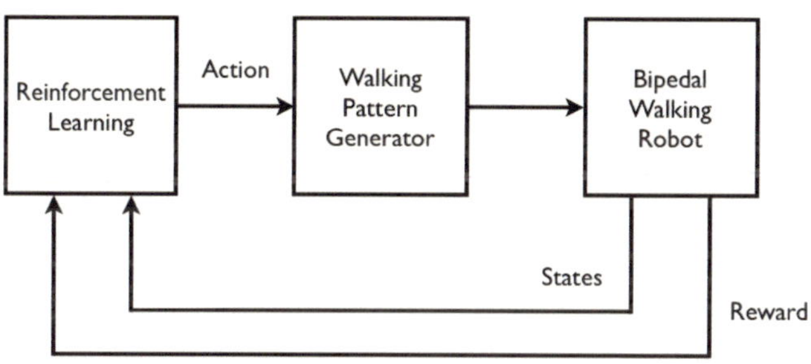

Fig. 17. Diagram of the biped walking pattern generation system

4.2.2 Coronal plane

The coronal plane motion is assumed to be weakly coupled to the sagittal plane motion. Thus, states for the sagittal plane motion are not considered in the coronal plane motion learning implementation. Regarding the inverted pendulum model, the linear dynamic equation for the inverted pendulum can be described as follows:

$$\ddot{y} = \frac{g}{l} y \tag{7}$$

Eq. 4-3 can be integrated to show the relationship between $\dot{y}$ and y:

$$\frac{\dot{y}^2}{2} = \frac{gy^2}{2l} + C \tag{8}$$

Here, C is the integration constant, which is termed the orbital energy [41]. Given the velocity y and the position y of the system at any instant, C can be computed. Eq. 4-4 defines the relationship between y and y for all other times before the next support exchange event. When C is greater than zero, the mass approaching the vertical plane that passes through the pivoting point will be able to travel across the plane. When C is less than zero, the mass will not be able to travel across the vertical plane. Instead, it reverts to its original direction of travel at some instant. If y and y are given, it is possible to predict stability using C. Thus, y (the velocity of the torso) and y (the position of the torso with respect to the support foot) are used as states.

As a simple sine function is used in the walking pattern of the coronal plane motion and because the step period is given, the only unknown parameter is the amplitude (sway amount) of the sine function. Thus, the amplitude is used as the action.

The reinforcement learning system for the coronal plane motion adopts a reward that only issues a punishment value when a failure state is encountered:

$$r = \begin{cases} 0 & for\,\Theta_l < \theta < \Theta_u \\ R & failure(otherwise) \end{cases} \tag{9}$$

Here, θ is the roll angle of the torso, which is bounded by Θ_l and Θ_u and R is a negative constant (punishment). It is important to note that the failure condition based on θ is checked at all times.

5. Experiment

To test the logic of the reinforcement learning algorithm, several experiments were carried out. It was assumed that the sagittal plane and the coronal plane motion are weakly related; for this reason, the coronal plane motion was considered first. The reinforcement learning agent for the coronal plane motion learns the stable walking pattern parameter. Based on this learning agent, the reinforcement learning agent for the sagittal plane motion is added. First, step walking was tested, and forward walking tests were performed after these tests.

5.1 Step walking

As step walking does not contain forward motion, it is similar to coronal plane motions.

Hence, the reinforcement learning agent for coronal plane motions learns stable parameters first through step walking experiments. Following this, the learning agent for the sagittal plane motion is added. To determine the suitable parameters, the learning agent for the coronal plane motion learns the sway amount. The update rate α and learning rate γ are set to 0.2 and 0.8, respectively, and the e-greedy value is set to 0.2 initially. The e-greedy value converges to 0 as the learning agent finds the successful parameter.
The experimental conditions are shown in Table 5-1. As shown in Fig. 5-1, step walking converged in 17 of the trails. The parameter for step walking (the sway amount) converges to 0.055m; Fig. 5-2 shows the sideways (y) direction movement of the torso. The movement of the torso is very stable with the period motion. The roll angle of the torso is within 0.4 degree, except for the initial movement, and is very stable, as shown in Fig. 5-3. Fig. 5-4 shows the z-direction movement of the foot, and shows that the robot walks stably without falling down. Additionally, it shows that the DSP time is set to 10% of the step period.

5.2 Forward walking - 15 cm

Based on the learning agent for coronal plane motions that were fully trained through the step walking experiment, a forward walking experiment of 15cm was performed to find a stable forward walking pattern. The experiment conditions are shown in Table 5-2.

As shown in Fig. 5-5, the learning agent learns the proper parameters within 14 trials. The converged final velocity of the walking pattern is 0.3m/sec. The stable walking pattern is identical to that of the forward (x) direction movement of the torso; Fig. 5-6 shows that the walking pattern is very stable. Moreover, the pitch angle of the torso does not exceed 2 degrees (failure condition), as shown in Fig. 5-7. Earlier research used the motion of the swing leg for stable walking while in this research the walking pattern for the support leg is considered. Therefore, it is possible to place the foot in the desired position. Figs 5-8 and 5-9 represent the movement of the foot. As shown in Fig. 5-9, the foot is located in the desired position (0.15m).

5.3 Forward walking - 20 cm

To test the robustness of the reinforcement learning agent for sagittal plane motions, an additional forward walking test was performed. In this test, the learning agent determined a stable parameter within 21 trials, and the pitch angle of the torso was within 1.3 degrees while walking. Fig. 5-14 shows that the foot is placed in the desired position.

Step period	1.0 sec
Step length	0.0 m
Lift-up	0.06 m
DSP time	0.1 sec
Update rate	0.2
Learning rate	0.8
Initial e-greedy	0.2

Table 5. Experiment condition for step walking

Step period	1.0 sec
Step length	0.15 m
Lift-up	0.06 m
DSP time	0.1 sec
Update rate	0.2
Learning rate	0.7
Initial e-greedy	0.1

Table 6. Experiment condition for forward walking (15 cm)

Step period	1.0 sec
Step length	0.2 m
Lift-up	0.06 m
DSP time	0.1 sec
Update rate	0.2
Learning rate	0.8
Initial e-greedy	0.1

Table 7. Experiment condition for forward walking (20 cm)

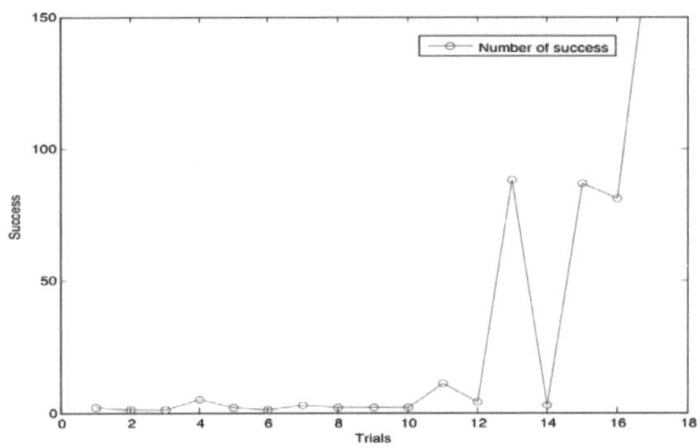

Fig. 18. Iteration and success in the Coronal plane motion

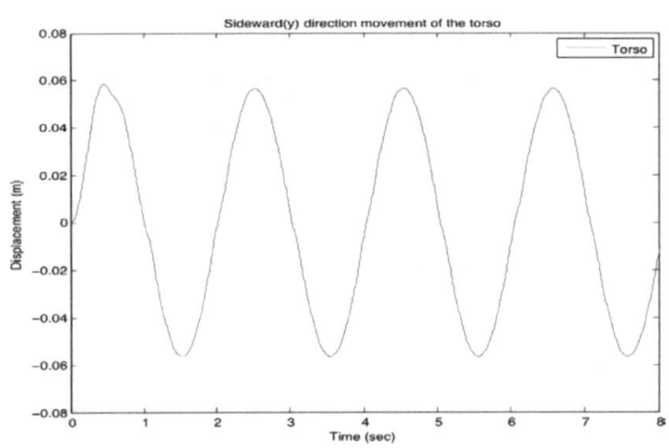

Fig. 19. Sideward(y) direction movement of the torso

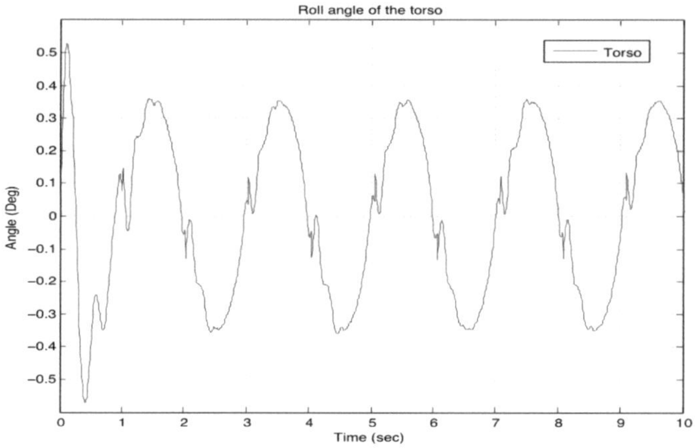

Fig. 20. Roll angle of the torso

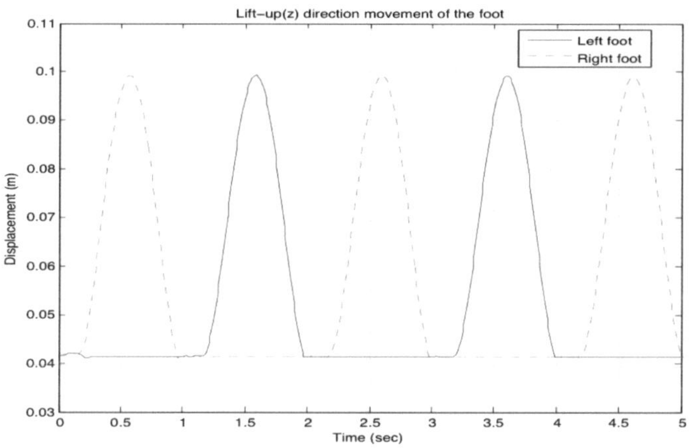

Fig. 21. Lift-up(z) direction movement of the foot

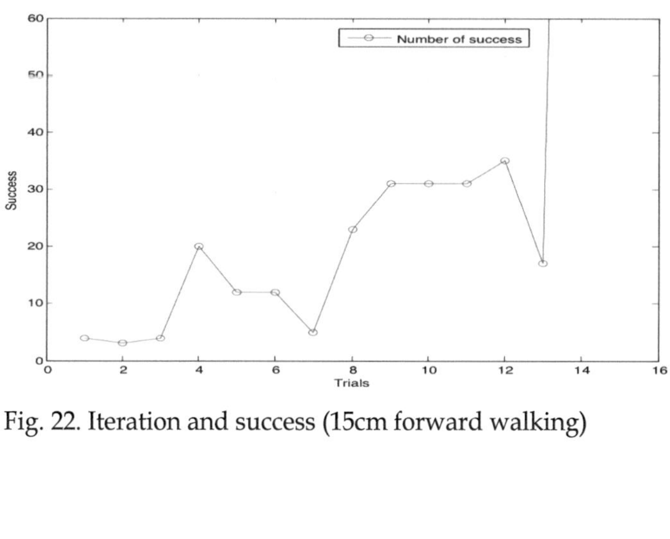

Fig. 22. Iteration and success (15cm forward walking)

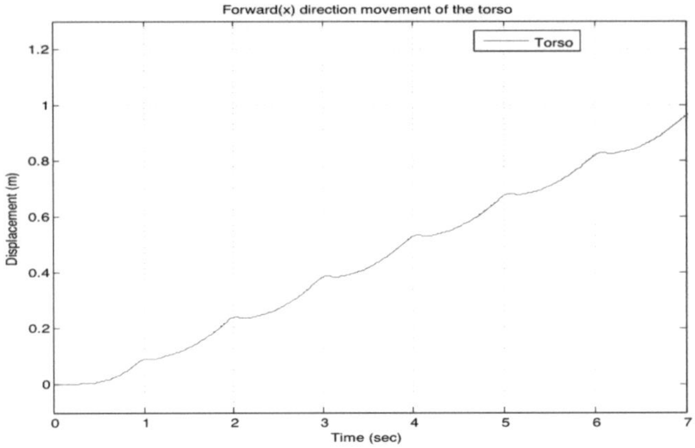

Fig. 23. Forward(x) direction movement of the torso (15cm forward walking)

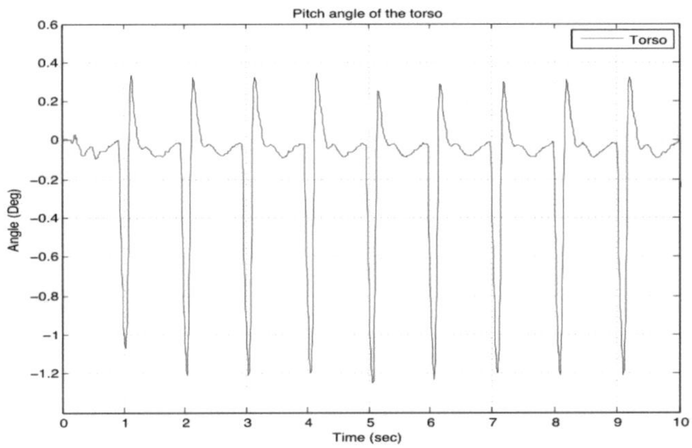

Fig. 24. Pitch angle of the torso (15cm forward walking)

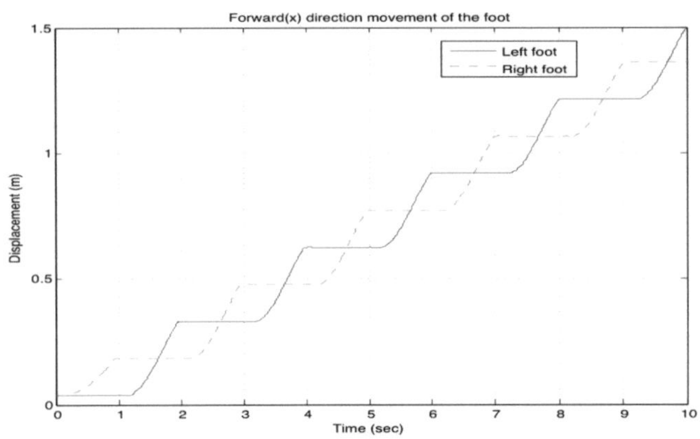

Fig. 25. Forward(x) direction movement of the foot (15cm forward walking)

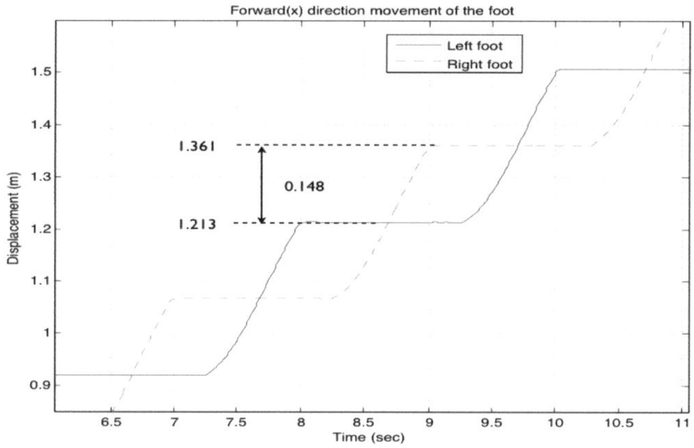

Fig. 26. Position of the foot (15cm forward walking)

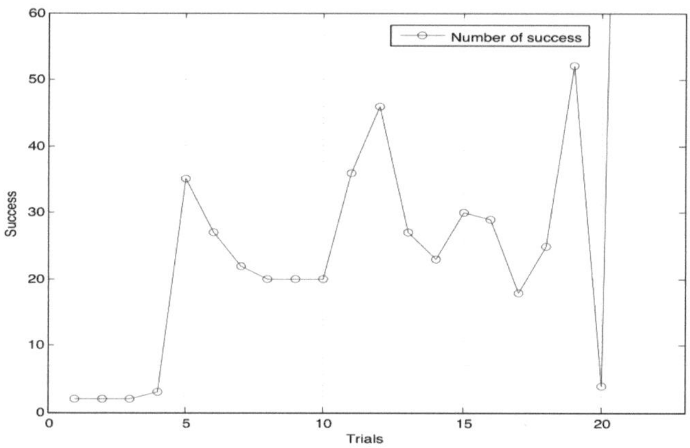

Fig. 27. Iteration and success (20cm forward walking)

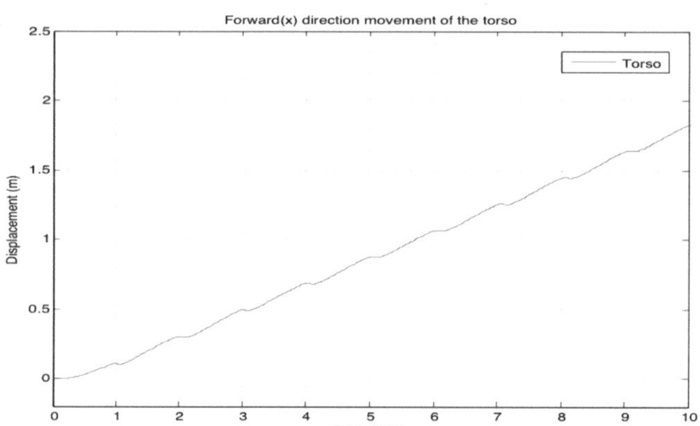

Fig. 28. Forward(x) direction movement of the torso (20cm forward walking)

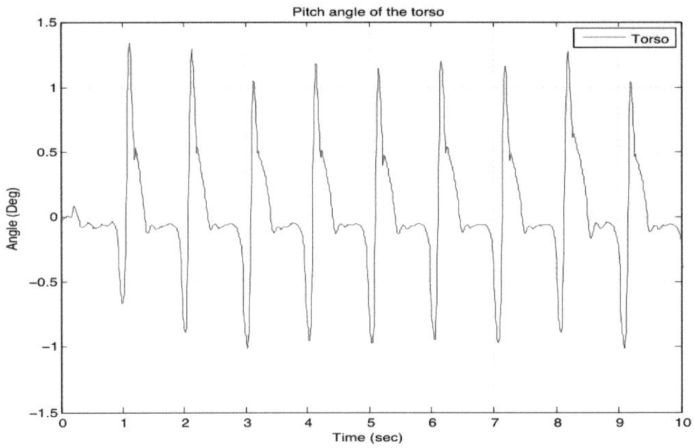

Fig. 29. Pitch angle of the torso (20cm forward walking)

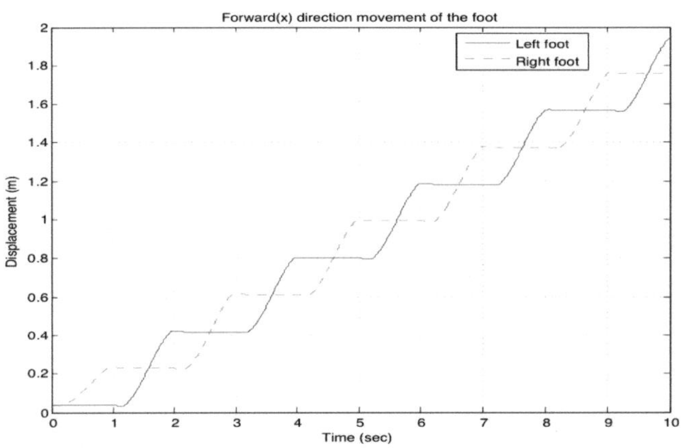

Fig. 30. Forward(x) direction movement of the foot (20cm forward walking)

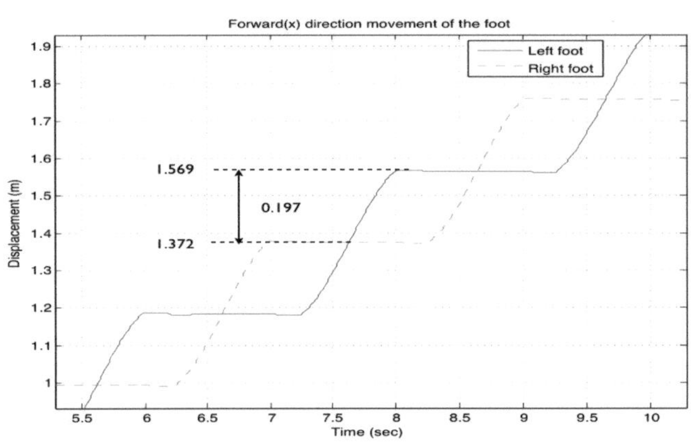

Fig. 31. Position of the foot (20cm forward walking)

7. Conclusion

The main purpose of this research is to generate the stable walking pattern. There are many methods about stable biped walking but these methods can be categorized into two groups. One is the 'inverted model based control method'. In this method, a simple inverted pendulum model is used as a biped walking model. Based on this model, a proper ZMP reference is generated and a ZMP feedback controller is designed to follow this reference. A second method is known as the 'accuracy model method'. This model requires an accurate model of the biped walking robot and its environment. In this method, a stable walking pattern is generated in advance based on the abovementioned accurate model and the biped walking robot follows this walking pattern without a ZMP feedback controller. One advantage of this method is that it allows control of the biped walking robot with a desired posture. Additionally, it does not require a ZMP controller. Each method has its own strength and weakness but in this research, the 'accuracy model method' is used.

However, a problem with the 'accuracy model method' involves difficulty in obtaining an accurate model of the robot and its environment, including such factors as the influence of the posture of the robot, the reaction force from the ground, and so on. Consequently, the generated walking pattern should be tuned by experiments. The generated walking pattern for a specific environment is sensitive to external forces, as this method does not include a ZMP controller. Also the tuning process takes much time and needs an expert.

In this research, reinforcement learning is used to solve this problem. Reinforcement learning is different from supervised learning, the kind of learning in most current research in machine learning, statistical pattern recognition, and artificial neural network. In point of view treating nonlinear problems which are finding optimal solution under given environment, supervised learning is similar to reinforcement learning. But this process is only possible to finding good solution if and if only there are good examples provided by external supervisor. But reinforcement learning learns this process without the external supervisor so if the system model is unknown or partially known, reinforcement learning will be good choice.

Reinforcement learning is based on trial-and-error methodology. It can be hazardous to apply a reinforcement learning system to an actual biped walking system before the learning system is trained sufficiently through many trials, as walking system likely has not been fully analyzed by the learning system. In particular, when such a system is inherently unstable, such as in the case of a biped walking robot, attention to detail is essential. Therefore, it is necessary to train a learning system sufficiently before applying it to a real system. For this reason, the HUBO simulator is developed. The HUBO simulator contains the physics engine, the learning system, the generalization module and other utility modules. The physics engine contains the environment model not only the robot model. So it is possible to simulate the interaction between the robot and its environment such as the ground. Its structure is modulated and simple, it is easy to update or add its functions or algorithm.

In this research, it is developed the stable biped walking pattern generation algorithm using reinforcement learning based on the HUBO simulator. Unlike former researches, the walking pattern of the support leg is considered. Existing researches use the motion of the swing leg for stable biped walking and extra controllers are needed for controlling the motion of the support leg. But because the algorithm developed in this research is for the walking pattern of the support leg and the motion of the swing is determined by the given specifications, extra controllers are not needed and overall control structure is very simple. Also this algorithm generates the stable biped walking pattern automatically, supervisors or experts are not needed.

Also algorithms developed by former researches were limited to the planar planed robot system, but the algorithm developed through this research considers the sagittal and the coronal plane motion. Former researches considered the sagittal plane motion only and the coronal plane motion was neglected. In this research, it is assumed that the sagittal and the coronal plane motion are weakly coupled. So the reinforcement learning systems for the each plane are trained separately and after the sufficient learning process, two learning systems are combined.

Through several experiments, it is validated the performance of the stable biped walking pattern generation algorithm. Several experiments are accomplished using the HUBO simulator and proper states and reward function for stable biped walking are founded. The algorithm is converged stably and its performance is superior and convergence time is faster than existing researches.

Although this research does not contain the experiment that contains the real system, the logic of the algorithm is tested and verified using the HUBO simulator. And the performance of the algorithm is distinguishable compare to the existing researches. But it is necessary to test or apply the algorithm to the real system and following research will contain this. Also the algorithm is tested to forward walking only but in the following research various motions such as side walking and asymmetric biped walking will be tested.

<table>
<tr><th>Contributions</th><th>Future work</th></tr>
<tr><td>The HUBO simulator is developed which contains the physics engine, the learning system, the generalization module and other utility modules.</td><td>It is needed to apply and test the algorithm to the real system.</td></tr>
<tr><td>Proper states and reward function are founded for the reinforcement learning system.</td><td rowspan="3">It is necessary to test various motions such as side walking and asymmetric biped walking.</td></tr>
<tr><td>The third order polynomial walking pattern for the motion of support leg is generated using reinforcement learning.</td></tr>
<tr><td>In contrast to former researches, the sagittal and coronal plane motion is considered.</td></tr>
</table>

Table 8. Contributions and future work

8. Reference

A. A. Frank and M. Vokobratovic, *'On the Gait Stability of Biped Machine'*, IEEE Transactions on Automatic Control, December, 1970.

Ill-Woo Park, Jung-Yup Kim, Jungho Lee, and Jun-Ho Oh, *'Mechanical Design of the Humanoid Robot Platform, HUBO'*, Journal of Advanced Robotics, Vol. 21, No. 11, 2007.

K. Hirai, *'Current and Future Perspective of Honda Humanoid Robot'*, Proc. IEEE/RSJ International Conference on Intelligent Robots and Systems, p500-p508, 1997.

Y. Sakagami, R. Watanabe, C. Aoyama, S. Matsunaga, N. Higaki and K. Fujimura, *'The Intelligent ASIMO: System Overview and Integration'*, Proc. IEEE/RSJ International Conference on Intelligent Robots and Systems, p2478-p2483, 2002.

K. Kaneko, S. Kajita, F. Kanehiro, K. Yokoi, K. Fujiwara, H. Hirukawa, T. Kawasaki, M. Hirata and T. Isozumi, *'Design of Advanced Leg Module for Humanoid Robot Project of METI'*, Proc. IEEE International Conference on Robotics and Automation, p38-p45, 2002.

A. Takanishi, M. Ishida, M. Yamazaki and I. Kato, *'The Realization of Dynamic Walking by the Biped Walking Robot WL-10RD'*, ICRA 1985, 1985.

Jung-Yup Kim, Ill-Woo Park, and Jun-Ho Oh, *'Walking Control Algorithm of Biped Humanoid Robot on Uneven and Inclined Floor'*, Journal of Intelligent and Robotic Systems, Accepted, 2006.

K. Nagasaka, Y. Kuroki, S. Suzuki, Y. Itoh and J. Yamaguchi, *'Integrated Motion Control for Walking, Jumping and Running on a Small Bipedal Entertainment Robot'*, Proc. IEEE International Conference on Robotics and Automation, p648-p653, 2004.

Jung-Yup Kim, *'On the Stable Dynamic Walking of Biped Humanoid Robots'*, Ph. D Thesis, Korea Advanced Institute of Science and Technology, 2006.

KangKang Yin, Kevin Loken and Michiel van de Panne, *'SIMBICON: Simple Biped Locomotion Control'*, ACM SIGGRAPH 2007, 2007.

Jung-Yup Kim, Jungho Lee and Jun Ho Oh, *'Experimental Realization of Dynamic Walking for the Human-Riding Biped Robot, HUBO FX-1'*, Advanced Robotics, Volume 21, No. 3-4, p461-p484, 2007.

Jun-Ho Oh, David Hanson, Won-Sup Kim, Il Young Han, Jung-Yup Kim, and Ill-Woo Park, *'Design of Android type Humanoid Robot Albert HUBO'*, in Proc. IEEE/RSJ Int. Conf. on Intelligent Robots and Systems, Beijin, China, 2006.

Jungho Lee, Jung-Yup Kim, Ill-Woo Park, Baek-Kyu Cho, Min-Su Kim, Inhyeok Kim and Jun Ho Oh, *'Development of a Human-Riding Humanoid Robot HUBO FX-1'*, SICE-ICCAS 2006, 2006.

Chew-Meng Chew and Gill A. Pratt, *'Dynamic Bipedal Walking Assisted by Learning'*, Robotica, Volume 20, p477-p491, 2002.

Hamid Benbrahim and Judy A. Franklin, *'Biped Dynamic Walking Using Reinforcement Learning'*, Robotics and Autonomous Systems, Volume 22, p283-p302, 1997.

Jun Morimoto, Gordon Cheng, Christopher Atkeson and Garth Zeglin, *'A Simple Reinforcement Learning Algorithm for Biped Walking'*, Proc. of the 2004 International Conference on Robotics & Automation, p3030-p3035, 2004.

E. Schuitema, D. G. E. Hobbelen, P. P. Jonker, M. Wisse and J. G. D. Karssen, *'Using a ControllerBased on Reinforcement Learning for a Passive Dynamic Walking Robot'*, Proc. of IEEE-RAS International Conference on Humanoid Robots, Tsukuba Japan, 2005.

Jong-Hwan Kim, Kui-Hong Park, Jun-Su Jang, Yong-Duk Kim, Bum-Joo Lee and Ki-Pyo Kim, *'Humanoid Robot HanSaRam: Schemes for ZMP compensation'*, Proc. of International Conference on Computational Intelligence, Robotics and Autonomous Systems, Singapore, 2003.

Dusko Katic and Miomir Vukobratovic, *'Control Algorithm for Biped Walking Using Reinforcement Learning'*, 2nd Serbian-Hungarian Joint Symposium on Intelligent Systems, 2004.

Naoto Shiraga, Seiichi Ozawa and Shigeo Abe, *'A Reinforcement Leaning Algorithm for Neural Networks with Incremental Learning Ability'*, Proceedings of International Conference Neural Information Processing, 2002.

William Donard Smart, *'Making Reinforcement Learning Work on Real robots'*, Ph. D. Thesis, Brown University, 2002.

Shuuji Kajita, Fumio Kanehiro, Kenji Kaneko, Kiyoshi Fujiwara, Kensuke Harada, Kazuhito Yokoi and Hirohisa Hiukawa, *'Biped Walking Pattern Generation by Using Preview Control of Zero-Moment Point'*, Proceedings of the 2003 IEEE International Conference on Robotics & Automation, p1620-p1626, 2003.

I. Kato, S. Ohteru, H. Kobayashi, K. Shirai and A. Uchiyama, 'Information Power Machine with Senses and Limbs', First CISM-IFToMM Symposium on Theory and Practice of Robots and manipulators, 1974.

Ill-Woo Park, Jung-Yup Kim and Jun-Ho Oh, *'Online Walking Pattern Generation and Its Application to a Biped Humanoid Robot-KHR-3(HUBO)'*, Journal of Advanced Robotics, 2007.

Richard S. Sutton and Andrew G. Barto, *'Reinforcement Learning: An Introduction'*, The MIT Press, 1998.

Lee, Y., Jung, T., *'Reinforcement Learning'*, Kyunghee Univ., 2005.

Watkins, C. J. C. H., *'Learning from Delayed Rewards'*, Doctoral Thesis, Cambridge University, 1989.

William Donald Smart, *'Making Reinforcement Learning Work on Real Robots'*, Ph .D Thesis, Brown University, 2002.

J. S. Albus, *'Theoretical and experimental aspects of a cerebellar model'*, PhD. Dissertation, University of Maryland, 1972.

J. S. Albus, *'Data storage in the cerebellar model articulation controller'*, Journal of Dynamic Systems, Measurement and Control, pp. 228-233, 1975.

J. S. Albus, *'Brains, behavior, and robotics'*, Peterborough, N.H.: Byte Books/McGraw-Hill, 1981.

Hirohisa Hirukawa, Fumio Kanehiro and Shuuji Kajita, *'OpenHRP: Open Architecture Humanoid Robotics Platform'*, Robotics Research: The Tenth International Symposium, Volume 6, p99-p112, 2003.

Rawichote Chalodhorn, David B. Grimes, Gabriel Maganis and Rajesh P. N. Rao, *'Learning Dynamic Humanoid Motion using Predictive Control in Low Dimensional Subspace'*, IEEE-RAS International Conference on Humanoid Robots Humanoids2005, 2005.

Rawichote Chalodhorn, David B. Grimes, Gabriel Maganis, Rajesh P. N. Rao and Minoru Asada, *'Learning Humanoid Motion Dynamics through Sensory-Motor Mapping in Reduced Dimensional Spaces'*, 2006 IEEE International Conference on Robotics and Automation, 2006.

C. Angulo, R. Tellez and D. Pardo, *'Emergent Walking Behaviour in an Aibo Robot'*, ERCIM News 64, p38-p39, 2006.

L.Holh, R. Tellez, O. Michel and A. Ijspeert, *'Aibo and Webots: simulation, wireless remote control and controller transfer'*, Robotics and Autonomous Systems, Volume 54, Issue 6, p472-p485, 2006.

Wolff, K., and Nordin, P., *'Learning Biped Locomotion from First Principles on a Simulated Humanoid Robot using Linear Genetic Programming'*, Genetic and Evolutionary Computation GECCO 2003, 2003.

Russel Smith, *'www.ode.org/ode.html'*, 2007.

Wolff, K., and Nordin, P., *'Evolutionary Learning from First Principles of Biped Walking on a Simulated Humanoid Robot'*, The Advanced Simulation Technologies Conference 2003 ASTC'03, 2003.

Olivier Michel, *'Webots: Professional Mobile Robot Simulation'*, International Journal of Advanced Robotic System, Volume 1, Number 1, p39-p42, 2004.

Shuuji Kajita, Kazuo Tani and Akira Kobayashi, *'Dynamic walk control of a biped robot along the potential energy conserving orbit'*, IEEE International Conference on Intelligent Robots and Systems, p789-p794, 1990.

J. K. Hodgins, M. H. Raibert, *'Biped Gymnastics'*, The International Journal of Robotics Research, Volume 9, No. 2, 1990.

M. H. Raibert, *'Hopping in Legged Systems-Modeling and Simulation for the Two Dimensional One Legged Case'*, IEEE Transactions on Systems, Man and Cybernetics, Volume SMC-14, No. 3, 1984.

J. Yamaguchi , E. Soga, S. Inoue and A. Takanishi, *'Development of a Bipedal Humanoid Robot-Control Method of Whole Body Cooperative Dynamic Biped Walking'*, Proc. of the ICRA 1999, p368-p374, 1999.

M. Gienger, *'Toward the Design of a Biped Jogging Robot'*, Proc. of ICRA 2001, p4140-p4145, 2001.

T. Sugihara, Y. Nakamura and H. Inoue, *'Realtime Humanoid Motion Generation through Manipulation based on Inverted Pendulum Control'*, Proc. of the ICRA 2002, p1404-p1409, 2002.

J. Pratt, P. Dilworth and G. Pratt, *'Virtual Model Control of a Bipedal Walking Robot'*, Proc. of the ICRA 1997, p1476-p1481, 1997.

A. Takanishi, M. Tochizawa, H. Karaki and I. Kato, *'Dynamic Biped Walking Stabilized With Optimal Trunk and Waist Motion'*, Proc. IEEE/RSJ International Workshop on Intelligent Robots and Systems, p187-p192, 1989.

A. Takanishi, H. Lim, M. Tsuda and I. Kato, *'Realization of Dynamic Biped Walking Stabilized by Trunk Motion on a Sagittally Uneven Surface'*, Proc. IEEE International Workshop on Intelligent Robots and Systems, p323-p330, 1990.

Q. Li, A. Takanishi and I. Kato, *'A Biped Walking Robot Having a ZMP Measurement System Using Universal Force-Moment Sensors'*, Proc. IEEE/RSJ International Workshop on Intelligent Robots and Systems, p1568-p1573, 1991.

Y. Ogura, Y. Sugahara, Y. Kaneshima, N. Hieda, H. Lim and A. Takanishi, *'Interactive Biped Locomotion Based on Visual / Auditory Information'*, Proc. IEEE International Workshop on Robot and Human Interactive Communication, p253-p258, 2002.

N. Kanehira, T. Kawasaki, S. Ohta, T. Isozumi, T. Kawada, F. Kanehiro, S. Kajita and K. Kaneko, *'Design and Experiments of Advanced Leg Module(HRP-2L) for Humanoid Robot(HRP-2) Development'*, Proc. IEEE/RSJ International Conference on Intelligent Robots and Systems, p2455-p2460, 2002.

G. Pratt and M. M. Williamson, *'Series Elastic Actuators'*, Proc. IEEE/RSJ International Conference on Intelligent Robots and Systems, p399-p406, 1995.

Jerry E. Pratt, *'Exploiting Inherent Robustness and Natural Dynamics in the Control of Bipedal Walking Robots'*, Ph. D. Thesis, Massachusetts Institute of Technology, 2000.

T. Ishida, Y. Kuroki and T. Takahashi, *'Analysis of Motion of a small Biped Entertainment Robot'*, Proc. IEEE/RSJ International Conference on Intelligent Robots and Systems, p144-p147, 2004.

Claude F. Touzet, *'Neural Network and Q-Learning for Robotics'*, Proceedings of International Joint Conference on Neural Network, 1999.

Richard S Sutton, *'Generalization in Reinforcement Learning: Successful Examples Using Sparse Coarse Coding'*, Advances in Neural Information Processing System, Volume 8, p1038-p1044, 1996.

R. Matthew Kretchmar and Charles W. Anderson, *'Comparison of CMACs and Radial Basis Functions for Local Function Approximation in Reinforcement Learning'*, Proceedings of International Conference on Neural Network, 1997.

Juan Carlos Santamaria, Richard S. Sutton and Ashwin Ram, *'Experiments with Reinforcement Learning in Problems with Continuous State and Action Space'*, COINS, p96-p88, 1996

Jun Morimoto, Jun Nakanishi, Gen Endo, Gordon Cheng, Christopher G. Atkeson and Garth Zeglin, *'Poincare-Map based Reinforcement Learning for Biped Walking'*, Proc. of the 2005 International Conference on Robotics & Automation, 2005.

9

The Reaction Mass Pendulum (RMP) Model for Humanoid Robot Gait and Balance Control

Sung-Hee Lee
University of California, Los Angeles
Los Angeles, CA, U.S.A
sunghee@cs.ucla.edu
Ambarish Goswami
Honda Research Institute
Mountain View, CA, U.S.A
agoswami@honda-ri.com

1. Introduction

Next generation humanoids are expected to successfully coexist within human environments. This imposes very difficult challenges to the robot controller in the form of complex and flexible gait planning, truly dynamic movements, balance maintenance under unexpected environmental forces and disturbances. Manual programming of every gait and balance strategy is an extremely tedious proposition and is not practically implementable. Formulation and implementation of generic autonomous behavior, however, need a deep intuitive understanding of the fundamental humanoid dynamics.

In this regard, reduced biped models, such as the different variations of the inverted pendulum models (Kajita et al, 1992; Kajita et al, 2002; Sugihara & Nakamura, 2003; Komura et al., 2005), have been very beneficial. These models allow us to ignore the movements of the individual limbs of the humanoid, and instead, focus on two important points - the center of pressure (CoP) and the center of mass (CoM) - and the line joining them. It is customary to employ a reduced model during the planning and analysis stage and map the planned control strategy into the usual joint-level controller of the full humanoid for an implementation-ready control law. By focusing attention to the fundamental aspects of humanoid dynamics, such models open the way to new classes of control laws, which would otherwise be difficult or impossible to conceive.

A limitation of the above-mentioned reduced models is that they represent the entire humanoid body only as a point mass and do not characterize the significant centroidal moment of inertia of the humanoid body (except (Komura et al., 2005a)). The centroidal moment of inertia is a property of the distributed masses of the robot limbs (head, arms, legs, etc). We have earlier demonstrated that a humanoid's state of balance is closely related to its rotational equilibrium which, in turn, is dependent on its angular momentum rate change (Abdallah & Goswami, 2005). The centroidal moment of inertia directly contributes to the centroidal angular momentum and its rate change. Direct manipulation of momenta is becoming a reasonable, and sometimes preferable, way to control a robot (Kajita et al., 2003; Vermeulen et al., 2006; Hofmann, 2005). The Reaction Mass Pendulum (RMP) model (Lee & Goswami, 2007), which we describe in this paper, is expected to be useful for these controllers.

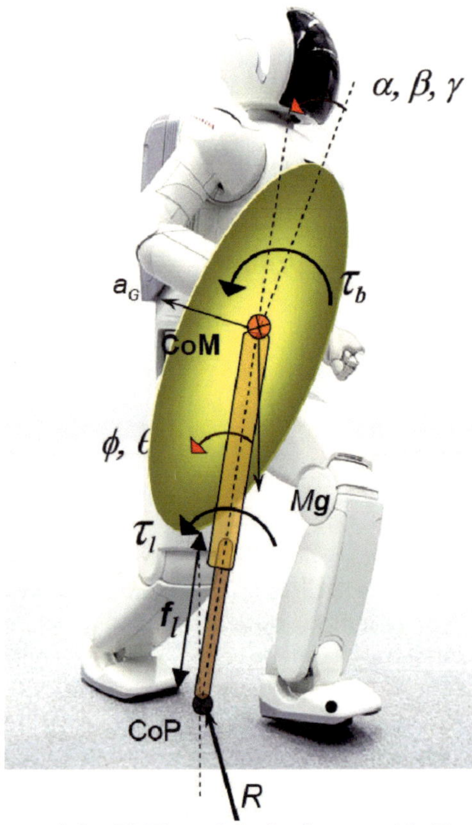

Fig. 1: Conceptual diagram of the RMP model of a humanoid. The RMP consists of a "leg" connecting the robot's CoM and CoP. The reaction mass ellipsoid, signifying the aggregate spatial inertia of the robot at its CoM, sits atop the leg. As the robot moves, the shape, size and orientation of the ellipsoid changes in a manner described in Section 2.

As shown in Fig. 1, an RMP consists of two components, a "leg" that joins the CoP and the CoM, and an ellipsoidal "body" - the abstracted reaction mass - that characterizes the inertia of the entire robot projected at the CoM. As the robot moves in space, so does the RMP, resulting in a movement of the CoP and CoM. All limb movements of the robot affect its centroidal moment of inertia, which is captured by the changing shape, size and orientation of the ellipsoidal reaction mass.

The rest of this chapter is as follows: We first provide detailed introduction and insights to the mathematical preliminaries used in this chapter (Section 2), and then derive the equations for the RMP model of a humanoid (Section 3) as well as the description of the parameters and properties of the mechanical realization of RMP model (Section 4). Finally, we present demonstrations of the application of inertia shaping technique (Section 5) followed by conclusions and future work (Section 6).

2. Mathematical Background

In this paper, we have used a Lie Group based approach (Murray et al., 1994) to derive the details of the RMP model. In this section we introduce preliminary geometric quantities and use them to define the spatial inertia of a single rigid body. Spatial inertia is a critical quantity in our model and we later generalize this concept to a multi-body system.

2.1 Geometric Preliminaries

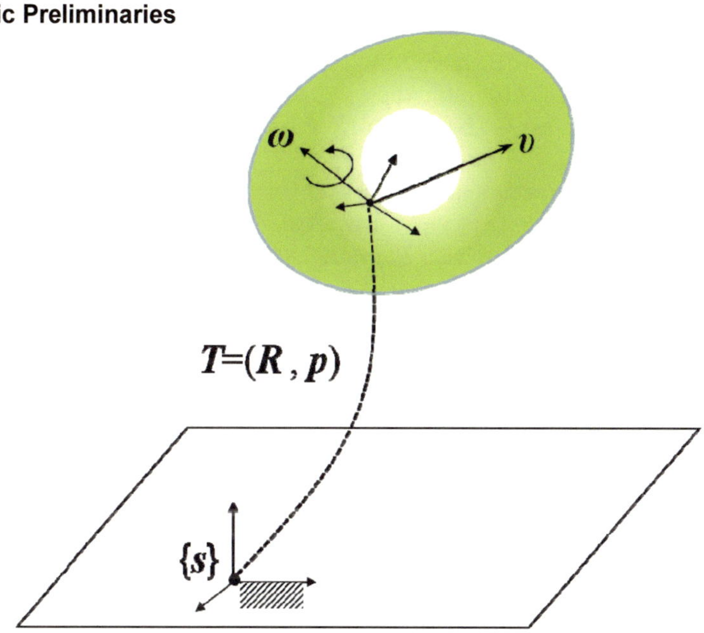

Fig. 2: A rigid body moving in space. $\boldsymbol{R}$ and $\boldsymbol{p}$ are the orientation and position of the body frame with respect to a spatial frame {s}, respectively. $\boldsymbol{\omega}$ and $\boldsymbol{v}$ are its angular and linear velocities expressed with respect to the body frame.

Let us consider a moving rigid body such as in Fig. 2. $\boldsymbol{T} = \begin{bmatrix} \boldsymbol{R} & \boldsymbol{p} \\ \boldsymbol{0} & 1 \end{bmatrix} \in SE(3)$ denotes the homogeneous transformation matrix of the body frame with respect to a spatial frame $\{s\}$, any frame fixed in space. When the body frame is moving in space, its *spatial velocity*[1] expressed with respect to the body frame (hence called *body velocity*) is defined as a twist

[1] Per (Featherstone, 1987), we use a term "spatial" to the definitions that combine angular and linear properties. For instance, a spatial velocity is the combination of the angular and linear velocities. Likewise, we will use a spatial force (the combination of torque and linear force), spatial inertia (rotational inertia and mass), and spatial momentum (angular and linear momenta). Note that in (Murray et al., 1994), the spatial velocity has different meaning; it refers to a velocity expressed with respect to a spatial frame, as opposed to the

$$\hat{\mathbf{v}} = \boldsymbol{T}^{-1}\dot{\boldsymbol{T}} = \begin{bmatrix} \boldsymbol{S}(\boldsymbol{\omega}) & \boldsymbol{v} \\ \mathbf{0} & 0 \end{bmatrix}, \quad (1)$$

which is an element of se(3), the Lie algebra of SE(3), and expressed as a 4× 4 matrix where $\boldsymbol{\omega}$ and $\boldsymbol{v}$ are its angular and linear velocities. The quantities $\boldsymbol{\omega}$ and $\boldsymbol{v}$ are expressed with respect to the instantaneous body frame. $\boldsymbol{S}(\boldsymbol{\omega})$ denotes the skew-symmetric matrix representation of $\boldsymbol{\omega} = [\omega_x, \omega_y, \omega_z]^T$; i.e.,

$$\boldsymbol{S}(\boldsymbol{\omega}) = \begin{bmatrix} 0 & -\omega_z & \omega_y \\ \omega_z & 0 & -\omega_x \\ -\omega_y & \omega_x & 0 \end{bmatrix}. \quad (2)$$

Note that, when multiplied with a vector, the skew-symmetric matrix yields the cross product, i.e., $\boldsymbol{S}(\boldsymbol{\omega})\boldsymbol{v} = \boldsymbol{\omega} \times \boldsymbol{v}$. Although represented as a 4×4 matrix, a twist has only 6 components; we can express a twist $\hat{\mathbf{v}}$ as a 6 dimensional vector $\mathbf{v} = [\boldsymbol{\omega}^T, \boldsymbol{v}^T]^T$ for convenience.

For all $\mathbf{s} \in$ se(3), exp ($\hat{\mathbf{s}}$) is an element of SE(3) and there exists a closed-form formula of the exponential map exp: se(3) → SE(3) (see Murray et al., 1994). When $\mathbf{s} \in$ se(3) and $q \in \mathbb{R}$ refer to a screw parameter of a joint and its joint angle, exp ($\hat{\mathbf{s}}q$) represents the transformation that is made by the joint motion.[2]

Coordinate transformation of the twist is achieved by the so-called *adjoint mapping*. Given $\boldsymbol{T} \in$ SE(3) and some $\mathbf{g} = [\boldsymbol{\rho}^T, \boldsymbol{\varphi}^T]^T \in$ se(3), the adjoint mapping $\mathrm{Ad}_{\boldsymbol{T}}$: se(3) → se(3) is defined as $\mathrm{Ad}_{\boldsymbol{T}}\hat{\mathbf{g}} = \boldsymbol{T}\hat{\mathbf{g}}\boldsymbol{T}^{-1}$, or in matrix form as

$$\mathrm{Ad}_{\boldsymbol{T}}\mathbf{g} = \begin{bmatrix} \boldsymbol{R} & \mathbf{0} \\ \boldsymbol{S}(\boldsymbol{p})\boldsymbol{R} & \boldsymbol{R} \end{bmatrix} \begin{bmatrix} \boldsymbol{\rho} \\ \boldsymbol{\varphi} \end{bmatrix}. \quad (3)$$

The spatial velocity $\mathbf{v}$ with respect to $\{s\}$ is given by ${}^{s}\mathbf{v} = \mathrm{Ad}_{\boldsymbol{T}}\mathbf{v}$.[3] Another useful operator that we use in this chapter is the Lie bracket $\mathrm{ad}_{\mathbf{v}}$: se(3) → se(3) and it occurs when $\mathrm{Ad}_{\boldsymbol{T}}$ is differentiated. The Lie bracket is defined as $\mathrm{ad}_{\hat{\mathbf{v}}_1}\hat{\mathbf{v}}_2 = \hat{\mathbf{v}}_1\hat{\mathbf{v}}_2 - \hat{\mathbf{v}}_2\hat{\mathbf{v}}_1$, or in matrix form

$$\mathrm{ad}_{\mathbf{v}_1}\mathbf{v}_2 = \begin{bmatrix} \mathbf{S}(\boldsymbol{\omega}_1) & \mathbf{0} \\ \mathbf{S}(\boldsymbol{v}_1) & \mathbf{S}(\boldsymbol{\omega}_1) \end{bmatrix} \begin{bmatrix} \boldsymbol{\omega}_2 \\ \boldsymbol{v}_2 \end{bmatrix}. \quad (4)$$

---

body velocity, which is expressed with respect to a body frame. However, we continue to use the term spatial frame to mean a ground reference frame, a frame rigidly attached to the ground.

[2] For instance, $\mathbf{s} = (\boldsymbol{a}^T, \mathbf{0}^T)^T$ for a revolute joint, where a unit vector $\boldsymbol{a} \in \mathbb{R}^3$ is the joint axis with respect to the body frame, and $\mathbf{s} = (\mathbf{0}^T, \boldsymbol{a}^T)^T$ for a prismatic joint, where $\boldsymbol{a}$ is the axis of translation.

[3] Left superscript s indicates the symbol is expressed in a spatial frame $\{s\}$. Likewise, we will use left superscript 0 and g to indicate a spatial frame that coincides with the base frame and a frame located at CoM of the humanoid robot, respectively. No left superscript is used when a symbol is expressed in the body frame.

One can easily verify that $\mathrm{Ad}_{\boldsymbol{T}}^{-1} = \mathrm{Ad}_{\boldsymbol{T}^{-1}}$ and $\mathrm{ad}_{\mathbf{v}}\mathbf{v} = \mathbf{0}$.
The spatial force $\mathbf{f} = [\boldsymbol{m}^T, \boldsymbol{f}^T]^T$ is an element of se*(3), the dual space of se(3), where $\boldsymbol{m} \in \mathbb{R}^3$ and $\boldsymbol{f} \in \mathbb{R}^3$ represent a moment and linear force, respectively. Corresponding to the adjoint mappings in se(3) space, the *dual adjoint mappings* $\mathrm{Ad}_{\boldsymbol{T}}^* : \mathrm{se}^*(3) \rightarrow \mathrm{se}^*(3)$ and $\mathrm{ad}_{\mathbf{g}}^* : \mathrm{se}^*(3) \rightarrow \mathrm{se}^*(3)$ are also defined and, in matrix form, they are the transposes of $\mathrm{Ad}_{\boldsymbol{T}}$ and $\mathrm{ad}_{\mathbf{g}}$; i.e.,

$$\mathrm{Ad}_{\boldsymbol{T}}^* = \mathrm{Ad}_{\boldsymbol{T}}^T, \qquad (5)$$
$$\mathrm{ad}_{\boldsymbol{T}}^* = \mathrm{ad}_{\boldsymbol{T}}^T.$$

The spatial force with respect to $\{s\}$ is given as ${}^{s}\mathbf{f} = \mathrm{Ad}_{\boldsymbol{T}^{-1}}^* \mathbf{f}$.[4]

2.2 Spatial inertia of a single rigid body

The spatial inertia of a body represents its aggregate inertial property by combining its translational mass and rotational inertia.
When a body frame is located at the CoM of a rigid body, its kinetic energy takes the following simple form,

$$E_k = \frac{1}{2} m\, \boldsymbol{v}^T \boldsymbol{v} + \frac{1}{2}\, \boldsymbol{\omega}^T \bar{\boldsymbol{I}}_{CoM} \boldsymbol{\omega}, \qquad (6)$$

where m is the mass, $\bar{\boldsymbol{I}}_{CoM} \in \mathbb{R}^{3\times 3}$ is the rotational inertia matrix, and $\boldsymbol{v}$ and $\boldsymbol{\omega}$ are linear and angular velocities. We use the subscript *CoM* to stress that the reference frame is at the CoM. In terms of its spatial velocity, the kinetic energy is expressed as follows:

$$E_k = \frac{1}{2} \mathbf{v}_{CoM}^T \boldsymbol{I}_{CoM}\, \mathbf{v}_{CoM}, \qquad (7)$$

where $\boldsymbol{I}_{CoM} = \begin{bmatrix} \bar{\boldsymbol{I}}_{CoM} & \mathbf{0} \\ \mathbf{0} & m\mathbf{1} \end{bmatrix}$ is called the spatial inertia of the rigid body. Now we consider an arbitrary body frame that is not located at the CoM. Let $\boldsymbol{G} = (\boldsymbol{R}, \boldsymbol{c}) \in \mathrm{SE}(3)$ denote a transformation matrix from this (arbitrary) body frame to a frame at the CoM where $\boldsymbol{c}$ is the position of the CoM. Since the kinetic energy is coordinate-independent,

$$E_k = \frac{1}{2} \mathbf{v}_{CoM}^T \boldsymbol{I}_{CoM}\, \mathbf{v}_{CoM} = \frac{1}{2} \mathbf{v}^T \boldsymbol{I}\, \mathbf{v}. \qquad (8)$$

Substituting the relation $\mathbf{v} = \mathrm{Ad}_{\boldsymbol{T}} \mathbf{v}_{CoM}$, we can derive the structure of the spatial inertia with respect to an arbitrary body frame:

$$\begin{aligned} \boldsymbol{I} &= \mathrm{Ad}_{\boldsymbol{G}^{-1}}^* \boldsymbol{I}_{CoM}\, \mathrm{Ad}_{\boldsymbol{G}^{-1}} \\ &= \begin{bmatrix} \boldsymbol{R}\bar{\boldsymbol{I}}_{CoM}\boldsymbol{R}^T - m\boldsymbol{S}(\boldsymbol{c})^2 & m\boldsymbol{S}(\boldsymbol{c}) \\ -m\boldsymbol{S}(\boldsymbol{c}) & m\mathbf{1} \end{bmatrix} \\ &= \begin{bmatrix} \bar{\boldsymbol{I}} & m\boldsymbol{S}(\boldsymbol{c}) \\ -m\boldsymbol{S}(\boldsymbol{c}) & m\mathbf{1} \end{bmatrix}, \end{aligned} \qquad (9)$$

4 Note that $\mathrm{Ad}_{\boldsymbol{T}^{-1}}^* = \begin{bmatrix} \boldsymbol{R} & \boldsymbol{S}(\boldsymbol{p})\boldsymbol{R} \\ \mathbf{0} & \boldsymbol{R} \end{bmatrix}$.

where $\bar{\boldsymbol{I}} = \boldsymbol{R}\bar{\boldsymbol{I}}_{CoM}\boldsymbol{R}^T - m\boldsymbol{S}(\boldsymbol{c})^2$ is the rotational inertia matrix with respect to the body frame. Similarly, one can verify that the coordinate transformation of the spatial inertia is accomplished by pre and post multiplying the adjoint matrices. For instance, the spatial inertia with respect to the spatial frame is as follows:

$$^{s}\boldsymbol{I} = \mathrm{Ad}^{*}_{\boldsymbol{T}^{-1}}\, \boldsymbol{I}\, \mathrm{Ad}_{\boldsymbol{T}^{-1}}. \tag{10}$$

Note that $\bar{\boldsymbol{I}}$ and $\boldsymbol{I}$ are both symmetric positive definite matrices.
Having defined the spatial inertia, we can define the *spatial momentum* of the rigid body as an element of se*(3) as follows:

$$\boldsymbol{h} = \begin{bmatrix}\boldsymbol{k}\\ \boldsymbol{l}\end{bmatrix} = \boldsymbol{I}\mathbf{v} \in \mathrm{se}^*(3), \tag{11}$$

where $\boldsymbol{k}$ and $\boldsymbol{l}$ are the angular and linear momenta, respectively. Substituting (9) into (11), we can derive the expressions of the angular and linear momenta; $\boldsymbol{k} = \bar{\boldsymbol{I}}\boldsymbol{\omega} + m\boldsymbol{c}\times\boldsymbol{v}$ and $\boldsymbol{l} = m(\boldsymbol{v} - \boldsymbol{c}\times\boldsymbol{\omega})$. Note that, since the spatial momentum is an element of se*(3), coordinate transformation of the spatial momentum is achieved through the dual adjoint mapping. For instance, the spatial momentum with respect to $\{s\}$ is $^{s}\boldsymbol{h} = \mathrm{Ad}^{*}_{\boldsymbol{T}^{-1}}\, \boldsymbol{h}$.

3. Generating the RMP Model of a Humanoid

We generate the RMP model of a humanoid by extending the concept of spatial inertia to an articulated chain. In this section we derive the necessary equations to exploit the concept of composite rigid body (CRB) inertia and show how it relates to the momentum of a humanoid. In the next section we will outline a mechanical realization of the RMP.

3.1 CRB inertia of a humanoid robot

We assume that a humanoid robot model consists of $n+1$ links with the base link, usually the pelvis, indexed as 0 (Fig. 3).
Let $\boldsymbol{T}_0 \in SE(3)$ denote the transformation matrix of the body frame of the base link (*base frame* hereafter) and $\boldsymbol{q} = (q_1, \dots, q_n)^T \in \mathbb{R}^n$ the joint angle vector of the robot, then $\boldsymbol{\Theta} = (\boldsymbol{T}_0, \boldsymbol{q})$ defines the position and orientation of the humanoid robot. Subsequently, $\dot{\boldsymbol{\Theta}} = (\mathbf{v}_0, \dot{\boldsymbol{q}})$ will denote the body velocity of the base frame and joint velocities.[5]
$\boldsymbol{T}_i$ denotes the transformation matrix from the spatial frame to the body frame of link i. For a non-base link, $\boldsymbol{T}_i$ $(i > 0)$ is determined by $\boldsymbol{T}_0$ and the joint angles $\boldsymbol{q}$, i.e., $\boldsymbol{T}_i = \boldsymbol{T}_0\, \boldsymbol{G}_i(\boldsymbol{q})$, where $\boldsymbol{G}_i = \boldsymbol{T}_0^{-1}\boldsymbol{T}_i$ is the transformation matrix from the base frame to link i. Note that $\boldsymbol{G}_i$ does not depend on $\boldsymbol{T}_0$ and is entirely determined by $\boldsymbol{q}$. For simplicity, we assume that, except for the base link, each link is connected to its parent link by a 1-DOF joint. Then

[5] Note that $\dot{\boldsymbol{\Theta}}$ is a slight abuse of notation because $\mathbf{v}_0$ is not $\dot{\boldsymbol{T}}_0$, but $\boldsymbol{T}_0^{-1}\dot{\boldsymbol{T}}_0$. In this representation, $\boldsymbol{\Theta}$ is not generalized coordinates because a 4×4 matrix $\boldsymbol{T}_0$ has only 6 DOFs. Another, more common way to represent the configuration of a robot is using 6 numbers (3 for the orientation and 3 for the position) instead of using a 4×4 matrix $\boldsymbol{T}_0$. An advantage of this representation is that the configuration vector is generalized coordinates as its dimension is same as the DOFs of the robot. However, since this uses 3 numbers to represent the orientation, it has singularities at certain configurations.

$\boldsymbol{T}_i = \boldsymbol{T}_{p(i)}\boldsymbol{H}_i e^{\hat{\mathbf{s}}_i q_i}$ holds for $i = 1 \dots n$ where $p(i)$ denotes the parent link of link i, $\boldsymbol{H}_i \in \mathrm{SE}(3)$ is the transformation from $p(i)$ to i at $q_i = 0$, and $\mathbf{s}_i \in \mathrm{se}(3)$ is the screw parameter of the joint.

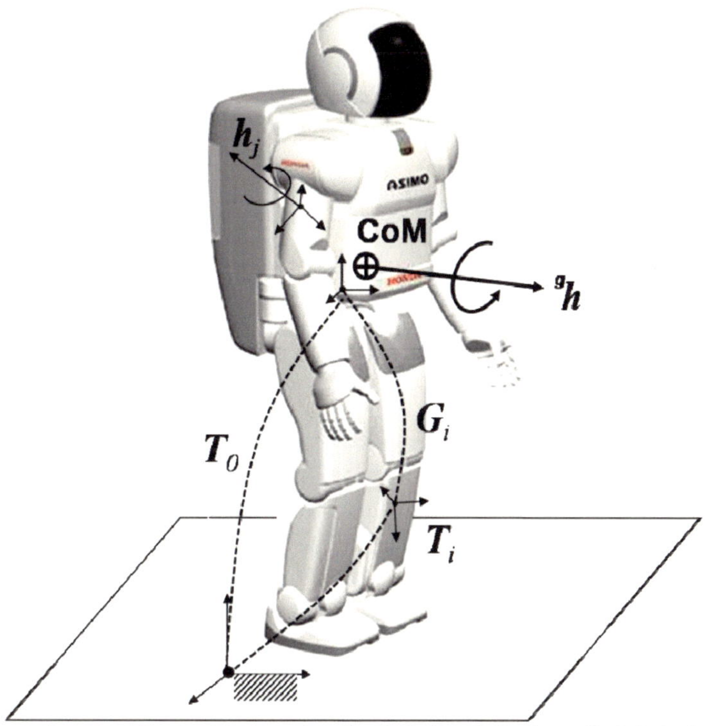

Fig. 3: T_i denotes the transformation matrix of a link i and the base link is indexed as 0. G_i is the transformation of link i as seen from the base frame and it is determined by the joint angles. The centroidal momentum of a humanoid robot ${}^g\boldsymbol{h}$ is computed as the sum of the momenum of each link $\boldsymbol{h}_i$ expressed with respect to the CoM.

The composite rigid body (CRB) inertia (Walker & Orin, 1982) of a humanoid robot is its instantaneous spatial inertia, assuming that all of its joints are frozen. It has the same structure as the spatial inertia of a single rigid body (Eq. 9). CRB inertia is identical to the so-called locked inertia, a term that is used in geometric mechanics (Ostrowski, 1999). Mathematically, CRB inertia of a humanoid with respect to its base frame is expressed as follows:

$$\begin{aligned} {}^0\boldsymbol{I} &= \sum {}^0\boldsymbol{I}_i \\ &= \sum \mathrm{Ad}^*_{G_i^{-1}} \boldsymbol{I}_i \, \mathrm{Ad}_{G_i^{-1}} \end{aligned} \tag{12}$$

Next, we show how CRB inertia is related to the momentum of a humanoid robot. To this end, we first define the *link Jacobian* $\boldsymbol{J}_i$ of a link i that is similar to the manipulator Jacobian as follows;

$$J_i = [\mathrm{Ad}_{G_i^{-1}},\ J_{i,q}], \quad (13)$$

$$J_{i,q} = [J_{i,1}, \dots, J_{i,n}] \in \mathbb{R}^{6\times n}, \quad (14)$$

where $$\hat{J}_{i,j} = T_i^{-1}(\partial T_i/\partial q_j) = G_i^{-1}(\partial G_i/\partial q_j).$$

As in Section 2.1, $\hat{J}_{i,j}$ is the twist, a 4 × 4 matrix while $J_{i,j}$ denotes its 6 dimensional vector form. Note that $J_{i,j}$ can be computed efficiently in a recursive manner as follows;

$$J_{i,j} = \mathrm{Ad}^{-1}_{H_i e^{\hat{s}_i q_i}} J_{p(i),j} + s_i \delta_{i,j} \quad \text{for} \quad i = 1 \dots n, \quad (15)$$

where $\delta_{i,j}$ is the Kronecker delta function and $J_{0,j} = \mathbf{0}$. In (13) and (14), $\mathrm{Ad}_{G_i^{-1}}$ and $J_{i,q}$ are the Jacobians due to the change of the base frame and joint angles respectively. Using the link Jacobian, we can decompose the velocity of $\{i\}$ into the sum of the velocity due to the base link and the one due to the joint velocities;

$$\mathbf{v}_i = J_i \dot{\Theta} = \mathrm{Ad}_{G_i^{-1}} \mathbf{v}_0 + J_{i,q} \dot{q}\,. \quad (16)$$

The spatial momentum h of the humanoid robot is the sum of spatial momentum of each link. The one with respect to the base frame [6] is

$$\begin{aligned} {}^0h &= \sum_i {}^0h_i = \sum_i {}^0I_i\, {}^0\mathbf{v}_i \\ &= {}^0I\mathbf{v}_0 + \textstyle\sum_i \mathrm{Ad}^*_{G_i^{-1}} I_i J_{i,q} \dot{q} \\ &= {}^0I\,(\mathbf{v}_0 + \mathbf{A}\dot{q}) \end{aligned} \quad (17)$$

where $\mathbf{A}(q)$ is the so-called *mechanical connection* (Ostrowski, 1999). As can be seen in (17), the CRB inertia contributes directly to the spatial momentum of a robot via the mechanical connection.

While the CRB inertia can be expressed with respect to any frame, it is particularly interesting to express it at the CoM since it is related to the centroidal angular momentum (Fig. 3). The CRB inertia expressed at the humanoid CoM is called the *centroidal* CRB inertia. To denote this we use ${}^gI$ in the RMP model.

3.2 Equimomental ellipsoids

The association of the rigid body inertia to an ellipsoid is well known and has been thoroughly exploited in physics and engineering (Crandall et al., 1982). As a straight-forward extension we determine the ellipsoid associated with the centroidal CRB inertia of an articulated chain. Instead of using the kinetic energy ellipsoid, which is traditionally described with an inertia, we derive the *equimomental* ellipsoid corresponding to a CRB inertia.

Two inertias are said to be equimomental if their moments of inertia about any arbitrary axis are equal (Beer & Johnson, 1984). The equimomental ellipsoid of a rigid body is an

[6] More precisely, a spatial frame that instantaneously coincides with the base frame.

ellipsoid with a uniform density set as the mean density of the body and having the same rotational inertia about any arbitrary axis as that of the rigid body. Kinetic energy ellipsoid characterizes the torque needed to rotate the body *about* an axis whereas the equimomental ellipsoid reflects the mass distribution *along* an axis.

We prefer the uncommon equimomental ellipsoid over the well-known kinetic energy ellipsoid because the shape of the former approximates the mass distribution of the overall multi-body system. For example, the ellipsoid for an upright humanoid will be long and narrow, thereby rendering some gross geometric resemblance to the humanoid. This is not the case for the kinetic energy ellipsoid.

Let $(\sigma_1, \sigma_2, \sigma_3)$ denote the eigenvalues of the rotational inertia, and (a_1, a_2, a_3) denote the semi-axes of the equimomental ellipsoid. From the relationships $\sigma_i = m(a_j^2 + a_k^2)/5$ and $m = 4\pi a_1 a_2 a_3 \rho/3$, where ρ is the mean density, we can derive the following:

$$a_i = \left(\frac{15}{8\pi\rho}\right)^{1/5} \frac{(-\sigma_i+\sigma_j+\sigma_k)^{2/5}}{\{(\sigma_i-\sigma_j+\sigma_k)(\sigma_i+\sigma_j-\sigma_k)\}^{1/10}} \tag{18}$$

for $i, j, k = 1 \ldots 3$ and $i \neq j \neq k$.

3.3 Simulation results

Fig. 4: Snapshots of HOAP2 robot performing Sumo-style motion superposed with corresponding RMP models. The reaction mass geometry undergoes significant changes during this motion.

The process of mapping a humanoid to its corresponding RMP involves the computation of 1) CoM, 2) CoP, and 3) centroidal CRB inertia, using robot kinematic and dynamic parameters, as well as motion data.[7] Refer to the Appendix for the computation of the CoP. We simulated the Fujitsu HOAP2 biped model, for which the parameters for some dramatic movements are available (Cominoli, 2005). The simulation is implemented using Webots (www.cyberbotics.com), a commercial simulation software.

Fig. 4 shows snapshots of HOAP2 executing Sumo-style movements. Notice the significant changes in the shape, size and orientation of the reaction mass ellipsoid[8] as the robot moves through different phases of its motion. Since the robot→RMP is a mapping to a lower dimension, different poses of the robot, at least theoretically, may get mapped to the same RMP.

4. Properties and Parameters of RMP

We have now shown how a humanoid robot can be reduced to an RMP. In this section we will discuss the realization of a mechanical model of the RMP. The RMP is the generalized 3D version of the 2D *reaction wheel pendulum* which has been studied before (Astrom et al., 2001; Olfati-Saber, 2001; Spong et al., 2001). A reaction wheel pendulum can be constructed by attaching an actuated reaction wheel to a rigid rod. The inclination angle of the pendulum can be controlled by controlling the angular acceleration of the reaction wheel. The reaction wheel, which is also called an inertia wheel, is one of a number of standard momentum exchange devices that are used to control satellite orientation (Sidi, 1997).

4.1 Description of RMP

The present work can be identified with those of [18, 19], where the benefit of a reaction mass feature of the humanoid as a mean to stabilize lateral biped dynamics is indicated. The current work is closest in spirit to the recently introduced inverted pendulum model with angular momentum properties (AMPM) (Komura et al., 2005 (a; b)). We seek to propose a physical model characterizing angular momentum.

The RMP *mathematical* model discussed here is not to be confused with the actual placement of a *physical* reaction mass device for the control of humanoid balance, as was done in (Mayer et al., 2005).

The 3D reaction mass may have continuously variable spatial inertia. At any given configuration of the robot, the centroidal CRB inertia can be reduced to an ellipsoid. This is modeled, as shown in Fig. 5 by three pairs of point masses linearly actuated along the three principal orthogonal directions of the ellipsoid. Along each axis k, the distance between the point masses is $2r_k$. The masses of each pair are always equi-distant from the ellipsoid center. The CoM of the ellipsoid is therefore always fixed at its center. The six point masses can have equal mass, i.e., $m = M/6$, so that they sum up to total mass of the humanoid robot. The distance between the masses depends on its corresponding rotational inertia, as each axis generates a moment of inertia mr_k^2.

---

[7] Refer to Appendix for equations to compute CoP.

[8] Reaction mass ellipsoid is synonymous with the equimomental inertia ellipsoid derived from the centroidal CRB inertia matrix.

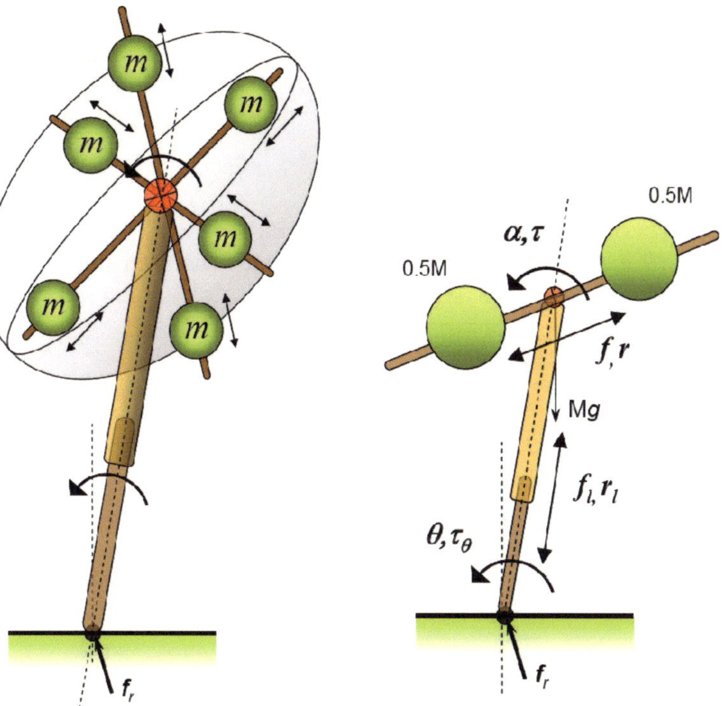

Fig. 5: (left) Conceptual mechanical realization of the 3D RMP. The gyrating ellipsoid is dynamically equivalent to three pairs of equal point masses at different radial distances that are actuated to slide on their linear tracks. The overall frame consisting of the three pairs of mutually perpendicular linear tracks form the skeleton which can be rotationally actuated in three DOFs. (right) 2D Reaction Wheel Pendulum Model. The distance between the two point masses is $\mathbf{2r}$.

Physical description	Generalized coordinates (forces)	
	2D	3D
Radial distances of three pairs of point masses forming the ellipsoid and their actuation on linear tracks	r (f)	r_1, r_2, r_3 (f_1, f_2, f_3)
Orientation angles of the ellipsoid body and their actuation	α (τ)	α, β, γ (τ_1, τ_2, τ_3)
Leg length and its actuation	r_l (f_l)	r_l (f_l)
Leg orientation angles and their actuation	θ (τ_θ)	θ, ϕ (τ_θ, τ_ϕ)
CoP position and ground reaction force	x_{CoP} (R_x, R_y)	x_{CoP}, y_{CoP} (R_x, R_y, R_z)

Table 1: Generalized variables of RMP. 2D reaction wheel pendulum model has 5 DOFs whereas 3D RMP has 11 DOFs.

The radial movement of the point masses only affects the shape and size of the ellipsoid. When $r_k = 0$ for $k = 1,2,3$ the ellipsoid reduces to a point mass and the RMP reduces to a 3D inverted pendulum. The list of all eleven generalized coordinates and nine generalized forces are listed in Table. 1.

4.2 2D Reaction wheel pendulum model

The 2D version of the RMP is equivalent to a reaction wheel pendulum, for which a realization is shown in Fig. 5(right). The generalized coordinates and generalized forces for this model are (θ, α, r_l, r) and $(\tau_\theta, \tau, f_l, f)$, respectively. The total mass of the pendulum is $0.5M + 0.5M = M$, whereas its rotational inertia about CoM is $\bar{I} = Mr^2$. The kinetic energy E_k and the potential energy E_p of the system are as follows,

$$E_k = \frac{1}{2}M\left(\dot{r}_l^2 + \dot{r}^2\right) + \frac{1}{2}Mr_l^2\dot{\theta}^2 + \frac{1}{2}Mr^2(\dot{\theta}^2 + \dot{\alpha}^2), \tag{19}$$

$$E_p = Mgr_l\sin\theta.$$

From the above relations, the equations of motion of this model are derived using Lagrangian techniques.

$$f_l = M\ddot{r}_l - Mr_l\dot{\theta}^2 + Mg\sin\theta \tag{20}$$

$$f = M\ddot{r} - Mr(\dot{\theta} + \dot{\alpha})^2 \tag{21}$$

$$\tau_\theta = Mr_l^2\ddot{\theta} + Mr^2(\ddot{\theta} + \ddot{\alpha}) + 2Mr_l\dot{r}_l\dot{\theta} + 2Mr\dot{r}(\dot{\theta} + \dot{\alpha}) + Mgr_l\cos\theta \tag{22}$$

$$\tau = Mr^2(\ddot{\theta} + \ddot{\alpha}) + 2Mr\dot{r}(\dot{\theta} + \dot{\alpha}). \tag{23}$$

The ground reaction force to the reaction mass pendulum is $\boldsymbol{f}_r = M(\ddot{\boldsymbol{r}}_l - \mathbf{g})$. We can also relate the rate of change of the angular momentum in terms of the generalized coordinates; i.e., $\tau = \dot{k}_G$ and $\tau_\theta = \dot{k}_P + Mgr_l\cos\theta$, where the angular momentum at CoM, $k_G = \bar{I}(\dot{\theta} + \dot{\alpha})$ and the one at CoP, $k_P = k_G + Mr_l^2\dot{\theta}$.

As known, a reaction wheel pendulum can have interesting dynamics. For example, if we set $\tau_\theta = 0$, $\dot{\theta} = \ddot{\theta} = 0$, then we can compute τ that keeps $\theta = \theta_c$ stationary, i.e., $\tau = -Mgr\cos\theta_c$. The torque creates an angular acceleration $\ddot{\alpha}$ which cannot continue indefinitely due to robot joint limits. However, the example showcases the situation where the robot "leg" can be in static stability while the CoM ground projection is outside of the support base.

5. Inertia Shaping: an RMP-Based Controller

A humanoid robot has a large number of DOFs: for example, the Fujitsu HOAP2 robot has 25 DOFs and the Honda humanoid Asimo has 27 DOFs. In order to kinematically transform an RMP back to a humanoid robot, one needs to generate a map from the 11 dimensional RMP space to the much larger robot kinematics space. A unique mapping will need additional constraints, such as in the form of desired hand or foot position.

In this section, we introduce the inertia shaping technique, an iterative method to compute joint angles to create the desired CRB inertia. To this end, we first derive the relationship between the joint angles and the CRB inertia of a humanoid robot. We call this the CRB inertia Jacobian. Specifically, since the number of independent joint angles of a humanoid robot differs according to its ground contact configuration due to the geometric constraints induced by the ground-robot contact, we formulate the CRB inertia Jacobian per each ground contact case.

5.1 CRB inertia Jacobian

Since the CRB inertia has the form of a matrix, its Jacobian should be a multi-dimensional tensor. Therefore, we string out the nonzero elements of a matrix of the CRB inertia so as to simplify the form of its Jacobian. The "strung out" vector corresponding to the spatial inertia matrix is $\check{\boldsymbol{I}} = (\check{\bar{\boldsymbol{I}}}^T, m\boldsymbol{r}^T)^T \in \mathbb{R}^9$ and $\check{\bar{\boldsymbol{I}}} = (I_{xx}, I_{xy}, I_{xz}, I_{yy}, I_{yz}, I_{zz})^T$. The first six elements correspond to rotational inertia and the last three elements to CoM multiplied by its mass.

In order to express the relation between a small change in CRB inertia and that of the generalized coordinates, we define the *CRB inertia Jacobian* $\boldsymbol{J}_I$ such that

$$\delta^s\check{\boldsymbol{I}} = \boldsymbol{J}_I\delta\boldsymbol{\Theta}. \tag{24}$$

Later on, the CRB inertia Jacobian will be used for solving the inverse problem, i.e., finding generalized coordinates corresponding to the given CRB inertia. In the following sections, we use $\boldsymbol{I}$ in the spatial frame without using superscripts. The CRB inertia Jacobian is decomposed into two parts, $\boldsymbol{J}_I = [\,\boldsymbol{J}_{I,0}\ \ \boldsymbol{J}_{I,q}]$, where $\boldsymbol{J}_{I,0} \in \mathbb{R}^{9\times 6}$ and $\boldsymbol{J}_{I,q} \in \mathbb{R}^{9\times n}$ map the motion of the base frame and the joint angles, respectively, to the rate of change of the CRB inertia; i.e.,

$$\delta\check{\boldsymbol{I}} = \boldsymbol{J}_{I,0}(\boldsymbol{T}_0^{-1}\delta\boldsymbol{T}_0) + \boldsymbol{J}_{I,q}\delta\boldsymbol{q}\,. \tag{25}$$

Specifically,

$$\boldsymbol{J}_{I,0} = \left(\boldsymbol{J}_{T_0,1}, \cdots, \boldsymbol{J}_{T_0,6}\right), \text{where } \boldsymbol{J}_{T_0,i} = \dot{\check{\boldsymbol{I}}}\,|_{\mathbf{v}=e_i,\dot{\boldsymbol{q}}=\mathbf{0}}$$

$$\boldsymbol{J}_{I,q} = \left(\boldsymbol{J}_{q_1}, \cdots, \boldsymbol{J}_{q_n}\right), \text{where } \boldsymbol{J}_{q_i} = \partial\check{\boldsymbol{I}}/\partial q_i.$$

Using the relations $\partial \mathrm{Ad}_{\boldsymbol{G}_i}/\partial q_j = \mathrm{Ad}_{\boldsymbol{G}_i}\mathrm{ad}_{\boldsymbol{J}_{i,j}} = \mathrm{ad}_{{}^s\boldsymbol{J}_{i,j}}\mathrm{Ad}_{\boldsymbol{G}_i}$ where $\hat{\boldsymbol{J}}_{i,j} = \boldsymbol{G}_i^{-1}\partial\boldsymbol{G}_i/\partial q_j$ [9] as defined in (14), we can derive analytical expression for $\dot{\boldsymbol{I}}$ and $\partial I/\partial q_j$; i.e.,

$$\dot{\boldsymbol{I}}|_{\dot{q}=0} = -\mathrm{ad}^*_{{}^s\mathbf{v}_0}\,{}^0\boldsymbol{I} - {}^0\boldsymbol{I}\,\mathrm{ad}_{{}^s\mathbf{v}_0} \tag{26}$$

$$\partial\boldsymbol{I}/\partial q_j = -\textstyle\sum_{i=1}^n\left(\mathrm{ad}^*_{{}^s\boldsymbol{J}_{i,j}}\boldsymbol{I}_i + \boldsymbol{I}_i\mathrm{ad}_{{}^s\boldsymbol{J}_{i,j}}\right). \tag{27}$$

[9] One can easily prove the relations by using $\hat{\boldsymbol{J}}_{i,j} = \begin{bmatrix} \boldsymbol{R}_i^T\frac{\partial\boldsymbol{R}_i}{\partial q_j} & \boldsymbol{R}_i^T\frac{\partial\boldsymbol{p}_i}{\partial q_j} \\ \boldsymbol{0} & 0 \end{bmatrix}$ where $\boldsymbol{G}_i = (\boldsymbol{R}_i, \boldsymbol{p}_i)$. Also, note that $\frac{d}{dt}\mathrm{Ad}_{\boldsymbol{T}} = \mathrm{Ad}_{\boldsymbol{T}}\mathrm{ad}_{\mathbf{v}} = \mathrm{ad}_{{}^s\mathbf{v}}\mathrm{Ad}_{\boldsymbol{T}}$ where $\hat{\mathbf{v}} = \boldsymbol{T}^{-1}\dot{\boldsymbol{T}}$.

Note that the CRB inertia Jacobian includes the CoM Jacobian, i.e., the mapping from the rate of change of generalized coordinates to that of the CoM position. If we partition $\boldsymbol{J}_I$ into $\boldsymbol{J}_I = [\boldsymbol{J}_{\bar{I}}^T\ \boldsymbol{J}_G^T]^T$ where $\boldsymbol{J}_G$ consists of the bottom 3 rows of $\boldsymbol{J}_I$, we get

$$m\dot{\boldsymbol{r}}_G = \boldsymbol{J}_G\dot{\boldsymbol{\Theta}}. \tag{28}$$

$\boldsymbol{J}_G$ maps the generalized velocity to the linear momentum of the system. In fact, it is the same as the CoM Jacobian scaled by the total mass.

5.2 CRB inertia Jacobian of humanoid

If the humanoid robot is not in contact with any external environment, (25) completely describes the CRB inertia Jacobian. Otherwise, however, geometric constraints arise among the generalized coordinates, and it is advantageous to describe the CRB inertia Jacobian in terms of independent coordinates.

Let us assume $\boldsymbol{q}$ comprises $\boldsymbol{q} = (\,\boldsymbol{q}_r^T\ \boldsymbol{q}_l^T\ \boldsymbol{q}_t^T)^T$, where $\boldsymbol{q}_{\{r,l\}} \in \mathbb{R}^6$ are joint angle vectors for right and left legs, respectively, and $\boldsymbol{q}_t$ is for the rest of joints. We also decompose $\boldsymbol{J}_I$ accordingly; i.e.,

$$\boldsymbol{J}_I = \begin{bmatrix} \boldsymbol{J}_{\bar{I}_0} & \boldsymbol{J}_{\bar{I}_r} & \boldsymbol{J}_{\bar{I}_l} & \boldsymbol{J}_{\bar{I}_t} \\ \boldsymbol{J}_{G_0} & \boldsymbol{J}_{G_r} & \boldsymbol{J}_{G_l} & \boldsymbol{J}_{G_t} \end{bmatrix}. \tag{29}$$

We describe CRB inertia Jacobian for each ground contact case.

5.2.1 Free floating

When a humanoid robot is floating in the air, all components of the generalized coordinates are independent. Rewriting (25), we get

$$\delta\check{\bar{\boldsymbol{I}}} = \boldsymbol{J}_{\bar{I}_0}(\boldsymbol{T}_0^{-1}\delta\boldsymbol{T}_0) + [\boldsymbol{J}_{\bar{I}_r}\ \boldsymbol{J}_{\bar{I}_l}\ \boldsymbol{J}_{\bar{I}_t}]\delta\boldsymbol{q} \tag{30}$$

$$m\delta\boldsymbol{r}_G = \boldsymbol{J}_{G_0}(\boldsymbol{T}_0^{-1}\delta\boldsymbol{T}_0) + [\boldsymbol{J}_{G_r}\ \boldsymbol{J}_{G_l}\ \boldsymbol{J}_{G_t}]\delta\boldsymbol{q} \tag{31}$$

Note that since we compute the CRB inertia with respect to a spatial frame, the configuration of the base frame affects the CRB inertia Jacobian. If we were only interested in the local "shape" of inertia, we need to transform the reference frame to the base frame, in which case the CRB inertia Jacobian is wholly determined by joint angles.

5.2.2 Single support by left or right foot

Let us suppose the humanoid robot is supported by one foot link, left foot link for example, which is stationary with respect to the ground. Then we can describe the constraint as follows,

$$\boldsymbol{T}_l^{-1}\delta\boldsymbol{T}_l = 0 \tag{32}$$

where $\boldsymbol{T}_l$ is the transformation matrix for the left foot link. Eq. 32 constrains that the left foot link should not move. From $\boldsymbol{T}_l = \boldsymbol{T}_0\,\boldsymbol{G}_l(\boldsymbol{q})$, where $\boldsymbol{G}_l(\boldsymbol{q})$ expresses the forward kinematic relation between the base frame and the left foot link, we can derive the following relation by plugging the relation into (32),

$$T_0^{-1}\delta T_0 = -\mathrm{Ad}_{G_l} J_l \delta q\,. \tag{33}$$

Defining J_l^* such that $\mathrm{Ad}_{G_l} J_l \delta q = J_l^* \delta q_l$, the CRB inertia Jacobian is written with respect to the joint angles,

$$\delta \breve{\bar{I}} = [J_{\bar{I}_r} \quad (J_{\bar{I}_l} - J_{\bar{I}_0} J_l^*) \quad J_{\bar{I}_t}]\delta q, \tag{34}$$

$$m\delta r_G = [J_{G_r} \quad (J_{G_l} - J_{G_0} J_l^*) \quad J_{G_t}]\delta q\,. \tag{35}$$

5.2.3 Double support

When both feet are stationary to the ground, we have an additional constraint $J_l^*\delta q_l = J_r^*\delta q_r$, where J_r^* of the right foot link corresponds to J_l^* of the left foot. This constraint yields $\delta q_r = J_r^{*-1} J_l^* \delta q_l$. Therefore,

$$\delta \breve{\bar{I}} = \left[\{J_{\bar{I}_l} + \left(J_{\bar{I}_r} J_r^{*-1} - J_{\bar{I}_0}\right) J_l^*\} \quad J_{\bar{I}_t}\right] \begin{bmatrix} \delta q_l \\ \delta q_t \end{bmatrix} \tag{36}$$

$$m\delta r_G = \left[\{J_{G_l} + \left(J_{G_r} J_r^{*-1} - J_{G_0}\right) J_l^*\} \quad J_{G_t}\right] \begin{bmatrix} \delta q_l \\ \delta q_t \end{bmatrix}. \tag{37}$$

5.3 Inertia shaping

An interesting application of our RMP modeling approach is what we call *inertia shaping* of an articulated chain (Lee & Goswami, 2007). Inertia shaping is a high-level approach to precisely control the aggregate kino-dynamic characteristics of an articulated chain by controlling its CRB inertia. Given a desired CRB inertia I_d of the robot, the inertia shaping controller seeks to determine the proper configuration of the robot that attains it. This can be posed as an inverse kinematics problem with the desired CRB inertia constraints. Since we have derived the CRB inertia Jacobian, the inverse kinematics problem can be solved by any suitable optimization algorithm. One solution will be to iteratively update the desired joint angles using pseudo-inverse of the centroidal CRB inertia Jacobian (24), i.e.,

$$\delta \Theta_I = J_I^{\dagger} \delta(\breve{I}_d - \breve{I}\,), \tag{38}$$

where Θ_I is the vector of independent generalized coordinates and $J_I^{\dagger} = J_I^T (J_I J_I^T)^{-1}$. This is the simplest form of inertia shaping that does not consider additional constraints such as obstacles in the environment or self-collision.

For the inertia shaping algorithm to be applied to actual humanoid robots, more sophisticated algorithm can be developed. For example, we may need to incorporate inverse kinematics problems into the inertia shaping process. The CRB inertia Jacobian provides useful information on the rate of change of the CRB inertia as a function of generalized coordinates, which is necessary in any iterative gradient-based algorithms for computing optimal joint angles to achieve desired inertia.

Fig. 6 presents three examples of inertia shaping on a non-contacting Asimo-like floating in space (say, a humanoid astronaut). The robot is given three different commands, shown in series a, b, and c, respectively, to try to match its own CRB inertia to a desired CRB inertia. Starting from an initial configuration, the robot moves its joints such that the cost function, considered to be the Frobenius norm of the difference between the two inertia matrices, is minimized. Specifically, we deal with the inertia with respect to the base frame, hence controlling joint angles is sufficient.

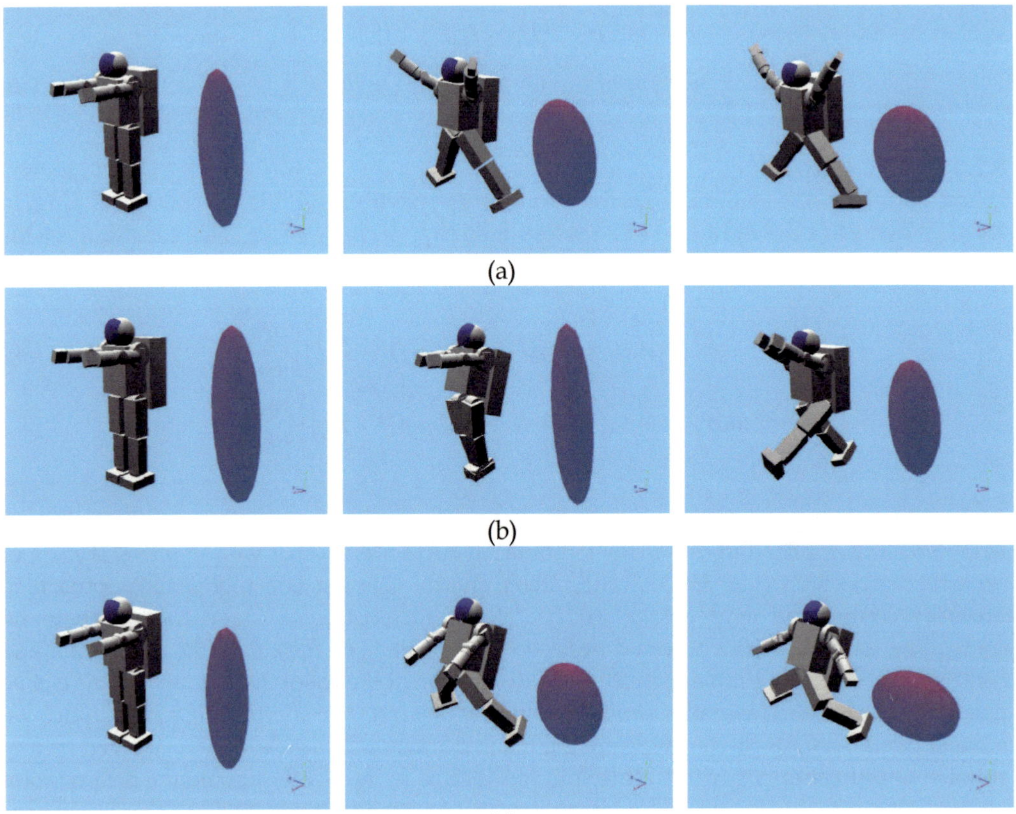

(a)
(b)
(c)

Fig. 6: Demonstration of the inertia shaping technique on a non-contacting biped robot floating in space. A floating robot has no CoP, so the RMP reduces to simply the ellipsoidal reaction mass.

In Fig. 6 (a) the desired inertia components along all three axes are equal and large. Hence the robot tries to "expand" in all directions. In Fig. 6 (b) the desired inertia in Y-component (vertical) is big, and the other components are very small. In Fig. 6 (c) robot tries to make its inertia in X and Z-components large.

This simulation demonstrates the important point of effectively controlling a complex biped with a very simple control law. While the robot model has 27 dofs, the control law deals with only three variables which are the three diagonal elements of the robot's rotational inertia.

In Fig. 7, we specify only desired CoM (Eq. 37) while keeping desired rotational inertia (Eq. 36) unconstrained. The inertia shaping algorithm computes optimal joint angles that creates specified CoM. In this example, we kept the joint angles for the arm and head fixed and moved only leg joints. Since we keep the ground projection of CoM within the support polygon, the humanoid robot maintains balance even in relatively extreme pose (e.g., bottom right figure).

Fig. 7: Given desired CoM position, the inertia shaping algorithm computes optimal joint angles to achieve the goal. In this demonstration, desired CoM is specified such that it moves gradually backward (top), forward (middle), and downward (bottom). Blue line segments connect CoP to CoM and radius of the green circle represents rotational inertia at CoM about Z-axis which is pointed towards the reader.

6. Conclusions and Future Work

We have introduced the reaction mass pendulum (RMP) model of a humanoid robot. The RMP model contains an actuated ellipsoidal reaction mass to explicitly model the robot's angular momentum. The ellipsoid represents the composite rigid body (CRB) inertia of the robot computed at its CoM. The reaction mass is an addition to the existing inverted pendulum humanoid models that only consider a point mass, and is also a mechanical realization of the AMPM model that accounts for the presence of centroidal angular momentum. The RMP is an instantaneous 3D capture of the aggregate kinematics and dynamics of a general humanoid robot. As a lower-dimensional ($n = 11$) dynamic equivalent of a high dof humanoid it lends itself to more probing analysis for dynamics and planning.

We presented the technique of inertia shaping, which can be thought of as a kino-dynamics-based higher-level control for humanoid. We have provided detailed formulations for each ground contact configuration and demonstrated the successful application to a free floating case as well as the CoM control in double support case.
RMP is introduced mainly as an analysis tool. For it to be a worthwhile successor to the very useful linear inverted pendulum (LIPM) or AMPM models, one need to develop an algorithm to compute desired trajectory of RMP. Our mechanical realization of the RMP model can be useful for developing such control laws. Also, we feel that analyzing RMP motion of human walking motion may provide a useful insight for this.
Once the planning is performed, one need to formulate control laws to apply desired RMP motion to humanoids. Inertia shaping can be used for this process, but more factors such as joint limits, collision with external objects, and prioritization of joints would have to be considered for a humanoid robot to perform various tasks while interacting with the environments (e.g., walking while holding a cup with the hand). This work is ongoing.

7. Acknowledgement

We appreciate Kangkang Yin for correcting errors in the original paper (Lee & Goswami, 2007). Ambarish appreciates fruitful discussions with Jerry Pratt, David Orin and especially with Stefano Stramigioli.

8. Appendix

8.1 Computation of CoP and ground reaction force/moment

Since the CoP is an important ground reference point for humanoid balance, equations to compute ground reaction force/moment and CoP are provided in numerous literatures (e.g., Abdallah & Goswami, 2005). However, for completeness we derive equations to compute CoP and ground reaction force/moment again using the notations used in this chapter.
The CoP is the point where the horizontal components of the moment generated by the ground reaction force (GRF) about the point vanish. The stationary humanoid robot can maintain balance when CoP is inside the supporting polygon given by the contact of feet and the ground. Since the GRF is unilateral, CoP is always in the support polygon. Assuming gravity $\boldsymbol{g}$ and GRF are the only external forces, GRF is $\boldsymbol{f}_r = M(\ddot{\boldsymbol{r}}_G - \boldsymbol{g})$ where $\boldsymbol{r}_G = \frac{1}{M}\sum m_i \boldsymbol{T}_i \boldsymbol{c}_i$ is the CoM of a humanoid robot.
From the relations between the rate of change of angular momentum about the CoM $\dot{\boldsymbol{k}}_G$ and the moment about the same point $\boldsymbol{m}_G$; i.e., $\dot{\boldsymbol{k}}_G = \boldsymbol{m}_G = (\boldsymbol{r}_P - \boldsymbol{r}_G) \times \boldsymbol{f}_r + \boldsymbol{m}_r$ where $\boldsymbol{r}_P$ is CoP, and assuming the ground normal $\boldsymbol{n}$ is parallel to the gravity vector and locating spatial coordinate frame on the ground (i.e., $\boldsymbol{r}_P \perp \boldsymbol{n}$), we can compute the CoP and the vertical component of $\boldsymbol{m}_r$:

$$\boldsymbol{r}_P = \frac{1}{\boldsymbol{n}\cdot\boldsymbol{f}_r}\boldsymbol{n}\times\left(\dot{\boldsymbol{k}}_G + \boldsymbol{r}_G \times \boldsymbol{f}_r\right)$$

$$\boldsymbol{n}\cdot\boldsymbol{m}_r = \boldsymbol{n}\cdot\left(\dot{\boldsymbol{k}}_G + (\boldsymbol{r}_G - \boldsymbol{r}_P)\times\boldsymbol{f}_r\right).$$

9. References

Kajita, S.; Yamaura, T. & Kobayashi, A. (1992). Dynamic walk control of a biped robot along the potential energy conserving orbit, *IEEE Transactions on Robotics and Automation*, vol. 8, no. 4, pp. 431-438

Kajita, S.; Kanehiro, F.; Kaneko, K.; Fujiwara, K.; Yokoi, K. & Hirukawa, H. (2002). A realtime pattern generator for biped walking, *IEEE International Conference on Robotics and Automation (ICRA)*, pp. 31-37, Washington, DC, 2002

Sugihara, T. & Nakamura, Y. (2003). Variable impedant inverted pendulum model control for a seamless contact phase transition on humanoid robot, *IEEE International Conference on Humanoid Robots (Humanoids 2003)*, 2003

Komura, T.; Nagano, A.; Leung, H. & Shinagawa, Y. (2005a). Simulating pathological gait using the enhanced linear inverted pendulum model, *IEEE Transactions on Biomedical Engineering*, vol. 52, no. 9, pp. 1502-1513, September 2005

Abdallah, M. & Goswami, A. (2005). A biomechanically motivated two-phase strategy for biped robot upright balance control, *IEEE International Conference on Robotics and Automation (ICRA)*, pp. 3707-3713, Barcelona, Spain, April 2005

Kajita, S.; Kanehiro, F.; Kaneko, K.; Fujiwara, K.; Harada, K.; Yokoi, K. & Hirukawa, H. (2003). Resolved momentum control: Humanoid motion planning based on the linear and angular momentum, *IEEE/RSJ International Conference on Intelligent Robots and Systems*, pp. 1644—1650, Las Vegas, NV, USA, 2003

Vermeulen, J.; Verrelst, B.; Vanderborght, B.; Lefeber, D. & Guillaume, P. (2006). Trajectory planning for the walking biped "Lucy", *The International Journal of Robotics Research*, vol. 25, no. 9, pp. 867—887, 2006

Featherstone, R (1987). *Robot Dynamics Algorithms*, Kluwer Academic Publishers

Murray, R. M.; Li, Z. X. & Sastry S. S. (1994). *A Mathematical Introduction to Robotic Manipulation*, CRC Press

Ostrowski, J. (1999). Computing reduced equations for robotic systems with constraints and symmetries, *IEEE Transactions on Robotics and Automation*, vol. 15, no. 1, pp. 111--123, 1999

Crandall, S. H.; Karnopp, D.; Kurtz, E. F. & Pridmore-Brown, D. C. (1982). *Dynamics of Mechanical and Electromechanical Systems*, Krieger Publishing Company

Beer F. P. & Johnson, E. R. (1984). *Vector Mechanics for Engineers: Dynamics*, McGraw-Hill Book Company, New York

Cominoli, P. (2005). *Development of a physical simulation of a real humanoid robot*, Master's thesis, Swiss Federal Institute of Technology (EPFL), Lausanne

Sidi, M. J. (1997). *Spacecraft Dynamics and Control*, Cambridge University Press, New York

Astrom, K.; Block, A. & Spong, M. (2001). *The Reaction Wheel Pendulum*, Lecture Notes, Ed. Downloadable from: http://decision.csl.uiuc.edu/~spong/main.htm

Olfati-Saber, R. (2001). Stabilization of a flat underactuated system: the inertia wheel pendulum, *40th Conference on Decision and Control*, pp. 306--308, Orlando, FL, December 2001

Spong, M. W.; Corke, P. & Lozano, R. (2001). Nonlinear control of inertia wheel pendulum, *Automatica*, vol. 37, pp. 1845--1851, February 2001

Kuo, A. (1999). Stabilization of lateral motion in passive dynamic walking, *International Journal of Robotics Research*, vol. 18, no. 9, pp. 917--930, September 1999

Pratt, J. (2000). *Exploiting inherent robustness and natural dynamics in the control of bipedal walking robots*, Ph.D. dissertation, MIT

Komura, T.; Leung, H.; Kudoh, S. & Kuffner, J. (2005b). A feedback controller for biped humanoids that can counteract large perturbations during gait, *IEEE International Conference on Robotics and Automation (ICRA)*, pp. 2001--2007, Barcelona, Spain, 2005

Mayer, N. M.; Farkas, F. & Asada, M. (2005). Balanced walking and rapid movements in a biped robot by using a symmetric rotor and a brake, *International Conference on Mechatronics and Automation*, Niagara Falls, Ontario, Canada, July 2005

Walker, M. W. & Orin, D. (1982). Efficient dynamic computer simulation of robotic mechanisms, *ASME Journal of Dynamic Systems, Measurement, and Control*, vol. 104, pp. 205-211, September 1982

Lee, S.-H. & Goswami, A. (2007). Reaction Mass Pendulum (RMP): An explicit model for centroidal angular momentum of humanoid robots, *IEEE International Conference on Robotics and Automation (ICRA)*, pp. 4667-4672, Rome, Italy, 2007

Hofmann, A. (2005). *Robust Execution of Bipedal Walking Tasks from Biomechanical Principles*, Ph.D. dissertation, MIT

10

Neurophysiological models of gaze control in Humanoid Robotics

Luigi Manfredi, Eliseo Stefano Maini & Cecilia Laschi
ARTS Lab (Advanced Robotics Technology and Systems Laboratory,
Scuola Superiore Sant'Anna
Italy
manfredi@ieee.org

1. Introduction

Thanks to the improvements in mechanical technology, it is currently possible to design robotic platforms that are increasingly similar to humans (Laschi et al., 2008; Kaneko, 2004; Kuffner et al., 2005). However, the increasing robot complexity (i.e. presence of many degrees of freedom, non linear actuation and complex geometries), requires more sophisticated control models and heavier computational burden. The development of humanoid robot is a very relevant issue in robotic research especially when one considers the challenges related to the actual implementation of a humanoid robot both in terms of mechanics and control system.
However these research efforts are justified considering that, an actual humanoid robot is regarded as a fundamental tool for neuroscience and, at the same time, neuroscience can be exploited as an alternative control solution for the design of humanoid robots (Kawato, 2000).
In this chapter, the neurophysiological models for gaze (i.e. the line of sight) shift control will be discussed and their implementation on a head robotic platform is presented. In particular the rapid movement of the gaze and the issues related to the eye-head coordination were investigated from neurophysiologic and robotics points of view. In neurophysiology the rapid movement of the gaze is known as *saccadic*. This movements are also classified either as *head-restrained* visual orienting movement or *head-free* visual orienting movement (Barnes, 1979; Bizzi et al., 1971; Bizzi, 1972; Guitton andVolle 1987; Guitton, 1992; Goossens and Van Opstal 1997).
The neurophysiologic models that will be discussed here are the visual mapping of superior colliculus and the independent gaze control model presented by Goossens and colleagues for the eye-head coordinated motion (Goossens & Van Opstal, 1997).
In the case of visual colliculus mapping, the input is a visual image that is mapped from camera image to the superior colliculus. Conversely, for the gaze model control, the input data are the two angular deviations (i.e. horizontal and vertical) that may be used to define the gaze shift amplitude and the movement orientation. The eye-head saccadic model

presented here derives from the one proposed by Goossens and Van Opstal with the aim of controlling both the eye and head movements.
The experimental setup was designed to test two typologies of results. One is the capability of reproducing the human behaviour both in terms of movement and velocity profile (i.e. velocity peak and movement duration for eye, head and gaze related to the displacement). Secondly, stability and response time have been measured in order to assess the ability of the model to control a redundant robotic platform.
The reported results show that, the visual target is properly achieved and its stability is maintained with low residual oscillation. Some limits have been noted in the beginning of the head movement when the amplitude of the movement is small. However, the obtained results are similar to humans (Goossens and Van Opstal, 1997). We are aware that higher repeatability and accuracy may be achieved by using the traditional control; however the residual errors of movements are lower or equal to those reported in neurophysiologic studies. Moreover, the impressive human like appearance of the motion obtained with this model can encourage its use in the design of more "social robot".
Future work could include the implementation of a dynamic model and the inclusion of an automatic learning phase of models gains.

2. Physiological models

The objective of studying physiological behaviour is to formulate a model description of the target - *black-box model* - (defined as $G(s)$, where s is the Laplace variable). This is a classic model identification problem.
Usually, a general physiological system is composed of many sub-models interacting with each other. Therefore, in order to be able to identify and exploit a particular sub-model, it needs to be isolated. The physiological models can be divided in two main categories: *qualitative*, and *quantitative*.
The *qualitative models*, often formulated by biologists, provide a description of the model behaviour by means of input-output relations.
Conversely, the *quantitative models* provide the mathematical formulations of the physiologic model. In this case, the system can be simulated and tested. Finally, real data can be compared with results in order to assess the ability of the model to describe the real system.
In recent years, the increasing computational power allows the simulation of more complex mathematical models, giving rise to more effective investigations. Accordingly, mechanical and electronic improvements have given the opportunity of building robotic platforms more similar to biological systems. Therefore, in these days we assist the birth of a new area research at the bridge of robotics and neuroscience. Its main goal is to use real robotic platforms to (1) validate the investigated model, and (2) use the proposed model to control a complex and redundant platform. The latter, can be used as an alternative way to the traditional control approach.
The classical control theory gives the possibility of being more accurate when the problem is mathematically identified. Indeed, the physiological approach is less accurate and repeatable. However, the particularity of physiological approach is the capability of generalizing and controlling a system with slow sensory response. Many physiological studies suggested that prediction is used in many tests for compensating slow sensory response (Wolpert and Kawato 1998).

There are two types of control based prediction:

- using motor commands to achieve a desired output;
- predicting the output obtained from the given motor command.

In the first case, the input of the prediction model is the desired task output, and it provides the motor commands required to accomplish the task. In this case, the prediction is made by identification of the inverse system model ($\hat{G}(s)^{-1}$).

In the second case, the input is the motor commands, and the prediction model provides the future output of the system.

In this case the prediction is made by identifying the model of the system: $\hat{G}(s)$.

Investigation of such problems is inherently complex hence a multidisciplinary group is fundamental; this should include biologists, physiologists, neuroscientists and engineers.

3. Ocular movements

In human as in primates, vision is based on images with a space-variant resolution. The photosensitive area of the eye is characterized by a non-uniform density of photo receptors (Green, 1970). The central part of the retina within 5 deg, is known as *fovea*, which has much higher resolution than the periphery. Thus, the target object needs to fall within the *fovea*. This implies the execution of many movements in eyes and head but in general also torso and limbs. The peripheral part has a resolution good enough to perceive visual stimuli, and the low quantity of information allows fast processing and especially fast visual-motor reaction. Both, the two eyes and the head compose the vision system. The *gaze* is defined as the eye position relative an external or spatial frame of reference.

How the brain transforms two-dimensional visual signals into multi-dimensional motor commands, and subsequently how it constrains the redundant degrees of freedom, are fundamental problems in sensory-motor control.

The common ocular movements are the following that are controlled by different part of the brain (Henn, 1993).

- *Microsaccades*: they are small, jerk-like, involuntary eye movements, similar to miniature versions of voluntary saccade.
- *Vestibulo-ocular reflex (VOR)*: is a reflex eye movement that stabilizes images on the retina during head movement, by producing an eye movement in the opposite direction of the head movement, thus preserving the image on the center of the visual field. It is a very reactive movement, with a latency (time between the head and the eye movement) of only 14 ms, guided by the vestibular system response, which detects head motion and posture. The VOR does not depend on the visual input and works even in total darkness or when the eyes are closed. It is considered as slow movement.
- *Optokinetic reflex (OKR)*: allows the eyes to follow objects in motion when the head remains stationary (and thus the VOR is not elicited) and it is activated when the image of the world slips on a large part of the retina; this reflex is based on the visual information and the latency is longer (100–150 ms) than in the VOR.
- *Saccadic movements*: are fast, ballistic, simultaneous movements of both eyes in the same direction. They can reach 800 - 1000 °/s, in a time of 30–120 ms (Bahill et al., 1975; Baloh et al., 1975);

- *Smooth pursuit*: is the ability of the eyes to smoothly follow a moving object. It is a slow movement.
- *Vergence movement*: is the simultaneous movement of both eyes in opposite directions to obtain or maintain single binocular vision. It is slower and can take up to 1s.

All movements of the eyes are in the same direction. Only the vergence makes an opposite movement of the eyes.
With regard to the neck, human motion is pretty complex, but three main movements can be identified (Zangemeister et al., 1981):

- Flexion/extension (i.e., ventral or dorsal flexion of the neck).
- Lateral bending (i.e., right and left flexion of the neck).
- Axial rotation (i.e., right and left rotation of the neck on the vertical axis).

Table 1 reports data on performance of human eye and neck movement, in terms of ranges of motion of cervical spine segments (ROM), maximum speed (VMax) and maximum acceleration (AMax) (Zatsiorsky, 1998).

Joint	ROM (deg)	*V*Max (deg/s)	*A*Max (deg/s2)
Neck lateral bending	−27/+27	100	1000
Neck flexion/extension	−45/+54	100	1000
Neck axial rotation	−58/+58	200	3000
Eye pitch	−40/+40	700	12000
Eye yaw	−45/+45	700	12000

Table 1. Human eye-head data performance: amplitude (ROM), maximum speed (VMax) and maximum accelerations (AMax).

Therefore, both the eyes and head can rotate about three mutually perpendicular axes: horizontal rotations about a vertical axis, vertical rotations about a horizontal axis, and torsional rotations about a naso-occipital axis.

4. Gaze movements

The saccadic and head-eye saccadic movement can be driven by different stimuli: visual, auditory, tactile or olfactory. The stimulus coordinates need to be converted in the retina reference system. In the case of visual stimulus, the retinal error drives the movements.
The main problem is how the saccadic system maps from two-dimensional (2-D) of retinal target error in to 3-D rotation of the eye and head. Indeed, the eye has 3 degree of freedom (DOFs) and another 3 DOFs for the head (excluding head translational movements). Therefore, the mapping is from 2-D to 9-D and it must preserve eyes and head mechanical constrains. The most important constrain is that the *eye-in-head* must obey Listing law (Radau et al., 1994) and the *head-on-torso* must obey Donders law (Radau et al., 1994). However there is a problem of redundancy that the brain solves in slightly different ways for the eyes and the head (Bernstein, 1967; Crawford et al., 2003).

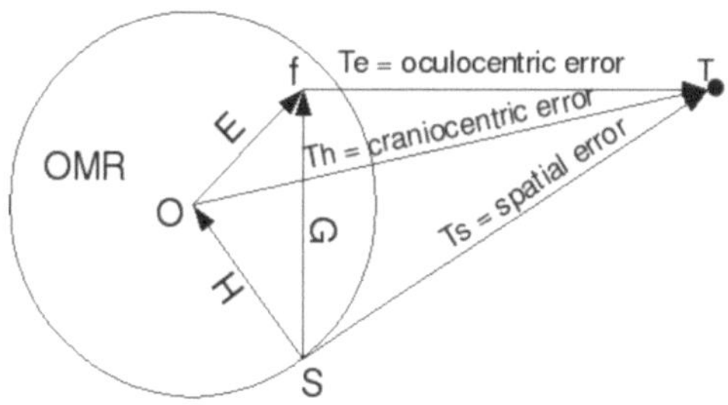

Fig. 1. Reference frame coordinates. T is the target stimulus, S, O, f are respectively the external reference frame (inside oculomotor range OMR), head end eye reference system. G represent *eye-in-space*, H *head-in-space*, E *eye-in-head*. **Te** is the *target-in-eye*, **Th** *target-in-head* and **Ts** *target-in-space*.

If the target is indicated with **T**, the craniocentric error is **Th,** spatial error **Ts** and oculocentric error **Te.** The eye displacement is **Te=Th-E**.
Two main approaches are followed to investigate the gaze control model:

1. head-restrained visual orienting movements: *eye saccadic movements;*
2. head-free visual orienting movements: *eye- head saccadic movements;*

Actually, the movement is more complex because also the torso and limbs interact with the movement.
This kind of control is obtained by transforming position into motor activation (in time) which is a function of time. This is the so called *spatio-temporal transformation* (STT) (Sparks, 2002).

4.1. Head-restrained visual orienting movements: *saccadic movements.*

When the head is fixed, the gaze is performed by rapid eye movements that can be accurately predicted, if the amplitude and direction are known, in term of duration and peak velocity (Bahill et al., 1975; Baloh, 1975). The relationship between duration and saccadic magnitude is linear (VanGisberg and Johnson, 1984). The latency saccadic period depends on the stimulus (source and intensity Bell et al. 2006).
Moreover, when this criterion is not respected, often a neuromechanical deficit is present (Westheimer and Brail, 1973; Zee et al., 1976).

$$V = \alpha(1 - e^{-\frac{x}{\beta}}) \qquad (1)$$

Equation (1) shows the relation between eye velocity and eye amplitude where V is the maximum velocity, x the saccadic amplitude α and β are the parameters of the exponential model which can be calculate as the minimum of the square error between the model and the data (Baloh et al., 1975; Collewijn et al., 1988).
An example of saccadic movement is shows in Fig. 2

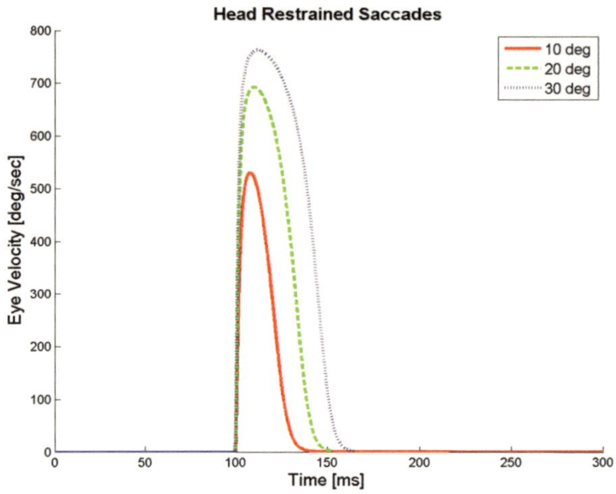

Fig. 2. Velocity profile of saccadic movement. The delay of activation strictly depends on stimulus source and intensity.

Many saccadic models have been proposed with the aim of understanding how a more simple system acts, e.g. an eye, in order to understand how a more complex system (e.g. a limb, which use multiple joints and operate with varying loads) is controlled.

The saccadic system is under a continuous feedback control. The main question is what the source of the feedback is. Cannot be visual because the retina signal is too slow compared to the saccadic dynamics. Many hypothesis and many models have been formulated (Robinson, 1975).

Many parts of the brain are involved in transforming the visuo-information provided by the retina in to the motor-neuron activation (STT). The main problem is to understand which the control strategy is; whether it is feed-back or feed-forward based, considering that the movement is very fast and the signal transmission is slow.

One of the first models formulated by Robinson (Robinson, 1975) supposes that the saccadic movements work as non ballistic systems actuated by a feed-back loop. The feed-back is based on an efferent copy of the output of the saccadic burst generator, which is able to performance of on-flying correction of the trajectory.

Van Gisbergen and colleagues (Gisbergen et al., 1985) studied the dynamic of oblique saccades. They proposed that, in order to obtain a straight trajectory, the times needed to complete the two components of the movements are identical.

Tweed and Vilis (Tweed and Vills, 1985) extended the previous model by introducing of different gain constants for horizontal and vertical saccades, due to the probably different 3rd order dynamics of the eye muscles. In this model, it may be possible to have different dynamic calibration.

Grossberg and Kuperstein (Grossberg and Kuperstein, 1986), expanded the models of Robinson and colleagues by introducing a neural network model to actuate the antagonist eye muscles in the horizontal plane, instead of classic control approach.

4.2. Head-free visual orienting movements: *head-saccadic movements.*

When the gaze shift becomes large, the eye and the head work together. The oculomotor systems transformation the object position into target reference system and then in head reference system (Goossens and Van Opstal, 1997; Freedman and Sparks, 2000).
Many models of these transformations were formulated.
Bizzi and colleagues (Bizzi et al. 1971, Bizzi et al. 1972) proposed the so-called *oculocentric hypothesis*. According to this model, the head-free gaze is like the head-fixed gaze, independently of the occurrence and size of the concomitant head movement. The vestibulo-ocular reflex (VOR) is switched-off, and cancels any contribution of the head to the gaze. Several experiments showed that the VOR is more or less switched-off during saccades (Laurutis and Robinson, 1986; Pelisson et al., 1988; Lefevre et al. 1992). Further experiments validated this hypothesis for gaze shifts smaller than 10 degrees.
Differently from the head-fixed, in which the hypothesis of eye movements with local feedback of current position is accepted (Robinson, 1975), in the head-free model, it is supposed that the gaze motor-error drive the oculomotor system. According to this so called *gaze feed-back hypothesis*, the gaze saccades can be maintained even if the VOR is suppressed during the movement.
The main question is *what is the source of the feedback?*
As described in Fig. 3 two hypotheses are being explored. In the first hypothesis the gaze error, after decomposition in eye-error and head-error drives both, eye and head (Fig. 3-A) (Goossens and Van Opstal, 1997) . Differently in the second architecture, there exists only a local feedback of the eye-position whereas the head act in open loop (Fig. 3-B) (Freedman and Sparks, 2000; Freedman and Sparks, 1997).

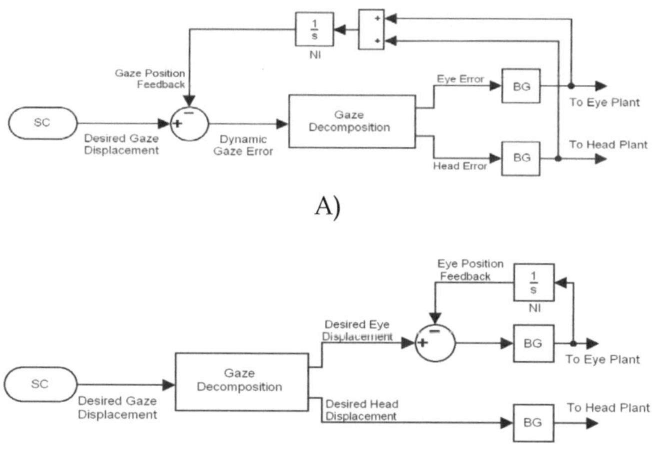

Fig. 3. Two main hypothesis of gaze control architecture. **A.** The gaze error, after decomposition, drives both eye and head (Goossens and Van Opstal, 1997). **B.** After gaze displacement decomposition, a local feedback is present only in the eye. To identify the action of the feedback, several tests with external disturbance have been made (Freedman and Sparks, 2000). In both architecture (A-B) the system is robust to external disturbance.

Note that, in **B** architecture, the external disturbance is compensated making use of head and neck reflexes.

4.3.The collicular mapping of the visual field

Many models of collicular mapping have been proposed (see review of Girard and Berthoz 2005).
A critical structure in the gaze control circuitry is the superior colliculus of the midbrain. Indeed its deeper layers contain a retinotopically organized motor map in which each site is thought to encode a specific gaze saccade vector.
The way in which the mapping of the visual field occurs onto the SC was experimentally measured in primates by Robinson (Robinson, 1972) and subsequently mathematically modeled by Ottes and colleagues (Ottes et al., 1986). The equations introduced in (Ottes et al., 1986) for mapping the retinotopic coordinates (R,θ) onto the Cartesian plane (X,Y) of *the colliculus are*:

$$\begin{cases} X = B_x \ln\left(\sqrt{\dfrac{R^2 + 2AR\cos(\theta) + A^2}{A}}\right) \\ Y = B_y \arctan\left(\dfrac{R\sin(\theta)}{R\cos(\theta) + A}\right) \end{cases} \quad (2)$$

Where a = 3 deg, B_X 1.4 mm and B_Y = 1.8 mm. The expression of this mapping is rather complicated but it may be reformulated using complex logarithm in the following form:

$$\begin{cases} \dfrac{X}{B_x} + i\dfrac{Y}{B_Y} = \ln\left(\dfrac{z+A}{A}\right) \\ z = \alpha + i\beta \end{cases} \quad (3)$$

where α and β represent respectively the azimuth and the elevation of the saccade

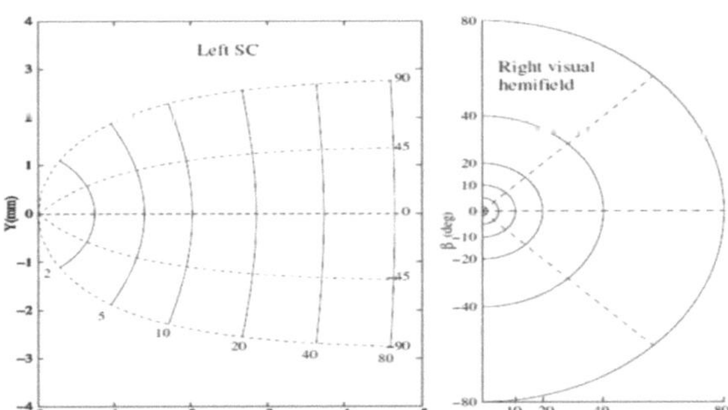

Fig. 4 The visual mapping from retina to colliculus.

The two right half sides of the retinas of the two eyes are projected onto the right SC, in the right side of the brain, while the two left half sides of the retinas of the two eyes are projected on the left SC.
Equation (3) has been use for collicular mapping software implementation.

5. Robotics application of gaze control

Eyes provide the ability to perceive and interact in unconstrained dynamics environments. This capability has been investigated in robotics as active vision. Several new researches have born to deep investigate this capability.
Many robotics applications have the scope to implement only fast eye movements useful for industrial applications and commercial vision systems. In this case relevant are the mechanical performance and the very fast control loop able to perform very fast movements.
From the viewpoint of social interaction, the robot has to give more social acceptability (Shibata, 2004).
Takanishi Lab of Waseda University, have developed a humanoid robot - WE-3RII - to mimic facial expressions, and saccadic movements have been implemented, with the same delay of human ones, with the purpose to facilitate social interaction (Takanishi et al., 1999).
Instead, studying the modeling and validation of neurophysiological models gives two possibilities: to understand how the models are correctly implemented and how it can be possible to control a robotic platform.
Clark and colleagues (Clark, 1998) have implemented, in a simple pan-tilt camera, a spatial attention mechanism with related saccadic camera motion. They have implemented a non programmed saccadic movement based on the pattern activity of the attention mechanism, based on winner-take-all model.
Ferrell (Ferrell, 1996) has implemented a saccadic system based on Self Organizing Maps (SOM) on the Cog platform (Brooks et al., 1999). By using feedback error-learning (model formulated by Kawato (Kawato et al., 1987))
Bruske and colleagues (Bruske et al., 1997) have proposed an adaptive saccade model for a 4 DOFs binocular camera head. The goal was not to emphasize the biological model, but to improve a real-time architecture based on a small neural network.
Panarei and colleagues have implemented image stabilization by using a bioinspired solution. It was implemented by using two sensors, inertial and visual (like vestibular and optokinetic in nature). The relations between sensors and system control are obtained by the process of learning of a neural network.
Berthouze and Kunihoshi (Berthouze & Kuniyoshi, 1998) have demonstrated the learning by neural network of sensory-motor coordination (i.e. VOR and OKR) by a 4 DOFs robot head.
Manfredi and colleagues (Manfredi et al., 2006) have implemented an eye-saccadic model in an anthropomorphic robotic head. The model has been implemented in the robot in order to validate it.
Preliminary experiments on eye-head coordination models were already performed by Maini and colleagues (Maini et al., 2008) on a anthropomorphic robotic head.
Many works have not remarked the robotic implementation, which contribute for investigating the validity of the implemented model. The main issue is how the robotic

platform can be used to validate the neurophysiological model. Often, a simulation is not complete enough to investigate the real behaviour of the studied model.
In many cases, the robotic platform is non adequate because it has less DOFs or less performance respect to the real biological platform.
In the present work two main points are emphasized: (1) the performances of the robotic head (Laschi et al. 2008) in terms of dynamics are quite similar to human data available from literature (Zatsiorsky, 1998); in this way it is possible to duplicate the neurophysiological models; (2) investigating the validity of the proposed model (according to 2-D independent gaze control hypothesis formulated by Goossen and Van Opstal (Goossens & Van Opstal, 1997)) to control a redundant robotic platform and to obtain results comparable to literature data.

6.Experimental Validation

6.1.Overall system

Fig. 5 represents the overall system architecture. The fast gaze shifts were evoked by presenting a visual stimuli to the robot that executed the task according to the following step: selection and detection of the color of interest; mapping of the image onto the collicular; estimation of the stimulus center of mass. The elevation and the azimuth of the stimulus were then passed to the eye-head coordination model that generated a velocity profile with a sampling time of 3 ms.

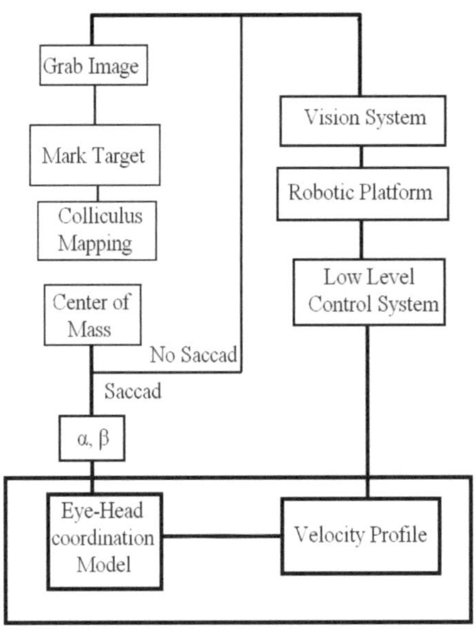

Fig. 5. Overall system architecture. After traditional color edge detection, the selected target is mapped in the superior colliculus. Then, the centre of mass of the mapped stimulus is detected and, it is send to the eye-head model which provides the velocity profile. The communication protocol is the TCP/IP. On the robotic head platform, the server application receives the data with a sample time of 3 ms.

6.2. The robotic platform and the vision system.

The robotic platform used for validating the system has been composed of 7 DOFs (Fig. 6). It is composed of two parts: a neck and a head. The neck has 4 DOFs respectfully 2 in the lower part (roll and lower-pitch) and last two in the upper part (upper pitch and yaw). The head has 3 DOFs, 1 common for the two eyes (pitch) and the last, one for each eye (yaw left and right). All the system has been controlled by an embedded PC where a server application has been implemented. The main characteristics of this hardware are the capability to perform the dynamics quite similar of human head. The robotic head performances are show in Table 2.

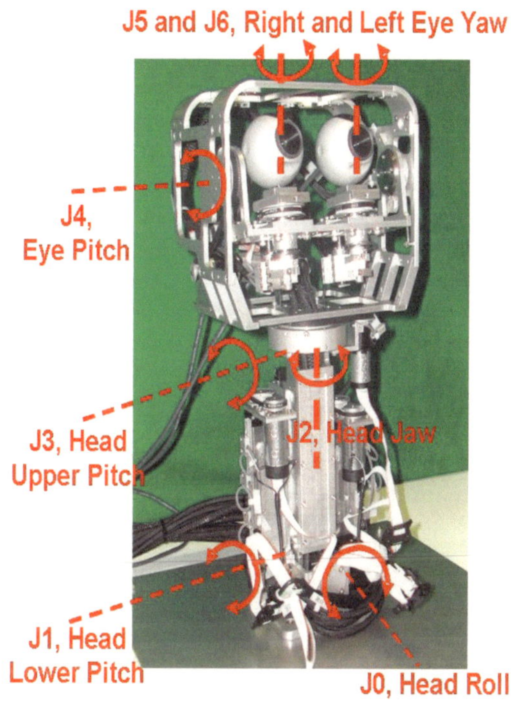

Fig. 6. Robotic platform composed of 7 DOFs. The neck has 4DOFs, two in the lower part (roll and lower-pitch) and two on the upper part (upper-pitch and yaw). The eyes have 3 DOFs, one common pitch and 1 yaw for each eye.

JOINT	ROM [deg]	V_{MAX} [deg/s]	A_{MAX} [deg/s^2]
Roll	± 30	25	200
Lower pitch	± 25	20	200
Yaw	± 100	170	750
Upper pitch	± 30	120	750
Eye pitch	± 47	600	4,500
Right Eye Yaw	± 45	1,000	10,000
Left Eye Yaw	± 45	1,000	10,000

Table 2. Robotic head ranges of motion (ROM), maximum speed (VMAX) and acceleration (AMAX).

Each eye has one commercial USB webcam, with the resolution of 640x480 pixel and 30 frame per second.
The robot is controlled using a PID (proportional, integrative and derivative) implemented in a DSP (Digital Signal Processor) motion controller (MEI 104/DSP-400 by Motion Engineering Inc, USA). Two boards are present, one for 4 axes and another for 3 axes. A software limit switch prevents mechanical injury. The axis control board is connected by PC104 bus in a Pentium IV PC, 3.0 GHz.
A *server application*, implemented in C++, receives and sends the command to control the robotic platform by using TCP/IP protocol. Therefore, the model implemented as *client application*, can run in the same PC or on other.
For more detail refer (Laschi et al., 2008).

6.3. Implementation of the Superior Colliculus mapping

It is worth noticing that, a straight use of this transformation for obtaining the collicular mapping may introduce some relevant ambiguities in real-world applications. Indeed, the CCD cameras have a uniform discrete pixel distribution whereas the equation (3) and (2) are defined on a continuous domain. Therefore, it is obvious that the logarithmic function will map the discrete points of the retinotopic plane in such a way that will spread the pixels pertaining to the central part of the collicular map. Conversely, in the peripheral part of the collicular map the density of pixels will be increased with a subsequent loss of information. To avoid this problem a void collicular map has been started from and for each pixel in the collicular map has been back-projected the coordinates onto the retinotopic plane making use of the equation (3) as reported in (Girard & Berthoz, 2005). Obviously, the results (as X_c, Y_c, coordinate in camera reference system) will be, in general, not integer hence with no physical representation on the CCD matrix. To solve this ambiguity we associated to the virtual point X_c, Y_c an intensity value that was calculated using a weighted interpolation of the surrounding four connected integer pixels. Finally, it has imposed the latter intensity value to the original point of the collicular map and we iterated this procedure for each point on the collicular plane. An example of the resulting well smoothed collicular maps is reported in Fig. 7.

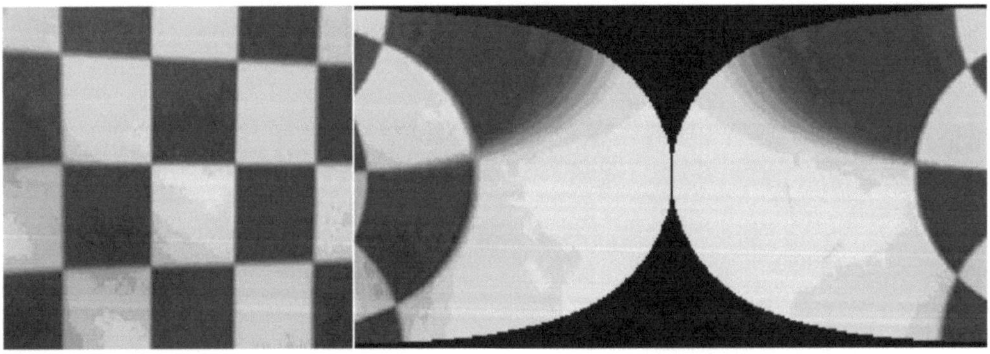

Fig. 7. Collicular mapping according to equation (3) with interpolated uniform resolution. **Left**: the image acquired by the camera; **Right**: the two halves of the image mapped with the collicular geometry expressed by equation. (3).

6.4. Stimulus detection

Visual stimuli are detected in the collicular image, as obtained after collicular mapping, by means of a color-based threshold algorithm. Specifically, the detection of the stimuli is performed with a simple procedure of color recognition based on a target color manually indicated.

The procedure makes use of the HSV (hue, saturation and value) color representation, instead of the classical RGB (red, green, blue), in order to improve the robustness with respect to light variations. After the detection of the selected color, the centroid of the object is calculated and the coordinates on the collicular plane are fed to the model, in order to produce the velocity profiles for controlling the motion of the eyes, similarly to what happens in humans.

In Fig. 8 is a show an example of stimulus detected after mapping and before gaze performed and in

Fig. 9 after gaze performed.

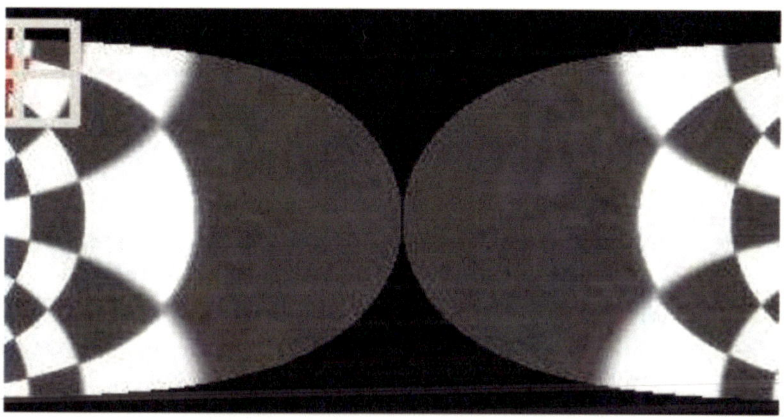

Fig. 8. Collicular mapping of the detected stimulus: the centroid of the object is marked with a targeting square on the collicular plane.

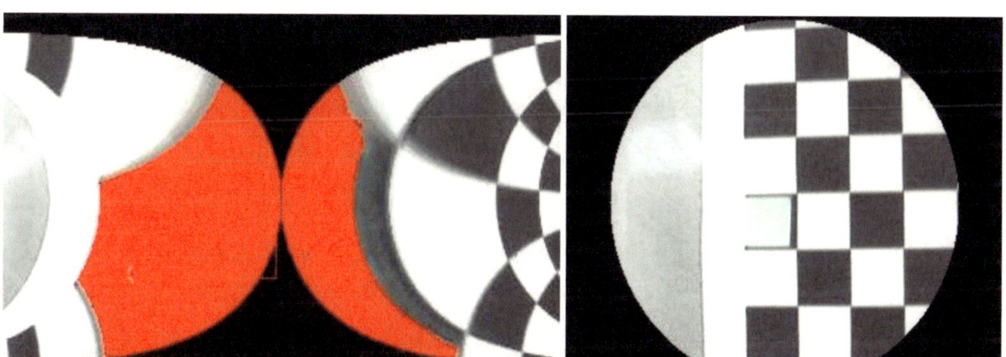

Fig. 9. The collicular mapping of the foveated object (left) and the visual field of the robot after the saccadic movement (right).

6.5. Model implementation

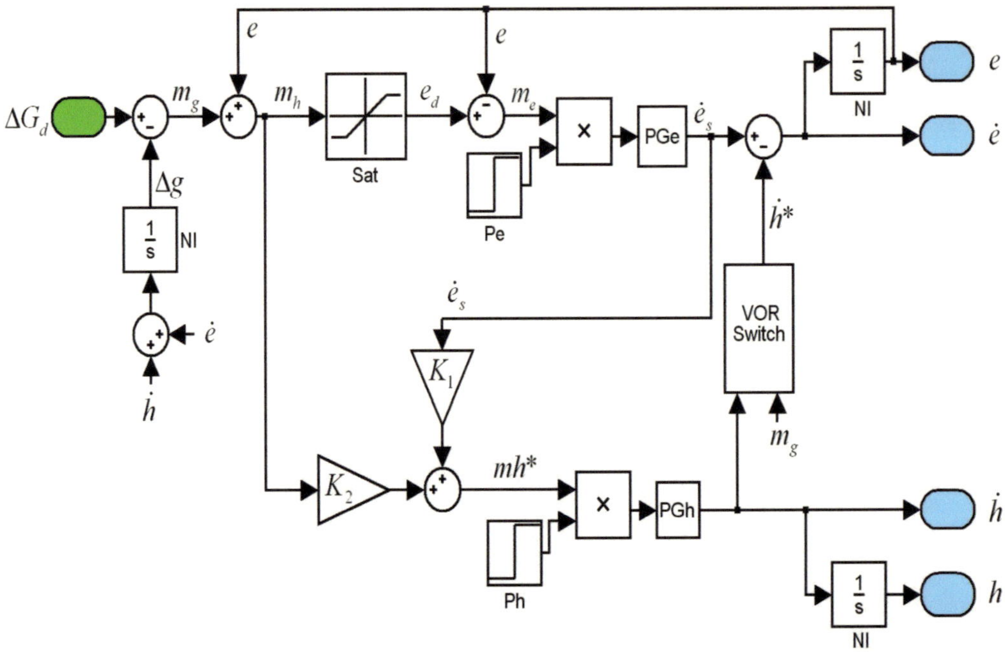

Fig. 10. The independent 2-D gaze control model as proposed in Goossens and Van Opstal (Goossens and Van Opstal, 1997), implemented in Matlab Simulink real-time. It is based on current gaze displacement feedback (modified after Guitton and Volle 1987 (Guitton & Volle, 1987)). ΔG_d is the target amplitude provided by superior colliculus (SC). Semicircular canals (SCC) provide the head velocity $\dot{H}$. The PG_h and PG_e blocks represents the head and eye pulse generator. **Sat** block, identifies the eye constraint. K_1 and K_2 define the head displacement decomposition. **VOR-Switch** represents the VOR behaviour.

A desired gaze displacement ΔG_d is given as input to the model and compared to current gaze displacement Δg (*eye-in-head* + *head-in-space*). The gaze displacement is obtained as integration (*rst* reset the integration of $\dot{g}$ after each saccadic movement) of eye ($\dot{e}$) and head ($\dot{H}$) velocity (provided by semicircular canals SCC - **VOR-Switch**). The residual displacement, is converted in head-motor error m_h. Initially **me** and **mh** are identical to the oculocentric and craniocentric target coordinates (**Te** and **Th**, see Fig. 1). The **Sat** block is to limit the eye mechanical constrain. Then, the desiderate eye position (e) is compared to current eye position and driving the eye as local feedback (as Robinson's local feedback model Robinson1975). Than, it is compared to current eye position. The head velocity, provide from SCC, after a partial attenuation ($\dot{h}^*$), interacts downstream from the eye pulse generator. In our implementation, because robotic platform fixed, the SCC is the actual velocity provided by axis control board. The eye and head movements can be controlled by separate gating mechanisms as possibility to test difference delay likes visual and auditory source.

The **VOR-Switch** block, has been implemented for deep investigate of its role. Until now, it is not clear how the VOR is implicated in the head-saccadic movement. It is know, from literature data, that it is weakened or shut off in the direction of the saccade. But it is not clear if it is shut-off in the direction of the *gaze* (*eye-in-space*) or *head-in-space* or *eye-in-head* motor error.

The model implemented is based as different timing:

1. **t=0** : the ΔG_d value is send to the system:

The input signal ΔG_d drives the system. The onset of the eye and head movement depends as source (Bell et al., 2006). The time can be defined by setting **Pe** and **Ph** delay variables.

2. **t=Te+** : the eye starts to move:
3.

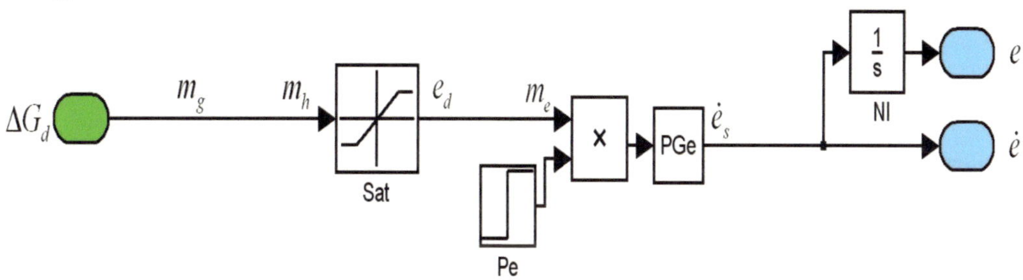

Fig. 11 Block diagram of the model at the time **t=Te+**

The eye movement starts onset of **Pe** block. The system is shown in Fig. 11.The eye has not received any feedback and it starts to move as open-loop system. The **PGe** block (eye pulse generator) is activated and provides the eye velocity. This is sent to the robotic platform, and it starts to move according to mechanical dynamics. The eye constrains is defined in **Sat** block.

4. **t>Te** : the feedback eye starts:

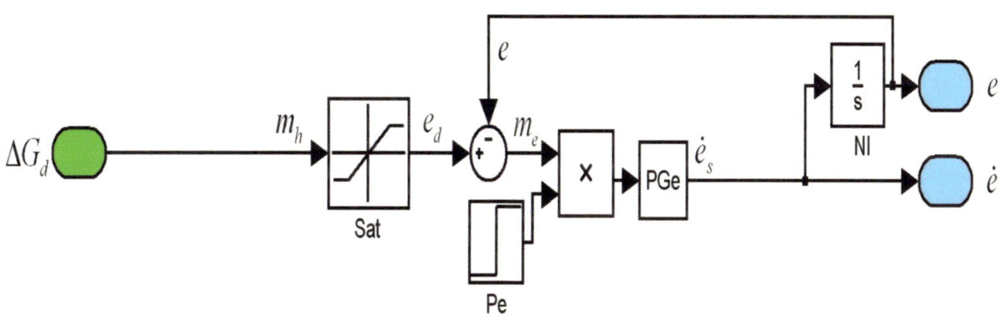

Fig. 12. Block diagram of the model at the time **t>Te**

In this phase (see Fig. 12), the eye starts to move as eye-saccadic movements according to Robinsons' local feedback model (Robinson, 1975). The ΔG_d, after saturation block (**Sat**, according to mechanical constraint), is sent to the **PGe** block which provides the velocity command to the eye. The eye perform the gaze shifting as own system, without head. In this condition, the eye starts to move according to maximum velocity performance. **PGe** block has been implemented concerning traditional used model (Becker & Reinhart , 1990) shown in equation (4). The equations (4) show the system equation.

$$\begin{cases} \dot{e} = PG_e(m_e) = K \cdot (1 - e^{-(\frac{m_e}{10})}) \\ me = e_d - e \\ ed = Sat(m_h) \end{cases} \quad (4)$$

5. **t=Th+** : the head starts to move:

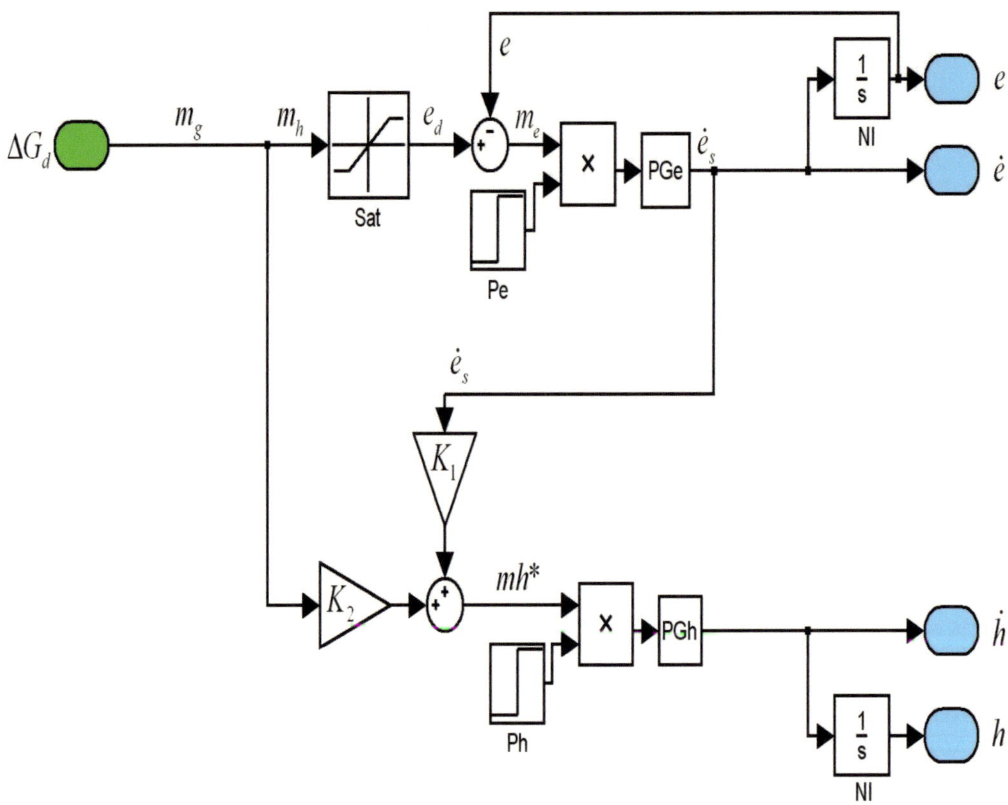

Fig. 13. Block diagram of the model at the time **t=Th+**

The mh^* is sent to the PG_h block (head pulse generator) and the velocity command $\dot{h}$ is sent to the robotic platform. The head starts to move according to mechanical dynamics. As shown in equation (5), the head has to perform the residual displacement error, until not performed by the eye. The block diagram is shown in Fig. 13.

$$\begin{cases} \dot{h} = PG_h(mh^*) \\ mh^* = K_1 \cdot \dot{e}_S + K_2 \cdot \Delta G_d \end{cases} \qquad (5)$$

6. **t>Th** : the feedback head starts:

At this stage, the head is already started and the system is going on. The block diagram is shown in Fig. 12.

$$\begin{cases} \dot{h} = PG_h(mh^*) \\ mh^* = K_1 \cdot \dot{e}_S + K_2 \cdot m_h \\ m_h = e + m_g = \Delta G_d - h \\ m_g = \Delta G_d - \Delta g = \Delta G_d - (e+h) \\ \Delta g = e + H \end{cases} \qquad (6)$$

In conclusion, the results of the model are strictly dependent on the given gain. Also the low level control (i.e. capability to fix the velocity profile provided by the model) influences the behaviour.

7.Experimental trial

The experiments and the setting of the system gain have been done to imitate the data reported in the Goossens and Van Opstal work (Goossens and Van Opstal, 1997).
The experiments were performed by positioning a visual stimulus (i.e. a green ball) in the robot's field of view and by recording (from the encoders) the actual positions of the involved joints. In the starting position, the eyes and the head were aligned. The position of the stimulus was then varied in order to span the entire visual field of the robot. In order to test the effectiveness of the model on a wide spectrum of movements the amplitude of the gaze shifts required to foveate the target was varied from 5 deg up to 35 deg for horizontal saccades. Several different trials were performed for each tested amplitude.
The model has been tested (1) to investigate the robot performance and compare them to physiological results and, (2) to test the capability of the model to control a redundant robotic platform using a neurophysiological model. Fig. 10 shows a typical example of head-saccadic movement.

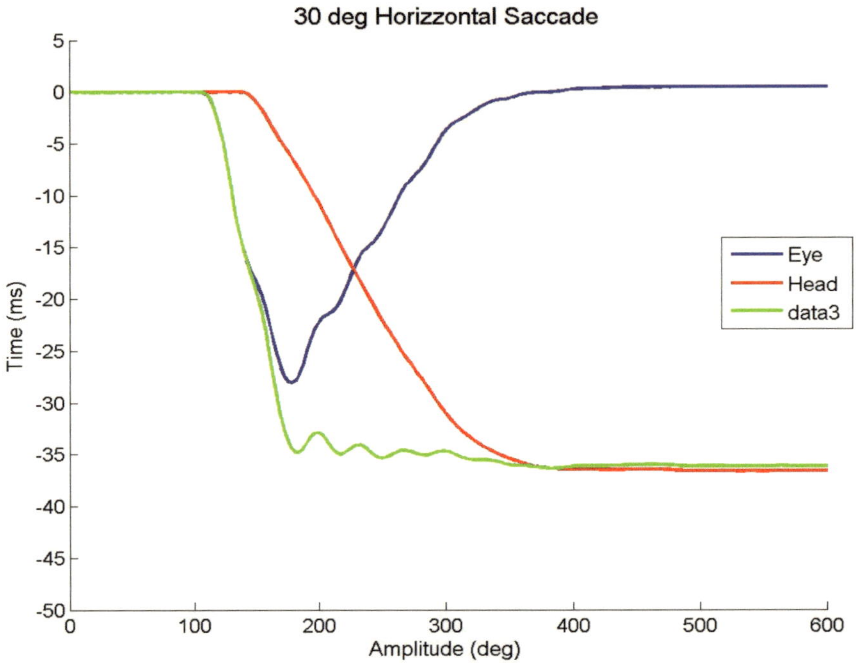

Fig. 14. Example of robotic head execution of 30 deg head-saccadic movement.

In the reported data, the zero time point is calculate from the onset of the **Pe** delay block.
To test the system performance, the following outcome measures were calculated:

Timing measures:

Gaze:

- **TGs10**: time to achieve 10% of final **gaze** displacement;
- *duration* **TGD**: time to achieve 90% of final **gaze** displacement;
- *delay time* **TGri**: time to achieve 50% of final displacement;
- *settling time* **TGst**: time to reduce the gaze oscillation below the 5% of the final displacement;
- **Tg** *gaze duration*: time between the gaze movement onset, **TGs10**, and the target achievement **TGs90**.

Head:

- **THs10**: time to achieve 10% of final **head** displacement;
- *duration* **THD**: time to achieve 90% of final **head** displacement;
- *delay time* **THri**: time to achieve 50% of final displacement;
- *settling time* **THst**: time to reduce the head oscillation below the 5% of the final displacement;
- **Th** *head duration*: time between the gaze movement onset, **THs10**, and the target achievement **THs90**.

Eye:

- **TED**: time to change the eye velocity sign (eye duration).

Velocity peak measures:

- **PVG**: gaze peak velocity;
- **PVH**: head peak velocity;
- **PVE**: eye peak velocity.

8.Results

The obtained results are shown for horizontals movements in table 4.

Fig. 15 shows the main sequence relation for the gaze, eye and head movements.
The main results are shown in Figure 16, 17.
Reported data are intended to investigate to different behaviors:

(1) accordance with the data provided from Goossens and Van Opstal study (Goossens and Van Opstal, 1997);

(2) capability to control the redundant robotic platform.

The first point is discussed in reference to the available data of eye and head coordination as reported by Goossens and Van Opstal study (Goossens and Van Opstal, 1997).
The second point has been tested, using traditional methods: timing response, residual oscillation for eye, head and gaze.

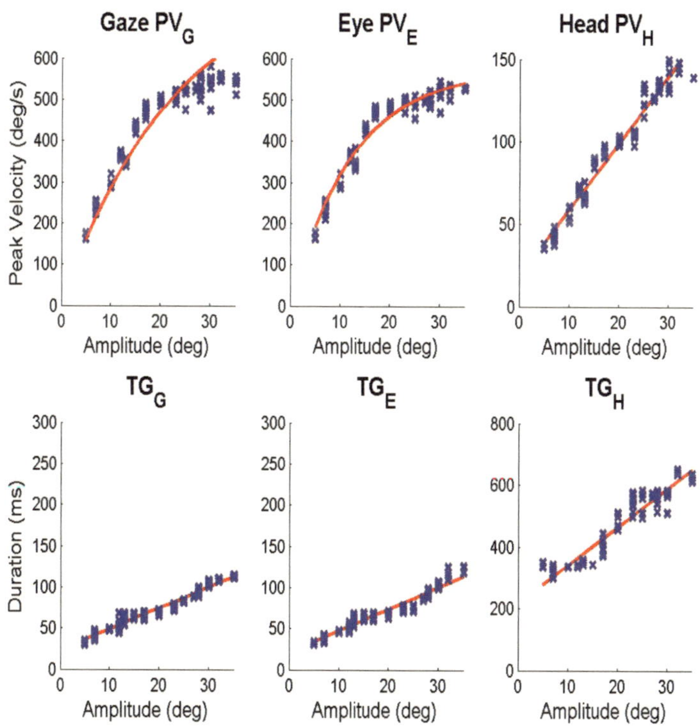

Fig. 15. Saccadic kinematics. Duration and peak velocity as a function of amplitude for gaze, eye and head. **PVG, PVE, PVH** are the peak velocity of the gaze, eye and head. **TGG, TGE,**

TGH are the movements duration of the gaze, eye and head. Goossens and Van Opstal obtained close results (Goossens and Van Opstal, 1997).

	Duration	Peck Velocity
Gaze:	$TG_G = 22.92 + 2.56 \cdot A$	$PV_G = 610 \cdot (1 - e^{-(\frac{A}{13.32})})$
Eye:	$TG_E = 20.06 + 2.67 \cdot A$	$PV_E = 573 \cdot (1 - e^{-(\frac{A}{12.25})})$
Head:	$TG_H = 216 + 12.34 \cdot A$	$PV_H = 18.04 + 4.05 \cdot A$

Table 3. Fitting equations of linear and exponential model for peak velocity and duration related to displacement. The eye and gaze data are fitted with exponential equation because the maximal velocity has been achieved.

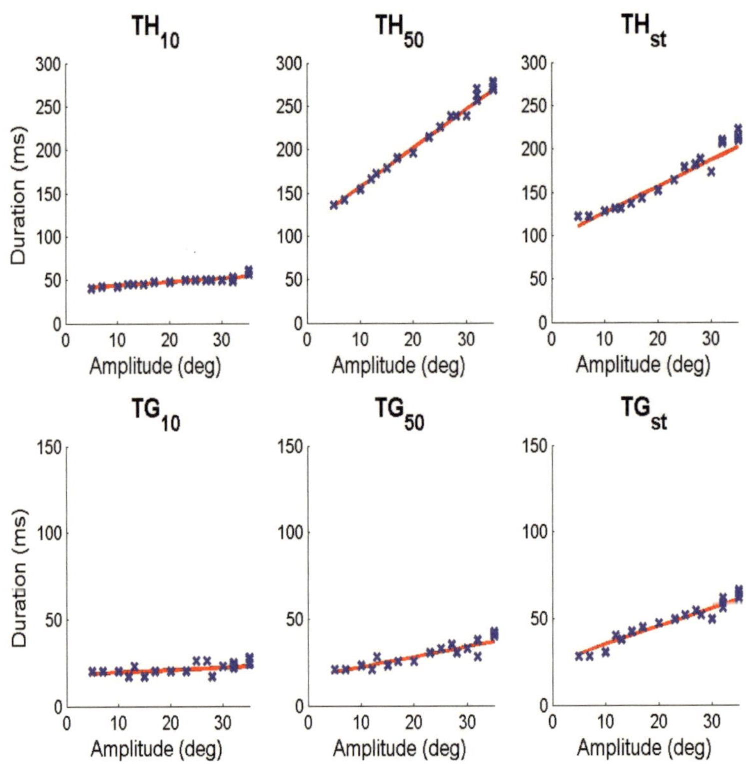

Fig. 16. Saccadic timing. Duration of **TH10**, **TH50**, **THST**, **TG10**, **TG50** and **TGST** as a function of amplitude for head and gaze.

	T10	T50	TST
Gaze:	$TG_{10} = 18.1 + 0.13 \cdot A$	$TG_{50} = 16.6 + 0.58 \cdot A$	$TG_{ST} = 24.2 + 1.05 \cdot A$
Head:	$TH_{10} = 39.5 + 0.4 \cdot A$	$TH_{50} = 111 + 4.46 \cdot A$	$TH_{ST} = 95 + 3.04 \cdot A$

Table 4. Fitting equations of linear and exponential model for peak velocity and duration related in related to displacement.

9. Discussion

This work present a robotic implementation of a neurophysiological model of rapid orienting gaze shifts in humans, with the final goal of model parameters validation and tuning. The quantitative assessment of robot performance confirmed a good ability to foveate the target with low residual errors around the desired target position. Furthermore, the ability to maintain the desired position was good and the gaze fixation after the saccadic movement was executed with only few oscillations of the head and eye. This is because the model required a very high dynamic.

9.1. Robotic point of view

The head and eye residual oscillations increase linearly with increasing amplitude. In Fig. 16 is evident that the residual gaze oscillation is less than head. This is explained with the compensation introduced by the eye oscillations which compensate the gaze which becomes more stable.

We explain these findings by observing that the accelerations required to execute (or stop-and-invert) the movement are very high especially for the eye movement. Even if the robotic head was designed to match the human performances (in terms of angle and velocities) in its present configuration it is still not capable produce such accelerations. This is particularly evident for the movement of the eye because the motor has to invert its rotation when the fixation point is first achieved.

With respect to the timing of the movement it has been found that the results of the experiments are in close accordance to the data available on humans (Goossens and Van Opstal, 1997). The same conclusion may be drawn for the shapes of the coordinated movement that can be directly compared to the typical examples reported in Fig. 14.

Figure 16, 17 show that the model is capable of providing inadequate control of the redundant platform.

The system response is very fast, due to the robotic head platform design.

TGst time take into account the problem of eye-head coordination and the very high acceleration.

The head is voluntarily delayed less than 30 millisecond after eye movement, according to human physiology, by means of **Ph** block (Goossens and Van Opstal ,1997).

9.2. Neurophysiological point of view

A typical robotic eye-head movement is shows in Fig. 14.

Very similar kinematic profile can be observed in neurophysiological studies on human (CITE CON PROFILI). The model drives the gaze by means of the global error displacement. Therefore, if an external disturbance has been applied, the oculomotor system reacts by compensating the gaze error.
The VOR plays an important role for the architecture behaviour. Therefore, a VOR block has been created to build different behaviour inside.
The obtained timing is in good accordance with human physiology reported in Goossens.
In the presented model, different eye dynamics may be obtained by modification of **PGe** block.
If the head is disabled by shutting off the **PGh** block, can be also possible to implement an eye-saccadic movement within the head-restrain paradigm.
In this case the model is in good accordance with Robinsons' local feedback model (Robinson, 1975)
Concerning the relation within peak velocity and gaze amplitude, it may be noted the close similarity between gaze and eye. This is explained with the preponderant role of the eye.
Conversely, the head velocity is linear because of the slow response as compared with the eye peak velocity.
As for the gaze duration, eye, head and gaze show a linear relation with gaze amplitude.
The different duration are explaining with the different dynamic.
The equation (7) shows the relation between peak velocity and displacement both for the gaze and eye. This is in exponential form in which V_p is the peak velocity, A the movement amplitude, and the constant to be fitted are V_M (the maximal eye velocity) and A_0 (that define how the system become saturated).

$$V_p = V_M \cdot (1 - e^{-(\frac{A}{A_0})}) \tag{7}$$

In the first part, below 400 deg/sec, the relation is almost linear, but, for increasing velocity it tends to be saturated for the achieved of limit velocity. The gaze is quite similar to the eye because, in this range, the eye plays the main role in the movements. Conversely for the head, the peak velocity is linear, because the different inertial properties of the head.

10. Conclusion

In our vision these results encourage the accomplishment of focused experimental trials in which the robot behaviour can help investigate the accuracy of the model (e.g. by varying the loop-gains) - and to possibly revise it - with respect to human behaviour in head-saccade execution.

This work may be considered the first preliminary attempt to reproduce the neurophysiologic model of human gaze control on an actual robotic artifact. However the obtained results are in close agreement with reported results on human's experiments both in terms of peak velocity and gaze duration (Goossens and Van Opstal, 1997). Moreover,

from a robotics point of view, this model has proved to perform an adequate control of a redundant robotic platform.
Nevertheless we are aware that future research activities should be devoted to the possible improvement and amelioration of the proposed model. Specifically, we envisage two strategic lines of research. On one hand, supplementary research activities should be dedicated to a better understanding on how the VOR interacts with the saccadic movement. Indeed, as already remarked, the VOR is known to be weakened or shut off in the direction of the saccade, but it is not clear if it is shut-off in the direction of the gaze (eye-in-space) or head-in-space or eye-in-head motor error.
On the other hand, we remark that the acquisition of coordinated motion of the eye and the head is a developmental issue. For instance, in newborns gaze shifts towards auditory of visual stimuli are absent or improperly accomplished until the development of appropriate strategies which occurs at about 6-9 months of age. This observation leads to the conclusion that -in humans- the model internal parameters should be learned by execution. Accordingly, a possible improvement of the model may include the introduction of adequate artificial neural network for learning the internal model parameters. This may be accomplished -for instance- by minimizing the error existing between actual and target kinematics until convergence of the network.

11. References

Bahill, A.; Clark, M. & Stark, L. (1975). The main sequence, a tool for studying human eye movements, *Math Biosciences*, Vol. 24, (1975), pp. 191-204.

Baloh, R. (1975). Quantitative measurement of saccade amplitude, duration, and velocity, *Neurology*, Vol. 25, No. 11, (1975), pp. 1065-1070.

Baloh, R.; Konrad, H.; Sills, A. & Honrubia, V. (1975). The saccade velocity test, *Journal Neurology*, Vol. 25, (1975), pp. 1071-1076.

Barnes, G. (1979). Vestibulo-ocular function during co-ordinated head and eye movements to acquire visual targets, *Journal Physiol*, Vol. 287, (1979), pp. 127-147.

Becker, W. & Reinhart, J. (1990). Human oblique saccades: Quantitative analysis of the relation between horizontal and vertical components, *Vision Research*, Vol. 30, No. 6, (1990), pp. 893-920.

Bell, A. H.; Meredith, M. A.; Opstal, A. J. V. & Munoz, D. P. (2006). Stimulus intensity modifies saccadic reaction time and visual response latency in the superior colliculus, *Experimental Brain Research*, Vol. 174, No. 1, (2006), pp. 53-59.

Bernstein, N. (1967), *The coordination and regulation of movements*, pp. 196, Oxford, UK, Pergamo, (1967)

Berthouze, L. & Kuniyoshi, Y. (1998). Emergence and Categorization of Coordinated Visual Behavior Through Embodied Interaction, *Autonomous Robots*, Vol. 5, No. 3, (1998), pp. 369-379.

Bizzi, E.; Kalil, R. & Morasso, P. (1972). Two modes of active eye-head coordination in monkeys, *Brain Research*, Vol. 40, (1972), pp. 45-48.

Bizzi, E.; Kalil, R. & Tagliasco, V. (1971). Eye-head coordination in monkeys, No. evidence for centrally patterned organization, *Science*, Vol. 173, (1971), pp. 452-454.

Brooks, R.; Breazeal, C.; Marjanovic, M.; Scassellati, B. & Williamson, M. (1999). The Cog Project: Building a Humanoid Robot, In: *Lecture Notes in Computer Science*, pp. 52-87,

Springer-Verlag.
Bruske, J.; Hansen, M.; Riehn, L. & Sommer, G. (1997). Biologically inspired calibration-free adaptive saccade control of a binocular camera-head, *Biological Cybernetics*, Vol. 77, No. 6, (1997), pp. 433-446.
Collewijn, H.; Erkelens, C. & Steinman, R. (1988). Binocular coordination of human horizontal saccadic eye movements, *The Journal of Physiology*, Vol. 404, No. 1, (1988), pp. 157-182.
Crawford, J. D.; Douglas, Tweed, B. & Vilis, T. (2003). Static ocular counterroll is implemented through the 3-D neural integrator, *Journal of Neurophysiol*, Vol. 90, (2003), pp. 2777-2784.
Freedman, E. & Sparks, D. (2000). Coordination of the eyes and head, No. movement kinematics, *Experimental Brain Research*, Vol. 131, No. 1, (2000), pp. 22-32.
Freedman, E. & Sparks, D. (1997). Eye-head coordination during head-unrestrained gaze shifts in rhesus monkeys, *Journal Neurophysiology*, Vol. 77, (1997), pp. 2328-2348.
Galiana, H. & Guitton, D. (1992). Central organization and modeling of eye-head coordination during orienting gaze shifts., *Annals of the New York Academy of Sciences*, Vol. 656, (1992), pp. 452-471.
Girard, B. & Berthoz, A. (2005). From brainstem to cortex, No. computational models of the saccade generation circuitry, *Progress in Neurobiology*, Vol. 77, No. 4, (2005), pp. 215-255.
Gisbergen, J. V.; Opstal, A. V. & Tax, A. (1985). Experimental test of two models for the generation of oblique saccades, *Experimental Brain Research*, Vol. 57, No. 2, (1985), pp. 321-336.
Gisbergen, J. V.; Opstal, J. V. & Ottes, F. (1984). Parametrization of saccadic velocity profiles in man, *in Theoretical and applied aspects of eye movement research*, Gale, A. G., Johnson, J., (Ed.) Elsevier Science, pp. 87-94, New York
Goossens, H. & Opstal, A. V. (1997). Human eye-head coordination in two dimensions under different sensorimotor conditions, *Experimental Brain Research*, Vol. 114, (1997), pp. 542-560.
Green, D. G. (1970). Regional variations in the visual acuity for interference fringes on the retina. *The Journal of Physiology*, Vol. 207, No. 2, (April 1970) 351–356.
Grossberg, S. & Kuperstein, M. (1986). *Neural dynamics of adaptive sensory motor control: ballistic eye movements.*, Elsevier, Amsterdam.
Guitton, D. & Volle, M. (1987). Gaze control in humans, No. eye-head coordination during orienting movements to targets within and beyond the oculomotor range., *Journal of Neurophysiol* , Vol. 58, (1987), pp. 496-508.
Henn, V. (1993). Neural Control of Saccadic Eye Movements, In: *Multisensory control of movement,* Berthoz, A., (Ed.), Oxford University Press, pp. 7-26, Oxford, UK
Kaneko, K.; Kanehiro, F.; Kajita, S.; Hirukawa, H.; Kawasaki, T.; Hirata, M.; Akachi, K. & Isozumi, T. (2004). Humanoid robot HRP-2, *Proceedings of Robotics and Automation,* pp. 1083-1090, ISBN: 0-7803-8232-3 , New Orleans, April 2005.
Kawato, M.; Furukawa, K. & Suzuki, R. (1987). A hierarchical neural-network model for control and learning of voluntary movement, *Biological Cybernetics*, Vol. 57, No. 3, (1987), pp. 169-185.
Laschi, C.; Patane, F.; Maini, E.; Manfredi, L.; Teti, G.; Zollo, L.; Guglielmelli, E. & Dario, P. (2008). An Anthropomorphic Robotic Head for Investigating Gaze Control,

Advanced Robotics, Vol. 22, No. 33, (2008), pp. 57-89.
Laurutis, V. P. & Robinson, D. A. (1986). The vestibulo-ocular reflex during human saccadic eye movements., *The Journal of Physiology*, Vol. 373, (1986), pp. 209-233.
Lefèvre, P.; Bottemanne, I. & Roucoux, A. (1992). Experimental study and modeling of vestibulo-ocular reflex modulation during large shifts of gaze in humans, *Experimental Brain Research*, Vol. 91, No. 3, (1992), pp. 496-508.
Manfredi, L.; Maini, E. S.; Dario, P.; Laschi, C.; Girard, B.; Tabareau, N. & Berthoz, A. (2006). Implementation of a neurophysiological model of saccadic eye movements on an anthropomorphic robotic head, *International Conference on Humanoid Robots, 2006 6th IEEE-RAS*, pp. 438-443, 1-4244-0200-X, Genova, December 2006.
Maini, E. S.; Manfredi, L.; Laschi, C.; Dario, P. (2008). Bioinspired velocity control of fast gaze shifts on a robotic anthropomorphic head, *Autonomous Robots*, Vol. 25, (2008), pp. 37-58.
Ottes, F.; Van Gisbergen J.A. & Eggermont, J. (1986). Visuomotor fields of the superior colliculus, No. a quantitative model, *Vision Research*, Vol. 26, (1986), pp. 857-873.
Pelisson, D.; Prablanc, C. & Urquizar, C. (1988). Vestibuloocular reflex inhibition and gaze saccade control characteristics during eye-head orientation in humans, *Journal of Neurophysiology*, Vol. 59, No. 3, (1988), pp. 997-1013.
Radau, P.; Tweed, D. & Vilis, T. (1994). Three-dimensional eye, head, and chest orientations after large gaze shifts and the underlying neural strategies, *Journal of Neurophysiology*, Vol. 72, (1994), pp. 2840-2852.
Robinson, D. (1972). Eye movements evoked by collicular stimulation in the alert monkey, *Vision Research*, Vol. 12, (1972), pp. 1795-1808.
Robinson, D. A. (1975). Oculomotor control signals, In: *Basic mechanism of ocular motility and their clinical implications*, Lennerstrand, G. D. & and Rita, P. B. (Ed.), pp. 337-374, Pergamon, Oxfork, UK, Pergamon, Oxford
Sparks, D. L. (2002). The brainstem control of saccadic eye movements, *Nature Reviews Neuroscience*, Vol. 3, (2002), pp. 952-964.
Takanishi, A.; Takanobu, H.; Kato, I. & Umetsu, T. (1999). *Development of the Anthropomorphic Head-Eye Robot WE-3RII with an Autonomous Facial Expression Mechanism,*. Proceedings of International Conference on Robotics and Automation, pp. 3255–3260, 0-7803-5180-0, Detroit, October 1999
Tweed, D.; Glenn, B. & Vilis, T. (1995). Eye-head coordination during large gaze shifts, *Journal Neurophysiol*, Vol. 73, (1995), pp. 766-779.
Westheimer, G. & Blair, S. (1973). Oculomotor defects in cerebellectomized monkeys., *Investigative Ophthalmology*, Vol. 12, No. 8, (1973), pp. 618-621.
Wolpert, D. M. & Kawato, M. (1998). Multiple paired forward and inverse models for motor control, *Neural Networks*, Vol. 11, No. 7-8, (1998), pp. 1317-1329.
Zangemeister, W.; Lehman, S. & Stark, L. (1981). Simulation of head movement trajectories: model and fit to main sequence, *Biological Cybernetics*, Vol. 41, No. 1, (1981), pp. 19-32.
Zangemeister, W. H. & Stark, L. (1981). Active head rotations and eye-head coordination, *Annals of the New York Academy of Sciences*, Vol. 374, No. 1, (1981), pp. 540-559.
Zatsiorsky, M. (1998). *Kinematics of Human Motion*, Human Kinetics Europe, United Kingdom.

Zee, D.; Optican, L.; Cook, J.; Robinson, D. & Engel, W. (1976). Slow saccades in spinocerebellar degeneration., *Archives of Neurology*, Vol. 33, No. 4, (1976), pp. 243-251.

11

Dynamic decision making for humanoid robots based on a modular task structure

Giulio Milighetti
Fraunhofer Institute IITB
Germany

1. Introduction

Within the last years a new challenging field of robotics has arisen in which assisting robots cooperate and interact with humans, both imitating their behaviour patterns and performing with them and/or in their environment diverse complex tasks.
Due to the ambitious tasks and the complexity of a time varying environment, these "humanoid" robots have to be provided with much more intelligence in comparison with the current industrial robots available on the market.
Thus, the robot has to dispose on the one hand of a large variety of dedicated sensors which cover the full range of human perception modalities and which collect information about the state of the robot and of the environment. On the other hand the control concept has to provide the capability of extracting, combining and analyzing this information in order to cope autonomously with these extreme conditions in the most appropriate way. Moreover, in order to be able to face a manifold of different possible situations, the robot has to manage specific human-like skills. That means it has to perceive, act, deliberate and communicate in a robust manner.

In order to develop advanced cognitive systems, various architectures have been proposed and validated in different application fields (Vernon et al., 2007). However, the implementation of such architectures in the robotics results often in a three-layer structure with a planning, an executive and a functional (control) level (Gat, 1997).
In spite of the many proposed approaches the communication problem between executive and planning is still an unsolved one because of the two different abstraction levels (Joyeux et al., 2007). In order to fill in this gap, some solutions have been developed merging executive and planning into a hybrid level. Following this kind of approach architectures like IDEA (Finzi et al., 2004) and CLARAty (Estlin et al., 2001) have been designed.

Within the Collaborative Research Center 588 "Humanoid Robots – Learning and Cooperating Multimodal Robots" (SFB588) a three-layer cognitive architecture has been presented for coping with the complexity of the challenging tasks that a humanoid robot is supposed to achieve (Burghart et al., 2005). Based on this reference a two-layer discrete-continuous control located in the two lowest levels of this architecture has been developed at Fraunhofer IITB in order to enable the robot to achieve successfully its goal also in a complex dynamical environment. Only by reconfiguring its plan (discrete control) and by adapting its control strategy (continuous control) the robot can react autonomously to external disturbances and to unexpected events occurred during the execution of its plan.

For choosing the optimal strategy to solve a task, different decision making algorithms driven by the actual situation have been proposed in the past, mainly in the form of reactive planning in mobile robotics (Finzi et al., 2004; Estlin et al., 2001). Most of them rely on the use of Neural Nets, Fuzzy methods or Bayes approaches (Gao & Zhou, 2005; Lundell et al., 2005; Schrempf & Hanebeck, 2005; Serrano, 2006).

The description of discrete-continuous hybrid systems makes often use of discrete models based on Petri nets or automata (Nenninger & Krebs, 1997; Antsaklis et al., 1998). Although the widening of these tools can be led back to applications in the production scheduling, Petri-nets are also used more and more in the field of robotics for modeling and planning complex tasks (Cao & Sanderson, 1994; Chang et al., 2004) as well as for their coordination and supervision (Asfour et al., 2004; Lehmann et al., 2006). Many applications can be found for example both in the area of unmanned vehicles (Palomeras et al.,2006) as well as of humanoid robots (Kobayashi et al., 2002).

The discrete-continuous control concept developed at Fraunhofer Institute IITB (Milighetti & Kuntze, 2006) utilizes a Fuzzy decision making algorithm supported by a modular task sequence based on Primitive Skills (PS) and modelled by Petri nets which considerably improves the autonomy and the flexibility of the robot. The efficiency of the proposed concept is demonstrated in a first experiment on a grasping task in which different vision and acoustic sensors can be deployed.

2. Control Architecture

The quality of a plan depends strongly on the accuracy of the available information about the environment. In a complex time-varying environment where a large variety of events can not be foreseen (e. g. dynamic obstacles) an off-lined scheduled sequence of actions leads often to a failure in the execution of the task. Therefore a discrete-continuous control has to assure that even in the presence of unpredicted conditions the plan is dynamically adapted in order to enable the robot to achieve the goal.

The basic structure of the discrete-continuous control concept developed for the supervision of the robot throughout its task is shown in Fig. 1. The state of the robot and of the environment in which it acts are captured with the help of proprioceptive (e.g. encoder) and exteroceptive sensors (e.g. camera, microphone, force-torque). A hierarchical two-level control is then responsible for the interpretation of such an information.

In the upper level, a discrete control processes the measurements coming from the sensors and uses them in order to generate diagnosis signals that contain quantitative information about the continuous state of the system (e.g. position of objects, sounds, forces). As a second step, this information serves for the identification of the discrete state or event

providing the qualitative information about the present situation (e.g. a contact has been established). By interpreting the acquired knowledge about both the continuous and discrete state of the system, a decision unit analyses the actual task sequence and if necessary adapts it or introduces new appropriate actions for overcoming unexpected situations. A fast online task adaptation and a transparent and efficient structure on Which supports an easy implementation of an online decision making algorithm, have been obtained by adopting a modular architecture based on the concept of Primitive Skills.
Every robot task can be divided into a discrete sequence of elementary actions (called Primitive Skills - PS), each with its own control strategy. A simple assembling process can for example be described by the following five PS:

PS1. Localisation of the part by means of an optical sensor (e.g. a stereo camera);
PS2. Approach to the object with a time optimal and collision-free trajectory;
PS3. Grasping the part by means of tactile and slip sensors;
PS4. Motion to the assembling point and establishment of the contact;
PS5. Assembling using force-torque sensors.

Once the decision unit has determined the PS sequence that has to be performed and the most appropriate controller to execute the currently active PS, in the lower hierarchy level the continuous control ensures an optimal system response and the attainment of the desired values by adapting the parameters of the employed control law.

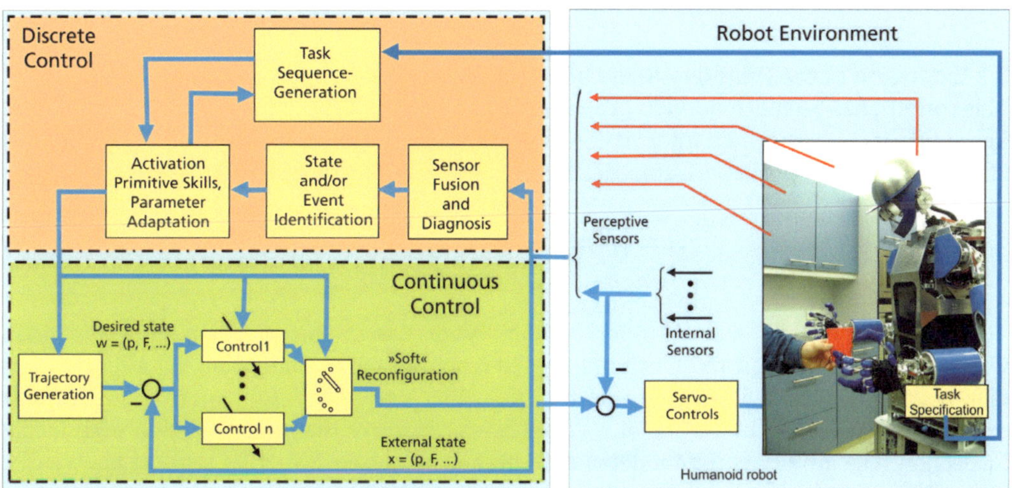

Fig. 1. Discrete-continuous control architecture

3. Task structure based on Primitive Skills and Petri Nets

3.1 Primitive Skills basic concept

A modular structure of a robot task offers numerous advantages. First of all a transparent action sequence is easy to understand and to be handled by the user. Moreover it offers a flexible plan which can be adapted online simply introducing new actions in the sequence and removing or adapting the existing ones. In addition, the presence of software-modules reduces the effort of the programmer since it maximizes the reusability of the code. Finally such a structure emphasises the discrete nature of the system, offering an optimal basis for the communication with a symbolic planner and supporting also the integration of learned processes, since many learning approaches are based on the decomposition of human actions into sequences of elementary actions (Pardowitz et al., 2007; Peters et al., 2007).

Following this modular approach different concepts have been proposed especially in the area of the manipulation (Hasegawa et al., 1992; McCarragher, 1996; Morrow & Khosla, 1997). In the present contribution an elementary action (Primitive Skill) has been defined starting from the tuple (1) introduced in (Thomas et al., 2003).

$$PS = \{HM, \tau, \lambda\} \tag{1}$$

The variable HM (Hybrid Motion) contains the information needed for the trajectory generation and for the control of the motion (e.g. desired values, active control-law and used parameters, control frame,...) in order to reach the PS-goal. The resources involved in the execution of the PS (e.g. sensors, tools, objects,...) and the associated commands (e.g. activate, open,...) are stored in the variable τ. The third variable λ represents the leave condition of the PS, which has to be fulfilled in order to consider the PS completed and to execute the next one (position reached, malfunctioning,...). More theoretical details about these three variables can be found in (Thomas et al., 2003).

3.2 Extension of the Primitive Skills concept

This concept has been extended in (Milighetti & Kuntze, 2006) by adding two more variables in order to include in the PS information about its actual capabilities and about the influence of the environment on its performance. This extension leads to a PS defined as

$$PS = \{HM, \tau, \lambda, E, a\} \tag{2}$$

The affinity a evaluates the a-priori suitability of the PS for achieving a given goal. Comparing two different PS which are able to reach the same final state, the affinity value tells how optimal is a PS with respect to the other one (e.g. faster, more accurate). For example a localization by means of a visual sensor is faster than a localization with tactile inspection. The affinity is time independent because it is based only on a-priori knowledge about the performance of each PS without taking into account the actual situation.
If PS_{opt} is the a-priori optimal choice in order to achieve a goal g (that means an affinity $a_{opt} = 1$), the affinity a_i of another PS_i compared with it can be expressed for example by:

$$a_i = \begin{cases} 0 & if\ g \notin \{goalset_i\} \\ 1 - k \cdot \dfrac{|p_i - p_{opt}|}{\max(p_i, p_{opt})} & if\ g \in \{goalset_i\} \end{cases} \tag{3}$$

with p_{opt} the optimal value of the comparison parameter (time, accuracy,...) and

$$0 < k < \max\left(p_i, p_{opt}\right) \tag{4}$$

The efficiency E on the contrary measures the current suitability of the PS by analyzing how good every resource needed for its execution is functioning with respect to the actual state of the system.
The efficiency $E_k(t)$ of a resource k can be decomposed into the following two components:

$$E_k(t) = av_k(t)\cdot q_k(t), \quad av \in \{0,1\}, \quad q \in [0,1] \tag{5}$$

where av_k is the availability of the k-th resource and q_k its quality. While the availability can assume only the values 0 (the resource is available) and 1 (the resource is not available), the quality can be estimated using the whole interval [0,1] with 1 indicating an optimal functional capability and 0 a not functioning resource.
In order to obtain the total efficiency of the PS, the efficiencies of the n involved resources are combined as follows:

$$E(t) = m\cdot\prod_{i=1}^{n_{comp}} \max_{j=1\ldots n_{coop}}\left(E_{ji}(t)\right) \tag{6}$$

The single efficiencies calculated by (6) can be adjusted in a n_{coop} x n_{comp} matrix, where n_{comp} is the maximum number of complementary resources and n_{coop} the number of the cooperative ones. With such a structure the efficiencies of the resources, which are working in cooperation (they provide the same kind of measurement) can be listed in each column (see equation (7)). Thus, according to equation (6), the total efficiency of the PS can be evaluated by multiplying the maxima of each column.

$$\uparrow_{coop}\begin{pmatrix} E_{11} & \cdots & E_{1j} & \cdots & E_{1n_{comp}} \\ \vdots & & \vdots & & 0 \\ E_{i1} & & \vdots & & \vdots \\ 0 & & \vdots & & \vdots \\ \vdots & & \vdots & & \vdots \\ 0 & \cdots & E_{n_{comp}j} & \cdots & 0 \end{pmatrix}_{\rightarrow comp} \tag{7}$$

In the PS4 of the example introduced in paragraph 2 (''Motion to the assembling point'') the assembling point can be identified by means of several fixed cameras while the establishment of the contact with the surface can be supervised with the help of a force-torque-sensor. In this case the efficiency of the PS is evaluated taking the maximum of the camera efficiencies and multiplying it with the efficiency of the force-torque-sensor.

A memory factor *m* has been introduced in (6) in order to consider a learning capability of the robot with respect to previous execution of the same PS. A break of the action before the goal has been achieved results in a decrement of *m*, a successful termination in its increment.

Since the efficiency measures the present performance capability of a PS, its value is actualized online during the whole robot task. In this way the discrete control has access at any times to the information that describes in a compact form the influence of the state of the system on the actions of the robot.

3.3 Task modeling by means of Petri Nets

In order to achieve the desired goal, the robot can plan the execution of different complex actions. The division of each action into a sequence of PS can result either from the segmentation used for learning the action (Pardowitz et al., 2007) or from automatic approaches (Bagchi et al., 2000). Thus, the complete robot task results in a chain of PS, which can be intuitively modelled by means of a Petri net associating every PS to a place. Following this approach, the exemplary assembly task presented in paragraph 2 can be described by the net of Fig. 2.
The marked place represents the actual discrete state of the system, that is the currently performed PS. A transition is activated once the leave condition λ of the executed PS has been fulfilled. By firing the transition the next PS in the net will be activated.

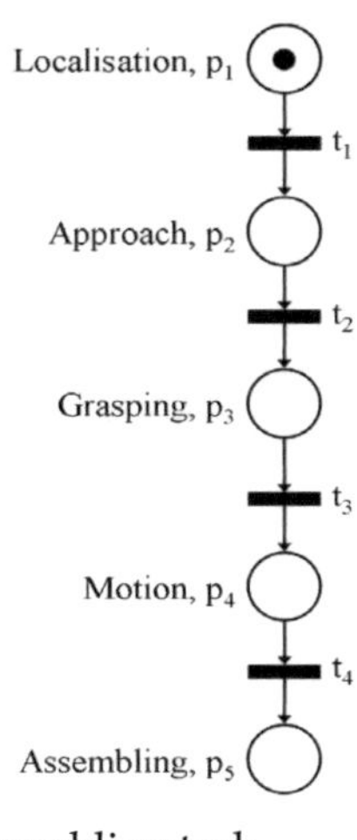

Fig. 2. Petri net modelling an assembling task

Of course the resulting net is not as simple as the one in Fig. 2 for every planned task. In the majority of the cases the robot has to face nets with different alternatives, which could result for example from the following situations:

- parallel tasks, that the robot has to perform;
- processes, that can be activated at any time during the execution of the main task;

- different available strategies for solving the given task or some of its sub-actions.

In order to resolve this kind of conflicts in the net (more than one transition firable at the same time) a decision making algorithm at higher level has to be implemented.

4. Decision Making Algorithm

4.1 Decision making algorithm

The decision making algorithm can be seen as the discrete control law of the robotic system. In fact, every time step the actual discrete state is compared with the desired optimal one and corrected in the case that it is not optimal with respect to the actual situation.

The optimal discrete state is the one which ensures the best performances, that is the PS with the highest efficiency (the PS with the currently best functioning resources) and the highest affinity (the a-priori optimal choice). The decision solving the conflict in the net can thus be made by taking the PS with the highest value of the utility function given by the product $E \cdot a$.

$$\max(E) \wedge \max(a) \Rightarrow \max(E \cdot a) \tag{8}$$

At every time step the efficiency of every PS is updated depending on the actual information about its resources and then used to make the optimal decision.

With this approach a decentralized decision making structure is obtained which relies on the measurements attached to every PS independently from its position in the net and thus unrelated to a particular net configuration or conflict (see Fig. 3). In this way the increase in complexity of the decision algorithm is negligible when the number of possible choices rises. Moreover, having most of the intelligence needed for the decision stored locally in every single PS results in an algorithm which works automatically also in case of self-generated task-nets.

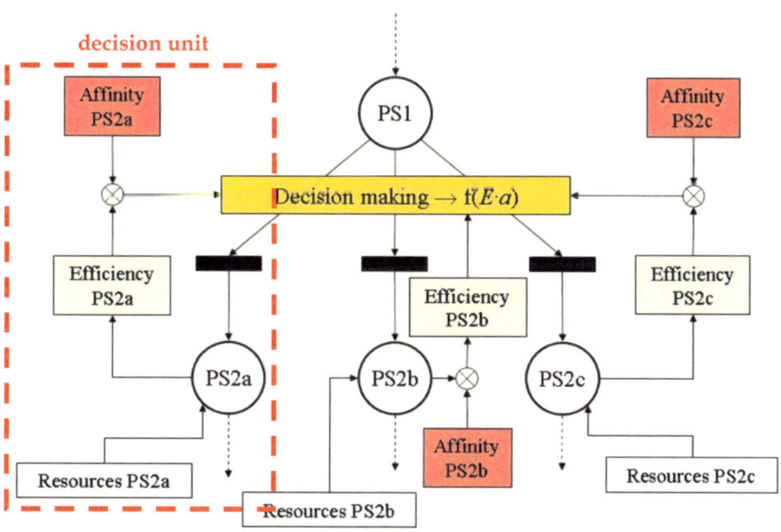

Fig. 3. Decentralized PS-based decision making structure

Comparing by (5) the different PS available in the next execution step, a local optimization problem can be solved finding the optimal action which has to be performed in the discrete task sequence. However, the time horizon of the decision can be extended considering the global net that describes the entire task and finding the optimal path from the currently active PS to the goal.
In order to do this, the arcs entering the k-th PS can be dynamically weighted with $1\text{-}E_k \cdot a_k$ obtaining a net where an arc with minimal weight corresponds to a PS with maximum utility ($E \cdot a=1$).
By using for example a slightly modified version of the Dijkstra algorithm a global optimal path can be evaluated every time step and used to extract the next PS avoiding in this way a deadlock in the task execution that could result by taking an optimal but local decision (see Fig. 4).

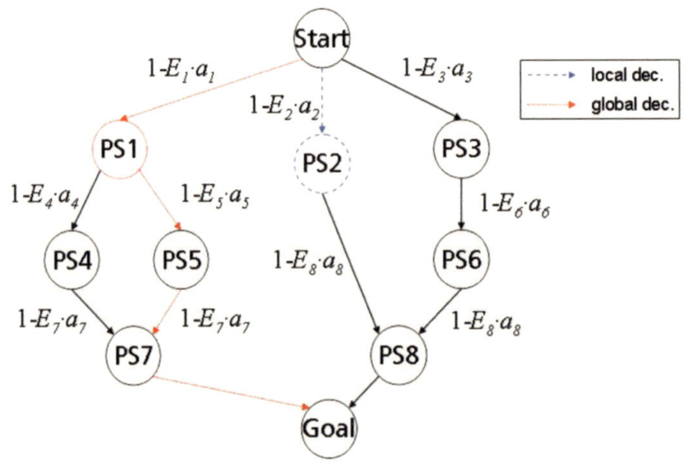

Fig. 4. Global vs. local decision making

4.2 Fuzzy-based efficiency evaluation

Equation (5) has shown that the value of each single efficiency is given by two different parameters:

- the availability *av* of the resource;
- the quality *q* of the resource.

Even if the estimation of the quality can be performed using any arbitrary method that returns a value in the interval [0,1], a fuzzy approach has been chosen. Thanks to this kind of approach, it is easier to transfer the human experience into the system, obtaining a more transparent and more comprehensible decision unit.

The fuzzy-based method for the quality evaluation can be clarified by taking as an example the PS1 introduced in paragraph 2, that is the localisation of an object by means of a stereo camera. In this case the two main resources involved are the camera and the object. In order to simplify the example it is supposed that the efficiency of the object is always constant and equal to one. Thus, the efficiency of the PS can be reduced to the efficiency of the camera only.

The evaluation of the efficiency can be carried out on the basis of three main factors:

- availability of the communication between sensor and robot (1 = available, 0 = not available);
- availability of an actual measurement (1 = received, 0 = no new measurement in the last *n* time steps);
- quality of the last measurement (1 = good and reliable, 0 = bad and/or unreliable).

The quality of a measurement is evaluated by taking into account three more factors that mostly influence a camera:

- luminosity of the environment;
- noise of the measurement;
- working range of the sensor.

The membership functions associated with each of these three factors are shown in Fig. 5.

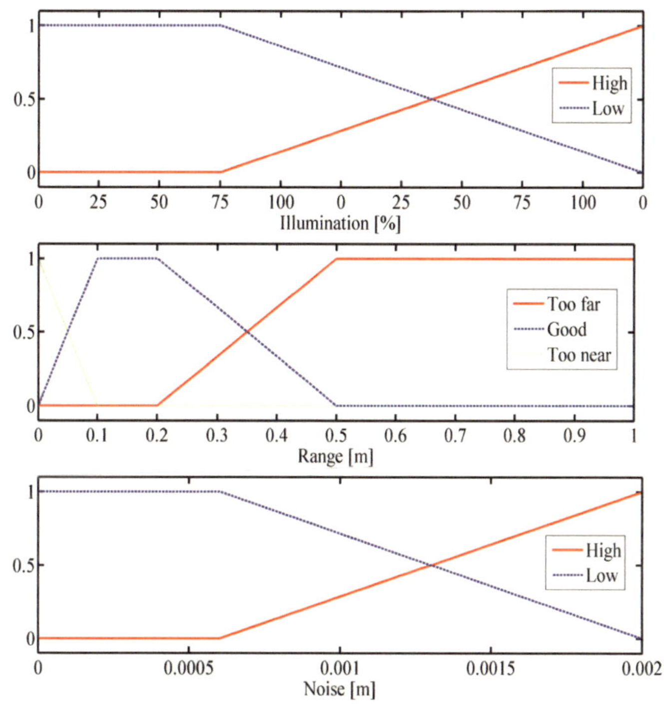

Fig. 5. Membership functions for the variables Illumination, Range and Noise

Once the values have been fuzzified they are evaluated with very intuitive rules like for example

If (*Noise is High*) or (*Range is Too far*) or
(*Illumination is Low*) then (*Quality is Low*)

(9)

If (*Noise is Low*) and (*Range is Good*) and
(*Illumination is High*) then (*Quality is High*)

After the defuzzification process a value of the quality between 0 and 1 is obtained and is weighted with the corresponding availabilities in order to estimate the value of the efficiency function needed in order to solve the conflict.

5. Experimental Results

5.1 Multi-sensor Robot Platform

The experimental test platform available at Fraunhofer IITB used for the development and investigation of the proposed control concept is shown in Fig. 6. It consists of two robot arms (A) each with 7 degrees of freedom (DoF), a 2DoF pan-tilt sensor head (B) and a five finger fluid hand (C).

For coping with a variety of interactive basic skills the robot is equipped with several redundant (cooperative) and complementary sensors. The head is equipped with a stereo camera able to track predefined objects and with an acoustic sensor (microphone array) able to determine the position of sound sources. Moreover, a miniaturized camera for accurately localizing objects at close range is integrated in the palm of the hand (see Fig. 7).

For the tactile inspection two force-torque sensors are mounted on the wrists (D) and the fingers of the gripper are equipped with tactile arrays and with a slip sensor able to detect the relative motions between end-effector and surfaces in contact with it.

Both cameras as well as the acoustic and slip sensor are connected to a dedicated computer where a first processing of the data takes place. The results are then sent via UDP/IP communication to the main computer where the robot control is implemented.

The different control programs have been developed in C++ under Windows. The control algorithms which have been successfully implemented and optimized on the presented test platform can be transferred and integrated in the common SFB demonstrator ARMAR with the help of the Modular Control Architecture (MCA2).

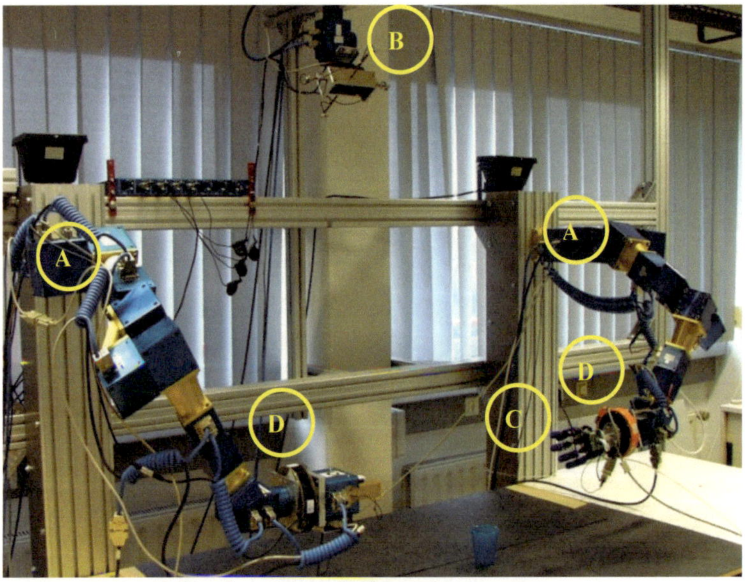

Fig. 6. Multi-sensor test and development platform

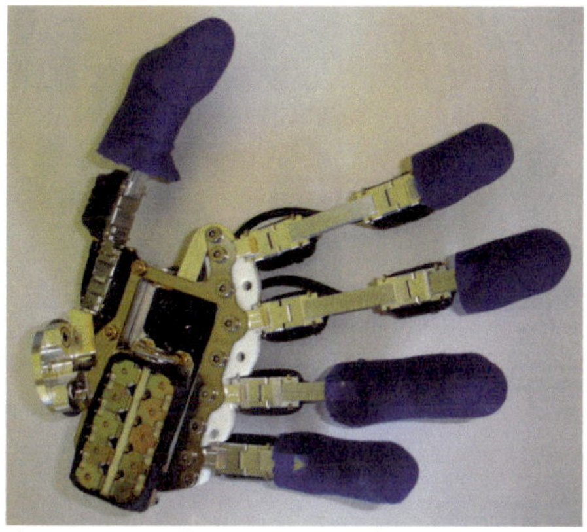

Fig. 7. Five finger hand with integrated miniaturised camera

5.2. Case study

In order to validate the control concept a case study typical for a kitchen environment has been considered. While the robot is performing a "pick and place" task to transport an object between two points A and B (e.g. taking different ingredients and putting them into a pan), the audio array hears a foreign sound. Thus, one of the three transitions associated with a sound event is triggered:

T_A: the carried object or a similar one has fallen down;

T_B: unknown (or uninteresting) sound;

T_C: an alarm (e.g. microwave) is ringing.

The robot has to cope with this unexpected situation without forgetting the initial task. In Fig. 8 the PS-based task structure in form of a pseudo Petri net describing this example is shown.

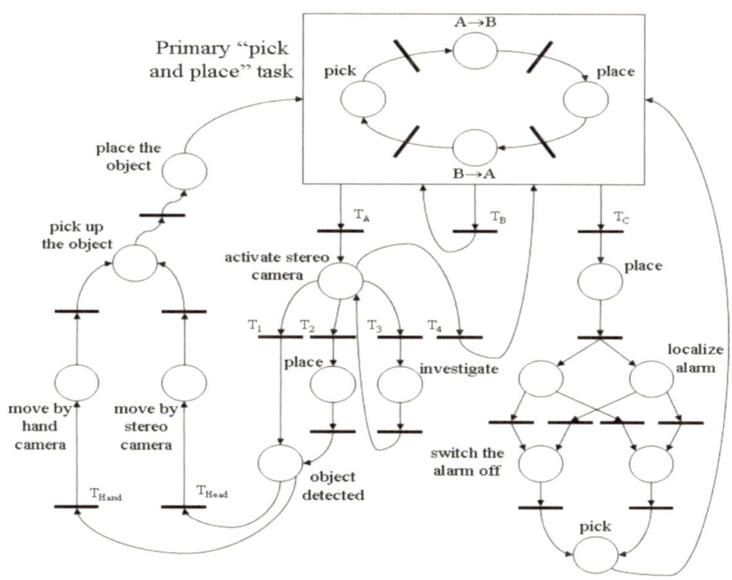

Fig. 8. Pseudo Petri net of a case study

In order to reduce the complexity of the implemented problem, only the first described situation will be discussed (i.e. an object falls down in the robot workspace). An example of the two angles representing the identified impact direction are shown in Fig. 9 (see (Milighetti et al., 2006) for more details).

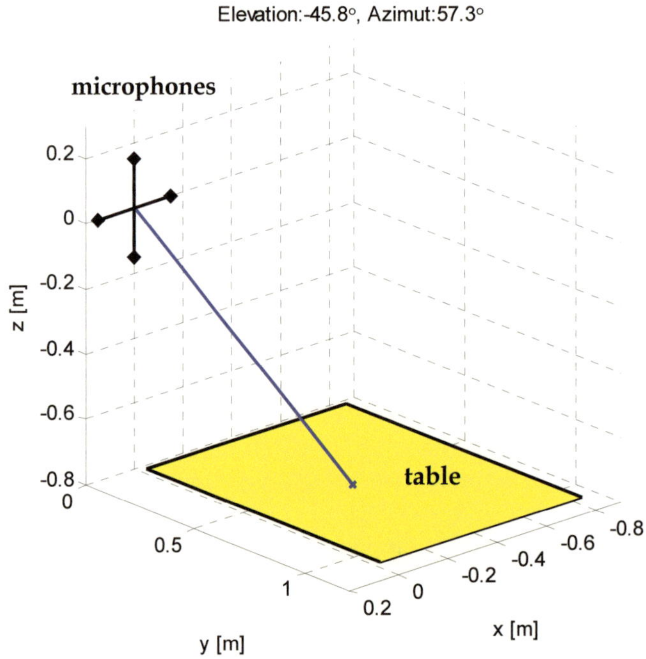

Fig. 9. Example of an audio localization

The robot stops immediately the primary "pick-and-place" task in order to activate a PS-sequence able to cope with the new situation.

First the stereo camera in the head is aligned with the sound direction in order to search for the fallen object. As shown in Fig. 8, four different events are possible at this point and can be distinguished by merging both audio and vision measurements and comparing them with the robot position.

Depending on the identified event, the following four transitions can be fired:

- T_1: the carried object has fallen down;
- T_2: another similar object has fallen down in the robot workspace;
- T_3: no object has been found in the field of view of the camera;
- T_4: the fallen object cannot be reached.

Two consistent measurements located in a region far from the actual working area of the robot are shown in Fig. 10. In this case it can be supposed that the impact was caused by a second object that can be picked up only after placing the carried one (T_2). In Fig. 11 instead the two measurements are inconsistent and a more accurate investigation (maybe enlarging the searching area) is required before a decision is made (T_3).

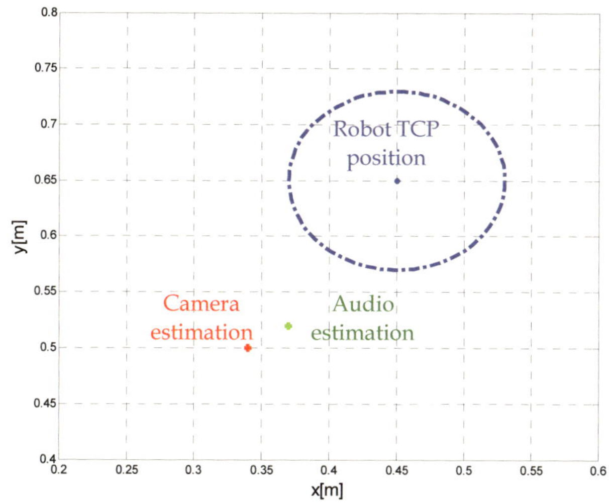

Fig. 10. Consistent audio and visual estimations

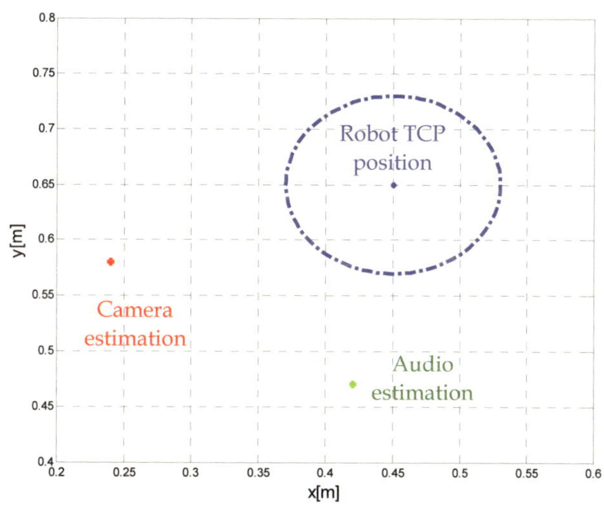

Fig. 11. Inconsistent audio and visual estimations

Once it has been determined which object has to be picked up, two different vision-based approach strategies can be adopted:

T_{Head}:the robot approaches the object using the measurements of the stereo camera in the head;

T_{Hand}:the robot approaches the object using the measurements of the camera integrated in the hand (see Fig. 7).

Finally, the object can be picked up and the primary "pick and place" task can be resumed.

During the execution of the approach phase it can be shown in detail, how the presented fuzzy-based decision making algorithm based on the evaluation of the PS-efficiency works. The affinities of the two considered PS are defined as follows:

$a_{HandCamera} = 1$ (the most accurate)

$a_{StereoCamera} = 0.8$

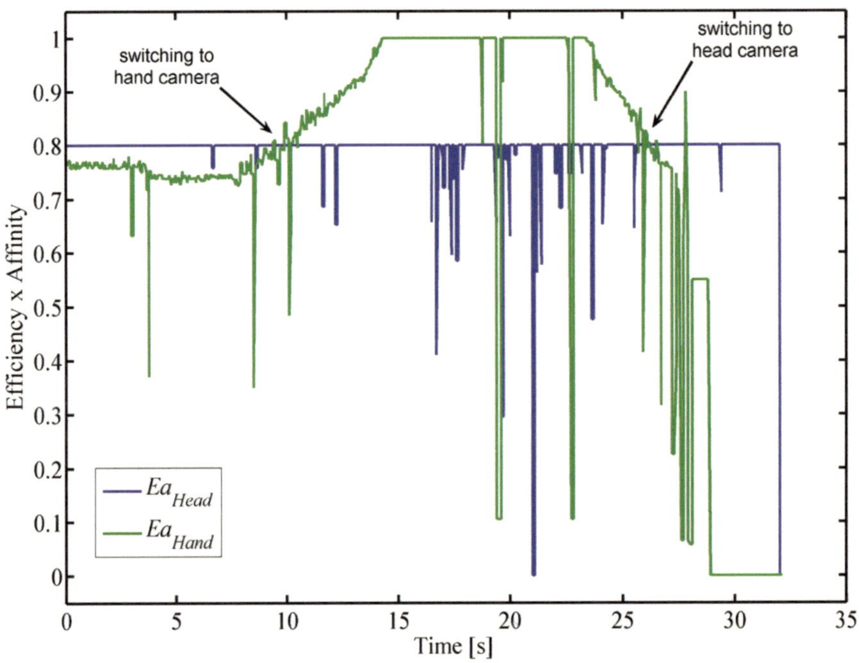

Fig. 12. Utility of the two PS "move by head camera" and "move by stereo camera" during an approach phase

Fig. 12 shows the efficiencies of the two used sensors during a normal approach towards the goal already weighted with the corresponding affinities. Considering for the other employed resources (robot arm and object) a maximal efficiency, the visualised results represent the total utilities of the two PS.

Both efficiencies are influenced by the noise in the measurements and by some false or completely missed measurements (i.e. at ca. 20 seconds for the hand camera or at 21 seconds for the head camera).

Except for these variations, the efficiency of the head camera remains constant during the analyzed time interval because the object is not moving and therefore the working range of the camera is not changing. Only at the end of the task, while the hand is grasping the object, the head camera is no longer able to localize it and its efficiency sinks to zero (at ca. 32 seconds).

On the contrary, the closer the hand camera comes to the object, the better its working range becomes and its efficiency grows accordingly till the optimal value of 1 has been reached.

At a certain instant (at ca. 23 seconds), the hand camera is too close to the object and its efficiency begins to sink once again. In the last grasping phase, shortly before the object has been reached, the localization is no longer reliable as shown also by the extreme fluctuations in the efficiency.

On the basis of the calculated efficiencies, the robot can switch between the two different choices depending on the actual situation, always activating the control strategy based on the currently optimal sensor.

In the presented example the approach is started closing in the control loop the stereo camera. In the central phase of the task the hand camera provides measurements with a higher quality and therefore the PS using it is the best choice. In order to avoid a blind grasping phase at the end of the approach (where the hand camera is no more able to localise correctly the object) the robot has to switch back again to the head camera.

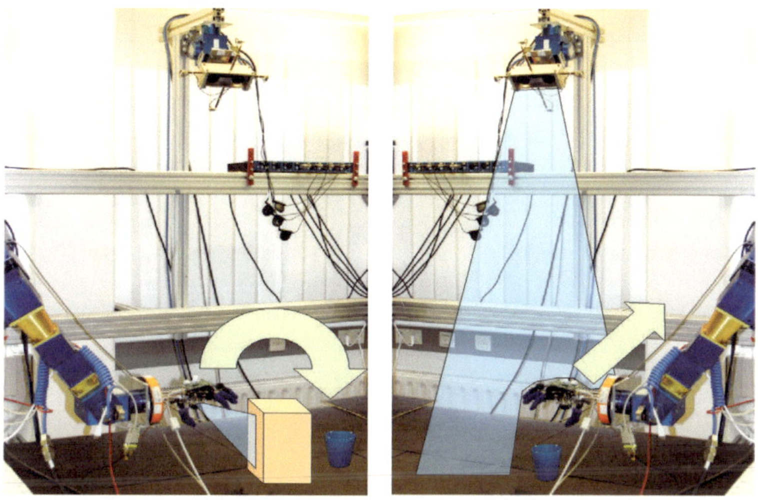

Fig. 13. Unexpected situations during the task execution: occlusion of
a) hand camera b) stereo camera

Also in case of unexpected events like for example the presence of an obstacle (Fig. 13a) or the occlusion of the view due to the motion of the arm (Fig. 13b), the values of the efficiencies of the two sensors and of their associated PS can be used in order to activate some correction in the plan. In the two presented situations, a vertical and a lateral motion can be respectively executed for overcoming the obstructions.

In Fig. 14 and Fig. 15 the efficiencies of the two PS and the cartesian trajectory of the robot TCP during a scenario with several occlusions are respectively shown.

Firstly, four occlusions of the hand camera have been simulated. The correspondence between the lowering of the availability (in this time interval no measurements are received) and the value of the efficiency is clearly observable. The robot reacts with vertical motions until a new measurement is available. Switching off the hand camera (at ca. 45 seconds) leads the robot to choose an execution by means of the stereo camera, although its optimal range was not yet reached.

Finally, the stereo camera was shortly occluded three times and the robot moves laterally until it has again free field of view. Some isolated peaks in the two efficiencies are visible as in the previous experiment.

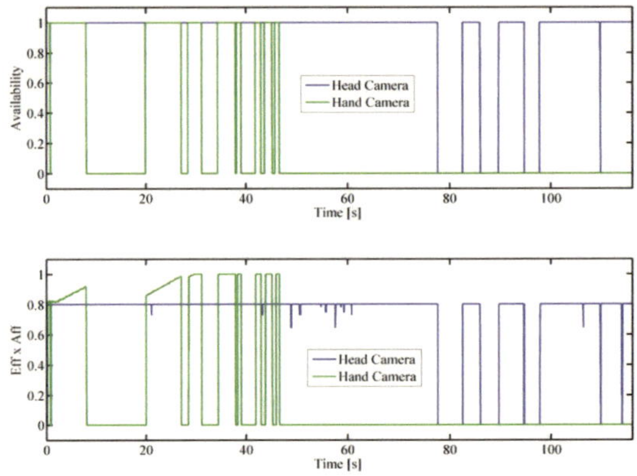

Fig. 14. Estimated efficiencies during an approach with several occlusions of the cameras

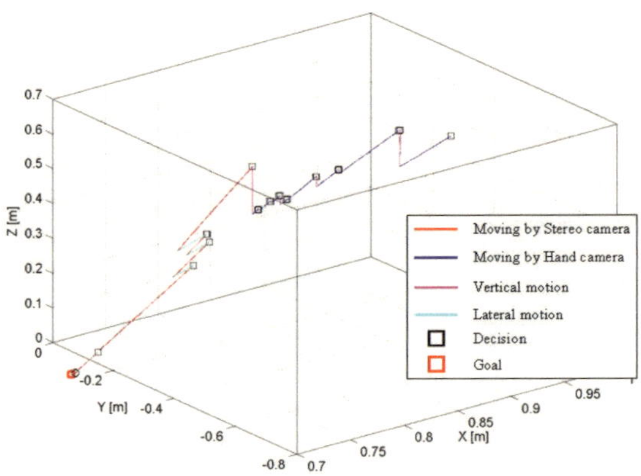

Fig. 15. Cartesian trajectory during an approach with several occlusions of the cameras

6. Conclusions

A multi-sensor based discrete-continuous control concept able to supervise complex robot tasks in a time varying environment has been presented. A flexible and transparent task architecture has been developed using the concept of Primitive Skills (PS). Every PS can be associated with a place of a Petri net which models the discrete structure of the task.

Through its multi-sensor perception the robot is able to identify failures, unexpected events or circumstances during the execution of the task. A fuzzy approach which models the human knowledge has been investigated in order to give the robot the intelligence to choose always the optimal configuration and control strategy according to the actual situation.

The efficiency of the proposed concept has been demonstrated by first experiments involving a grasping process by means of different visual and acoustic sensors. The achieved results persuasively show that the proposed approach is still valid for more complex tasks.

7. References

Antsaklis, P.J.; Koutsoukos, X.D.; Zaytoon, J. (1998). On Hybrid Control of Complex Systems: A Survey, *Proceedings of the 3rd International Conference ADMP'98, Automation of Mixed Processes: Dynamic Hybrid Systems*, March 1998.

Asfour, T.; Ly, D.N.; Regenstein, K.; Dillmann, R. (2004). Coordinated Task Execution for Humanoid Robots, *Proceedings of the 9th International Symposium on Experimental Robotics*, Marina Mandarin, Singapore, June 2004, pp. 1-10.

Bagchi, S. ; Biswas, G. ; Kawamura, K. (2000). Task Planning under Uncertainty using a Spreading Activation Network, *IEEE Transaction on Systems, Man, and Cybernetics*, vol. 30, no. 6, pages 639-650, November 2000.

Burghart, C.; Mikut, R.; Stiefelhagen, R.; Asfour, T.; Holzapfel, H.; Steinhaus, P.; Dillmann, R. (2005). A cognitive architecture for a humanoid robot: a first approach, *Proceedings of the IEEE-RAS International Conference on Humanoid Robots*, Tsukuba, Japan, 5-7 December, 2005.

Cao, T.; Sanderson, A.C. (1994). Modeling of Sensor-Based Robotic Task Plans Using Fuzzy Petri Nets, *Proceedings of the Fourth International Conference on Computer Integrated Manufacturing and Automation Technology*, 10-12 October 1994.

Chang, Z.-H.; Bian, X.-Q.; Shi, X.-C. (2004). Autonomous Underwater Vehicle: Petri Net Based Hybrid Control of Mission and Motion, *Proceedings of the Third International Conference on Machine Learning and Cybernetics*, 26-29 August 2004.

Estlin, T.; Volpe, R.; Nesnas, I.; Mutz, D.; Fisher, F.; Engelhardt, B.; Chien, S. (2001). Decision-making in a robotic architecture for autonomy, *Proceedings of the 6th International Symposium on Artificial Intelligence, Robotics and Automation in Space* (iSAIRAS), Montreal, Canada, June 18-22, 2001.

Finzi, A.; Ingrand, F.; Muscettola, N. (2004). Model-based executive control through reactive planning for autonomous rovers, *Proceedings of the IEEE International Conference on Intelligent Robots and Systems* (IROS), Sendai, Japan, September 28 - October 2, 2004.

Gat, E. (1997). On three-layer architectures, *Artificial Intelligence and mobile robots*, D. Kortenkamp, R. P. Bonnasso, and R. Murphy, MIT/AAAI press, 1997.

Gao, M.; Zhou, M. (2005). Control Strategy Selection for Autonomous Vehicles in a Dynamic Environment, *Proceedings of the IEEE International Conference on Systems, Man and Cybernetics*, 10-12 October 2005.

Hasegawa, T. ; Suehiro, T. ; Takase, K. (1992). A Model-Based Manipulation System with Skill-based Execution, *IEEE Transaction on Robotics and Automation*, vol.8, no. 5, pages 535-544, October 1992.

Kobayashi, K.; Nakatani, A.; Takahashi, H.; Ushio, T. (2002). Motion planning for humanoid robots using timed Petri net and modular state net, *Proceedings of the IEEE International Conference on Systems, Man and Cybernetics*, 6-9 October 2002.

Joyeux, S.; Alami, R.; Lacroix, S. (2007). A Software component for simultaneous plan execution and adaptation, *Proceedings of the IEEE International Conference on Intelligent Robots and Systems* (IROS), San Diego, CA, USA, October 29 - November 2, 2007.

Lehmann, A.; Mikut, R.; Asfour, T. (2006). Petri Nets for Task Supervision in Humanoid Robots, *Proceedings of the 37th International Symposium on Robotics* (ISR 2006), München, May; 2006.

Lundell, M.; Jinpeng Tang; Nygard, K. (2005). Fuzzy Petri Net for UAV Decision Making, *Proceedings on the 2005 International Symposium on Collaborative Technologies and Systems*, 15-20 May 2005.

Milighetti, G.; Kuntze, H.-B. (2006). On the Discrete-Continuous Control of Basic Skills for Humanoid Robots, *Proceedings of the International Conference on Intelligent Robots and Systems IROS 2006*, Beijing, China, 9-15 October 2006.

Milighetti, G. ; Emter, T. ; Kuntze, H.-B. ; Bechler, D. ; Kroschel, K. (2006). Primitive-Skill Based Supervisory Control of a Humanoid Robot Applied to a Visual-Acoustic Localization Task, *Proceedings of the Robotik 2006*, Munich, Germany, 15th -17th May 2006.

McCarragher, B.J. (1996). Task primitives for the discrete event modeling and control of 6-DOF assembly tasks. *IEEE Transactions on Robotics and Automation*, vol. 12, no. 2, pages 280–289, April 1996.

Morrow, J.D. ; Khosla, P.K. (1997). Manipulation Task Primitives for Composing Robot Skills. *Proceedings of the IEEE International Conference on Robotics and Automation*, pages 3354–3359, Albuquerque, Mexico, April 1997.

Nenninger, G.; Krebs, V. (1997). Modeling and Analysis of Hybrid Systems: A New Approach Integrating Petri Nets and Differential Equations, *Proceedings of the Joint Workshop on Parallel and Distributed Real-Time Systems*, 1997.

Palomeras, N.; Carreras, M.; Ridao, P.; Emili, H. (2006). Mission control system for dam inspection with an AUV, *Proceedings of the 2006 IEEE/RSJ International Conference on Intelligent Robots and Systems*, 9-15 October 2006.

Pardowitz, M.; Zöllner, R.; Knoop, S.; Dillmann R. (2007). Incremental Learning of Tasks From User Demonstrations, Past Experience, and Vocal Comments, *IEEE Transactions on Systems, Man and Cybernetics*, Part B, Vol. 37, No. 2, pages 322-332, 2007.

Peters, J.; Schaal, S.; Schoelkopf, B. (2007). Towards Machine Learning of Motor Skills, *20.Fachgespräch Autonome Mobile Systeme (AMS)*, Informatik aktuell, Springer, Kaiserslautern, October, 2007.

Schrempf, O.C.; Hanebeck, U.D. (2005). A generic model for estimating user intentions in human-robot cooperation, *Proceedings of the Second International Conference on Informatics in Control, Automation and Robotics (ICINCO)*, 14-17 September 2005.

Serrano, N. (2006). A Bayesian Framework for Landing Site Selection during Autonomous Spacecraft Descent, *Proceedings of the 2006 IEEE/RSJ International Conference on Intelligent Robots and Systems*, Beijing, China, 9-15 October 2006.

Thomas, U.; Finkemeyer, B.; Kröger, T.; Wahl, F.M. (2003). Error-Tolerant Execution of Complex Robot Tasks Based on Skill Primitives, *Proceedings of the IEEE International Conference on Robotics and Automation*, Taipei, Taiwan, May 2003.

Vernon, D.; Metta, G.; Sandini, G. (2007). A survey of artificial cognitive systems: implications for the autonomous development of mental capabilities in computational agents, *IEEE Transaction on Evolutionary Computation*, Vol. 11, No. 2, April 2007.

12

The Design of humanoid Robot Arm based on Morphological and Neurological Analysis of Human Arm

Yongseon Moon[1], Nak Yong Ko[2] and Youngchul Bae[3]
[1] *School of Information Communication Sunchon National Universit, Suncheon,5407-42.Korea, moon@sunchon.ac.kr*
[2] *Department of Control and Instrumentation Engineering, Chosun University, Gwangju,550-759,Korea, nyko@chosun.ac.kr*
[3] *Division of Electrical·Electronic Communication·Computer Engineering, Chonnam National University, Yeosu,550-745, Korea, ycbae@chonnam.ac.kr*

Abstract - **In this paper, we analyzed and verified possibility and validity of overcoming present limitations of humanoid robot by using morphological and neurological analysis of human arm. Through design, implementation and performance evaluation of humanoid robot arm, we will be verifying applicability and effectiveness of humanoid robot arm system based on SERCOS network that fulfills the concept of opening, networking and modularizations that are progressive direction of future robot.**
Index Terms - humanoid robot, UML, ISO 15745, SERCOS Communication

1. Introduction

At first appearance of the industrial robots during the 1960s, the concepts for the usage of the robots were only as manipulator in which to perform pre-ordered commands. After 20 years later during the 1980s, together with the appearance of microprocessor, it marked the beginning of the intense research of the field of robot. As the research progressed, robots were recognized not only as simple action performer but as a machine that have diverse and variety of purposes and usages.

As the technology and recognition of robots improved in a variety of ways, humans began to relate themselves with robots; hence, these robots are called humanoid robot in which they resemble appearance of human and imitate their behaviour. There are two representative humanoid robots that have currently developed. They are Japanese Honda's *ASIMO* which is well known as superior to any other humanoid robots and KAIST's *HUBO*.

However, these robots have some limiting factors. First, they have limited control performance due to DC motor usage for joint actuator and difficulty in constant repairing and replacement of joint that are caused by wearing down of brush and commutator. These factors cause limited life time of robots, which in turn brings economic burden. Second, current robot control networks are RS-232[2], USB(Universal serial bus)[3], CAN(Control area network)[4], Ethernet[5] which cannot sufficiently implemented the performance of humanoid robots. Third, current humanoid robot technology is focused on production and implementation of skills without proper mechanism for opening methodology of development for analysis, design, implementation, and integration for robot development.

In this paper, we analyzed and verified possibility and validity of overcoming present limitations of humanoid robot by using morphological and neurological analysis of human arm. Through design, implementation and performance evaluation of human robot arm, we will be verifying applicability and effectiveness of humanoid robot arm system based on SERCOS network that fulfills the concept of opening, networking and modularization that are progressive direction of future robot.

2. The Development Methodology of Humanoid Robot Arm

A. Open human robot development methodology

In order to establish a generalized opening technology system for the analysis, design, integration, and implementation of the humanoid robot, we proposed a new mechanism of development using generalized method of robot arm that is core make up component of the arm. Fig. 1 show open development methodology for humanoid robot arm based on ISO15745.

Opened humanoid robot development methodology begins from analysing morphological and neurological structure of human arm and through yielding of current developed humanoid robot arm model. Then, we develop a design integration model of humanoid robot arm using opening ISO15745 standardization criterion and definition. In ISO15745 standard, regulation is made by using UML (Unified modelling language) which is standard model of orientation in designing the humanoid robot arm. Humanoid robot arm designing integration class that has been finally deducted by using UML through another standardized transformation process for profile description is described as a XML(Extensible Markup Language) schema. Final design profile is prepared through instance of document of schema and process. These humanoid robot arm design profile that has been going through these processes is used as standard manual for robot arm development [6].

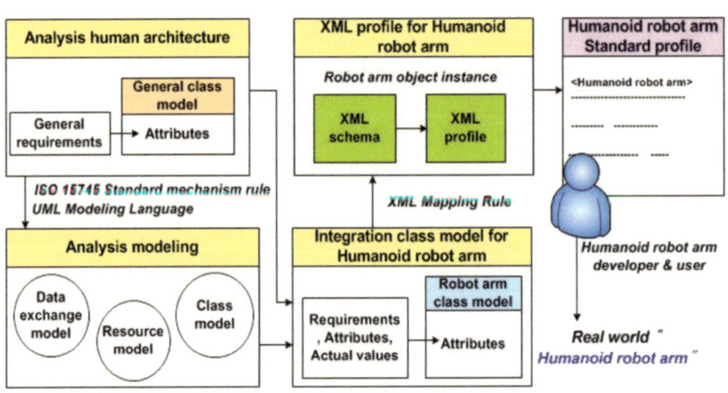

Fig. 1. ISO 15745 based open development methodology for humanoid robot arm development

B. Morphological analysis of human arm

In this paper, we applied mapping concept of human morphological structure for implementation of robot arm that is similar to structure and movement to that of human arm. Fig.2 show human -robot morphological mapping.

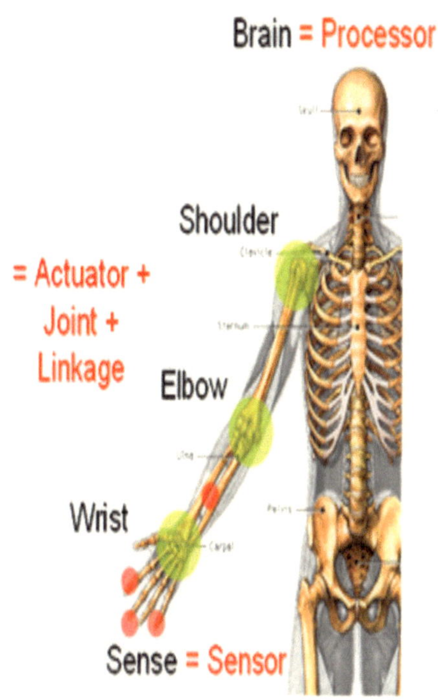

Human	Robot
Brain	Processor
Shoulder, Elbow, Wrist	Actuator + Joint + Linkage
Sense	Sensor

Fig. 2. Human-robot morphological mapping

C. Neurological analysis of structure of human arm

Through neurological analysis of structure of human [7], we define control network and processor structures for implementation of humanoid robot arm. Humans perform movement of muscles and acquire internal and external information through each organ based on nerve activity. These nerves can be viewed as robot's network in functional sense. Furthermore, according to human neurological classification, nervous system is classified in brain, spinal cord, sensorimotor. These can be matched with functional module of processor of robot control. Fig.3 shows a Human-robot neurological mapping.

3. Modeling of Humanoid Robot Arm

Based on requirement of robotic arm through neurological and morphological analysis, we can develop designing model for humanoid robot arm.

Table 1 present basic requirement for implementation of humanoid robot arms currently developed national and international wide through analysis of structure of humanoid robot arm and human morphology.

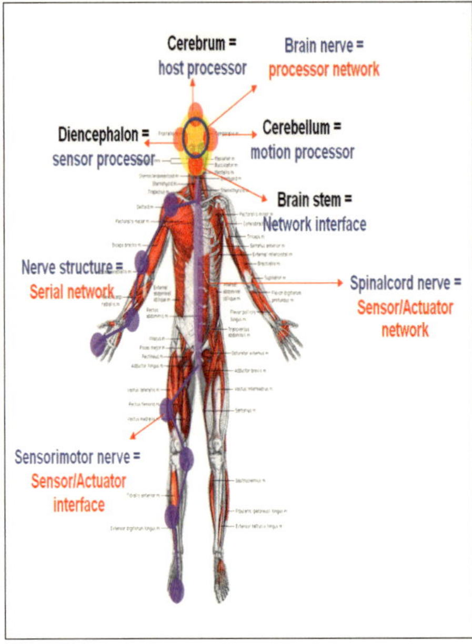

Human	Robot
Cerebrum	Host processor
Cerebellum	Motion processor
interbrain	Sensor processor
Brain stem	Network interface
Spinal nerve	Sensor/Actuator network
Sensori-motor nerve	Sensor/Actuator interface

Fig. 3. Human-robot neurological mapping

Item	Requirement	Reason
Arm type	Module structure	- Simplification of arm structure and wire problem solving
Degree of freedom	5□7	- Optimal behavior implementation based on human structure
Actuator	AC servo motor	- High precision control and everlasting use of joint
Control network	High speed network	- High speed data processing
	Motion control network	-Advance performance of motion control and multi axes synchronization
	Serial network	- Human nerve system, cable wire remove, and modularization implementation

Table 1. The requirements for implementation of humanoid robot arms

A. Humanoid robot arm resource integration model

ISO15745 standard represents an integration model called AIF(Application Integration Model) which defines the opening development system. AIF integration model is subdivided into a process integration model, a data exchange model, and a resource integration model. Among these detailed models, we standardized them as resource integration mode to define external structure of system l.

Through this paper, we defined external structure and configuration of humanoid robot arm using resource integration model of ISO15745. (Fig. 4)

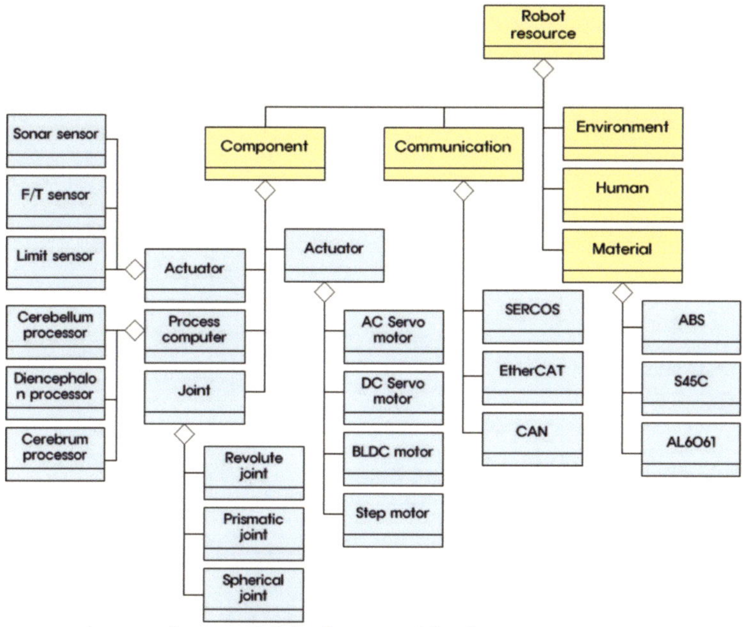

Fig. 4. The resource integration model for humanoid robot arm

B. Humanoid robot arm data exchange model

By using humanoid robot arm that behaves in ISO15745 standard that are based on human neurological and morphological transaction structure, we presented behaviour of internal system for movement of humanoid robot in the Fig. 5. [6]

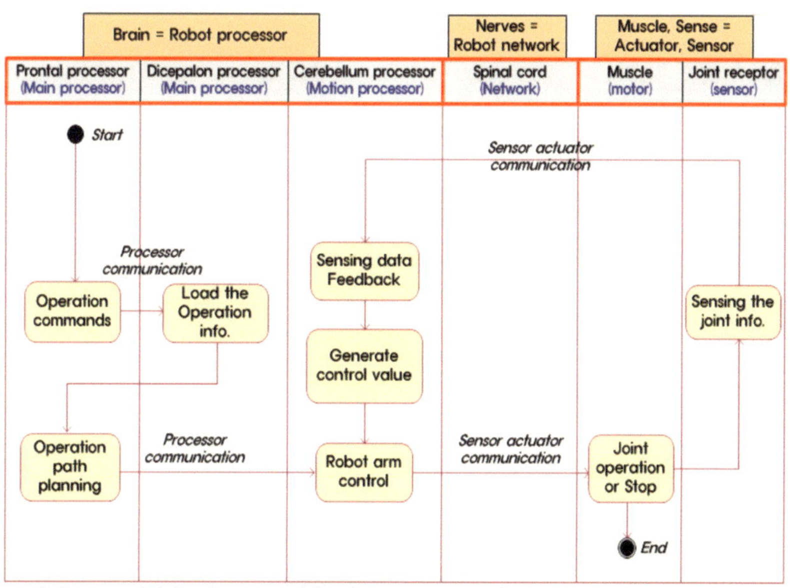

Fig. 5. The activity model for humanoid robot arms motion

C. Humanoid robot arm integration model

Through analysis of human arm movement and control component, the final designed model of humanoid arm that are based on human structure like Fig. 5, can be created.

Fig 6 shows the integration class model of the humanoid robot arm structure that will be developed.

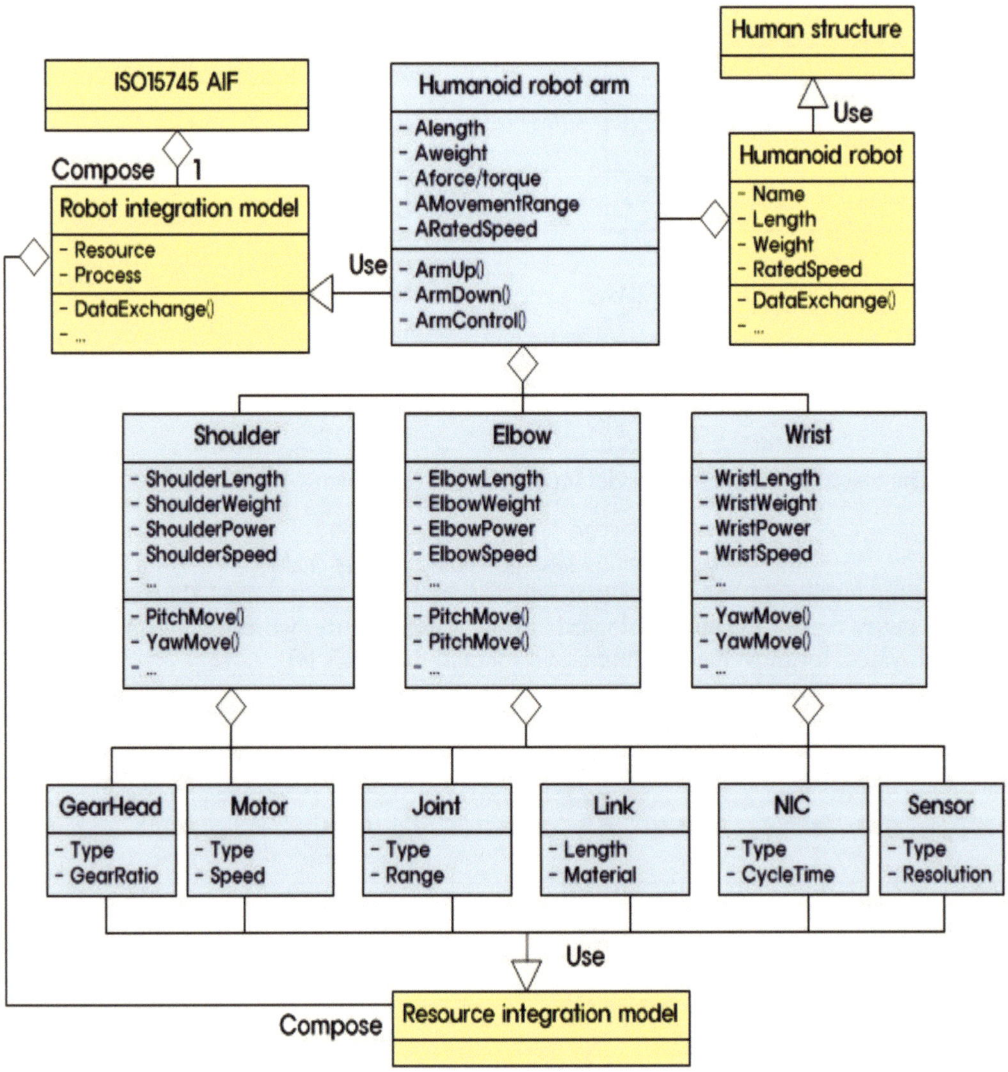

Fig. 6. Integration class model for humanoid robot arms

D. Humanoid robot arm design profile

In order to write up XML standard profile for humanoid robot arm, we must define DTD(Data type definition), schema that defines basic structure of XML documents. Fig. 7 represents standard profile of XML of humanoid robot arm that has been object instance based on XML schema structure.

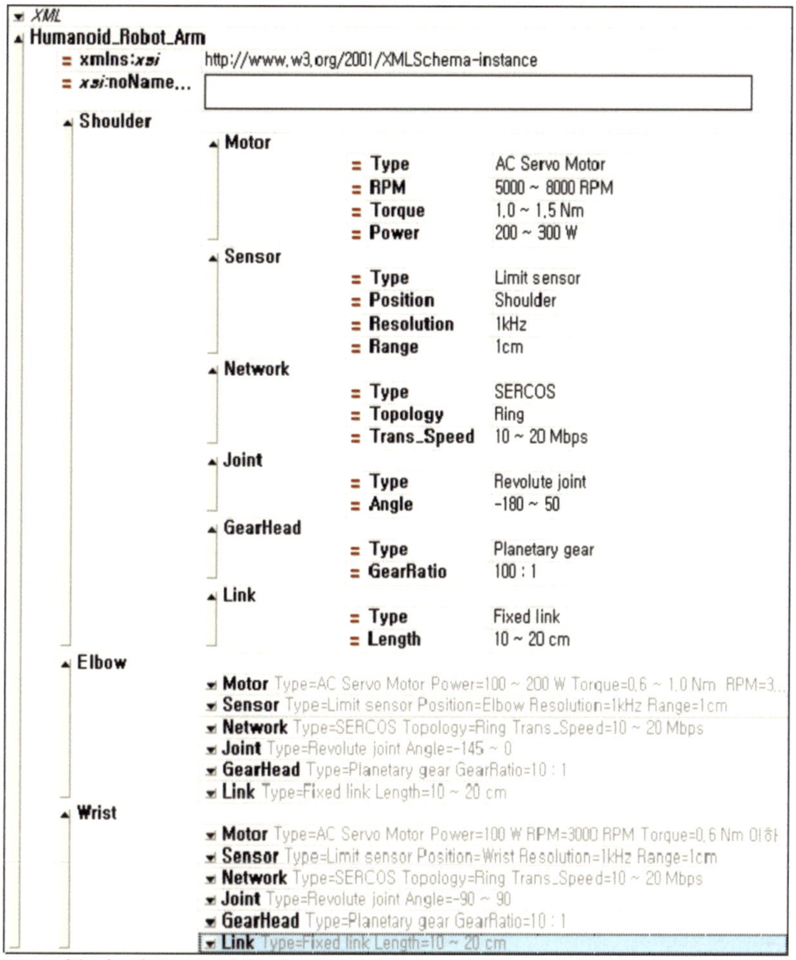

Fig. 7. XML profile for humanoid robot arm

4. Sercos Based Robot Arm Design Analysis

A. Humanoid robot arm structure and system

The degree of freedom of humanoid robot arm joints consist of total 5 degree of freedoms excluding 2 degrees of freedom from total 7 DOF of humans which are radial/ulnar and flection/extension. Moreover, humanoid robot arm control network and control system will consist of robotic arm control system that uses motion control SERCOS communication and high precision AC servo motors which can overcome problems that current humanoid robot system have.

Fig 8 represents concept of degrees of freedom and structure of humanoid robot arm that will be implemented to humanoid robot.

In conceptual diagram of Fig 8, among total seven degrees of freedom that humans have, except two degrees of freedom of wrist that have very little influence in motion control, the five degrees of freedom should be implemented including shoulder 3 degrees of freedom, elbow 1 degrees of freedom, wrist 1 degrees of freedom. Control motions that are related with the movement may deal the rest of 2 degrees of freedom.

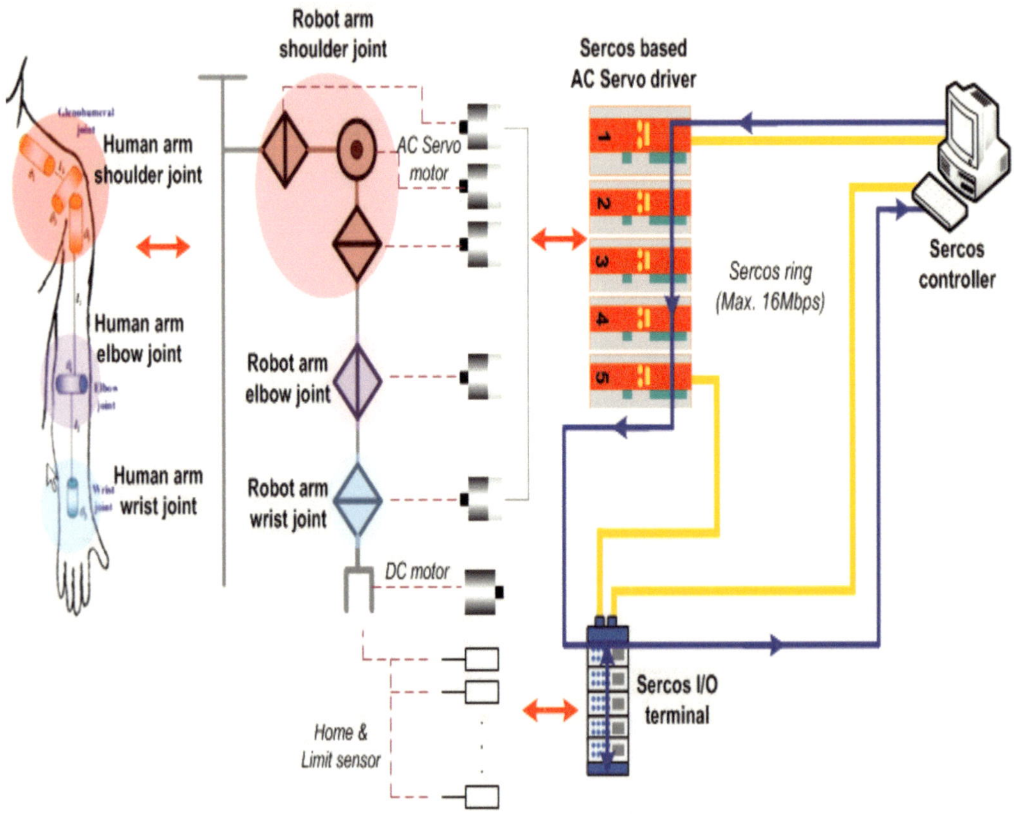

Fig. 8. The control system configuration of the humanoid robot arms

Unlike usual robot control method in microprocessor environment, SERCOS network based on humanoid robot arm control system in Fig. 8 uses SoftPLC(Software Programmable Logical Controller) technology which is the control method of constructing false image of continuous control environment in PC and uses mater/slave control method that is the method of continuous control through SERCOS communication.

The transfer of control data through SERCOS communication supports high speed transmission that speeds up to 16Mbps, and it is possible to support precise motion of humanoid robot arm through standardized motion function block and to realize precision motion and synchronized control of separated humanoid robot joints.

Current International Motion Association provides standardized structure and function blocks for PTP motion control. In this paper, we tried to control the motion of humanoid robot arm by applying standardized PTP (point to point) control system and interpolation control.

Fig. 9 represents basic movement structure of state chart model that defines movement of standardized PTP motion control method.

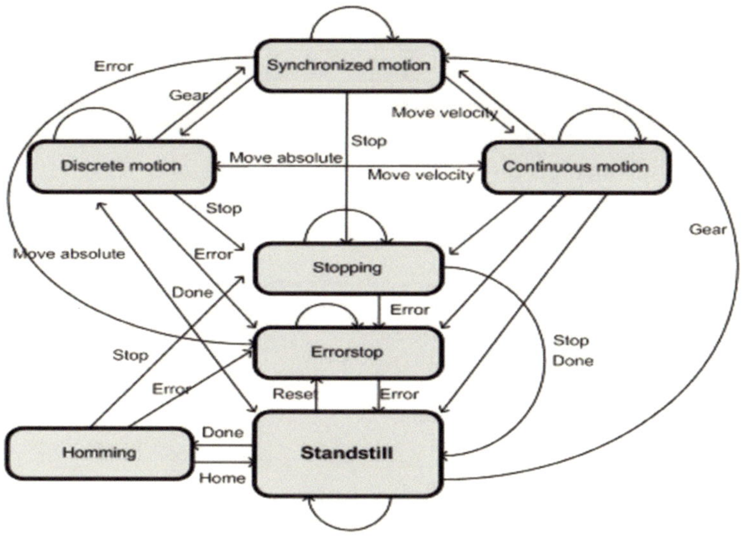

Fig. 9. The configuration drawing for the standard control function block and the condition

Interpolation control is the method for path driving of mutual interpolation of variety of axes in space that has been proposed to overcome disadvantage of PTP control method. Interpolation control, instead of control over discontinuous section, was optimal for continuous section.

Moreover, it is possible to make continuous section during discontinuous section using interpolation control like in Fig. 10.

In order for interpolation control to work properly for humanoid robot arm in various environments, detailed processes of motion interrupter establishment and control should be implemented additionally.

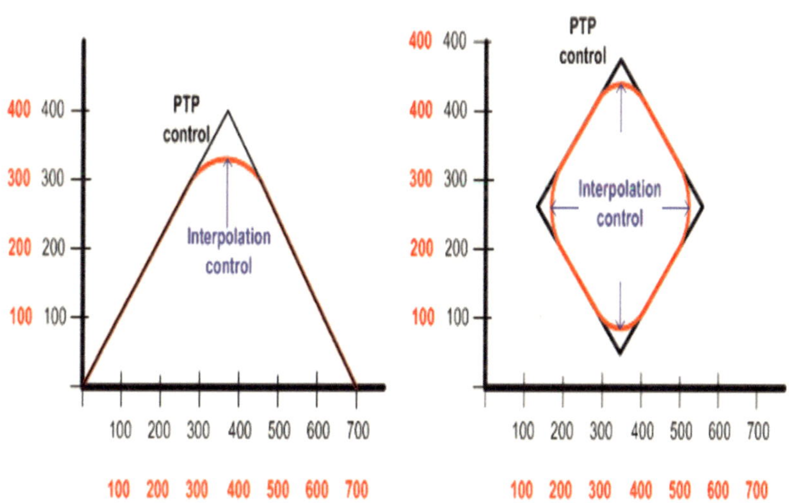

Fig. 10. Motion trace from interpolation control and PTP control

5. Humanoid Robot Arm Implementation and Evaluation

The humanoid robot arm in the right side of Fig. 11 was implemented with international standard motion control SERCOS network and AC servo control system. Each arm links machinery module structure that is possible to attached and remove.
The measurement and calculation for kinematics of matrix of humanoid robot arm motion tracing has been implemented through visual C++. Fig. 11 also shows the experimental diagram for humanoid robot arm motion control and calculation for tracing of rotational angel with input condition by using kinematics model.

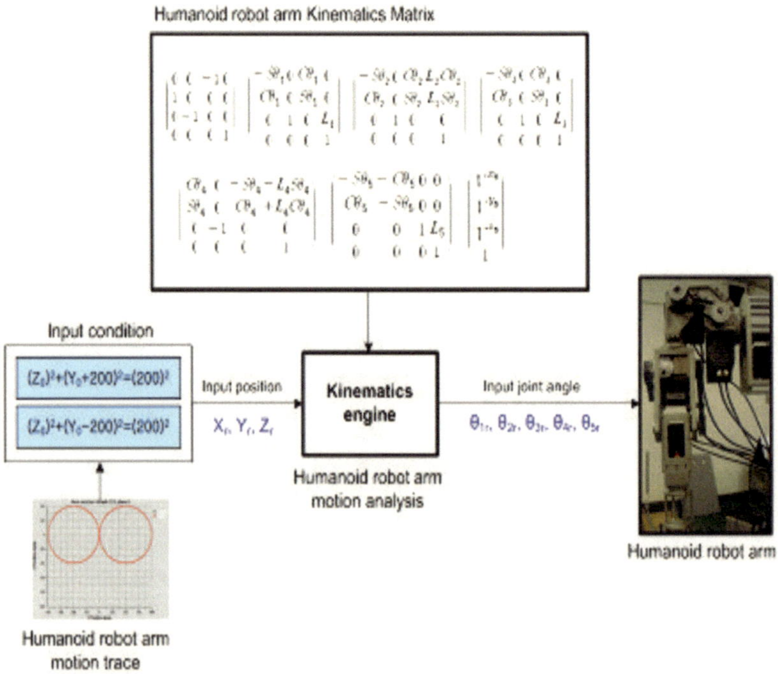

Fig. 11. The structure of calculating the information for the join rotation of humanoid robot arm

By using 3 dimensional space coordinating calculation of robot for input of kinematics matrix for rotational angle of 5 joints of humanoid robot arm measured during motion trace used as the method of measuring motion trace of humanoid robot arm, we used method of yielding real motion trace value of humanoid robot arm.
In Fig 11, we performed performance evaluation for circle motion control of humanoid robot arm using control performance, motion trace measurement, and calculation structure. In this paper, we used double circle trace for the purpose of humanoid robot arm motion control. Fig. 12 represents structure of double circle trace to be inputted for circle motion control of humanoid robot arm.
In Fig. 13, rotational angle information that will be used as humanoid robot arm was produced by input of position value of circle trace of desired formula for kinematics matrix.

Since induced input joint rotational angle is dependent on kinematics matrix, in case of induction of real kinematics matrix does not occur properly, we must use properly induced kinematics matrix. Otherwise, it will cause totally difference output of desired motion due to production of false input information.

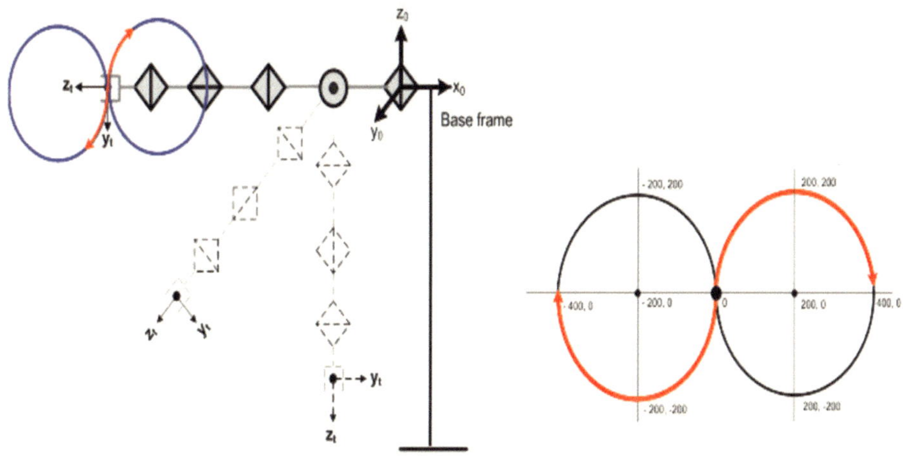

Fig. 12. The outline of the circle motion control trace setting for the humanoid robot arms

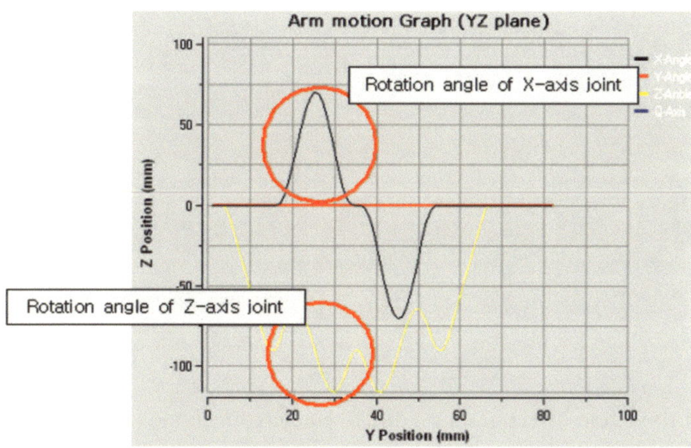

Fig. 13. Circle control input degree trace

As we know input angle trace information from Fig. 14, we know it can be formed by driving two joints that are circular traces. Furthermore, we created one circular trace through the rotational angel trace of cosine type to determine direction of motion trace and the rotational angel trace of sine type to determine position of motion trace.

Fig. 14 shows repeated motion trace and normal circular trace during the 5 repeating motion trace as a PTP output trace for double circular motion with input rotation angel.

As we know through the circular control trace of PTP from Fig. 14, the repeated position precision of trace is superior to position repeating precision trace itself. However, precision of position tracking for actual desired circular motion is fairly poor in the PTP control case of humanoid robot arm.

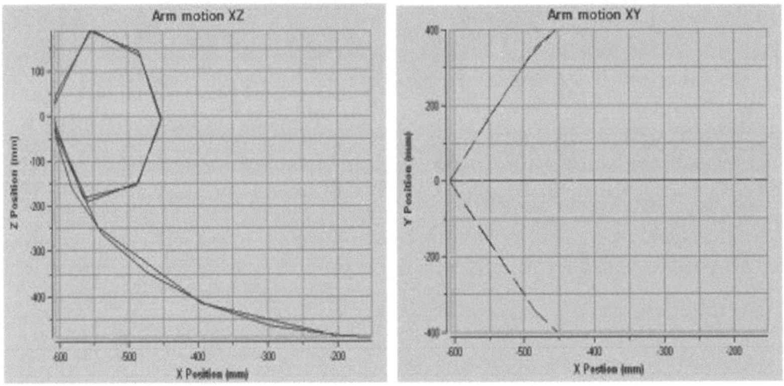

Fig. 14. Result of the circle PTP control trace from plane (XZ, XY)

In this paper, in order to solve PTP control problem of low precision, we applied interpolation control which is an optimal control for continuous section in the 3D space. Fig. 15 shows repeated motion trace and normal circular trace during the 5 repeats as an interpolation control input trace for double circular motion with input rotational angel.

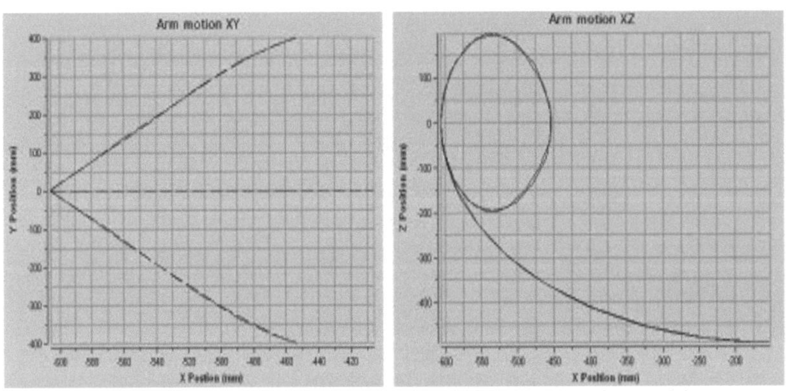

Fig. 15. Result of the circle interpolation control trace from plane (XZ, XY)

We can see the interpolation control which has smoother trace and high precision that is superior to PTP control. There are a small error found between the actual trace and the interpolation control trace. The comparison of PTP and interpolation control represented in Fig. 14 and Fig. 15.

Fig. 16 shows measurement of trace for error between the interpolation control and actual circular control. Since the errors that occur in each section of real control sampling are different, we used mean error for each section and evaluated tracking error that occur in peak location dot tracking for error section and repeating error that represents floating ratio that occurs between sections. Calculated error value from measurement is represented in Fig 16.

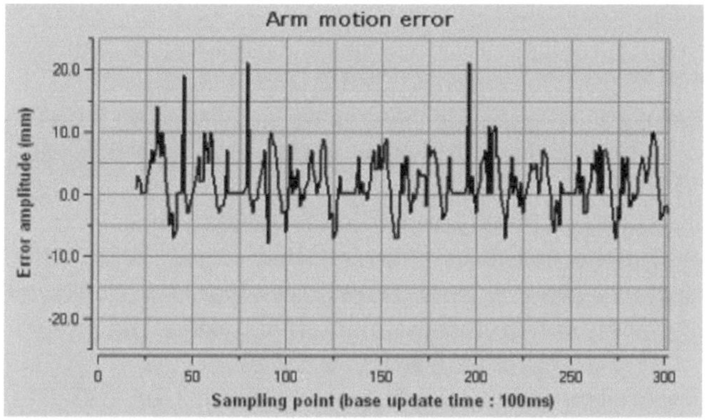

Fig. 16. Circle control error output (motion 5 times repetition error)

Table 2 represent the motion error measurement data of the humanoid robot arm. Through the 5 controls, we found that mean position trace error is about 4.112mm and mean position repetition error is about 0.135mm in the Table 2.

Times	Sampling time(ms)	Sum of error(mm)	Mean error(mm)
1	50 Point (5s/0.1s(100ms)	204mm	204/50 = 4.08mm
2	50 Point (5s/0.1s(100ms)	209mm	209/50 = 4.18mm
3	50 Point (5s/0.1s(100ms)	199mm	199/50 = 3.98mm
4	50 Point (5s/0.1s(100ms)	205mm	205/50 = 4.1mm
5	50 Point (5s/0.1s(100ms)	211mm	211/50 = 4.22mm
Total	Mean trace error : 4.112mm Mean poison repetition error : (0.1+0.2+0.12+0.12)/4 = 0.135mm		

Table 2. The motion error measurement data of the humanoid robot arms

6. Conclusion

In this paper, we have presented the implementation and performance evaluation for SERCOS based humanoid robot arm by using morphological and neurological analysis of human arm. Moreover, we reviewed the possibility of application of these robot arms. First, we proposed robot development methodology of open architecture based on ISO15745 for "opening of humanoid robot." Then, we verified the method of implementation of humanoid robot arm and its application to the real world.

We have implemented robot arm using SERCOS communication and AC servo motor for high precision motion control; in addition, we got a mean position trace error of 4.112mm and mean position repetition error of 0.135mm as a control performance.

7. References

Karl Williams, "*Build Your Own humanoid Robots*", Tab Bookks, 2004.

Christopher E. Strangio, "The RS232 Standard - A Tutorial with Signal Names and Definitions" 1993~2006. "Universal Serial Bus Specification", Compaq, Intel, Microsof, NEC, 998.

Robot Bosch Gmbh, "*CAN Specification v2.0*", Bosch, 1991.

John F. Shoch, "*An Introduction to the Ethernet Specification*", ACM SIGCOMM Computer Communication Review volume 11, pp 17-19, New York, USA, 1981.

ISO TC 184/SC 5, "*ISO 156745* - Industrial automation system and integration Part1, 1999.

David G. Amaral, "Anatomical organization of the central M. King, B. Zhu, and S. Tang, "Optimal path planning," *Mobile Robots*, vol. 8, no. 2, pp. 520-531, March 2001.

Rainer Bischoff, Volker Graefe, "*HREMES-a Versatile Personal Robotic Assistant*", IEEE-Special Issue on Huamn Interactive Robots for Psychological Enrichment, pp. 1759-1779, Bundeswehr University Munich, Germany.

In A. Zelinsky, "*Design Concept and Realization of the humanoid Service Robot HERMES*", Field and Service Robotics, London, 1998

Rainer Bischoff, "*HERMES-A humanoid Mobile Manipulator for Service Task*", International Conference on field and Service Robots, Canberra, December 1997.

Rainer Bischoff, "*Advances in the Development of the humanoid Service Robot HERMES*", Second International Conference on field and Service robotics, 1999.

H. Netter MD, "*Atlas of Human Anatomy, Professional edition*", W.B Saunders, 2006. Function blocks for motion control, "PLCopen-Technical Committee2", 2002.

13

6-DOF Motion Sensor System Using Multiple Linear Accelerometers

Ryoji Onodera and Nobuharu Mimura
Tsuruoka National College of Technology, Niigata University
Japan

1. Introduction

This chapter describes a multi degrees of freedom (hereafter, "DOF") motion sensor system in 3D space. There are two major areas where multiple-DOF motion sensors are needed. The first involves vehicles, helicopters or humanoid robots, which are multi-input multi-output objects, each having 3 DOF in all translational and rotational directions. Thus, an accurate measurement of all 6 DOF motions is needed for their analysis and control. The second area involves considering not only the translational components, but also the rotational components for 1-DOF linear motion on ground since this type of motion is always affected to a certain extent by motion along other axes, and mixed signals of translational and rotational components are detected in the case of using 1-DOF motion sensors.
So far, inertial navigation systems (INS) with highly accurate gyro sensors and accelerometers have been used in the field of rocketry and aerodynamics. However, it is difficult to install such systems combined with gyroscopes and accelerometers on small robots due to their large size, weight and cost. In recent years, small vibration gyro systems and micro-machine gyro systems have been developed by using MEMS (micro-electro-mechanical system) technology. However, the sensor units which utilize these systems are not small, weighing about 6 to 10 kilograms, and in addition there are some problems with measurement accuracy and stability.
In the case of multiaxial sensors, it is known that the specific problem of non-linear cross effect arises, which means that other axial components interfere with the target one, and the measured values differ from the true values due to this effect. As a result, the cross effect reduces the measurement accuracy. In addition, as a result of increasing the effect by the gain and offset errors of each axis, stability is lost. Therefore, sensor calibration is required for suppressing the cross effect, as well as for decreasing the gain and offset errors. However, in general, gyro systems cannot be calibrated in isolation, which needs special equipments, such as an accurate rotary table, for generating a nominal motion. Thus, a proper calibration of the multiaxial sensors by the user is extremely difficult.
In the present work, we propose a newly developed 6-DOF motion sensor using only multiple accelerometers, without the gyro system. The advantage of using accelerometers is that they can be calibrated with relative ease, using only the gravitational acceleration without any special equipment. So far, we have performed several experiments using the

prototype sensor, and observed rapid divergence followed by the specific cross effect in the multiaxial sensors. Therefore, in this chapter, we investigate these two problems of divergence and cross effect. Regarding the divergence, we analyzed the stability based on the geometric structure of the sensor system. Furthermore, we analyzed the cross effect with respect to the alignment error of the linear accelerometer, and proposed a relatively easy calibration method based on the analytical results. Finally, we investigated the proposed system and methods in an experiment involving vehicle motion, which is particularly prone to the cross effect, and demonstrated that this sensor system (i.e., the 6-DOF accelerometer) performs well.

2. Measurement principle and extension to multiple axes

First, we consider the acceleration which occurs at point *i* on a rigid body (Fig.1). We define a position vector for the moving origin of the body (Σ_b) relative to the reference frame (Σ_o) as ${}^o\boldsymbol{p}_b = [{}^o p_{bx} \quad {}^o p_{by} \quad {}^o p_{bz}]^T$, and the position vector of the point *i* relative to the body origin as ${}^o\boldsymbol{r}_i = [{}^o r_{ix} \quad {}^o r_{iy} \quad {}^o r_{iz}]^T$. Then, the position vector ${}^o\boldsymbol{p}_b$ from Σ_o is is represented as follows:

$${}^o\boldsymbol{p}_i = {}^o\boldsymbol{p}_b + {}^o\boldsymbol{r}_i. \tag{1}$$

When the rigid body revolves around the origin Σ_b, Eq.(1) is written as

$${}^o\dot{\boldsymbol{p}}_i = {}^o\dot{\boldsymbol{p}}_b + {}^o\dot{\boldsymbol{\omega}}_b \times {}^o\boldsymbol{r}_i. \tag{2}$$

Moreover, we can rewrite the equation by taking into consideration the gravitational acceleration as follows:

$${}^o\ddot{\boldsymbol{p}}_i = {}^o\ddot{\boldsymbol{p}}_b + \boldsymbol{g} + {}^o\dot{\boldsymbol{\omega}}_b \times {}^o\boldsymbol{r}_i + {}^o\boldsymbol{\omega}_b \times \left({}^o\boldsymbol{\omega}_b \times {}^o\boldsymbol{r}_i\right). \tag{3}$$

Equation (3) represents the acceleration of point *i* on the body. If one linear accelerometer is installed at point *i*, the accelerometer output a_i is

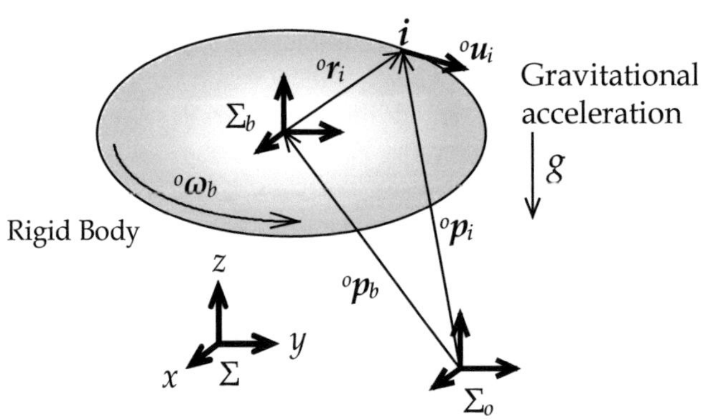

Fig. 1. Acceleration at point *i* on a rigid body

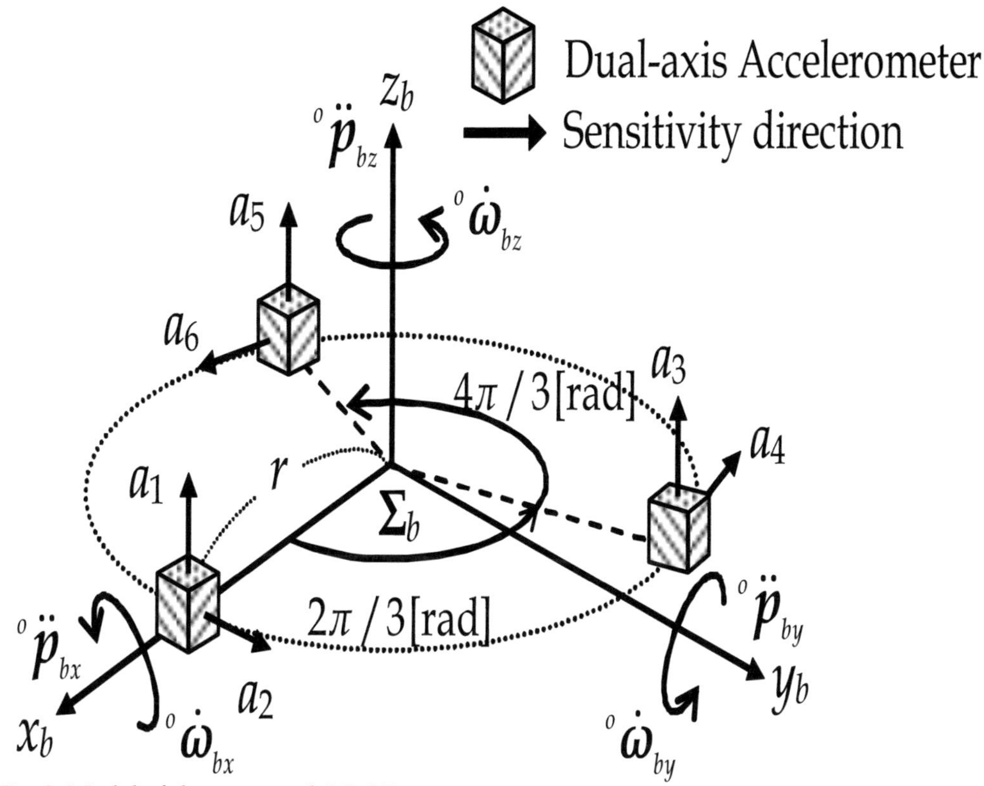

Fig. 2. Model of the proposed 6-DOF measurement system

$$\begin{aligned} a_i &= {}^o\boldsymbol{u}_i^{T\,o}\ddot{\boldsymbol{p}}_i \\ &= \begin{bmatrix} {}^o\boldsymbol{u}_i^T & {}^o\boldsymbol{u}_i^{T\,o}\boldsymbol{R}_i \end{bmatrix} \begin{bmatrix} {}^o\ddot{\boldsymbol{p}}_b + g \\ {}^o\dot{\boldsymbol{\omega}}_b \end{bmatrix} + {}^o\boldsymbol{u}_i^T\left\{{}^o\boldsymbol{\omega}_b \times \left({}^o\boldsymbol{\omega}_b \times {}^o\boldsymbol{r}_i\right)\right\}, \end{aligned} \tag{4}$$

where

$$ {}^o\boldsymbol{R}_i = \begin{bmatrix} 0 & -{}^o r_{iz} & {}^o r_{iy} \\ {}^o r_{iz} & 0 & -{}^o r_{ix} \\ -{}^o r_{iy} & {}^o r_{ix} & 0 \end{bmatrix}. \tag{5}$$

Here, ${}^o\boldsymbol{u}_i = [{}^o u_{ix} \quad {}^o u_{iy} \quad {}^o u_{iz}]^T$ is the sensitivity unit vector. In Eq.(3), the translational, rotational, centrifugal and gravitational accelerations are mixed, and cannot be separated using only one accelerometer output. In order to obtain 6-DOF acceleration data (i.e., data regarding translational and rotational motion), we need to resolve multiple acceleration signals, when more than six linear accelerometers are needed. Thus, in the six accelerometers, each output is given by (see Fig.2)

$$\begin{bmatrix} a_1 \\ a_2 \\ \vdots \\ a_6 \end{bmatrix} = {}^o\boldsymbol{R} \begin{bmatrix} {}^o\ddot{\boldsymbol{p}}_b + \boldsymbol{g} \\ {}^o\dot{\boldsymbol{\omega}}_b \end{bmatrix} + \begin{bmatrix} {}^o\boldsymbol{u}_1^T\{{}^o\boldsymbol{\omega}_b \times ({}^o\boldsymbol{\omega}_b \times {}^o\boldsymbol{r}_1)\} \\ {}^o\boldsymbol{u}_2^T\{{}^o\boldsymbol{\omega}_b \times ({}^o\boldsymbol{\omega}_b \times {}^o\boldsymbol{r}_2)\} \\ \vdots \\ {}^o\boldsymbol{u}_6^T\{{}^o\boldsymbol{\omega}_b \times ({}^o\boldsymbol{\omega}_b \times {}^o\boldsymbol{r}_6)\} \end{bmatrix}, \tag{6}$$

where

$${}^o\boldsymbol{R} = \begin{bmatrix} {}^o\boldsymbol{u}_1^T & -{}^o\boldsymbol{u}_1^T\,{}^o\boldsymbol{R}_1 \\ {}^o\boldsymbol{u}_2^T & -{}^o\boldsymbol{u}_2^T\,{}^o\boldsymbol{R}_2 \\ \vdots & \vdots \\ {}^o\boldsymbol{u}_6^T & -{}^o\boldsymbol{u}_6^T\,{}^o\boldsymbol{R}_6 \end{bmatrix}. \tag{7}$$

${}^o\boldsymbol{R}$ is a gain matrix which depends on the direction of the sensitivity vectors and position vectors. ${}^o\boldsymbol{R}$ is a known constant matrix. If ${}^o\boldsymbol{R}$ is a non-singular matrix, Eq.(6) can be written as

$$\begin{bmatrix} {}^o\ddot{\boldsymbol{p}}_b + \boldsymbol{g} \\ {}^o\dot{\boldsymbol{\omega}}_b \end{bmatrix} = {}^o\boldsymbol{R}^{-1} \left(\begin{bmatrix} a_1 \\ a_2 \\ a_3 \\ a_4 \\ a_5 \\ a_6 \end{bmatrix} - \begin{bmatrix} -0.03({}^o\omega_{bx}{}^2 - {}^o\omega_{by}{}^2) \\ 0.06\,{}^o\omega_{bx}\,{}^o\omega_{by} \\ 0 \\ {}^o\omega_{by}\,{}^o\omega_{bz} \\ -{}^o\omega_{bx}\,{}^o\omega_{bz} \\ 0 \end{bmatrix} \right), \tag{8}$$

where

$$[{}^o\ddot{\boldsymbol{p}}_b + \boldsymbol{g}] = \begin{bmatrix} {}^o\ddot{p}_{bx} - g\,{}^o\theta_{by} \\ {}^o\ddot{p}_{by} + g\,{}^o\theta_{bx} \\ {}^o\ddot{p}_{bz} + g \end{bmatrix}, \quad [{}^o\dot{\boldsymbol{\omega}}_b] = \begin{bmatrix} {}^o\dot{\omega}_{bx} \\ {}^o\dot{\omega}_{by} \\ {}^o\dot{\omega}_{bz} \end{bmatrix}. \tag{9}$$

Here, the terms consisting of angular velocities (${}^o\omega_{bx}, {}^o\omega_{by}, {}^o\omega_{bz}$) are the centrifugal acceleration components in Eq.(8). We tentatively refer to these terms as "CF terms". Furthermore, ${}^o\theta_{bx}$ and ${}^o\theta_{by}$ are the attitude angle of the x axis (roll) and the y axis (pitch), respectively. In addition, the distance r between Σ_b and each accelerometer is 0.06 m.

3. Stability analysis

In motion sensor systems, one of the most serious problems is the drift effect. Although it occurs for various reasons (e.g., vibration or environmental temperature fluctuations), if the solutions are obtained only from the accelerometer outputs (that is, if they are represented by an algebraic equation), they can be improved with relatively high accuracy by performing sensor calibration. However, as shown in Eq.(8) of this system, the z axis acceleration is resolved only by accelerometer outputs (a_1-a_6), while the x and y axis accelerations are resolved by outputs and CF terms with the cross effect in each other. In past studies, it has been indicated that these terms interact with each other, starting with the drift error. As a result, a rapid divergence of the solutions (i.e., the 6-DOF acceleration) caused by the cross effect in addition to the drift error has already been confirmed in past systems. Therefore, in order to analyze the stability of the proposed system in the same case, let us assume the measurement errors occurring as follows:

$$\begin{aligned} &{}^o\dot{\boldsymbol{\omega}}_b = {}^o\dot{\boldsymbol{\omega}}_{bn} + \Delta^o\dot{\boldsymbol{\omega}}_b, \\ &a_i = a_{in} + \Delta a_i \quad (i = 1,2,\ldots,6), \end{aligned} \tag{10}$$

where the suffix n represents a true value, and Δ represents the combined error with gain and offset error. Here, we substitute Eq.(10) into Eq.(8) and solve the equation for the error terms (Δ). As a result, we obtain the following second-order differential equations for the error:

$$\begin{cases} \Delta^o\ddot{\omega}_{bx}(n) - \dfrac{{}^o\dot{\omega}_{bzn}(n)}{{}^o\omega_{bzn}(n)}\Delta^o\dot{\omega}_{bx}(n) + {}^o\omega_{bzn}{}^2(n)\Delta^o\omega_{bx}(n) = F_x(\Delta a_i(n)) \\ \\ \Delta^o\ddot{\omega}_{by}(n) - \dfrac{{}^o\dot{\omega}_{bzn}(n)}{{}^o\omega_{bzn}(n)}\Delta^o\dot{\omega}_{by}(n) + {}^o\omega_{bzn}{}^2(n)\Delta^o\omega_{by}(n) = F_y(\Delta a_i(n)) \end{cases}, \tag{11}$$

where we assume that the above errors are negligibly small, and thus we obtain Eq.(11) by linear approximation.

$$\begin{aligned} &\Delta^o\boldsymbol{\omega}_b(n) << 0, \\ &\Delta^o\boldsymbol{\omega}_b(n)\cdot\Delta^o\boldsymbol{\omega}_b(n) \approx 0. \end{aligned} \tag{12}$$

Equation (11) is a Mathew-type differential equation, which is known to be intrinsically unstable, and therefore we analyzed Eq.(11) by using numerical calculation in order to clarify the stability or instability conditions of the 6-DOF sensor system.

The analytical result is shown in Fig.3, which shows the state of the amplitudes of the error terms ($\Delta^o\boldsymbol{\omega}_b$) for the frequency ratio $\omega_{x\,\mathrm{or}\,y}/\omega_z\,(\omega = 2\pi f)$ when time approached infinity. Here, ω_x, ω_y and ω_z represent the roll, pitch, and yaw frequencies, respectively. This result shows the error terms increase rapidly when the frequency ratio becomes even (i.e., $\omega_{x\,\mathrm{or}\,y}/\omega_z = 2.0, 4.0\,\mathrm{or}\,6.0\ldots$).

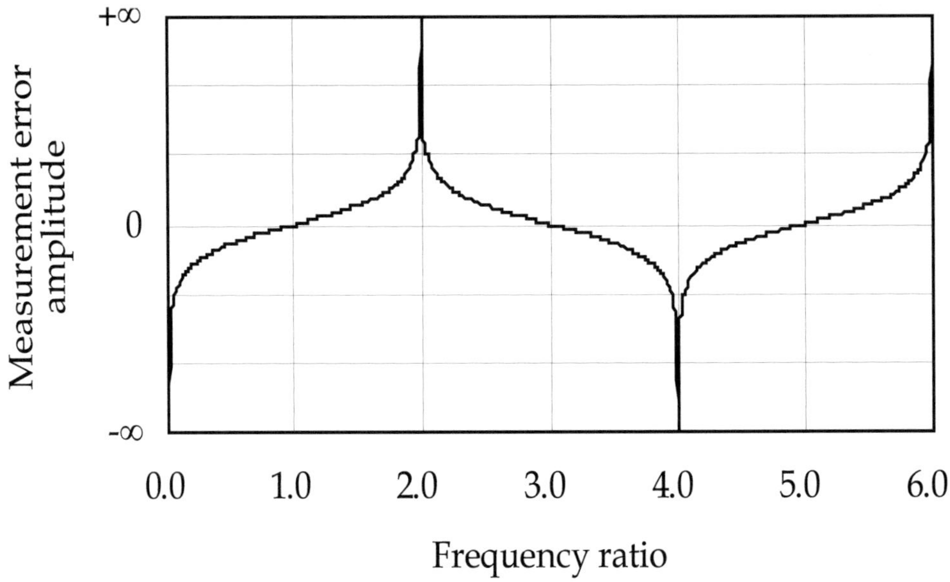

Fig. 3. Measurement error amplitude as plotted against the frequency ratio

Thus, it is considered that this system might become unstable when the roll or the pitch frequency become even for the yaw motion. However, in 3D motion of a rigid body, it is unlikely that the rigid body motion satisfies the above instability condition since the roll or pitch motion is generally synchronized with the yaw motion. This condition is likely to occur in specific cases, such as machinery vibration. Thus, conversely, it is unlikely that the system becomes unstable as a result of the above condition in a rigid body motion, such as the motion of a vehicle or an aircraft. However, the rapid divergence seen in Fig.3 is likely to occur when the system satisfies this condition due to measurement errors or the cross effect induced by the alignment error in this system. In the next section, we analyze the estimated alignment error and investigate a sensor calibration in order to minimize it.

4. Sensor calibration

4.1 Accelerometer error analysis

This sensor system obtains 6 DOF accelerations by resolving multiple linear acceleration signals. Therefore, the sensitivity vector (${}^{o}\boldsymbol{u}_i$) and the position vector (${}^{o}\boldsymbol{r}_i$) of the linear accelerometers must be exact. In the previous section, we investigated the case for each accelerometer without considering the alignment errors. Therefore, in this section, we investigate the accelerometer outputs a_{mi} by taking into account the error terms $\Delta{}^{o}\boldsymbol{u}_i$ and $\Delta{}^{o}\boldsymbol{r}_i$, which constitute the alignment errors of this sensor. Additionally, since accelerometers generally have offset errors, we assume the offset error to be Δa_{ofi}, in which case the accelerometer outputs with these alignment errors can be written as follows:

$$\begin{aligned} \boldsymbol{a}_m &= \left({}^o\boldsymbol{R}+\Delta^o\boldsymbol{R}\right)\begin{bmatrix} {}^o\ddot{\boldsymbol{p}}_b+\boldsymbol{g} \\ {}^o\dot{\boldsymbol{\omega}}_b \end{bmatrix}+\begin{bmatrix} \left({}^o\boldsymbol{u}_i^T+\Delta^o\boldsymbol{u}_i^T\right)\left[{}^o\boldsymbol{\omega}_b\times\left\{{}^o\boldsymbol{\omega}_b\times\left({}^o\boldsymbol{r}_i^T+\Delta^o\boldsymbol{r}_i^T\right)\right\}\right] \end{bmatrix}+\Delta\boldsymbol{a}_{of} \\ &= \left({}^o\boldsymbol{R}+\Delta^o\boldsymbol{R}\right)\begin{bmatrix} {}^o\ddot{\boldsymbol{p}}_b+\boldsymbol{g} \\ {}^o\dot{\boldsymbol{\omega}}_b \end{bmatrix}+\left(\boldsymbol{c}_{rvi}+\Delta\boldsymbol{c}_{rvi}\right)vec\left({}^o\boldsymbol{\Omega}_b^T\right)+\Delta\boldsymbol{a}_{of}, \\ \boldsymbol{a}_m &= \begin{bmatrix} a_{m1} & a_{m2} & \cdots & a_{m6} \end{bmatrix}^T, \\ \Delta\boldsymbol{a}_{of} &= \begin{bmatrix} \Delta a_{of1} & \Delta a_{of2} & \cdots & \Delta a_{of6} \end{bmatrix}^T, \end{aligned} \tag{13}$$

where

$$\begin{aligned} \left({}^o\boldsymbol{R}+\Delta^o\boldsymbol{R}\right) &= \left[{}^o\boldsymbol{u}_i^T+\Delta^o\boldsymbol{u}_i^T \quad -\left({}^o\boldsymbol{u}_i^T+\Delta^o\boldsymbol{u}_i^T\right)\left({}^o\boldsymbol{R}_i+\Delta^o\boldsymbol{R}_i\right)\right], \\ \left(\boldsymbol{c}_{rvi}+\Delta\boldsymbol{c}_{rvi}\right) &= \left[vec\left\{\left({}^o\boldsymbol{u}_i\,{}^o\boldsymbol{r}_i^T\right)\right\}^T\right]^T+\left[vec\left\{\left(\Delta^o\boldsymbol{u}_i\,{}^o\boldsymbol{r}_i^T\right)\right\}^T\right]^T+\left[vec\left\{\left({}^o\boldsymbol{u}_i\Delta^o\boldsymbol{r}_i^T\right)\right\}^T\right]^T+\left[vec\left\{\left(\Delta^o\boldsymbol{u}_i\Delta^o\boldsymbol{r}_i^T\right)\right\}^T\right]^T, \\ vec\left(\begin{bmatrix} a_{11} & a_{12} & a_{13} \\ a_{21} & a_{22} & a_{23} \\ a_{31} & a_{32} & a_{33} \end{bmatrix}\right) &= \left[a_{11} \;\; a_{21} \;\; a_{31} \;\; a_{12} \;\; a_{22} \;\; a_{32} \;\; a_{13} \;\; a_{23} \;\; a_{33}\right]^T, \\ {}^o\boldsymbol{\Omega}_b &= \left({}^o\boldsymbol{\omega}_b\times{}^o\boldsymbol{\omega}_b\times\right) \\ &= \begin{bmatrix} -\left({}^o\omega_{by}^2+{}^o\omega_{bz}^2\right) & {}^o\omega_{bx}{}^o\omega_{by} & {}^o\omega_{bx}{}^o\omega_{bz} \\ {}^o\omega_{bx}{}^o\omega_{by} & -\left({}^o\omega_{bx}^2+{}^o\omega_{bz}^2\right) & {}^o\omega_{by}{}^o\omega_{bz} \\ {}^o\omega_{bx}{}^o\omega_{bz} & {}^o\omega_{by}{}^o\omega_{bz} & -\left({}^o\omega_{bx}^2+{}^o\omega_{by}^2\right) \end{bmatrix}. \end{aligned} \tag{14}$$

In Eqs.(13) and (14), Δa_{ofi} and $\Delta^o\boldsymbol{u}_i$ are based on Table 1, and $\Delta^o\boldsymbol{r}_i$ is based on the accuracy of the finish for the base frame. The sensor frame was cut by NC machinery. Its machining accuracy is ± 0.1 [mm] and ± 0.1 [deg] or better in a length accuracy and an angular accuracy, respectively. The results of calculating the maximum range of each term are shown in Table 2 when the above performance and accuracy are considered. As a result, the terms including $\Delta^o\boldsymbol{r}_i$ can be omitted since they are negligibly small in comparison to the other terms, as seen clearly from Table 2. Thus, we should consider only accelerometer errors including $\Delta^o\boldsymbol{u}_i$, and the error terms (i.e., $\Delta^o\boldsymbol{R}$ and $\Delta\boldsymbol{c}_{rvi}$) can be written as follows;

$$\Delta^o\boldsymbol{R} = \begin{bmatrix} \Delta^o\boldsymbol{u}_1^T & -\Delta^o\boldsymbol{u}_1^T\,{}^o\boldsymbol{R}_1 \\ \Delta^o\boldsymbol{u}_2^T & -\Delta^o\boldsymbol{u}_2^T\,{}^o\boldsymbol{R}_2 \\ \vdots & \vdots \\ \Delta^o\boldsymbol{u}_6^T & -\Delta^o\boldsymbol{u}_6^T\,{}^o\boldsymbol{R}_6 \end{bmatrix}, \quad \Delta\boldsymbol{c}_{rvi} = \begin{bmatrix} \left[vec\left\{\left(\Delta^o\boldsymbol{u}_1\,{}^o\boldsymbol{r}_1^T\right)^T\right\}\right]^T \\ \left[vec\left\{\left(\Delta^o\boldsymbol{u}_2\,{}^o\boldsymbol{r}_2^T\right)^T\right\}\right]^T \\ \vdots \\ \left[vec\left\{\left(\Delta^o\boldsymbol{u}_6\,{}^o\boldsymbol{r}_6^T\right)^T\right\}\right]^T \end{bmatrix}. \tag{15}$$

Therefore, Eqs.(13) and (15) are sensor equations which include principal errors.

Measurement Range		±20 [m/s²]
Resolution (at 60 Hz)		0.02 [m/s²]
Operating Voltage Range		3 ~ 5 [V]
Quiescent Supply Current		0.6 [mA]
Temp. Operating Range		0 ~ 70 [degree Celsius]
Size		5×5×2 [mm]
Gain error	$\mid \Delta u_j \mid / \mid {}^j u_j \mid$	±20 [%]
Absolute alignment error	$\mid \Delta^j \alpha_j \mid / \mid {}^j u_j \mid$, $\mid \Delta^j \beta_j \mid / \mid {}^j u_j \mid$	±1.75 [%]
Offset error	$\mid \Delta a_{ofj} \mid / g$	±200 [%]

Table 1. Specifications of the dual-axis accelerometer chip (Analog Devices, Inc. Accelerometer ADXL202E)

Error factor	$\Delta^o \boldsymbol{R}$			
Error definition	$\dfrac{\lvert \Delta^o \boldsymbol{u}_i \rvert_{\max}}{\lvert {}^o \boldsymbol{u}_i \rvert}$	$\dfrac{\lvert \Delta^o \boldsymbol{u}_i^T \, {}^o \boldsymbol{R}_i \rvert_{\max}}{\lvert {}^o \boldsymbol{u}_i^T \, {}^o \boldsymbol{R}_i \rvert}$	$\dfrac{\lvert {}^o \boldsymbol{u}_i^T \, \Delta^o \boldsymbol{R}_i \rvert_{\max}}{\lvert {}^o \boldsymbol{u}_i^T \, {}^o \boldsymbol{R}_i \rvert}$	$\dfrac{\lvert \Delta^o \boldsymbol{u}_i^T \, \Delta^o \boldsymbol{R}_i \rvert_{\max}}{\lvert {}^o \boldsymbol{u}_i^T \, {}^o \boldsymbol{R}_i \rvert}$
Estimated value	0.251	0.251	0.242×10^{-2}	0.651×10^{-3}

Error factor	Δc_{rvi}		
Error definition	$\dfrac{\lvert vec\{(\Delta^o \boldsymbol{u}_i \, {}^o \boldsymbol{r}_i^T)^T\} \rvert_{\max}}{\lvert vec\{({}^o \boldsymbol{u}_i \, {}^o \boldsymbol{r}_i^T)^T\} \rvert}$	$\dfrac{\lvert vec\{({}^o \boldsymbol{u}_i \Delta^o \boldsymbol{r}_i^T)^T\} \rvert_{\max}}{\lvert vec\{({}^o \boldsymbol{u}_i \, {}^o \boldsymbol{r}_i^T)^T\} \rvert}$	$\dfrac{\lvert vec\{(\Delta^o \boldsymbol{u}_i \Delta^o \boldsymbol{r}_i^T)^T\} \rvert_{\max}}{\lvert vec\{({}^o \boldsymbol{u}_i \, {}^o \boldsymbol{r}_i^T)^T\} \rvert}$
Estimated value	0.251	0.247×10^{-2}	0.751×10^{-3}

Table 2. Estimated error values

4.2 Calibration method for sensor errors

In general, a sensor system is calibrated using a known reference input. In the proposed 6-DOF sensor, it is necessary to determine at least 42 ($6 \times 6 + 6$) components. However, in this system, as shown in Eqs.(13) and (15), the unknown errors constitute 24 components (i.e.,

$\Delta^o \boldsymbol{u} \in \boldsymbol{R}^{3\times 6}$ and $\Delta \boldsymbol{a}_{of} = \left[\Delta a_{of1}\ \Delta a_{of2} \cdots \Delta a_{of6}\right]^T \in \boldsymbol{R}^6$), and these components can be estimated by using only certain translational inputs (i.e., $\boldsymbol{\omega}_b = 0, \dot{\boldsymbol{\omega}}_b = 0$). Thus, Eq.(13) can be written as follows:

$$\boldsymbol{a}_m = \left({}^o\boldsymbol{R} + \Delta^o\boldsymbol{R}\right)\begin{bmatrix} {}^o\ddot{\boldsymbol{p}}_b + \boldsymbol{g} \\ 0 \end{bmatrix} + \Delta \boldsymbol{a}_{of}. \tag{16}$$

When the above equation is solved for $\Delta^o \boldsymbol{u}$ and $\Delta \boldsymbol{a}_{of}$, we obtain

$$\begin{gathered}\begin{bmatrix} {}^o\boldsymbol{A} & \boldsymbol{I}_{6\times 6} \end{bmatrix}\begin{bmatrix} \Delta^o \boldsymbol{u}_i \\ \Delta \boldsymbol{a}_{ofi} \end{bmatrix} = \boldsymbol{a}_m - {}^o\boldsymbol{U}{}^o\ddot{\boldsymbol{p}}_{gb}, \\ {}^o\boldsymbol{A} = \begin{bmatrix} {}^o\ddot{\boldsymbol{p}}_{gb}^{\ T} & & \\ & \ddots & \\ & & {}^o\ddot{\boldsymbol{p}}_{gb}^{\ T} \end{bmatrix}, \quad {}^o\ddot{\boldsymbol{p}}_{gb} = {}^o\ddot{\boldsymbol{p}}_b + \boldsymbol{g}, \\ \Delta^o \boldsymbol{u} = \left[\Delta^o \boldsymbol{u}_1^{\ T} \cdots \Delta^o \boldsymbol{u}_6^{\ T}\right]^T, \quad {}^o\boldsymbol{U} = \left[{}^o\boldsymbol{u}_1^{\ T} \cdots {}^o\boldsymbol{u}_6^{\ T}\right]^T.\end{gathered} \tag{17}$$

In case Eq.(17) is measured n times,

$${}^o\boldsymbol{B}_n \begin{bmatrix} \Delta^o \boldsymbol{u} \\ \Delta \boldsymbol{a}_{of} \end{bmatrix} = \begin{bmatrix} \boldsymbol{a}_m - {}^o\boldsymbol{U}{}^o\ddot{\boldsymbol{p}}_{gb1} \\ \boldsymbol{a}_m - {}^o\boldsymbol{U}{}^o\ddot{\boldsymbol{p}}_{gb2} \\ \vdots \\ \boldsymbol{a}_m - {}^o\boldsymbol{U}{}^o\ddot{\boldsymbol{p}}_{gbn} \end{bmatrix}, {}^o\boldsymbol{B}_n = \begin{bmatrix} \begin{bmatrix} {}^o\boldsymbol{A}_1 & \boldsymbol{I}_{6\times 6} \end{bmatrix} \\ \begin{bmatrix} {}^o\boldsymbol{A}_2 & \boldsymbol{I}_{6\times 6} \end{bmatrix} \\ \vdots \\ \begin{bmatrix} {}^o\boldsymbol{A}_n & \boldsymbol{I}_{6\times 6} \end{bmatrix} \end{bmatrix} \in \boldsymbol{R}^{6n\times 24}. \tag{18}$$

In Eq.(18), ${}^o\boldsymbol{B}_n$ becomes a square matrix when n=4, whose number is the least number of measurement times, i.e., the principal errors $[\Delta^o \boldsymbol{u}^T \quad \Delta \boldsymbol{a}_{of}^{\ T}]^T$ can be estimated using data for only four position.

4.3 A 6-DOF acceleration sensor system

A prototype of the 6-DOF accelerometer and the sensor specifications are shown in Fig.4 and Table 3, respectively. In this sensor, we used a dual-axis accelerometer ADXL202E (Analog Devices Co., Table 1), and installed microcomputer H8-3664 (Renesas Technology Co.) and a USB interface internally. This system captures signals from the accelerometers and transfers the data to the host computer at 60 Hz while performing integer arithmetic calculations in order to speed up the data processing.

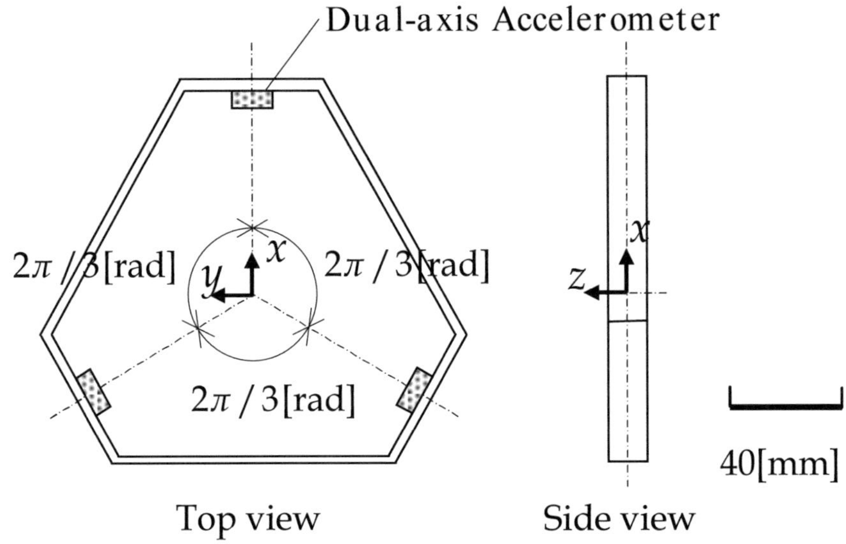

Fig. 4. Outline drawing of a prototype 6-DOF accelerometer

		Max	Typ	Resolution	Sampling
Translational	x	±19.6 [m/s²]	±9.8 [m/s²]	0.04 [m/s²]	60 [Hz]
	y				
	z				
Rotational	x	±490 [rad/s²]	±163 [rad/s²]	0.64 [rad/s²]	
	y				
	z				

Table 3. Specifications of the prototype 6-DOF accelerometer

4.4 Calibration results

Figures 5 and 6 show the measurement results as compared with the state before and after the calibration. This result took the measurement value when the gravitational acceleration was added for an arbitrary direction in the x-y plane of the sensor. Then, the z axis of the 6-DOF sensor was fixed in horizontally direction, and we rotated the sensor in the yaw direction. In addition, the difference between the values before and after the calibration was divided by the rated value (1G), and was thus represented as a dimensionless parameter. In Fig.5, the gain and offset error rates were reduced to 1% or less from about 2% and 4%, respectively. Then, the result from measuring the values along the x axis and the roll acceleration are shown in Fig.6. These accelerations should become essentially zero; however, the measurement values were perturbed by the offset error and the interference with the y axis the before calibration. After the calibration, the offset errors and the interference were reduced, and thus the availability of this calibration method was demonstrated.

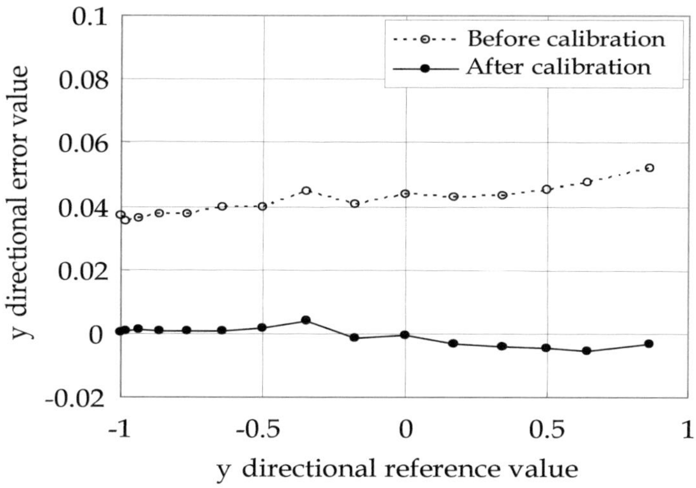

Fig. 5. *y* directional error characteristic before and after the calibration

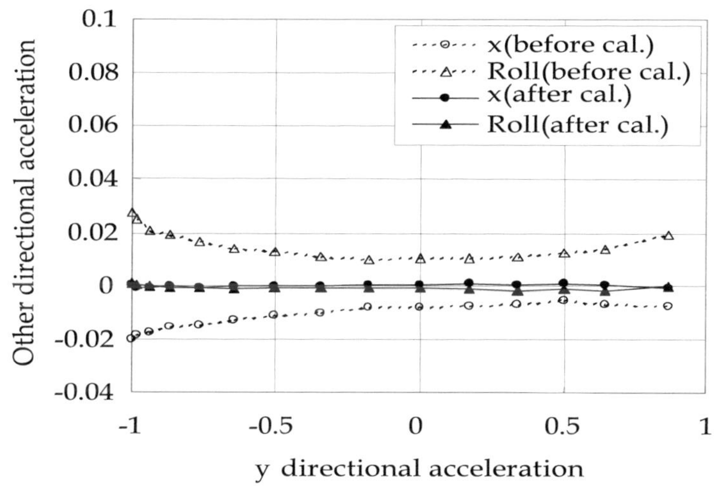

Fig. 6. Interference of the y directional acceleration to the other directions

5. Application to vehicle motion

5.1 Experimental task

In this section, we install the prototype sensor (see Table 3 and Fig.4) in a vehicle and investigate the applicability of the proposed methods to a relatively complex motion. The experimental task is the double lane change that involves the vehicle changing lane twice for a distance of about 2.5 m on a straight track. This task is known that detection of rotational components is relatively difficult because the influence of translational motion is large. Additionally, translational components are likely to become mixed with rotational components, which in turn is likely to induce the cross effect.

5.2 Experimental results

The experimental results are shown in Fig.7. In this experiment, we used a highly accurate triaxial rate sensor (Tamagawa Seiki Co.) which is unaffected by lateral motion as a reference. In the results, angular acceleration was transformed into angular velocity by numerical integration for the purpose of comparison with the rate sensor outputs. In addition, we applied a high-pass filter in order to suppress the drift error from emerging after integration. Figure 7 shows a comparison of the experimental results before and after the calibration in order to demonstrate the applicability of the above calibration method. Before performing the calibration, a large angular velocity was measured in the pitch direction, which should be almost zero, since the lateral components interfered with the rotational component. On the other hand, after performing the calibration, the cross effect was greatly reduced while at the same time the gain and the offset error were suppressed, and therefore the angular velocities were in good agreement with the rate sensor outputs. As a result, the applicability of this sensor system to a relatively complex motion was demonstrated.

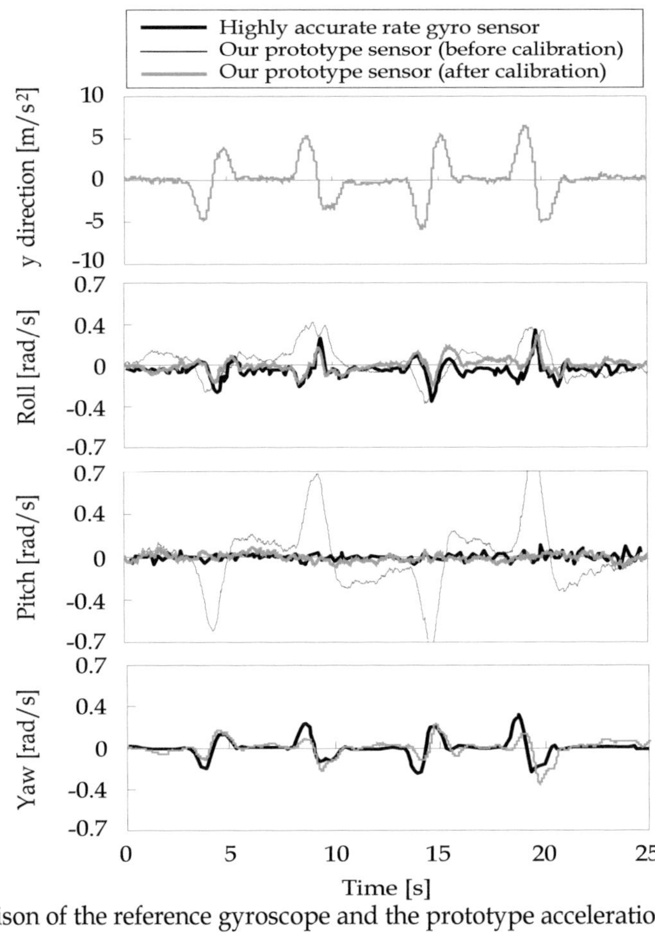

Fig. 7. Comparison of the reference gyroscope and the prototype acceleration sensor in double lane change experiment

6. Conclusion

This chapter described the stability and the error analysis of the proposed 6-DOF motion sensor system. A summary of the results is provided below.

1. This sensor system using multiple linear accelerometers can perform 6-DOF acceleration measurements by resolving each linear accelerometer outputs and CF terms, in other words, the terms consisting of angular velocities (see Eq.(8)). In order to clarify the instability conditions for this system, we analyzed the differential equation for the error by taking the measurement errors into account. As a result of analyzing the geometric instability conditions, the measurement errors increase rapidly when the roll or pitch frequency become even for the yaw motion and thus becomes unstable. However, it is unlikely for the system to become unstable under the influence of the above conditions in a rigid body motion, such as the motion of a vehicle or anaircraft.
2. The alignment errors of this sensor system consist of the position $(\Delta^o \boldsymbol{r}_i)$, sensitivity $(\Delta^o \boldsymbol{u}_i)$, and offset errors $(\Delta \boldsymbol{a}_{of})$. However, in this system, the principal errors are the offset and sensitivity errors since $\Delta^o \boldsymbol{r}_i$ is negligibly small (see Table 2). Thus, these errors can be estimated by using only four linear inputs (acceleration of gravity), and can be calibrated with relative ease and high accuracy for all 6 DOF.
3. We performed an experiment invoving a vehicle changing lanes twice (a double lane change task) using the prototype sensor, and the results demonstrated the applicability of the proposed system to relatively complex motion (see Fig.7).

7. References

Fraden, J. (2003). *Handbook of Modern Sensors,* AIP Press, page numbers (301-322), ISBN 0387007504

Padgaonkar,A.J, Krieger, K.W & King,A.I. (1975). Measurement of Angular Acceleration of a Rigid Body Using Linear Accelerometers. *Transaction of the ASME Journal of Applied Mechanics,* 42, page numbers (552-556)

K. Ohta, K. Kobayashi. (1994). Measurement of Angular Velocity and Angular Acceleration in Sports Using Accelerometers. *Transactions of the Society of Instrument and Control Engineers,* Vol. 30, No. 12, (December and 1994) page numbers (1442-1448), ISSN 04534654

K. Ohta, K. Kobayashi. (1995). Measurement of Angular Velocity and Angular Acceleration in Sports Using Extended Kalman Filter. *Transactions of the Society of Instrument and Control Engineers,* Vol. 31, No. 9, (September and 1995) page numbers (1265-1272), ISSN 04534654

T. Masuda, Y. Tanizawa. (2001). Measurement Experiment of the 6 DOF Acceleration Using Multiple Sensors. *Research Report of Industrial Reasearch Division, Mie Pref. Science & Techonology Promotion Center,* No.26, page numbers (117-120), ISSN 13478796

N. Mimura, R. Onodera. (2005). Stability Analysis of 6 DOF Acceleration Sensor System Using Multiple Accelerometers. *Transactions of the Japan Society of Mechanical Engineers Series C,* Vol. 71, No. 707, (July and 2005) page numbers (2218-2224), ISSN 03875024

N. Mimura, R. Onodera. (2006). Calibration Method for 6 DOF Acceleration Sensor Systems Using Multiple Accelerometers. *Transactions of the Japan Society of Mechanical Engineers Series C*, Vol. 72, No. 724, (December and 2006) page numbers (3798-3805), ISSN 03875024

R. Onodera, N. Mimura. (2006). A Stability of Six-Degrees-of-Freedom Acceleration Sensor System by Integration of Acceleration Signals (Considerations about a Dynamic Influence of Centrifugal Force Term in the Multi-Degree-of-Freedom Motion). *Transactions of the Japan Society of Mechanical Engineers Series C*, Vol. 72, No. 724, (December and 2006) page numbers (3789-3797), ISSN 03875024

R. Onodera, N. Mimura. (2008). Measurement of a Vehicle Motion Using a New 6-DOF Motion Sensor System –Angular Velocity Estimation with Kalman Filter Using Motion Characteristic of a Vehicle-. Journal of Robotics and Mechatronics, Vol.20, No.1, (February and 2008) page numbers (116-124), ISSN 09153942

14

Toward Intelligent Biped-Humanoids Gaits Generation

Nizar ROKBANI , Boudour AMMAR CHERIF and Adel M. ALIMI
nizar.rokbani@ieee.com , Boudour.ammar@gmail.com, adel.alimi@ieee.org
REGIM, Research Group on Intelligent Machines, National School of Engineering of Sfax, University Of Sfax, Bp W, Soukra road, 3038 Sfax, Tunisia.

Abstract

In this chapter we will highlight our experimental studies on natural human walking analysis and introduce a biologically inspired design for simple bipedal locomotion system of humanoid robots. Inspiration comes directly from human walking analysis and human muscles mechanism and control. A hybrid algorithm for walking gaits generation is then proposed as an innovative alternative to classically used kinematics and dynamic equations solving, the gaits include knee, ankle and hip trajectories. The proposed algorithm is an intelligent evolutionary based on particle swarm optimization paradigm. This proposal can be used for small size humanoid robots, with a knee an ankle and a hip and at least six Degrees of Freedom (DOF).

Keywords

Human gaits analysis, intelligent robotics, Humanoid robotics, Biped robots, evolutionary computing, particle swarm optimization.

1. Introduction

In near future humanoid robots will be asked to "live" and collaborate with humans, they have generally a friendly design with a great resemblance with us. Biped humanoids robots are expected to have increasing field of exploitations, they are naturally adapted to human like commodities and have the advantage to fit more the human's environment than other kind of robots; humanoids can coordinate tasks with humans workers (Arbulu & Balaguer, 2008) with human like comportments. Humanoid robotics issues include building machines with human like capabilities in motion control, gestures, and postures. Recently a lot of work is done in order to enhance their capacities in walking, fast walking and motion optimization (Xie et al, 2008) in a human like environment. The development of machine intelligence is crucial of the independence of these robots it is also crucial to the emergence of self learning attitudes. All these factors impulses the rapid growth of humanoid robotics.
Humanoid robots are legged robots with a limited number of two legs, such a limitation is important since it has direct impact on stability control policy. When two legs are used in a

dynamic walking process, the sustention polygon is limited to a single footprint. In natural human walking process the double support phase is rarely used (Rokbani et al, 2008).
In legged robotics gaits analyses are commonly used to understand the locomotion system. Gaits analysis is a methodology introduced by biomechanics scientists to analyze and enhance motion dynamics of humans and animals. The human walking is the result of a long learning process that begins in the childhood life. Humans naturally adapt their walking-steps to their environments in order to ensure their safety and avoid falling down; they naturally optimize their energy and maintain their anthropomorphic stand up while switching between the support and transfer phases or between static and dynamic walking. The walking process takes into consideration all the humans activities, they can walk while eating, drinking or simply responding to phone call.
The "IZIMAN" is a research project of the REGIM laboratory, "Research group on intelligent Machines", with some ambitious challenges aiming to propose intelligent architectures that are biologically inspired in order to offer to humanoid robots a form of self tutoring (Rokbani et al, 2007).
In second section of this chapter we will highlight briefly the main issues of bipedal and humanoid locomotion robotics; in section III we present our gaits capture experiments, note that the gaits are extracted to be used as a comparison material, in section IV, we propose a hybrid method for human like gaits generation. The last section, V, is reserved to discussions and further works.

2. Humanoid locomotion robotics

2.1. Humanoid and biped robots evolution

A humanoid robot is commonly assumed to be a biped otherwise some wheeled humanoid robots are proposed by researchers (Berns et al, 1999). Humanoid robotics includes all the aspects of human like machines such as: walking, grasping, emotion or cognition, etc. Humanoid locomotion robotics is interested only in the issues concerning gaits generation, stability control, walking energy optimization. Recently an increasing interest to humanoid robotics helps the development of more humanoids; small size humanoids are actually proposed for entertainments; some models can be used as low cost research validation platforms. One of the earliest projects in humanoid locomotion is the Waseda university, Japan where Kato and his team build their walker robot, WL1, since 1966. By 1984 a humanoid dynamic walker is proposed and called WL10 RD, this prototype is a 12(DOF)[1], walker robot. Wabian is a real humanoid robot having 52 (DOF), including arms and head and able to walk at a speed of 0.21 meter/second. On the other hand, some sophisticated humanoids robots are developed by Japanese industrial companies such as Sony with their QRIO humanoid robot and Honda with its well known ASSIMO. One of the most enhanced humanoid projects is certainly the HRP2 Japanese projects; this robot is comparable to ASSIMO, the HRP2 is a Human size robot and shape, with low energy needs and it is proposed free from the back-bag, used as battery container in ASSIMO. It is important to distinguish between full size humanoids and smaller ones. Full size humanoids such as HRP2 or ASSIMO have a size which is comparable to humans. Robots like Nimbro (Behnke

---

1 DOF : Degree Of Freedom

et al, 2007) or Qrio have sizes that not exceed 1m. Recently a French amazing humanoid robot, called Nao, was proposed while still under development. The Nao robot was the official platform for the IEEE Humanoid conference and Robocup competitions.

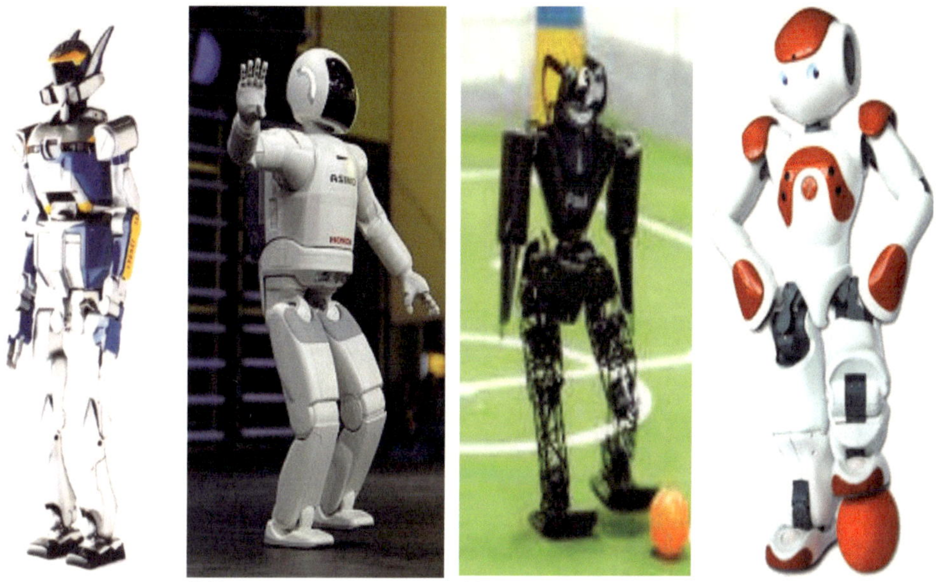

(a)[2] (b)[3] (c)[4] (d)[5]

Fig. 1. Humanoid robots: (a) and (b) are human size robots, (c) and (d) are middle size robots. (a) HRP2, (b) ASSIMO, (c) Nimbro soccer robot, (d) NAO robot.

In Europe, some research projects have been developed in France, Belgium, Deutschland and Italy. French projects are essentially focused on the locomotion system. Bip and rabbit are the most famous projects; the "Bip" robot is an autonomous walker carrying his control system while the rabbit robot is a platform for dynamic walking analysis and experiments (Azevedo et al, 2004) (Chemori, 2005). In Belgium the Lucy robot is also a walker robot; its particularity is that it uses pneumatic muscles as actuators. Johnnie is a Deutsch humanoid founded by Technical University of Munich. Recently new proposals are made in order to generalize the ZMP[6] approach, new intuitive methods such us the (RTC)[7] (Goobon, 2008) allowing to maintain biped walker stability on irregular floor or stepping over an obstacle while walking (Sabourin et al, 2008).

---

2 Source : http://www.generalrobotix.com/

3 Source : http://www.androidworld.com

4 Source : www.informatik.uni-freiburg.de

5 Source : www.aldebaran-robotics.com

6 ZMP : Zero Moment Point

7 RTC : Relative Trajectory Control

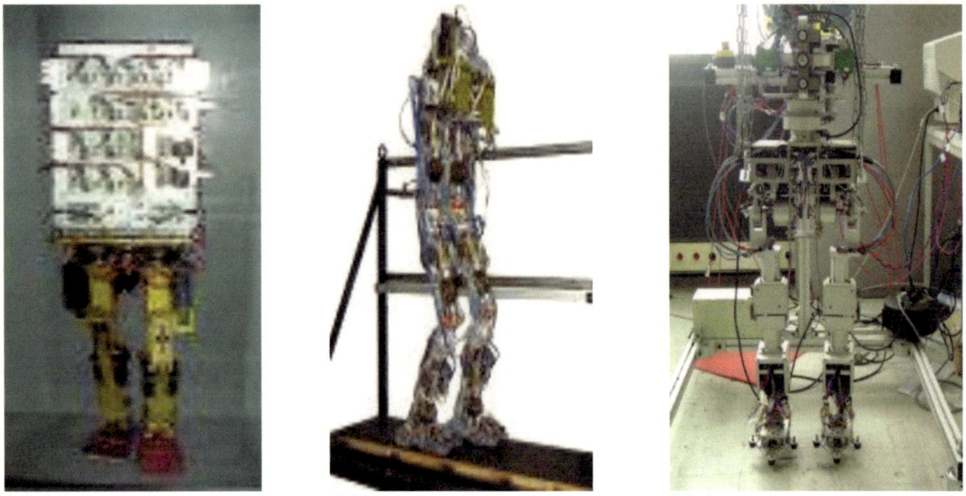

(a)[8] (b)[9] (c)[10]

Fig. 2. Humanoid locomotion systems or biped robots, (a) BIP robot, (b) Lucy robot, (C) Robian robot.

2.2. Description of the IZIMAN project

"IZIMAN" is a research project aiming to propose alternative intelligent solutions to humanoid robots. The IZIMAN architecture aims to propose a schema to help emergence of machine intelligence rather than to create an intelligent machine directly inspired from humans; the proposed architecture is a multi agent's one, composed by controls and decisions centers that have to collaborate to ensure the man-like tasks of the humanoid robot. A special focus is attributed to the learning processes; learning uses build in intelligent based on evolutionary, fuzzy or neuro-fuzzy algorithms. Human imitations are also considered as source of learning frames. Learning based on imitation needs efficient human detection and gestures tracking.

Assuming that the proposed control system is hybrid and integrates different types of controllers; a collaboration schema should be defined to avoid collateral risk of antagonist command signals. The collaboration schemas establish the hierarchy between the controllers and solve the multiplicity command problematic (Rokbani et al, 2007).

For the gaits generation of the biped locomotion system, we opted for a multi-swarms design, each swarm controls a joint. A collaboration schema insures the coherence on the generated joints trajectories called also gaits. This architecture will be detailed in section IV.

3. Human walking analysis

Human walking is a complex process that uses a 29 (DOF) locomotion system coupled with different neural centers including brain and spinal-reflex centers. More than 48 muscles

---

8 Source : http://www-lms.univ-poitiers.fr/article182.html

9 Source : http://lucy.vub.ac.be

10 Source : http://www2.cnrs.fr/presse/communique/410.htm

are used to coordinate foot motions and synchronize them with the upper body constrains. The joints of each foot perform elementary movements essentially flexion and extension; that includes the hip, the knee and the ankle. The walking is a result from the coordination of these simple movements. It is important to note that upper body is connected to the locomotion system trough the basin; the upper body includes a critical point assumed to be the center of mass, COM, and playing a key role in balance, stability and walking dynamics.

3.1. Description of human locomotion system

From an anatomic point of view, the lower trunk of humans is composed by a pelvic bone connecting two legs. A leg is composed by two vertical segments: the femur and the tibia. The femur is attached to the pelvic bone by the hip joint and to the tibia by the knee joint (Wagner & Carlier, 2002). Biomechanical description of the human like lower trunk shows that the ankle performs two rotations, the knee performs two rotations and the hip three rotations. Most biped construction use limited degrees of freedom compared to human locomotion system. Such a limitation reduces the dynamic of the walking but makes easier the stability control. Human motions result from flexions and extensions; these elementary actions are produced by muscles. Assuming the effectiveness of the muscles, pneumatic muscles proposed and had been used in biped robots such as Lucy biped. In human locomotion system flexion and extension of muscles produces rotations on the human joints. In a humanoid locomotion system these rotations depend on the skeleton structure and on the actuators specification; but whatever are the nature of the build skeleton and the used actuators most bipeds respect globally the human joints angular elongations limits.

3.2. Description of walking gaits

The human body is represented by segments and joints; this simplification is commonly used in biomechanics; the outline of a leg can be represented by seven anatomical markers for three joints (Winter, 1990), a segment is obtained by joining two consecutive markers. Considering the complexity of human walk it is common to use a limited number of states to represent a walking cycle; only key steps are detailed and assumed to be representative of overall a cycle. Wagner uses sixteen states (Wagner & Carlier, 2002) while Winter uses thirteen (Winter, 1990). In robotics the used steps are limited compared to biomechanics descriptions (azevedo & Heliot, 2005). This representation simplifies the computing of angular limits of the joints, from a gait analysis it becomes possible to establish an estimation of these limits.

To collect human gaits, we have used the classical marking methodology where participants are young humans with ages ranging from 20 to 24 years, the participants are students at the High Institute of Sport and Physical Education of Sfax, Tunisia. We have installed a walking scene of about 8 meters long, 2 meters large and 4 m height. The scene walls were covered by black, to simplify marker extraction. Experiments were conducted under supervision of biomechanics' walking specialists. Six participants are asked to work normally from a start point to the end of the scene, avoiding accelerations and decelerations. Two cameras were placed respectively at the left of the walkers and in front of them. We had market the hip, the knee, the ankle and the foot of each walker using white circular stickers with a diameter of 3 cm (centimeter), see figure 3. The walking cycles were captured using two numeric cameras placed respectively at 5m40 from the lateral plan and at 8m 25 from the frontal plan. The scene, a lateral walker and a frontal walker can be seen in respectively figure 3 (a), (b) and (c).

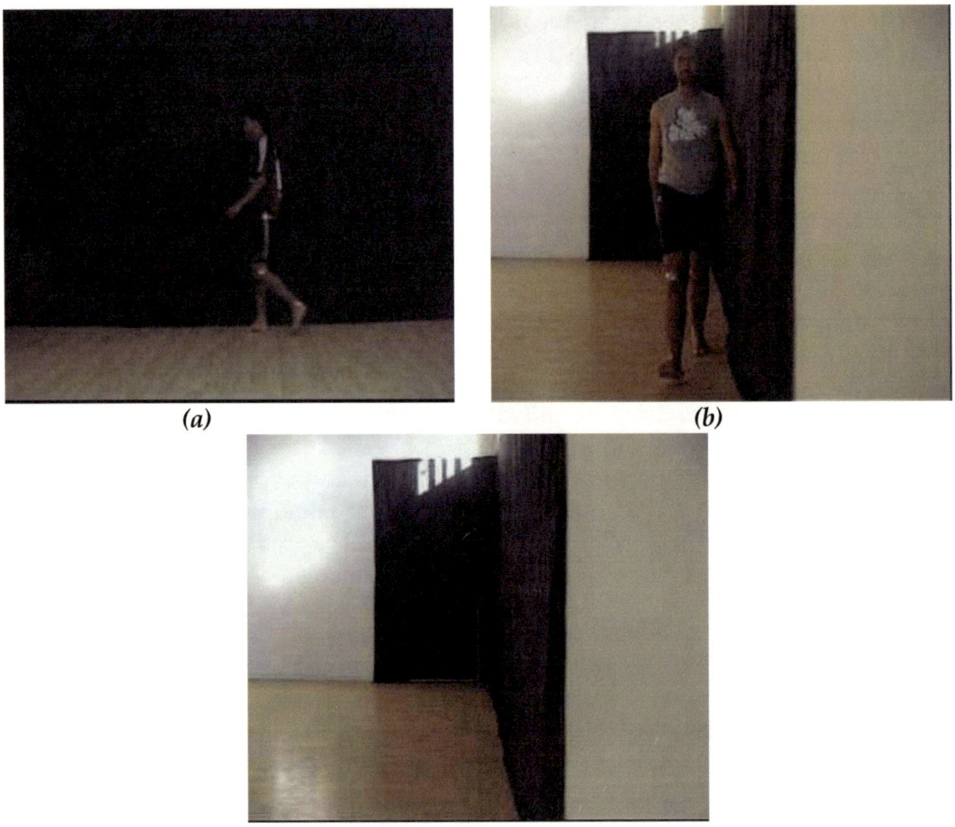

(a) *(b)*

(c)

Fig. 3. human gaits capture, (a) human walking on lateral plan, (b) human walking in frontal plan, (c) capture scene.

4. Hybrid approach to humanoid like gaits generation

4.1 Walker initialization

Since we are concerned by human inspired locomotion system, we used human anthropometry to establish an approximation for the humanoid body segments dimensions. In human anthropometry theses dimensions depend on sex, body build, race origin, etc. In our work, we are using an average length expressed in (Winter, 1990) and detailed by formulas (1) to (5). A humanoid segment-based representation is then built as it can be seen in figure 4(c). It is important to indicate that in this work the footprint is added just in order to simplify the static stability condition; in real applications a footprint introduces the problem to floor reaction.

$$fl = 0.152 * h \tag{1}$$

$$fb = 0.055 * h \tag{2}$$

$$tl = 0.246 * h \tag{3}$$

$$leg\_l = 0.53 * h \tag{4}$$

$$Int\_h = 0.191 * h \tag{5}$$

Where ***fl, fb*** represent respectively the foot length and the foot breadth, ***tl*** is the length of the segment knee to ankle; the ***leg_l*** term represents the leg length and ***Int_h*** is the inter-hip distance.
After walker skeleton initialization we introduce a new proposal to achieve gaits generation; instead of using classical kinematics modeling we introduce a swarm based search policy. A swarm of particles is placed in the corresponds joints of the walker, the swarm has to find a suitable new position for the joint assuming both forward stepping and avoiding the robot fall down.

In particle swarm optimization, as founded by Kennedy and Eberhart (Kennedy & Eberhart, 1995) (Dreo et al, 2003), the swarm is composed by a set of particles; each particle has its position, its velocity and its fitness function. Initially the particles are randomly initialized round a joint position; a search space is then allocated to the swarm. Particles have to find within the search space the new position, all the swarm will consider its actual position and the position of the best particle. A fitness function is used to evaluate the performance of each particle and to select the best one. The original formulation of PSO can be expressed through the following formulas:

$$v_{i(t)} = v_{i(t-1)} + c_1 * r_1 * \left[p_{lbest} - x_{i(t-1)} \right] + c_2 * r_2 * \left[p_{gbest} - x_{i(t-1)} \right] \tag{6}$$

$$x_{i(t)} = x_{i(t-1)} + v_{i(t)} \tag{7}$$

Where, $x_{i(t)}$ is the current position of the particle (i), $v_{i(t)}$ is the particle velocity;
c1 represents the moderation of particle personal contribution;
c2 represents the moderation of social contribution; (r1) and (r2) are random numbers within the interval [0.0, 1]. Note that p_{lbest} represents the best particle in the particle (i) neighborhood; p_{gbest} is the particle getting the best fitness function of the swarm. (t) is the current time step and (t-1) is the last time step.

The swarms are grouped in right and left foot sub-swarms. Since in walking cycle we have a support foot and an oscillating one the foot sub-swarms will moves alternatively; when a foot is assumed to be the support one, all its particles are in hunt mode. Collaboration is also very important, a forward valid step depends on the joints positions of the overall skeleton. To satisfy those constraints some collaboration connections are established between the joints swarms; the connections appear in figure 4(b). Really a connection means that the best position of a joint is communicated to the upper joint and the virtual COM in order to check and validate the stability of the forthcoming joints displacements.

4.2 Swarms processing details

The proposed architecture is composed by two sub-swarms; the sub-swarm(0), see figure 4 (b), represents the left foot and is composed by three main particles ***P0.1***, ***P0.2*** and ***P0.3*** respectively the hip, the knee and the ankle. The same organization is used for the right feet which is represented by the sub-swarm(1). These particles represent valid skeleton positions and are not supposed to move frequently, they are memories particles. Around each one of them, there is a limited set of search particles working in search space prospecting. A walking step includes seven coordinates representing the articulations and the body mass. To ensure the stability of the system we assign a memory-particle to each articulation. This particle will be moved only if its next position ensures a forward walking-step and the static stable attitude of the walker.

If we consider the particle ***p0.1***, see figure 4 (b), this particle is a memory and will not be authorized to move since its new position is not better than its actual one and since the new position do not insure stability. The search particles related to ***p0.1*** are initialized and iterates looking for the best position in their allocated search space. The local best position is elected according to both the local search strategy and with respect to global stability strategy of the walker. Global stability depends on whether the COM particle assumed here by ***P0.0*** is within the sustention polygon. ***P0.0*** represents the COM of the robot mass, it is a virtual particle and has no search swarm attached to it.

The swarms are conceived to work as follows: The memory particles are connected hierarchically from bottom to top, such a connection means that particle ***(p0.2)***, ***particle*** **2** of ***swarm 0*** communicates its position to both particles ***(p0.1)*** and ***(p0.3)***, it gathers also the respective positions of these particles. Particle ***(p0.3)*** has only the position of ***(p0.2)***. ***P0*** is a specific particle used to represent the body center of mass, *COM*, it's position is estimated using the best sub-swarms particles and using Eq(9)and (8). This particle is critic since the analysis of its coordinates will help to decide on whether the generated joints positions correspond to stable posture or not.

4.3 Stability control policy

Dynamic walking is the natural human attitude; the double support phases are very limited we move always on a single support. Static equilibrium walking can be described as a succession of standing stable positions, even in a simple support phase, if the walking is halted, during the cycle, the robot still standing up in an equilibrium position, Figure 5 illustrate the position of the COM during a walking cycle where simple support phases alternate with double support ones. For a humanoid it is easier to achieve a static equilibrium rather than a dynamic one. In a slow forward the static stability is assured if:

1) In a double support phase, the center of mass of the robot is placed in the center of the sustention polygon.

2) In a simple support phase, the center of mass COM is placed on the support foot. That condition insures that the robot will not fall down while it balances.

```
Initialize number of sub-Swarms.
Initialize number of particles of the sub-swarms.
Initialize n1, n2 = max iteration number1.
Initialize the joints memories M(i,j).
Begin
    Loop
Select the support leg
Initialize the oscillating leg sub-swarms:
        Sub-swarm(i)= {Si.0, Si.1, Si.2}
Initialize the local search swarms S(i,j)
Loop
lunch Sub-swarms(i)
         for i= 1 to number of sub-Swrams
        run sub-swarms()

       evaluate local bests : Gbest(i,j)
        Next (i)
Validate the step:
        Estimate COM(x,y)position
         Check the stability of the walker
M(i,j) = Gbest (s(i,j).
Change support leg
Transfer the particles dynamics ()
    Until (number of iteration = n1)
    Until (number of iteration = n2)
    End
```

(a)

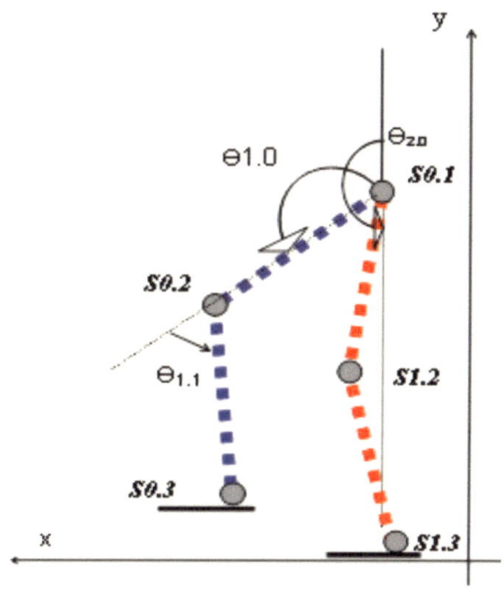

(c)

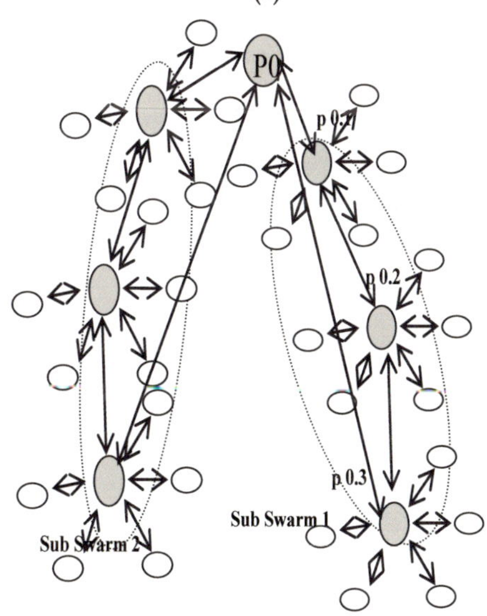

(b)

Fig. 4. (a) Pseudo-code of the proposed algorithm, (b) Sub swarms architecture and inference graph, the memories particles appears in grey, the search particles are white, (c) Side view of the simplified bipedal locomotion system.

From a biomechanics' point of view, the coordinates of the center of mass, COM, are estimated by Eq (9). If we consider a body with n segments (S_1, S_2, …,S_n) and if we assume that the weight of these segments are (m_1, m_2, …m_n) the center of mass can be estimated. In our case we need to estimate the coordinates of the CoM projection on the floor plan, which means that we need a pair of coordinates COM(x,y). The relative masses of the segments depend on the weight of the skeleton, on the nature and proportion of the used effectors and sensors; alternatively it can be directly inspired from human anthropomorphic studies even if human muscles still better than any industrial actuators.

$$x^i_{com} = \frac{\sum_{j=1}^{n\_Sw} m_j * x_j}{M} \quad (8)$$

$$y^i_{com} = \frac{\sum_{j=1}^{n\_Sw} m_j * y_j}{M} \quad (9)$$

Where :

(i) represents the iteration number,
n_Sw is the number of sub-Swarms in the proposed model, here ***n_Sw*** = 6, see figure 1(a). ***M*** is the mass of the locomotion system and m_j represents the mass of the segment (S_j).

If we assume that the robot has a footprint which is propositional to its locomotion-system dimensions, and if we assume that the footprint is rectangular, see subsection A, we deduce a simple representation of the sustention polygon in both double and single support phases, see Figure 5. The foot-print is supposed to be rectangular with the length ***fl*** and a breadth ***fb***, Eq(1) and (2). If only single support phases are used during the walking cycle the sustention polygon is limited to the segment joining ***p0.3*** (left ankle) to ***p1.3*** (right ankle), this is a constrained solution compared to the first one.

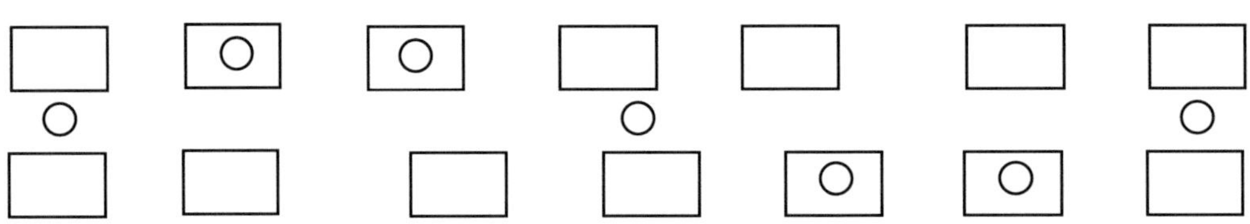

Fig. 5. A static walking cycle foot prints, COM (circle) projection on the footprints (rectangle).

4.4 Fitness functions

In our case we need both, local finesses functions in order to select local bests particles, and a global fitness function that allow us to select the best posture within those assuming stability. Local best particles are those minimizing the error expressed by Eq (10).

$$f_{sw(j)} = \left\| \overrightarrow{p^{s}{}_{i,j}\, p^{th}{}_{i,j}} \right\| \tag{10}$$

Where $f_{sw(j)}$ is the fitness function of the swarm (j), $p^{s}{}_{i,j}$ is the 3D point representing the memory particle of the joint (i,j); $p^{th}{}_{i,j}$ is the theoretical point corresponding to the joint position obtained by a direct kinematic solver.

$$p^{th}{}_{i,j} = \begin{bmatrix} x_{i-1} \\ y_{i-1} \\ z_{i-1} \end{bmatrix} + M^{sg}{}_{i} * M^{ft}{}_{i} * \begin{bmatrix} 0 \\ 0 \\ l \end{bmatrix} \tag{11}$$

Where $M^{sg}{}_{(i-1)i}$ and $M^{sg}{}_{(i-1)i}$ represent respectively the rotation matrices on the lateral and frontal plans they are expressed as fellows:

$$M^{sg}{}_{i} = \begin{bmatrix} \cos(\theta_{j,i-1},) & 0 & -\sin(\theta_{j,i-1}) \\ 0 & 1 & 0 \\ \sin(\theta_{j,i-1}) & 0 & \cos(\theta_{j,i-1}) \end{bmatrix} \tag{12}$$

Note that $\theta_{j,i-1}$ is the last stable position angle of the joint(i) that belongs to the sub-swarm(j) on lateral plan. The frontal plane rotation matrix for axis (Y,Z) is expressed by equation (13) :

$$M^{ft}{}_{i} = \begin{bmatrix} 1 & 0 & 0 \\ 0 & \cos(\alpha_{j,i-1}) & -\sin(\alpha_{j,i-1}) \\ 0 & \sin(\alpha_{j,i-1}) & \cos(\alpha_{j,i-1}) \end{bmatrix} \tag{13}$$

Where $\alpha_{j,i-1}$: is the last stable position angle of the joint (i) that belongs to the sub-swarm(j) on the frontal plan.

To evaluate global best positions and stability policy we try to minimize the following expression, the best pattern is that one who minimizes the distance between the projection of the COM and the Gravity center on the sustention polygon.

$$f_{itness} = \sqrt{\left(y_{com} - y_{polycenter}\right)^2} + \sqrt{\left(x_{com} - x_{polycenter}\right)^2}) \tag{14}$$

Note that $(x_{polycenter}, y_{polycenter})$ represent the coordinates of the sustention polygon center of gravity, they vary according to whether the robot is in single support or double support phase.

5. Discussions and Further Developments

In this paper we have briefly resumed main biped and humanoid locomotion research issues before introducing the IZIMAN project. We presented our experiments in human gaits captures, these gaits help us understanding the walking mechanism and are used as a comparative frame work to validate the simulated approaches; classical kinematics modeling is also used to generate joints trajectories but is not detailed in this paper. We essentially detailed our biologically inspired hybrid gaits generation methodology. The proposal is based on particle swarm optimization.

The joints extracted from the biomechanics experimentations are limited since only six walkers were involved in the experimentation process, in the future large scale gaits captures will be organized; even if this work had been done by earlier bipedal locomotion researches it still very instructive on the way walking and anthropomorphisms works.

Particle swarm optimization belongs to what is commonly assumed to be evolutionary computing; it is based on a quiet simple equation fast to compute with a low memory cost, they can be used as an alternative to mathematical equation solvers especially in non linear systems.

3D walking steps are shown in figure6 (a), while a lateral walking is represented in figure6 (b). The walking gaits of the COM can be observed in figure 6(c), evolutionary based ones appear in figure6 (d). The PSO based gait generator performs globally similar results to those obtained by classical modeling (Ammar, 2006) on the axes x and y, while the z-gaits are slightly different showing that the walker has a dissymmetric motion; one of the members oscillates longer than the other. In human normal walking attitude, a slight dissymmetry is also observed but it is not so important than that produced by the PSO approach. On the other hand and assuming that PSO is a non deterministic technique, only valid gaits should be saved in memory. The convergence of the algorithm does not insure that a complete walking cycle is always gathered..! A learning technique should be soon introduced to overcome this problem. The obtained gaits will be soon implemented as reference joint trajectories in a small size humanoid assembled in the REGIM laboratory using the BIOLOID expert robotics kit.

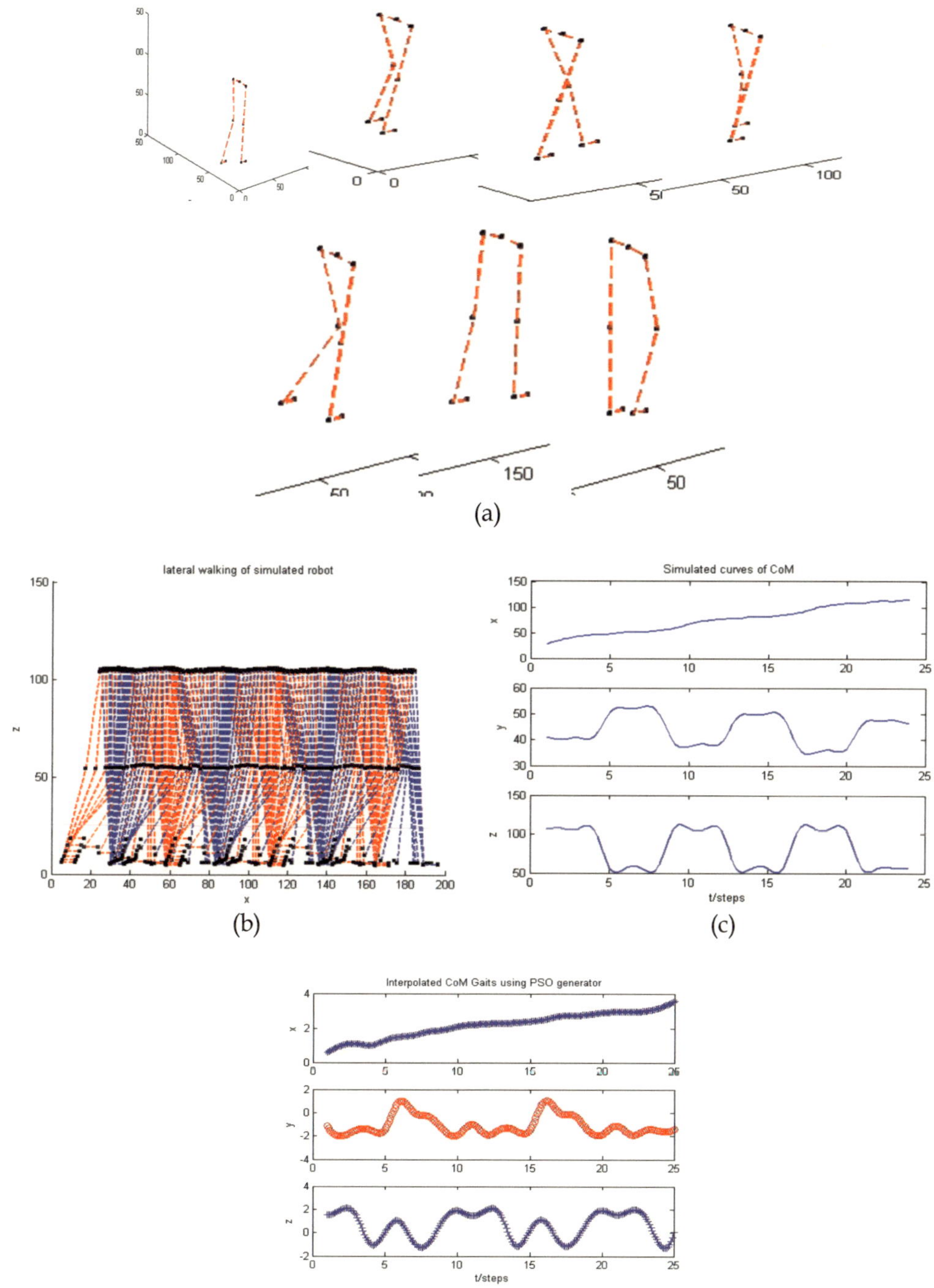

Fig. 6. Simulation results, (a) walking frames screen shots, (b) lateral gaits of segments femur and tibia from classical kinematics' simulation, (c) COM gaits from classical kinematics' simulation, (d) COM gaits from PSO proposal approach.

6. References

M.Arbulu and C.Balaguer, "Human-Humanoid robot cooperation in collaborative transportation tasks", in Proc of Clawar2008, Coimbra portugal 2008.

M.Xie, Z.W.ZHONG, L.Zhang, L.B.Xian, L.Wang, H.J.Yang, C.S.SONG and J.LI, "A deterministic way of planning and controlling biped walking of Loch Humanoid robot", In proc of Clawar2008, Coinbra, Portugal,2008.

N.Rokbani, A.M ALMI and B.ammar, "architectural proposal for an intelligent robotized humanoid iziman", in proc of iEEE conf on logistics and automation, xandho, china 2007.

N.Rokbani, E. H. Ben Boussada and A M.Alimi, "particle swarm optimization for humanoid walking-gaits generation", in proc of clawar 2008, coimbra portugal, 2008.

K.Berns, T. Asfour, and R.Dillman, "ARMAR-an anthropomorphic arm for humanoid service robot", IEEE Conf on Robotics and Automation, Volume 1, Issue , 1999 Page(s):702 - 707 ,1999

S.Behnke, M.Schreiber, J.Stückler, H.Strasdat, and K.Meier, "NimbRo TeenSize 2007 Team Description", In RoboCup 2007 Humanoid League Team Descriptions, Atlanta, July 2007.

C.Azevedo, P. Poignet and B.Espinau, "Artificial locomotion control: from human to robots", Robotics and Autonomous Systems, Vol 47, pp 203-204, 2004

A. Chemori, « Quelques contributions à la commande non linéaire des robots marcheurs bipèdes sous-actionnés », Thèse de l'INPG, Grenoble, 2005

C.Goobon , "simple intuitive method for a planar biped robot to walk", in proc of clawar 2008 conf, Coimbra, portugal 2008.

C. Sabourin, K.madani, W.yu and J.yan, "Obstacle avoidance strategy for biped robot based on fuzzy Q-learning", in proc of clawar 2008 conf, coimbra, portugal 2008.

D. A. Winter, «Biomechanics and motor control of human movement », Wiley-interscience Publication, NewYork, 1990.

J. Wagner and C. Carlier., « Biomécanique et physiologie du mouvement », Biomécanique de la marche revue de chirurgie orthopédique, p 69-73, 2002

C. Azevedo, R. Héliot, "Rehabilitation of Functional Posture and Walking: Coordination of healthy and Impaired Limbs", in: Journal of Automatic Control, 2005, vol. 15-Suppl., p. 11-15.

J.Kennedy , R.C.Eberhart , "Particle Swarm Optimisation", IEEE International Conference on neural Networks, p 1942-1948, 1995.

J.Dreo, A.Petrowski , P.Siarray , Taillard "methaheuristiques pour l'optimisation difficile", Eyrolles, Paris, 2003.

B Ammar CHERIF, « simulation of biped walking robot, IZIMAN », master thesis dissertation, ENIS 2006.

15

Humanoid Robot With Imitation Ability

WEN-JUNE WANG and LI-PO CHOU
National Central University & National Taipei University of Technology, Taiwan, ROC

1. Introduction

This chapter designs an intelligent humanoid robot that not only can walk forward, backward, turn left, turn right, walk sideward, squat down, stand up and bow smoothly, but also can imitate several human basic motions. The robot's structure is composed of 17 AI motors with some self- design acrylic sheets connections. The robot is like a human in that it also has two hands, two feet, and a head. The head is a web camera which serves as its eye of the robot. The eye can recognize the color marks pasted on the human body in any complex background. The robot can recognize and imitate human motions according to the relative positions of those marks. The imitated human motions include the various motions of the hand and the lower body, such as "raise hand", "Stand up", "Squat down", and "Stand on one foot". Furthermore, the robot can also imitate "walking forward", "walking backward" and "walking sideways". The webcam automatically rotates to search the marks pasted on the human when they move outside the robot's vision. Notably, the stability and balance of the robot should be maintained, regardless of the motion performed by the robot.

Humanoid biped robots have been widely studied. Those investigations always focus on keeping balance control and walking as smoothly as possible. Zero Moment Point (ZMP) concept has been used to implement the balance control for the biped robot (Erbatur et al., 2002), (Kim & Oh, 2004) and (Park & Chung, 1999). The paper (Kanehiro et al., 1996) developed a walking pattern generator and a gravity compensation function to enable the biped robot to walk and carry objects. (Grizzle et al., 2001) established the existence of a periodic orbit in a simple biped robot, and analyzed its stability properties. A biped robot has been designed in (Loffler et al., 2004) to achieve a dynamically stable gait pattern, allowing for high walking velocities. A walk control for biped robots, consisting of a feed forward dynamic pattern and a feedback sensory reflex, has also been proposed in (Huang & Nakamura, 2005).

(Sias & Zheng, 1990) proposed the number of degrees of freedom corresponding to robot motions. For instance, each foot should have four degrees of freedom at least for the basic walking of a biped robot and should have five degrees of freedom at least for walking up stairs and down stairs. A robot can turn and walk smoothly on the ground with six degrees of freedom per foot, and can walk with a large step given seven degrees of freedom per foot. Furthermore, a robot needs at least eight degrees of freedom per foot to walk like a human being. The above information is helpful for the robot designers when determining the number of degrees of freedom of a robot.

Conversely, many papers have discussed the interaction motions between robots and humans. Motion imitation is one of the interaction motions. (Nakaoka et al., 2005) used eight cameras to detect and recognize 32 marks on a dancer body such that the robot can imitate Japanese dance motions. (Zhao et al., 2004) applied six cameras to detect and recognize 38 marks to allow a robot to imitate humans in playing "TaiChi" gong-fu. (Tanco et al., 2005) designed a robot that can imitate hand motions of humans in real time by recognizing the skin of human's two hands. Moreover, a dance robot was developed in Japan in 2003 (Kosuge et al., 2003).

The organization of this chapter is as follow. Section 2 describes the mechanisms of the robot. Section 3 proposes the walking path planning method for the robot. Section 4 presents the motor torque control and timing arrangement for the robot. Section 5 describes the extraction of the markers pasted on the human's body and the motion imitation for the robot. Some experiments for the humanoid robot are provided in Section 6. The conclusion is discussed in the final section.

2. Mechanisms of the robot

The humanoid biped robot designed in this study is composed of 17 AI motors (AI-1001 and AI-601) and some self design acrylic sheet connections. The robot is 40 cm tall, 23 cm wide and 1.5 kg weight and is shown in Fig. 2.1. Each foot of the robot has five degrees of freedom comprising two degrees of freedom in the hip, one in the knee and two in the ankle. Each hand has three degrees of freedom consisting of two degrees of freedom in the shoulder and one in the elbow. There is also a degree of freedom in the neck and a camera on the head to serve as an eye. The motor on the neck can rotate the camera up and down. All AI motors have different ID numbers and are connected in series as shown in Fig. 2.2. The motors in highing load positions, namely ID-1~ID-14, are AI-1001 motors which has a maximum torque of 10Kg/cm. The motors in low loading positions, namely ID-15~ID-17, are AI-601 motors which has a maximum 6Kg/cm torque.

Fig. 2.1. The humanoid biped robot

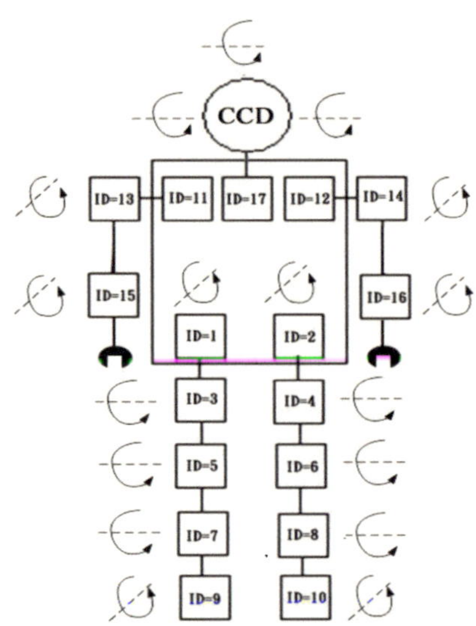

Fig. 2.2. All motors on the robot

The controller delivers command packets to the AI motor to control the motor actions. Those packets are the commands of Position Send, Position Read, Act Down, Power Down and 360° Rotation (see Fig. 2.3(a)). The motor returns response packets, containing the current and position data of the motor (see Fig. 2.3(b)), back to the controllers. The AI motor is connected to the personal computer (PC) via an RS-232 asynchronous serial communication link (Kim et al., 2000). The camera on the head of the robot is a webcam linked to the computer by USB 2.0 interface. The webcam can take 30 pictures/sec and has a resolution of 1280 × 960. Moreover, Borloand C++ Builder 6.0 is used to develop a human-machine interface. In summary, the robot is controlled by a PC with a webcam and the communication of RS-232. All hardware framework connection of the robot system is shown in Fig. 2.4.

1byte	1byte	1byte	1byte
Header	Data1	Data2	Checksum

Fig. 2.3(a). The command packets

1 byte	1 byte
Current	Position

Fig. 2.3(b). The response command packets

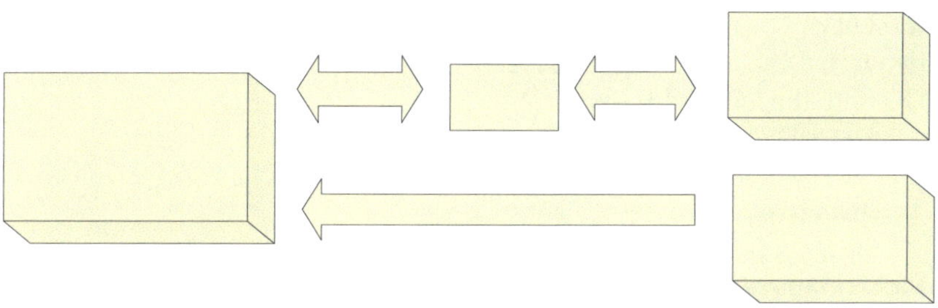

Fig. 2.4. The hardware framework connection of the robot system

3. Path planning for the basic walking

Walking path planning is important in ensuring that humanoid robot walks stably. Fig. 3.1 shows a walking path planning, namely the cycloid plan (Huang et al., 2001) & (Hodgins & Raibert, 1991)). In the figure, dotted lines denote the cycloid paths of the hip and the swinging ankle of a humanoid robot to perform a walking motion.

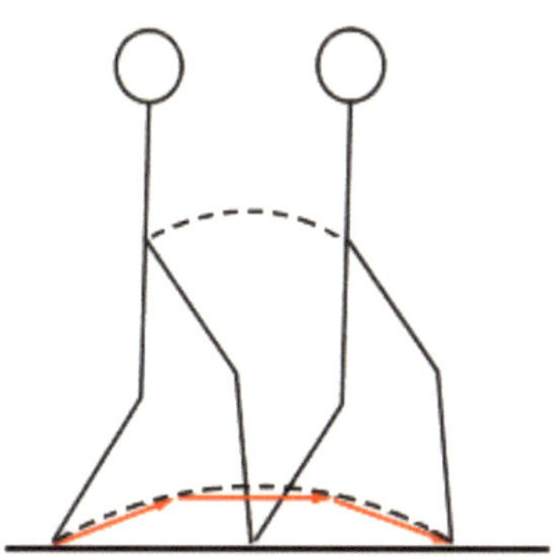

Fig. 3.1. The cycloid path

The cycloid paths shown in Fig. 3.2(a) can be obtained from equations in the papers (Hodgins & Raibert, 1991) and (Kurematsu et al., 1988). Any point on the cycloid paths can be solved from those equations and then the rotating angles of AI motors can be obtained by the inverse kinematics (Kim et al., 2000). However this approach is only suitable for walking and its calculation is very complicated. To simplify the calculation and design, the proposed method set four sampling points along a step path as shown by the gray points in Fig. 3.2(b). Further, one additional point (the white point in Fig. 3.2(b)) is added between the starting gray point and the second gray point to emphasize the smooth moving for the instant of off landing. These five sampling points are the reference points to establish the walking path.

Fig. 3.2. (a) The cycloid path of the swinging ankle (b) the sampling points selection

A walking path planning method is to construct a trajectory to be followed by the ankle or the hip. The next issue for the walking plan is to locate the center of gravity (COG). A basic walking motion of a robot can be divided into eight steps, in terms of the changes of COG:
Step 1: Stand with two feet; COG is between two feet.
Step 2: Move the COG onto the left (right) foot.
Step 3: Lift the right (left) foot and move forward.
Step 4: Land the right (left) foot on the ground.
Step 5: Move COG between two feet.
Step 6: Move COG on the right (left) foot.
Step 7: Lift the left foot and move forward.
Step 8: Land the left foot on the ground.

For a basic walking motion, the above eight steps can be seen as eight states. Fig. 3.3 is the series of walking motion, where the black point is the COG of the robot. The shifting of COG from Step 1 to Step 8 is presented in the figure. Fig. 3.4 shows the walking motion of a real humanoid robot.

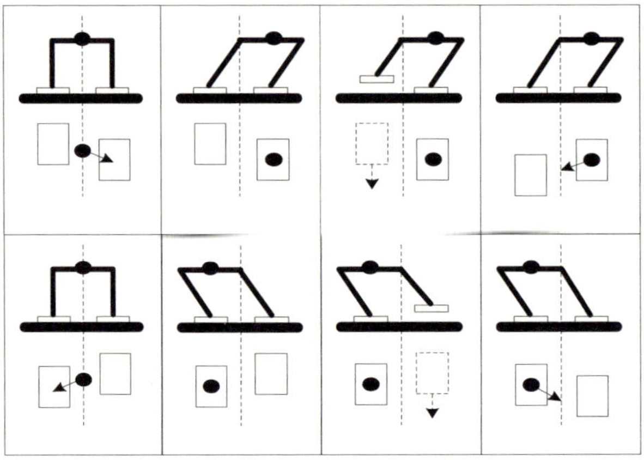

Fig. 3.3. A series of walking motions

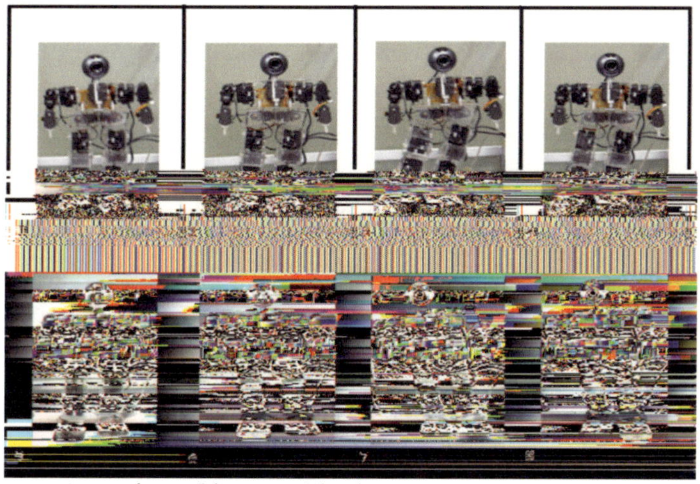

Fig. 3.4. Walking motions of a real humanoid robot

A humanoid robot walks following the above procedure. However, implementing these eight steps does not produce a smooth walking, since a motor with high torque that rotates a large degree directly and then generates an unpredictable inertia and momentum causing the robot to fall down or move unstably. Therefore, to improve the smoothness and stability of walking motion, the number of sample points between two successive states should be increased, ideally at least 10. Increasing the number of samples reduces inertia, thus enabling the robot to move smoothly and stably. However, the smoothness is ignored, if the robot needs additional momentum to achieve some motion at some instant (for instance, the instant of foot lifting off the ground). The number of samples can be reduced in this case. The above analysis for basic walking can also be applied to the other motions, such as stand up, squat down, hand up and down.

4. Motor torque control and timing arrangement

The torque control of each motor is also an issue when the robot is moving. The ankle always needs a large torque, because it has a very high loading. Furthermore, when the robot stands on one foot, the thigh needs a large torque too. In other words, each motor needs a proper torque corresponding to its motion. Therefore, each motor's torques should be properly given when the robot is performing a motion.

Additionally, the robot has many serially connected motors for which the timing arrangement is very important. It is known that the motor performing large degree rotation needs much more time than that motor performing small degree rotation. Therefore, time arrangement work is not only to arrange the order of motor operating, but also to set a certain delay time (waiting time) for the motor which performs small degree rotation such that the motor, which performs large degree rotation, can finish its operation before the next state starts.

After the moving path is planed and sample points are obtained, then the robot needs to follow the path and the sample points to move. Therefore, by trial and error, all proper data of the corresponding motors at each sample point are found and saved to perform the motion. The data contain the rotation degree, torque magnitude, and operation time for each motor. In summary, a basic walking motion involves torque control, path planning and timing arrangement for the motor operations.

5. Motions recognition and imitation

Before imitating a human motion, the robot has to recognize the motions exhibited by the imitated human (called Mr. A) in this work. An experiment was performed in which some specific marks with red and yellow color were pasted on the body of Mr. A. Two yellow marks were pasted on the stomach (or body) and right foot of Mr. A; and three red marks are pasted on two hands and left foot as shown in Fig. 5.1(a). The following motions were imitated: motions of lower part of the body, namely stand up, squat down, and stand on a single foot; and the motions of the hands, namely hand up and down, both hands lifting evenly, and hand curving. Sideways movement was also performed.

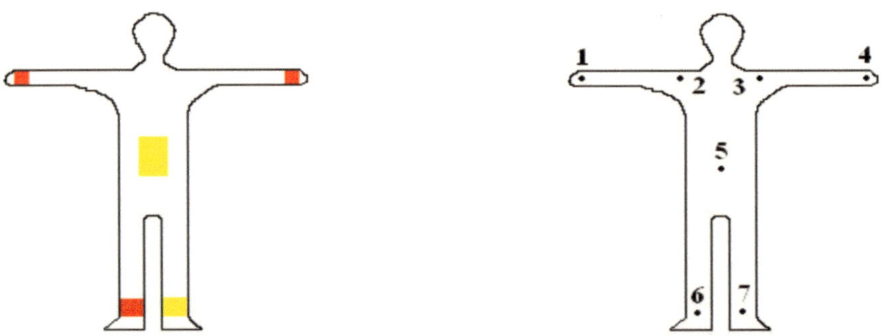

Fig. 5.1. (a) The five color marks (b). Seven key points

First, let us introduce the following procedure to get the seven key points of Mr. A.

Step 1: Capture the image of Mr. A with 320×240 pixels by robot's head camera.

Step 2: Transform RGB color model of the image to YUV color mode to reduce the effect of changes in brightness.

Step 3: Detect the five color marks on the body of Mr. A and calculate the center of each mark.

Step 4: Identify the given key points from the relative locations of five color marks on the body of Mr. A. These points are the stomach (or body), left hand, right hand, left foot and right foot.

Step 5: Based on the five key points in step 4, estimate the other two key points, left and right shoulders (see Fig. 5.1(b)).

Now let the coordinates of seven key points be shown in Table 5.1 and the motions of Mr. A as shown in Fig. 5.2 be imitated.

Positions	left hand	right hand	left foot	right foot	left shoulder	right shoulder	Stomach (body)
Notations	(x_{lh}, y_{lh})	(x_{rh}, y_{rh})	(x_{lf}, y_{lf})	(x_{rf}, y_{rf})	(x_{ls}, y_{ls})	(x_{rs}, y_{rs})	(x_b, y_b)

Table 5.1 Notations of seven key points

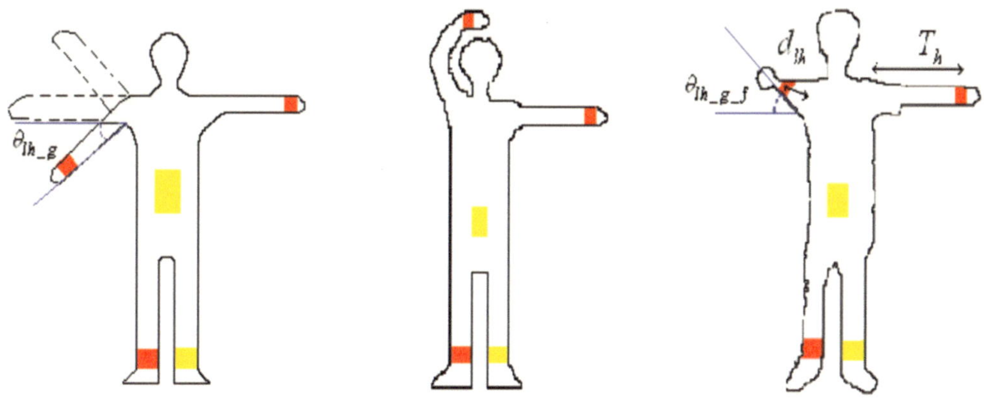

Fig. 5.2. (a) Hand up, down and lifting sideward; (b) Hand up and curving ; (c) Hand lifting forward.

The motions of Mr. A's hands in Fig. 5.2 can be recognized by the equations in the table.

Conditions	θ_{lh_g} or $\theta_{lh\_g\_f}$	Motions
$x_{lh} < x_{ls}$	$\theta_{lh\_g} = \tan^{-1}(\frac{y_{ls} - y_{lh}}{x_{ls} - x_{lh}})$	Hand up, down and lifting sideways
$x_{ls} < x_{lh}$ & $y_{lh} < y_{la}$	$\theta_{lh\_g} = \tan^{-1}(\frac{y_{ls} - y_{lh}}{x_{ls} - x_{lh}})$	Hand up and curving
$d_{lh} < T_h$	$\theta_{lh\_g\_f} = \cos^{-1}(\frac{d_{lh}}{T_h})$	Hand lifting forward

Table 5.2. Conditions of hands' motions

where θ_{lh_g} and $\theta_{lh\_g\_f}$ are the angles that left hand lifts sideways and forward, respectively. d_{lhs} and d_{rhs} are the distance from the left hand to left shoulder and from the right hand to right shoulder, respectively, where T_h is a threshold and d_{lhs} is defined as

$$d_{lhs} = ((x_{ls} - x_{lh})^2 + (y_{ls} - y_{lh})^2)^{1/2} \tag{1}$$

The motions of right hand can be figured out similarly.

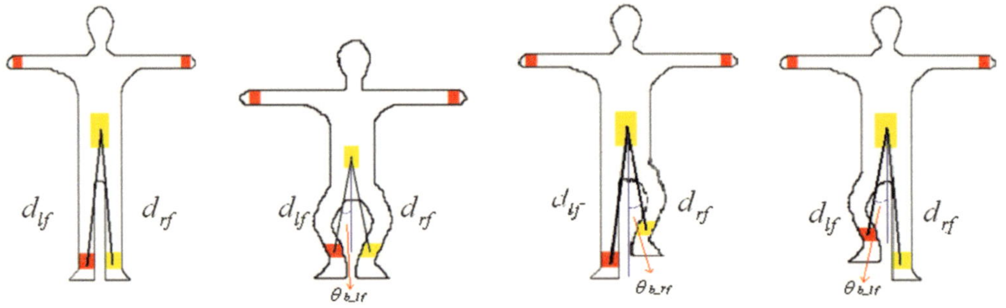

Fig. 5.3. (a) Stand up; (b) Squat down; (c) Stand on the left foot ; (d) Stand on the right foot.

The motions of Mr. A's feet as shown in Fig. 5.3 can be recognized by the equations in the following table.

Conditions	θ_{b_lf}	Motions
$d_{lf} > T_f$ & $d_{rf} > T_f$	$\theta_{b\_lf} = \tan^{-1}(\frac{x_b - x_{lf}}{y_b - y_{lf}})$ and $\Delta d = 0$	Stand up
$d_{lf} < T_f$ & $d_{rf} < T_f$	$\theta_{b\_lf} = \tan^{-1}(\frac{x_b - x_{lf}}{y_b - y_{lf}})$ and $\Delta d = 0$	Squat down
$d_{lf} > T_f$ & $d_{rf} < T_f$	$\theta_{b\_rf} = \tan^{-1}(\frac{x_{rf} - x_b}{y_b - y_{rf}})$ and $\Delta d = T_f - d_{rf}$	Stand on the left foot only
$d_{lf} < T_f$ & $d_{rf} > T_f$	$\theta_{b\_lf} = \tan^{-1}(\frac{x_b - x_{lf}}{y_b - y_{lf}})$ and $\Delta d = T_f - d_{lf}$	Stand on the right foot only

Table 5.3. Conditions of feet's motions

where θ_{b_lf} is the angle between body and the left foot, d_{lf} and d_{rf} are defined as follows:

$$d_{lf} = ((x_b - x_{lf})^2 + (y_b - y_{lf})^2)^{1/2}$$
$$d_{rf} = ((x_b - x_{rf})^2 + (y_b - y_{rf})^2)^{1/2} \qquad (2)$$

d_{lf} and d_{rf} are the distances from left foot to the body and right foot to the body, respectively. T_f is a threshold defined by the designer and $\Delta d = T_f - d_{lf}$ (or $\Delta d = T_f - d_{rf}$) is the heigh of one foot lifting. When Δd is large, that means the robot lifts one foot highly. When the robot recognizes the motions of human, then the robot should imitate those motions. However, the robot should also be able to switch smoothly from one motion to another motion. Fig. 5.4 shows a finite state machine representing the motion switching paths of the lower part of the robot's body.

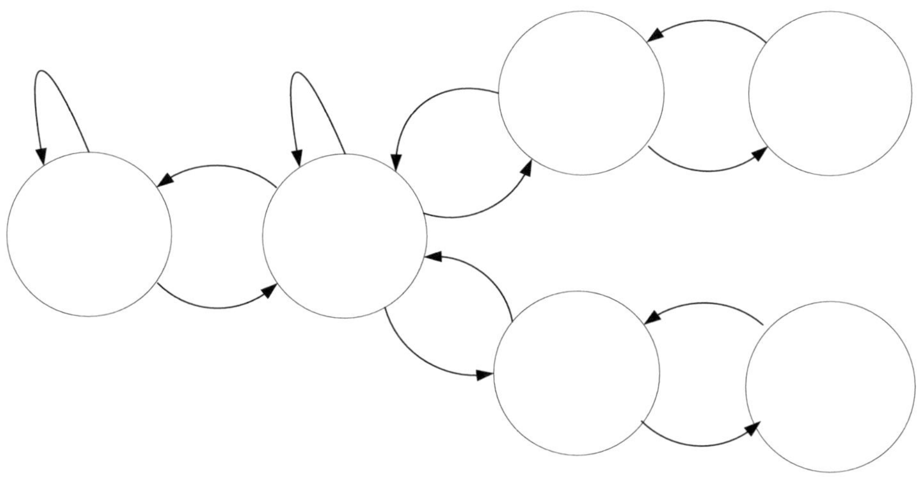

Fig. 5.4. Finite state machine for the switching of the robot's motions

For instance, suppose the robot stands on two feet initially (status 0). If the robot detects that Mr. A squate down, then it follows path 9 causing its status to change to status 5. If the robot then detect Mr. A stands on two feet again, then it follows path 10 to change its status to status 0 again. The robot can thus switch from status to status by following the corresponding path. Here, each path is planned in advance by following the method of Section 3.

The motions of moving sideways, forward and backward are now discussed. Since there is a yellow mark pasted on the body of Mr. A, the distance between the next and current locations of the mark's center can be adopted to recognize him moving sideways. For instance, $x_b(t+1) - x_b(t) < -T_b$ means Mr. A is moving left-ward; and $x_b(t+1) - x_b(t) > T_b$ means Mr. A is moving right-ward, where $x_b(t+1)$ and $x_b(t)$ are the next and current positions, respectively, of the yellow mark's center; further, T_b is a threshold. On the other hand, Mr. A moving forward or backward is recognized by the perceived size of the yellow mark on his body. If the mark looks smaller by the webcam of the robot at current status than at his previous status, then Mr. A moves backward. Conversely, if the mark appears larger at the current status than at the previous status, then Mr. A is moving forward.

Path 11

6. Experiment results

This section presents motion experiments for a humanoid robot. Part A presents some basic motions of a humanoid robot, namely walking forward and backward, turning right or left, moving sideways, squat down, stand up and bow. Part B presents the motion imitations of the robot. Data figures, photographs and detailed descriptions are presented in these two parts. Because the humanoid robot and the human face each other, the photographs are opposite to each other. For example, if Mr. A raises his right hand, the humanoid robot raises its left hand.

Part A: Basic motions

By using the path planning introduced in section 3, the humanoid robot practices the motions of forward, backward, turn right, turn left, stand up and squat down. For instance, Fig. 6.1 shows a series of pictures of two basic motions walking forward for two steps. Fig. 6.2 shows the motions of squatting down and standing up. The above six motions can be finished smoothly and stably following the path planning.

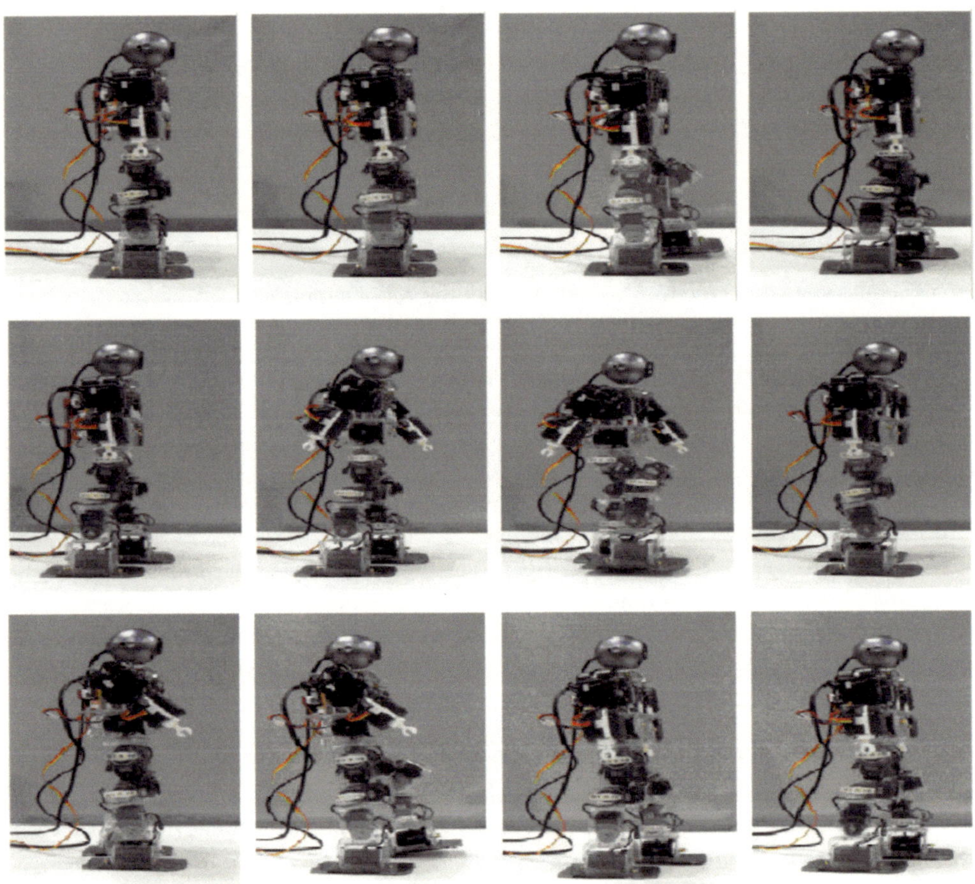

Fig. 6.1. A series of pictures of walking forward for the robot

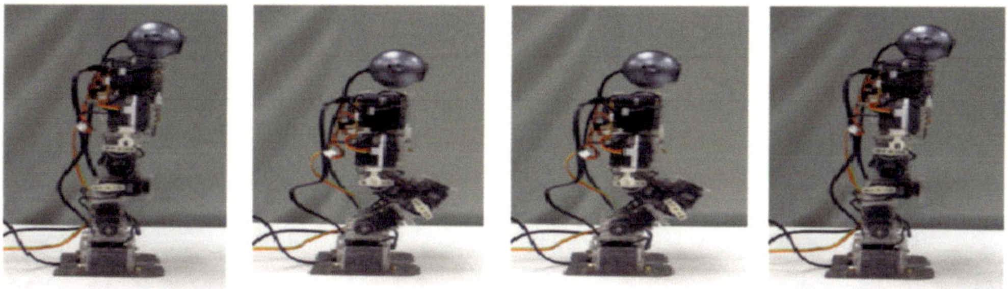

Fig. 6.2. The motions of squatting down and standing up

Part B: Motion imitation

The motions to be imitated in part B are hand up and down, hand lifting sideways, hand up and curving, and hand lifting forward. For instance, Fig. 6.3 shows a series of figures in which the robot's left hand is down initially and lifts sideways from the 1.7th second to the 3.3th second, and then puts down again at the 4.5th second. The left hand lifts again from the 5th second and keeps lifting from the 6th second to the 7.5th second and down at the 8th second. From Fig. 6.4, the humanoid robot is standing on two feet for the initial 7 seconds, standing on the right foot only during 7 -10 seconds, and standing on the left foot only during 15 -21 seconds; and then the robot squats down during 24 – 28 seconds and returns to stand on two feet after the 28th second. Experiment results indicate that the humanoid robot can switch its status when Mr. A changes his motions. Fig. 6.5 shows a series of pictures for several motion imitations, in which the humanoid robot imitate Mr. A's motions very well.

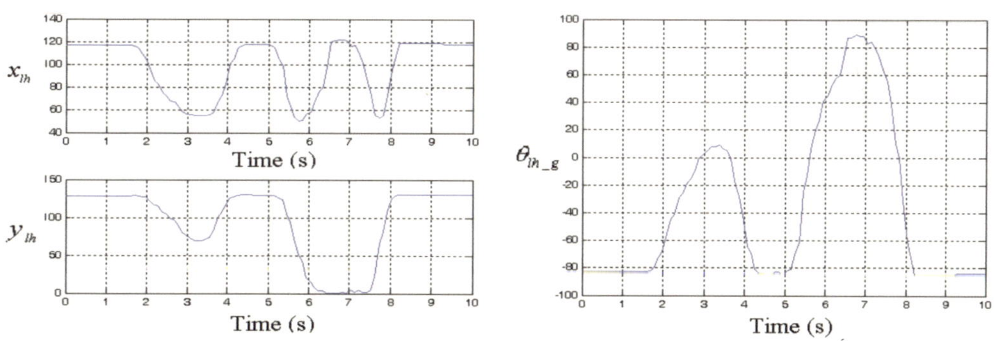

Fig. 6.3. A series of figures for the left hand lifting.

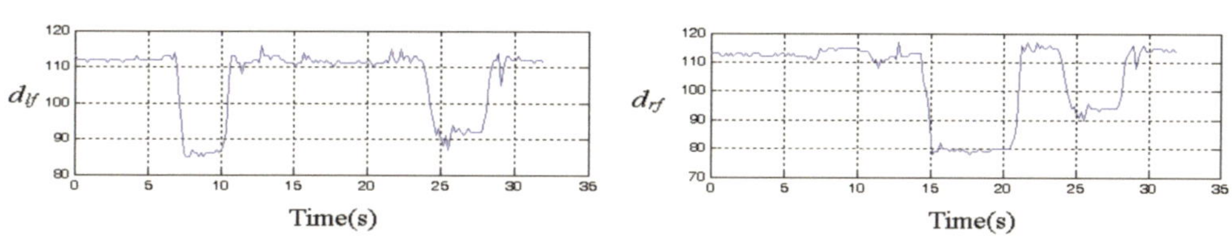

Fig. 6.4 The distances from the left foot and the right foot to the body.

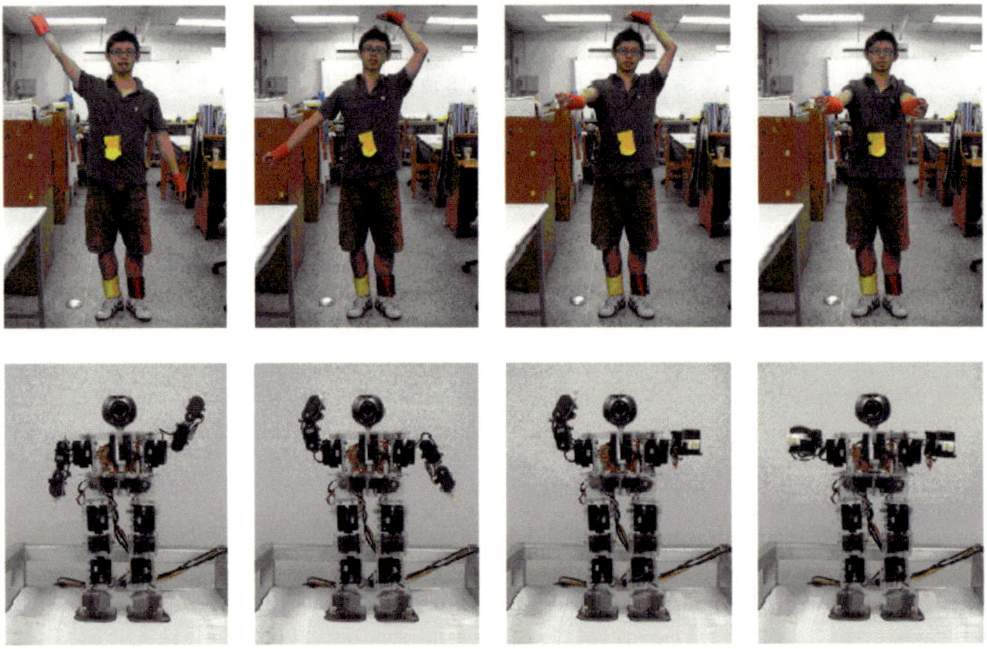

(a)

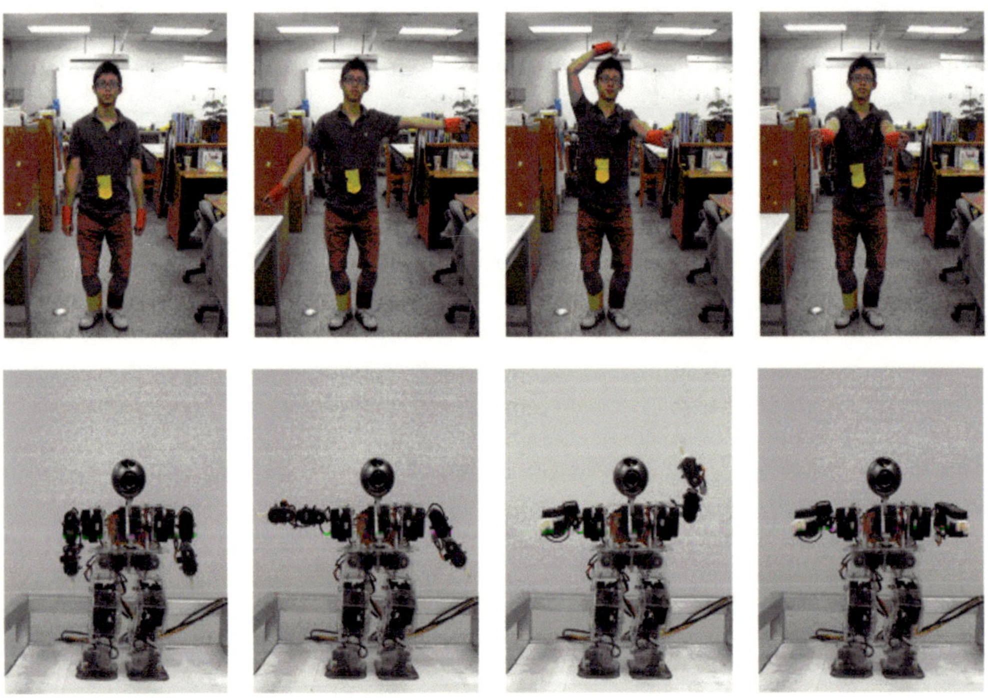

(b)

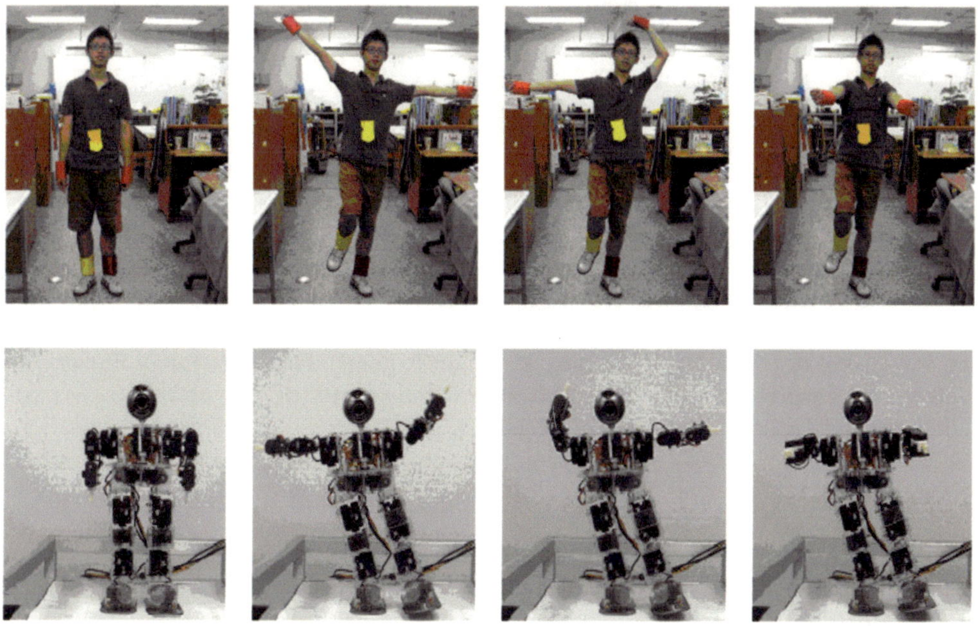

(c)

Fig. 6.5 Motion imitations (a) hand up, down, curving and lifting sideways; (b) squatting down; (c) Standing on one foot.

7. Conclusion

This chapter has proposed a scheme by which a humanoid robot can perform basic motions, and can imitate some human motions. The basic motions consist of walking forward, walking backward, turning left, turning right, walking sideways, squatting down, standing up and bowing. The imitated motions are hands up and down, squatting down, standing up, hand lifting sideways, hand up and curving, and hand lifting forward. The designed humanoid robot comprises 17 AI motors and one camera, and is controlled by a PC. The robot control strategy includes motion path planning, torque control, and timing arrangement. The experiments show that the robot works very well in both basic and imitated motions.

8. Reference

Erbatur, K.; Okazaki, A.; Obiya, K.; Takahashi, T. & Kawamura, A. (2002). **A** study on the zero moment point measurement for biped walking robots, *Proceedings of International Workshop on Advanced Motion Control*, pp. 431–436, Jul. 2002

Grizzle, J. W.; Abba, G. & Plestan, F. (2001). Asymptotically Stable Walking for Biped Robots: Analysis via Systems with Impulse Effects, *IEEE Transactions on automatic control*, Vol. 46, Jan. 2001 pp. 51 - 64, 0018-9286

Hodgins, J. K. & Raibert, M. H. (1991). Adjusting step length for rough terrain locomotion, *IEEE Transactions on Robotic and Automation*, Vol. 7, Jun. 1991 pp. 289–298, 1042-296X

Huang, Q. & Nakamura, Y. (2005). Sensory reflex control for humanoid walking, *IEEE Transactions on Robotic and Automation*, Vol. 21, Issue 5, Oct. 2005 pp. 977–984, 1552-3098

Huang, Q.; Yokoi, K. ; Kajita, S.; Kaneko, K.; Arai, H.; Koyachi, N. & Tanie, K. (2001). Planning Walking Patterns for a Biped Robot, *IEEE Transactions on Robotic and Automation*, Vol. 17, No. 3, Jun. 2001 pp. 280- 289, 1042-296X

Kanehiro, F.; Inaba, M. & Inoue, H. (1996). Development of a two-armed bipedal robot that can walk and carry objects, *Proceedings of IEEE/RSJ International conference on Intelligent Robots and Systems*, Vol.1, pp. 23–28, 1996

Kim, D. J.; Kim, K. I..; Zheng, Y. F.; Sun, Z. & Sun F. (2000). Design of Small Power Biped Robot by Load Sharing of Walking Gait, *Proceedings of IEEE International conference on Robotics and Automation*, Vol. 4, pp. 3359–3364, Apr. 2000

Kim, J. H. & Oh, J. H. (2004). Walking control of the humanoid platform KHR-1 based on torque feedback control, *Proceedings of IEEE International conference on Robotics and Automation*, Vol. 1, pp. 623–628, 2004

Kosuge, K.; Hayashi, T.; Hirata, Y. & Tobiyama, R. (2003). Dance partner robot - Ms DanceR, *Proceedings of IEEE/RSJ International conference on Intelligent Robots and Systems*, Vol. 3, pp. 3459-3464, 2003

Kurematsu, Y.; Kitamura, S. & Kondo, Y. (1988). Trajectory Planning and Control of a Biped Locomotive Robot-Simulation and Experiment, *Robotics and Manufacturing*, Recent Trends in Research, Education and Applications, M.Jamshidi(ed.), pp.65-72, *ASME Press*, 1988

Loffler, K.; Gienger, M.; Pfeiffer, F. & Ulbrich, H. (2004). Sensors and control concept of a biped robot, *IEEE Transactions on Industrial Electronics.*, Vol. 51, Oct. 2004 pp. 972-980, 0278-0046

Nakaoka, S.; Nakazawa, A.; Kanehiro, F.; Kaneko, K.; Morisawa, M. & Ikeuchi, K. (2005). Task model of lower body motion for a biped humanoid robot to imitate human dances?, *Proceedings of IEEE/RSJ International conference on Intelligent Robots and Systems*, pp. 3157–3162, 2005

Park, J. H. & Chung, H. (1999). ZMP compensation by online trajectory generation for biped robots, *Proceedings of IEEE International conference on Systems, Man, and Cybernetics*, Vol. 4, pp. 960–965, 1999

Sias, F. R.; Jr. & Zheng, Y. F. (1990). How Many Degrees-of-Freedom Does a Biped Need□, *Proceedings of IEEE International Workshop on Intelligent Robots and Systems*, Vol. 1, pp. 297–302, Jul. 1990

Tanco, L.; Bandera, J. P.; Marfil, R. & Sandoval, F. (2005). Real-time human motion analysis for human-robot interaction, *Proceedings of IEEE/RSJ International conference on Intelligent Robots and Systems*, pp. 1402–1407, 2005

Zhao, X. J.; Huang, O.; Peng, Z. & Li, K. (2004). Kinematics mapping and similarity evaluation of humanoid motion based on human motion capture, *Proceedings of IEEE/RSJ International conference on Intelligent Robots and Systems*, vol.1, pp. 840–845, 2004

16

Developing New Abilities for Humanoid Robots with a Wearable Interface

Hyun Seung Yang, Il Woong Jeong, Yeong Nam Chae, Gi Il Kwon and Yong-Ho Seo
AIM Lab, EECS, KAIST (Korea Advanced Institute of Science and Technology)
Republic of Korea

1. Introduction

A humanoid robot is a robot that has a similar appearance with humans. Although appearance is the one of the important features of humanoid robots, human-like behaviors and task solving abilities are more meaningful features in daily life. To make humanoid robots more useful, they should have the capacity to learn new abilities. However, today's humanoid robots are not smart enough to adapt to their working environments. In this paper, we propose a method to develop new abilities of humanoid robots on the basis of imitation, one of the most powerful methods to learn a new skill, not only for humans but also for humanoid robots.

To achieve this goal, we designed and implemented a wearable interface to teach humanoid robots via user demonstration. Magnetic markers and flex sensors are applied for capturing human motion. A head mount display and a microphone are used for communicating with the partner robot. A vision channel and speech channel are employed to communicate with the partner robot. The wearable system assists the user and enhances interaction with the robot through an intuitive motion imitating method and multimodal communication channels via which commands are given and sensory feedback is received.

A vision channel is used to complement the weakness in recognizing surrounding states. A human teacher can sense the environment by sharing the robot's sight using the head mount display and he or she can decide the most appropriate behavior for the observed situation. When the teacher concludes that there is no proper behavior among the robot's abilities, a new ability can be trained through demonstration by the teacher. We conducted a kinematic analysis on both a human user and a partner robot to overcome the joint difference problem when transferring motion from the human user to the robot. We used the humanoid robot AMIO as a test bed to validate the wearable interface and found that we could make the robot more intelligent with the wearable interface.

2. Related Work

Many researchers are trying to develop a novel way to enhance the abilities of humanoid robots. There have been some worthy researches on robotics which used motion capture

data to transfer human motion to a partner robot. Zhao et al., generated a new motion for a robot by mapping motion capture data to a robot model with employing similarity evaluation (Zhao et al., 2004). Nakazawa et al., proposed a method to generate new motions for humanoid robots by using motion primitives extracted from motion capture data (Nakazawa et al., 2002). Matsui et al., tried to minimize the motion difference between a human user and a partner robot by capturing motion from both the user and the robot (Matsui et al., 2005). Inamura et al., controls their robot by using voice command and motion capture data (Inamura et al., 2005). Kanzaki et al., used a wearable system instead of conventional motion capturing devices for generating new motions for humanoid robots. (Kanzaki et al., 2004).
Programming by demonstration (PbD) or learning by imitation is the one of the most popular approaches that a robot learns a new ability from an user's demonstration. Movement primitive based motion reproduction is a major research issue in PbD. Schaal et al., applied dynamic movement primitives for controlling actuators of their humanoid robot. (Schaal et al., 2004). Jenkins and Mataric proposed a system that extracts action units from motion capture data and they generated new motions for their robot based on action units. (Jenkins and Mataric, 2002). Stochastic motion modelling is another major research issue in PbD. Most researches adopted hidden Markov model (HMM) to handle the motion data as a time series data. Inamura et al. proposed a method to represent the original motion using HMM (Inamura et al., 2003). Calinon et al. used PCA, ICA and HMM for encoding the motion data (Calinon et al., 2005). Kwon et al. proposed a method to find the best combination of motion primitives using HMM (Kwon at el., 2006).
Motion capturing devices are very effective in terms of capturing precise motions from human user for generating humanoids' motion. But they are not appropriate for users to interact with a humanoid robot in daily life. Therefore we focused on enhancement of HRI using a wearable interface which is very easy to use in daily life. Motion primitive based motion reproduction methods have advantages of generalizing the task to be learned by a robot for later use but the methods cannot generate motions with variations. And HMM based motion generation methods need a lot of training data and a lot of time. So we propose a new method for extracting motion primitives from an user's demonstration and generating motions with variations for humanoid robots.

3. Wearable Interface

3.1. Wearable Interface Prototype

Motion capture devices are widely used for researches on human motions. But those devices are very expensive and difficult to use in daily life. Therefore we developed a wearable interface platform with some inexpensive sensors - inclinometer sensors, magnetic sensors and flex sensors. Our wearable interface is designed for the interaction between a human user and a partner robot in daily life, so it should be very easy to wear and easy to use. The prototype of our wearable interface platform is shown in Fig. 1.
A head mount display (HMD), PC Eye-Trek manufactured by Olympus Optical is used to interact with a partner robot. With the HMD, an user can share the sight of the robot. An earphone and a microphone are attached to the HMD for the interaction between the user and robot by speech channel.

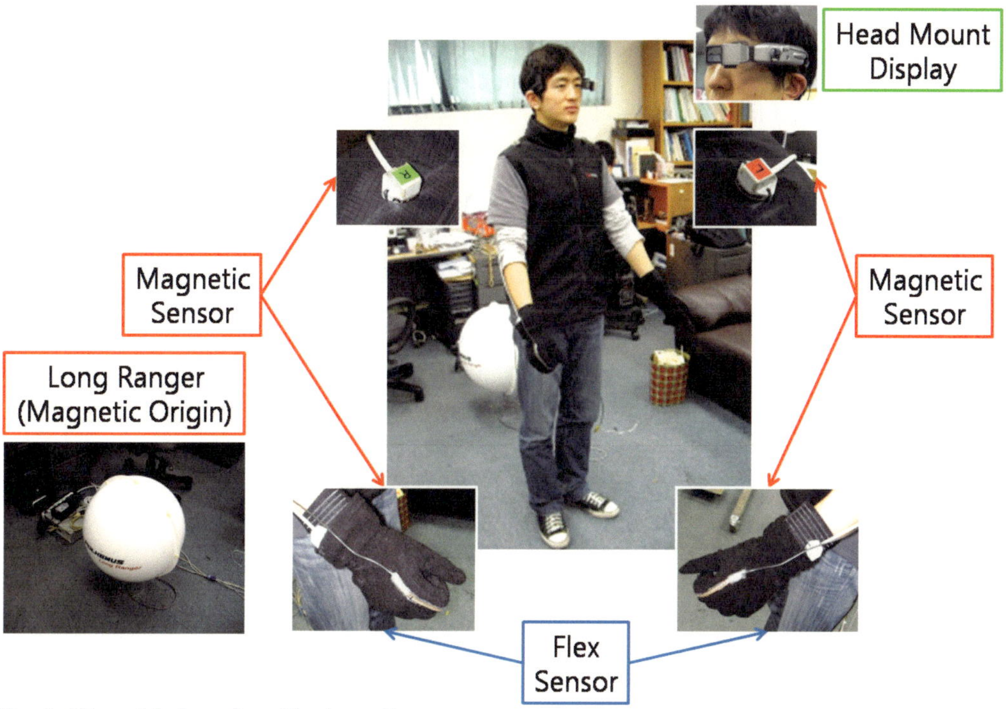

Fig. 1. Wearable Interface Platform Prototype

A Polhemus Fastrak system with a long range origin is used as magnetic sensor. Each magnetic marker can measure the 3-axis rotation angles and the distance from the origin. The effective workspace is 4.6 meters from the origin. The static accuracy for the position is 0.08 centimeters, and the static accuracy for the orientation is 0.15 degrees. Data acquisition rate is 40Hz with 4 markers, so it is fast enough to capture human motion. Four magnetic markers are used, two for shoulders and other two for wrists. By measuring the position and orientation of each shoulder and wrist, we can capture the human motion to generate a new behaviour for a humanoid robot. Two FLX-01 sensors manufactured by Abrams Gentile are attached to gloves for measuring bending degrees of human fingers. A MI-A3330LS 3-axis gyroscope sensor made by MicroInfinity is used to measure human head motion. Because our robot platform has a neck with 2 degrees of freedom, we measure the rotation angles of yaw axis and pitch axis. The sensors we used to develop our wearable interface are shown in Fig. 2.

Fig. 2. Sensors

3.2. Motion Capture using Wearable Interface

We used our wearable interface as a motion capturing device to extract motion data from the human user's demonstration. The motion of neck is measured by a gyroscope sensor attached to HMD and the motion of each hand is measured by flex sensor attached to gloves. Four magnetic sensors are used for gathering the position and orientation information of each shoulder and each wrist.

The motion of neck and hands can be applied directly to the robot because the degrees of freedoms which those sensor can measure are exactly same as the robot's ones. But in the case of arm motions, kinematic analysis must be done before applying arm motion to the robot because the control input and the measured information cannot be matched directly. The analytic solution for the inverse kinematics problem is hard to find and it makes multiple solutions. Therefore we tried to find a single solution by using geometrical information of the robot arms. After solving inverse kinematics problem, we can get 4 angle trajectories representing each arm motion. The angle trajectories of neck, hands and arms are used as the original motion data for the user's demonstration.

3.3. Human-Robot Interaction using Wearable Interface

A humanoid robot should have autonomy with a self-contained anthropomorphic body. it must have abilities to sense its surrounding environments. Furthermore, it must have sufficient intelligence to perform its task successfully. However, today's humanoid robots don't have enough abilities to have autonomy because the current limitations of robotic technologies. So we focused on the interaction with humanoids using our wearable interface with multi-modal communication channels to complement the limitation of humanoid robots' autonomy (Seo et al., 2007).

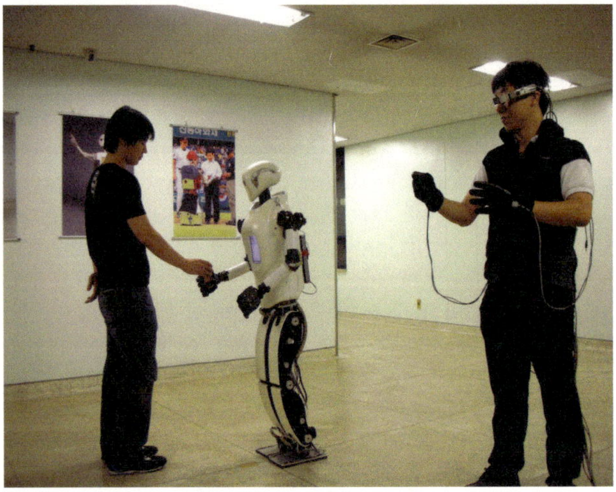

Fig. 3. HRI using the wearable interface

An user wearing the wearable interface gives commands to the partner robot through multi-modal communication channels - visual channel and speech channel. The user can see what the partner robot sees by sharing the scene from cameras in the robot's head. And then the user can choose the operation mode of robot - automatic and manual. The user can make the robot perform its task autonomously by sending speech command when the robot can solve that task by itself. If the task is too complex for the robot to solve by itself, the user can control the robot's movement manually by conducting demonstration for that task. Fig. 3 shows an example of human-robot interaction using our wearable interface.

4. A Humanoid Robot, AMIO

A humanoid robot AMIO is used as a partner robot in our experiments. AMIO is a humanoid robot developed in 2006 by Yang at el. (Yang at el. 2006). AMIO is designed to test biped walking algorithms and to imitate natural human motions. With a built-in Lithium-Polymer battery, it can operate its work up to 30 minutes without external source of electricity. AMIO has height of 150cm and weight of 45kg and it can walk at speed of 1km/h. It has a self-contained body, head, two arms with five finger hands, and two legs. The total degrees of freedom of the robot are 36 - 2 for neck, 5 for each arm, 6 for each hand, and 6 for each leg. The shape and the degrees of freedom of AMIO are shown in Fig. 4.
There exist two CCD cameras in its head and it can recognize human users and sense its surrounding environments using the stereo vision mechanism. AMIO can express its emotion using 3D characterized virtual face through a LCD screen in its chest, and it can recognize human speech command using a simple speech recognizer.

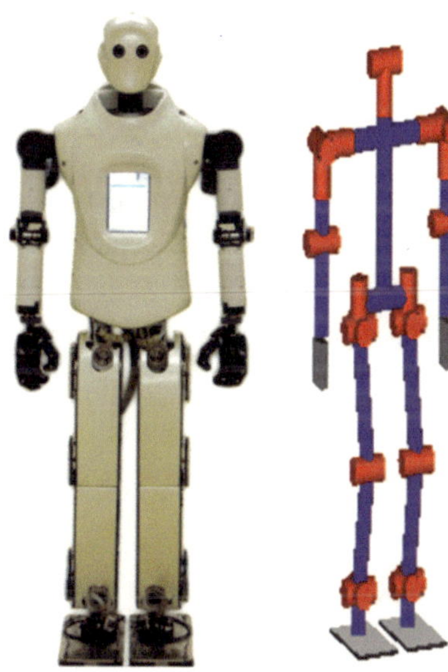

Fig. 4. A humanoid robot AMIO and its degrees of freedom

5. Developing New Abilities

5.1. Overview

We focused on generating new upper-body motions from user's demonstrations to enhance the robot's ability. It is intractable to generate new motions for a robot based on original motion data because the motion data is dense sequence of motion frames which is very hard to modify. Therefore we adopted the concept of motion primitives for representing motion sequences. The original demonstrated motion can be expressed with motion primitives and various new motions can be generated based on those motion primitives. For extracting motion primitives, we used a curve simplification algorithm and a clustering algorithm. After simplifying motion sequence, we conducted a clustering algorithm to cluster similar keyframe motions and we constructed motion primitives based on these motion clusters. The whole process of generating new motion is shown in Fig. 5.

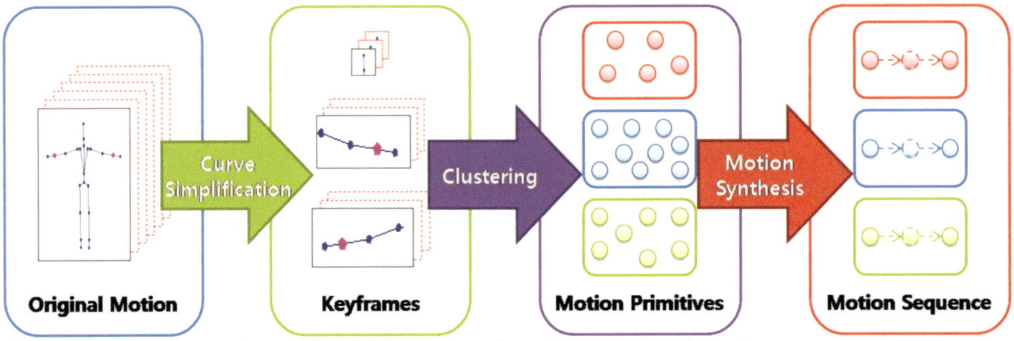

Fig. 5. Whole process for generating new motions from original motion

5.2. Motion Primitives

We simplified the original motion sequence using a curve simplification algorithm proposed by Lowe (Lowe, 1987). Lowe's algorithm converts the original curve into a set of vertices. We used the curve simplification algorithm for extracting keyframe motions from the original motion sequences. After then, we conducted a clustering algorithm to group the similar keyframe motions into the same cluster. The total number of cluster is unknown parameter, so we used adaptive resonance theory (ART) to cluster the keyframe motions (Alpaydin, 2004) (Bishop 2006). Each cluster is represented as a Gaussian distribution with its mean and variance, and it is a representative of similar keyframe motions. We used the clusters as motion primitives. The specific process of constructing motion primitives is shown in Fig. 6.

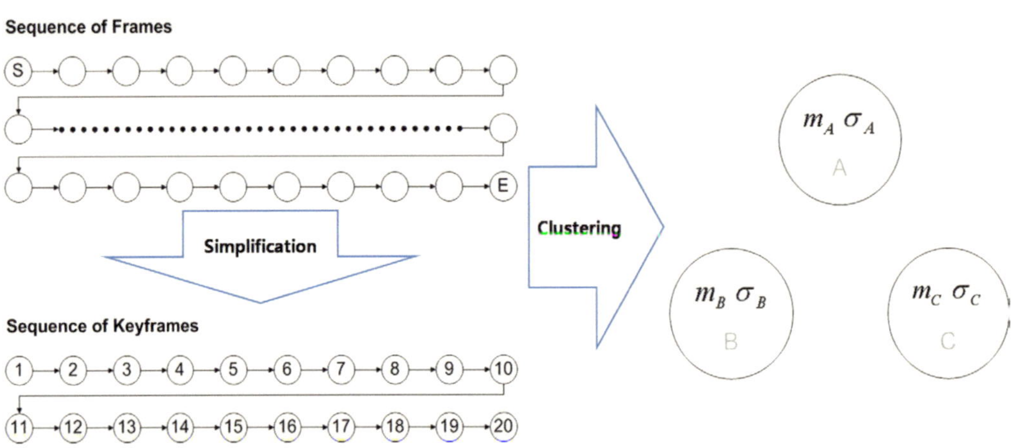

Fig. 6. Constructing motion primitives

5.3. Generating New Motions

The sequence of keyframe motions can be written as a sequence of motion primitives and we can calculate the transition probabilities among all the clusters. The transitions can be represented as a weighted directed graph. We used the graph as a model of a motion sequence for the user's demonstration. New motions can be generated by writing a new sequence with following the transitions of the model. We can create a new motion from multiple motion models. The transitions among multiple models can be only caused between the similar motion primitives and motion frame interpolations are conducted to complement the surprise changes in transitions between different motion models.

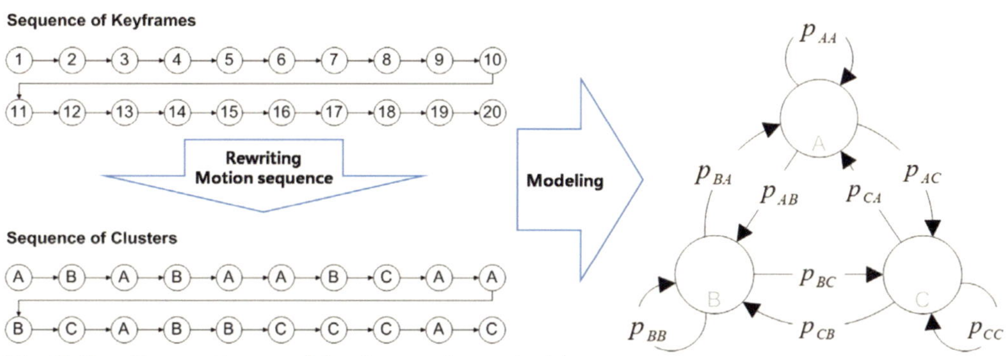

Fig. 7. Creating motion model using motion primitives

6. Examples

We tested our motion generation method using our humanoid robot, AMIO. After finishing the motion capture from the user demonstration, our system extracted motion primitives and made a motion model for the demonstrated motion. We implemented a simulator to check the validity of generated motion before applying to real robot platform like Fig. 8. An example of motion tracking using the wearable interface is shown in Fig 9.

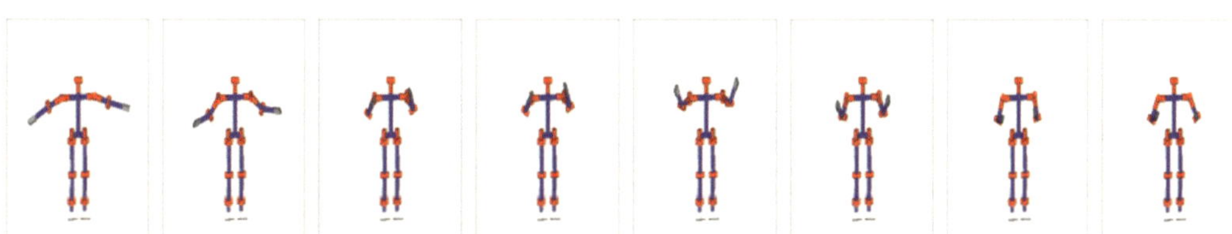

Fig. 8. An example of simulated motion

Two simple motions are demonstrated by an user - a simple heart drawing motion and a simple boxing motion. A heart drawing motion is used to test our wearable interface's motion capturing capability. The AMIO's heart drawing motion is shown in Fig. 10. And for generating new motions for AMIO, a basic boxing motion was demonstrated by an user and AMIO generated its modified motion based on the motion primitives from the demonstrated motion. Then the robot started to generate a new motion based on the motion model. The generated boxing motion for AMIO is shown in Fig. 11.

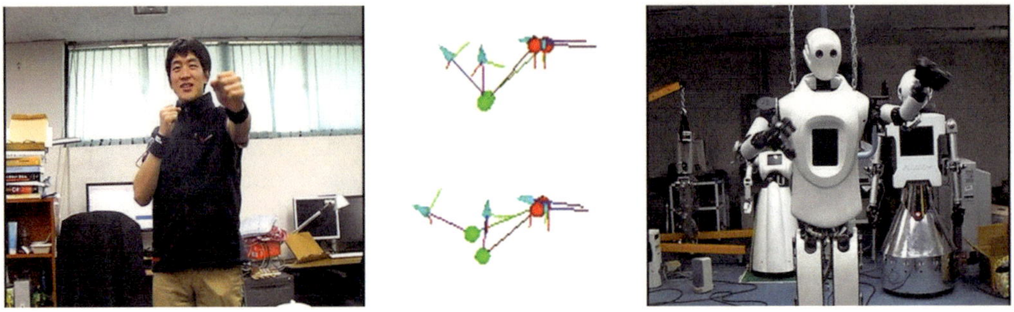

(a) (b) (c)

Fig. 9. Motion tracking from user demonstration

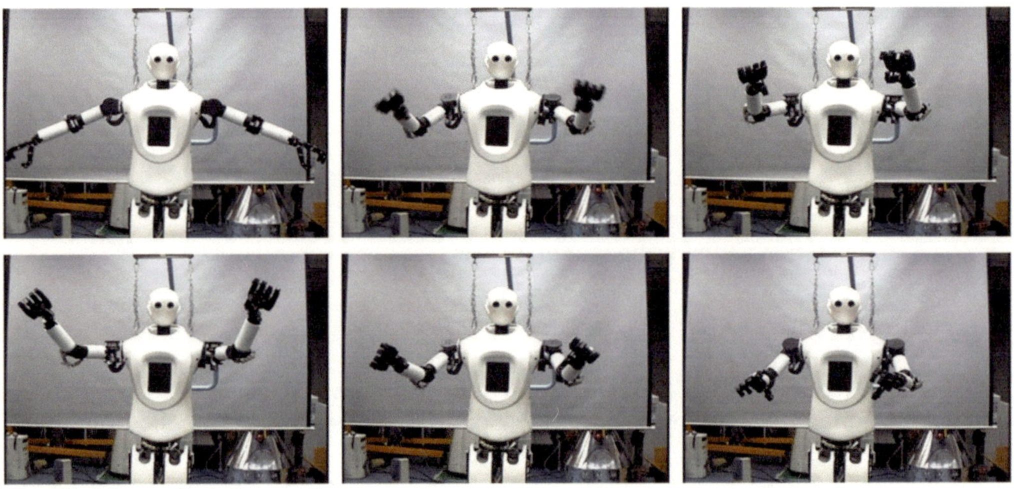

Fig. 10. A heart drawing motion

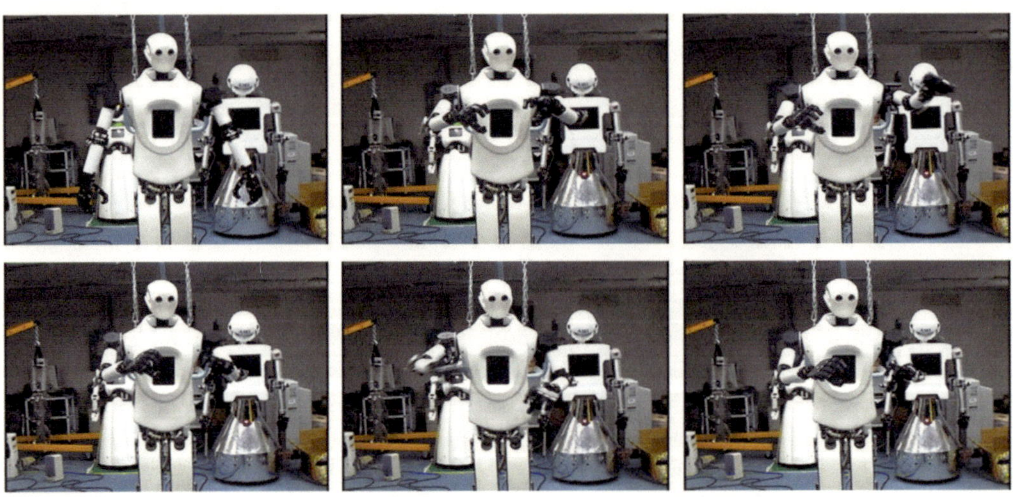

Fig. 11. A boxing motion

7. Conclusion

We focused on a method to enhance the abilities of humanoid robots by paying attention to the fact that imitation is the best way to learn a new ability. We designed and developed a wearable interface which is lightweight and presents multi-modal communication channels for interacting with robots. And we proposed a method to build motion primitives from the user demonstrated motion using curve simplification and clustering. A stochastic process is used for modelling motions and generating new motions. The stochastic model presents the way of generating various motions not monotonous repetition of demonstrated motions. And we tested our method by using a humanoid robot AMIO.

The limitations of our work are 1) limited working space of human user because our wearable interface uses magnetic sensors which can be operated near the origin sensor, and 2) the motions that we generated are not considering the meaning of the task. For the further work, we will replace the magnetic sensors with the other positioning sensors that do not have any spatial limitations. And for improving the intelligence of humanoid robots, defining task descriptors and extracting task descriptors from a demonstrated task are indispensable. We are planning to conduct a research on task description method for generating tasks with ease.

8. Acknowledgement

This research was supported by Foundation of Healthcare Robot Project, Center for Intelligent Robot, the Ministry of Knowledge Economy, and UCN Project, the Ministry of Knowledge Economy(MKE) 21st Century Frontier R&D Program in Korea and a result of subproject UCN 08B3-O4-10M

9. References

Alpaydin, E. (2004), *Introduction to Machine Learning*, The MIT Press, ISBN 0-262-012211-1, Cambridge, Massachusetts.

Bishop, C. (2006), *Pattern Recognition and Machine Learning*, Springer, ISBN 0-387-31073-8, Singapore.

Calinon, S. and Billard, A. (2005). Recognition and reproduction of gestures using a probabilistic framework combining PCA, ICA and HMM. *Proceedings of International Conference on Machine Learning*, pp. 105-112, 2005.

Inamura, T.; Tanıe, H. & Nakamura, Y. (2003). Keyframe compression and decompression for time series data based on the continuous hidden Markov model, *Proceedings of International Conference on Intelligent Robots and Systems*, pp. 1487-1492.

Inamura, T.; Kojo, N.; Sonoda, T.; Sakamoto, K.; Okada, K. & Inaba, M. (2005). Intent imitation using wearable motion capturing system with on-line teaching of task attention. Proceedings of Conference on Humanoid Robots, pp. 469-474, Toukuba, Japan.

Jenkins, O. C. and Mataric, M. J. (2002). Deriving action and behavior primitives from human motion data. *Proceedings of International Conference on Intelligent Robots and Systems*, pp. 2551-2556.

Kanzaki, S.; Fujumoto, Y.; Nishiwaki, K.; Inamura, T.; Inaba, M. & Inoue, H. (2004). Development of wearable controller with gloves and voice interface for humanoids whole-body motion generation, Proceedings of International Conference on Machine Automation, pp. 297-302, Osaka, Japan.

Lowe, D. (1987). Three-dimensional object recognintion from single two-dimensional images. *Artifical Intelligence.*

Nakazawa, A.; Nakaoka, S.; Ikeuchi, K. & Yokoi, K. (2002). Imitating human dance motions through motion structure analysis, Proceedings of International Conference on Intelligent Robots and Systems, pp. 2539-2544, Lausanne, Switzerland.

Schaal, S.; Peters. J.; Nakanishi, J. & Ijspeert, A. (2004). Learning movement primitives. *International Symposium on Robotic Research.*

Seo, Y.; Jeong, I. & Yang, H. (2007) Motion capture-based wearable interaction system, *Advanced Robotics*, pp. 1725-1741

Yang, H.; Seo, Y.; Chae, Y.; Jeong, I.; Kang, W. & Lee, J. (2006). Design and Development of Biped Humanoid Robot, AMI2, for Social Interaction with Humans, *Proceedings of IEEE-RAS Humanoids*, pp. 352-357, Genoa, Italy.

Zhao, X.; Huang, Q .; Peng, Z. & Li, K. (2004). Kinematics mapping and similarity evaluation of humanoid motion based on human motion capture, *Proceedings of Intelligent Robots and Systems*, pp. 840-845, Sendai, Japan.

17

Walking Gait Planning And Stability Control

Chenbo Yin, Jie Zhu and Haihan Xu
Nanjing University of Technology, School of Mechanical and Power Engineering

1. Introduction

Research on biped humanoid robots is currently one of the most exciting topics in the field of robotics and there are many ongoing projects. Because the walking of humanoid robot is complex dynamics inverse problem the pattern generation and dynamic simulation are extensive discussed. Many different models are proposed to simple the calculation. Many researches about the walking stability and pattern generation of biped robots are made using ZMP principle and other different methods.

Vukobratovic first proposed the concept of the ZMP (Zero Moment Point). Yoneda etc proposed another criterion of "Tumble Stability Criterion" for integrated locomotion and manipulation systems. Goswami proposed the FRI (Foot Rotation Indicator). As for the pushing manipulation, Harada researched the mechanics of the pushed object. Some researches mentioned that changes of angular momentum of biped robot play the key roles on the stability maintenance. However, there have been fewer researches on stability maintenance considering the reaction with external environment.

A loss of stability might result a potentially disastrous consequence for robot. Hence man has to track robot stability at every instant special under the external disturbance. For this purpose we need to evaluate quantity the danger extent of instability. Rotational equilibrium of the foot is therefore an important criterion for the evaluation and control of gait and postural stability in biped robots. In this paper by introducing a concept of fictitious zero-moment (FZMP), a method to maintain the whole body stability of robot under disturbance is presented.

2. Kinematics and dynamics of humanoid robot

Robot kinematics deals with several kinematic and kinetic considerations which are important in the control of robotic kinematics. In kinematic modeling of robots, we are interested in expressing end effector motions in terms of joint motions. This is the direct problem in robot kinematics. The inverse-kinematics problem is concerned with expressing joint motions in terms of end-effector motions. This latter problem is in general more complex. In robot dynamics (kinetics), the direct problem is the formulation of a model as a set of differential equations for robot response, with joint forces/torques as inputs. Such models are useful in simulations and dynamic evaluations of robots. The inverse-dynamics problem is concerned with the computation of joint forces/torques using a suitable robot model, with the knowledge of joint motions. The inverse problem in robot dynamics is directly applicable to computed-torque control (also known as feed forward control), and also somewhat indirectly to the nonlinear feedback control method employed here.

2.1 Representation of position and orientation

2.1.1 Description of a position

Once a coordinate system is established we can locate any point in the universe with a 3×1 position vector. Because we will often define many coordinate systems in addition to the universe coordinate system, vectors must be tagged with information identifying which coordinate system they are defined within. In this book vectors are written with a leading superscript indicating the coordinate system to which they are referenced (unless it is clear from context), for example, ${}^{A}P$. This means that the components of ${}^{A}P$ have numerical values which indicated distances along the axes of {A}. Each of these distances along an axis can be thought of as the result of projecting the vector onto the corresponding axis.
Figure 2.1 pictorially represents a coordinate system, {A}, with three mutually orthogonal unit vectors with solid heads. A point ${}^{A}P$ is represented with a vector and can equivalently be thought of as a position in space, or simply as an ordered set of three numbers. Individual elements of a vector are given subscripts $x, y,$ and z :

$$ {}^{A}P = \begin{bmatrix} p_x \\ p_y \\ p_z \end{bmatrix} \tag{1} $$

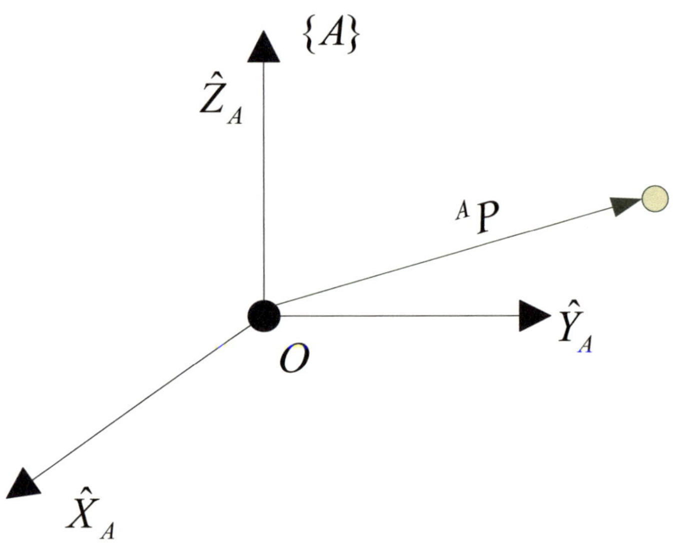

Fig. 1. Vector relative to frame example

In summary, we will describe the position of a point in space with a position vector. Other 3-tuple descriptions of the position of points, such as spherical or cylindrical coordinate representations are discussed in the exercises at the end of the chapter.

2.1.2 Description of an orientation

Often we will find it necessary not only to represent a point in space but also to describe the orientation of a body in space. For example, if vector ${}^AP$ in fig.2.2 locates the point directly between the fingertips of a manipulator's hand, the complete location of the hand is still not specified until its orientation is also given. Assuming that the manipulator has a sufficient number of joints the hand could be oriented arbitrarily while keeping the fingertips at the same position in space. In order to describe the orientation of a body we will attach a coordinate system to the body and then give a description of this coordinate system relative to the reference system. In Fig.2.2, coordinate system {B} has been attached to the body in a known way. A description of {B} relative to {A} now suffices to give the orientation of the body.

Thus, positions of points are described with vectors and orientations of bodies are described with an attached coordinate system. One way to describe the body-attached coordinate system, {B}, is to write the un it vectors of its three principal axes in terms of the coordinate system {A}.

We denote the unit vectors giving the principal directions of coordinate system {B} as ${}^A\hat{X}_B, {}^B\hat{Y}_B, and\, {}^A\hat{Z}_B$. When written in terms of coordinate system {A} they are called $\hat{X}_B, \hat{Y}_B, and\, \hat{Z}_B$. It will be convenient if we stack these three unit vectors together as the columns of a 3×3 matrix, in the order ${}^A\hat{X}_B, {}^B\hat{Y}_B, {}^A\hat{Z}_B$. We will call this matrix a rotation matrix, and because this particular rotation matrix describes {B} relative to {A}, we name it with the notation ${}^A_BR$. The choice of leading sub-and superscripts in the definition of rotation matrices will become clear in following sections.

$$ {}^A_BR = \left[{}^A\hat{X}_B \quad {}^A\hat{Y}_B \quad {}^A\hat{Z}_B \right] = \begin{bmatrix} r_{11} & r_{12} & r_{13} \\ r_{21} & r_{22} & r_{23} \\ r_{31} & r_{32} & r_{33} \end{bmatrix} \tag{2} $$

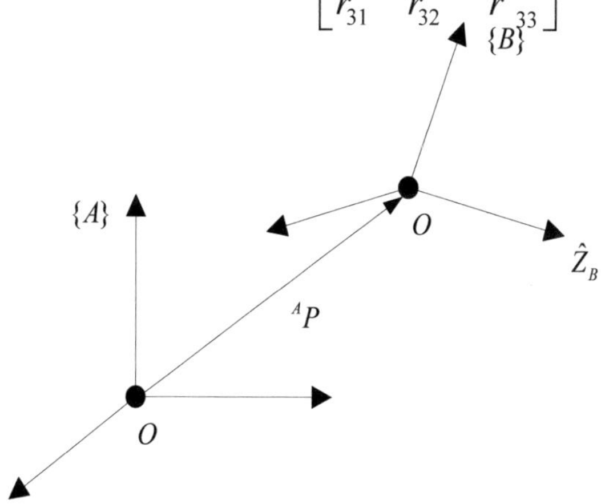

Fig. 2. locating an object in position and orientation

In summary, a set of three vectors may be used to specify an orientation. For convenience we will construct a 3×3 matrix which has these three vectors as its columns. Hence, whereas the position of a point is represented with a vector, the orientation of a body is represented with a matrix. In section 2.8 we will consider some other descriptions of orientation which require only three parameters.

We can give expressions for the scalars r_{ij} in (2.2) by nothing that the components of any vector are simply the projections of that vector onto the unit directions of its reference frame. Hence, each component of ${}^{A}_{B}R$ in (2.2) can be written as the dot product of a pair of unit vectors as

$$ {}^{A}_{B}R = \begin{bmatrix} {}^{A}\hat{X}_B & {}^{A}\hat{Y}_B & {}^{A}\hat{Z}_B \end{bmatrix} = \begin{bmatrix} \hat{X}_B \cdot \hat{X}_A & \hat{Y}_B \cdot \hat{X}_A & \hat{Z}_B \cdot \hat{X}_A \\ \hat{X}_B \cdot \hat{Y}_A & \hat{Y}_B \cdot \hat{Y}_A & \hat{Z}_B \cdot \hat{Y}_A \\ \hat{X}_B \cdot \hat{Z}_A & \hat{Y}_B \cdot \hat{Z}_A & \hat{Z}_B \cdot \hat{Z}_A \end{bmatrix} \tag{3} $$

For brevity we have omitted the leading superscripts in the rightmost matrix of (2.3). In fact the choice of frame in which to describe the unit vectors is arbitrary as long as it is the same for each pair being dotted. Since the dot product of two unit vectors yields the cosine of the angle between them, it is clear why the components of rotation matrices are often referred to as direction cosines.

Further inspection of (2.3) shows that the rows of the matrix are the unit vectors of {A} expressed in {B}; that is,

$$ {}^{A}_{B}R = \begin{bmatrix} {}^{A}\hat{X}_B & {}^{A}\hat{Y}_B & {}^{A}\hat{Z}_B \end{bmatrix} = \begin{bmatrix} {}^{A}\hat{X}_B^T \\ {}^{A}\hat{Y}_B^T \\ {}^{A}\hat{Z}_B^T \end{bmatrix} \tag{4} $$

Hence, ${}^{B}_{A}R$, the description of frame {A} relative to {B} is given by the transpose of (2.3); that is,

$$ {}^{B}_{A}R = {}^{B}_{A}R^T \tag{5} $$

This suggests that the inverse of a rotation matrix is equal to its transpose, a fact which can be easily verified as

$$ {}^{A}_{B}R^T \, {}^{A}_{B}R = \begin{bmatrix} {}^{A}\hat{X}_B^T \\ {}^{A}\hat{Y}_B^T \\ {}^{A}\hat{Z}_B^T \end{bmatrix} \begin{bmatrix} {}^{A}\hat{X}_B & {}^{A}\hat{X}_B & {}^{A}\hat{X}_B \end{bmatrix} = I_3 \tag{6} $$

Where I_3 is the 3×3 identity matrix. Henec,

$$ {}^{B}_{A}R = {}^{B}_{A}R^{-1} = {}^{B}_{A}R^T \tag{7} $$

Indeed from linear algebra we know that the inverse of a matrix with orthonormal columns is equal to its transpose. We have just shown this geometrically.

2.1.3 Description of a frame

The information needed to completely specify the whereabouts of the manipulator hand in Fig.2.2 is a position and an orientation. The point on the body whose position we describe could be chosen arbitrarily, however: For convenience, the point whose position we will describe is chosen as the origin of the body-attached frame. The situation of a position and an orientation pair arises so often in robotics that we define an entity called a frame, which is a set of four vectors giving position and orientation information. For example, in Fig.2.2 one vector locates the fingertip position and three more describe its orientation. Equivalently, the description of a frame can be thought of as a position vector and a rotation matrix. Note that a frame is a coordinate system, where in addition to the orientation we give a position vector which locates its origin relative to some other embedding frame. For example, frame {B} is described by ${}_{B}^{A}R$ and ${}^{A}R_{BORG}$, where ${}^{A}R_{BORG}$ is the vector which locates the origin of the frame {B}:

$$\{B\} = \{{}_{B}^{A}R, {}^{A}P_{BORG}\} \tag{8}$$

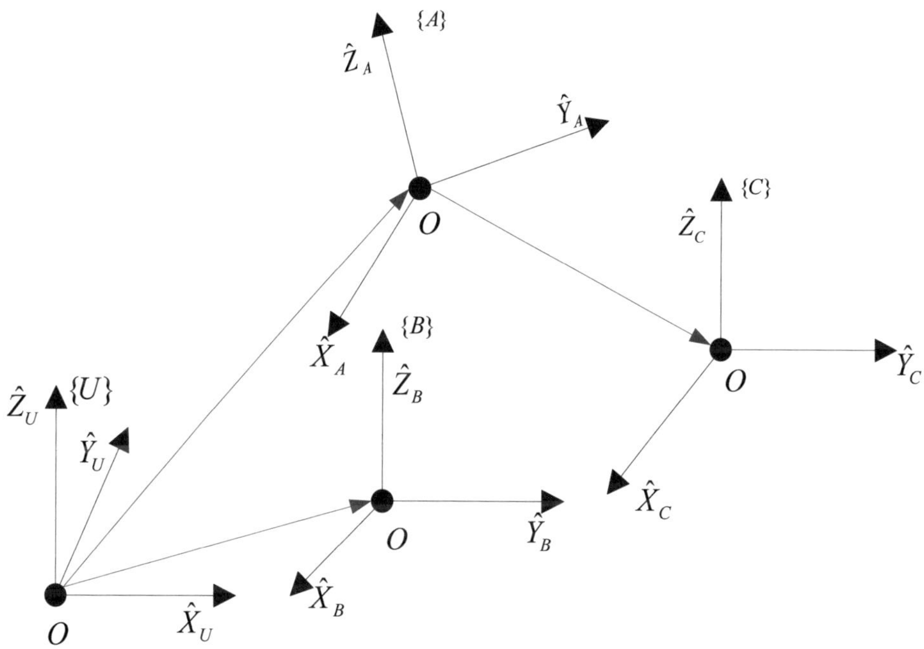

Fig. 3. Example of several frames

In Fig.2.3 there are three frames that are shown along with the universe coordinate system. Frames {A} and {B} are known relative to the universe coordinate system and frame {C} is known relative to frame {A}.

In Fig.2.3 we introduce a graphical representation of frames which is convenient in visualizing frames. A frame is depicted by three arrows representing unit vectors defining the principal axes of the frame. An arrow representing a vector is drawn from one origin to another. This vector represents the position of the origin at the head of the arrow in terms of the frame at the tail of the arrow. The direction of this locating arrow tells us, for example, in Fig.2.3, that {C} is known relative to {A} and not vice versa.

In summary, a frame can be used as description of one coordinate system relative to another. A frame encompasses the ideas of representing both position and orientation, and so may be thought of as a generalization of those two ideas. Position could be represented by a frame whose rotation matrix part is the identity matrix and whose position vector part locates the point being described. Likewise, an orientation could be represented with a frame. Whose position vector part was the zero vector.

2.2 Coordinate transformation

2.2.1 Changing descriptions from frame to frame

In a great many of the problems in robotics, we are concerned with expressing the same quantity in terms of various reference coordinate systems. The previous section having introduced descriptions of positions, orientations, and frames, we now consider the mathematics of mapping in order to change descriptions frame to frame.

Mappings involving translated frames

In Fig.2.4 we have a position defined by the vector ${}^{B}P$. We wish to express this point in space in terms of frame {A}, when {A} has the same orientation as {B}. In this case, {B} differs from {A} only by a translation which is given by ${}^{B}P_{BORG}$, a vector which locates the origin of {B} relative to {A}.

Because both vectors are defined relative to frames of the same orientation, we calculate the description of point P relative to {A}, ${}^{A}P$, by vector addition:

$$ {}^{A}P = {}^{B}P + {}^{A}P_{BORG} \tag{9} $$

Note that only in the special case of equivalent orientations may we add vectors which are defined in terms of different frames.

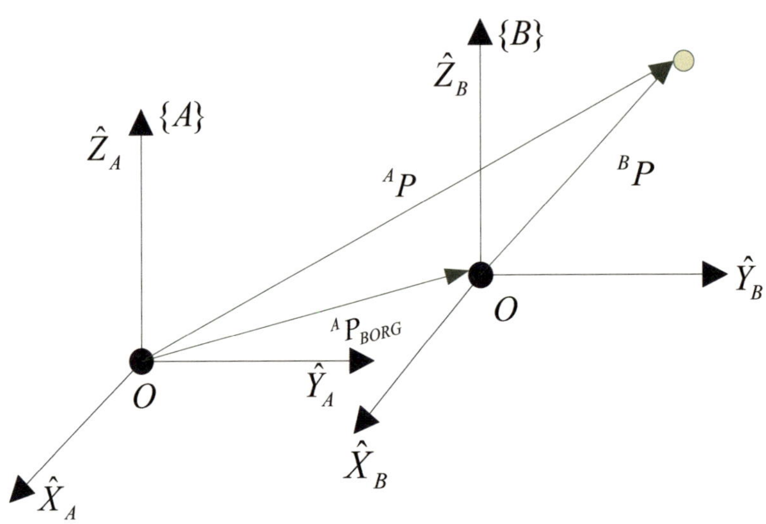

Fig. 4. Translational mapping

In this simple example we have illustrated mapping a vector from one frame to another. This idea of mapping, or changing the description from one frame to another, is an extremely important concept. The quantity itself (here, a point in space) is not changed; only its description in changed. This is illustrated in Fig.2.4, where the point described by ^{B}P is not translated, but remains the same, and instead we have computed a new description of the same point, but now with respect to system {A}.

We say that the vector $^{A}P_{BORG}$ defines this mapping, since all the information needed to perform the change in description is contained in $^{A}P_{BORG}$ (along with the knowledge that the frames had equivalent orientation).

Mappings involving rotated frames

Section 2.2 introduced the notion of describing an orientation by three unit vectors denoting the principal axes of a body-attached coordinate system. For convenience we stack these three unit vectors together as the columns of a 3×3 matrix. We will call this matrix a rotation matrix, and if this particular rotation matrix describes {B} relative to {A}, we name it with the notation $^{A}_{B}R$.

Note that by our definition, the columns of a rotation matrix all have unit magnitude, and further, these unit vectors are orthogonal. As we saw earlier, a consequence of this is that

$$^{A}_{B}R = {}^{B}_{A}R^{-1} = {}^{B}_{A}R^{T} \tag{10}$$

Therefore, since the columns of $^{A}_{B}R$ are the unit vectors of {B} written in {A}, then the rows of $^{A}_{B}R$ are the unit vectors of {A} written, in {B}.

So a rotation matrix can be interpreted as a set of three column vectors or as a set of three row vectors as follows:

$$^{A}_{B}R = \begin{bmatrix} {}^{A}\hat{X}_{B} & {}^{A}\hat{Y}_{B} & {}^{A}\hat{Z}_{B} \end{bmatrix} = \begin{bmatrix} {}^{B}\hat{X}_{A}^{T} \\ {}^{B}\hat{Y}_{A}^{T} \\ {}^{B}\hat{Z}_{A}^{T} \end{bmatrix} \tag{11}$$

As in Fig.2.5, the situation will arise often where we know the definition of a vector with respect to some frame, {B}, and we would like to know its definition with respect to another frame, {A}, where the origins of the two frames are coincident. This computation is possible when a description of the orientation of {B}, is known relative to {A}. This orientation is given by the rotation matrix $^{A}_{B}R$, whose columns are the unit vectors of {B} written in {A}.

In order to calculate ^{A}P, we note that the components of any vector are simply the projections of that vector onto the unit directions of its frame. The projection is calculated with the vector dot product. Thus we see that the components of ^{A}P may be calculated as

$$
\begin{aligned}
{}^{A}p_x &= {}^{B}\hat{X}_A \cdot {}^{B}P, \\
{}^{A}p_y &= {}^{B}\hat{Y}_A \cdot {}^{B}P, \\
{}^{A}p_z &= {}^{B}\hat{Z}_A \cdot {}^{B}P.
\end{aligned}
\tag{12}
$$

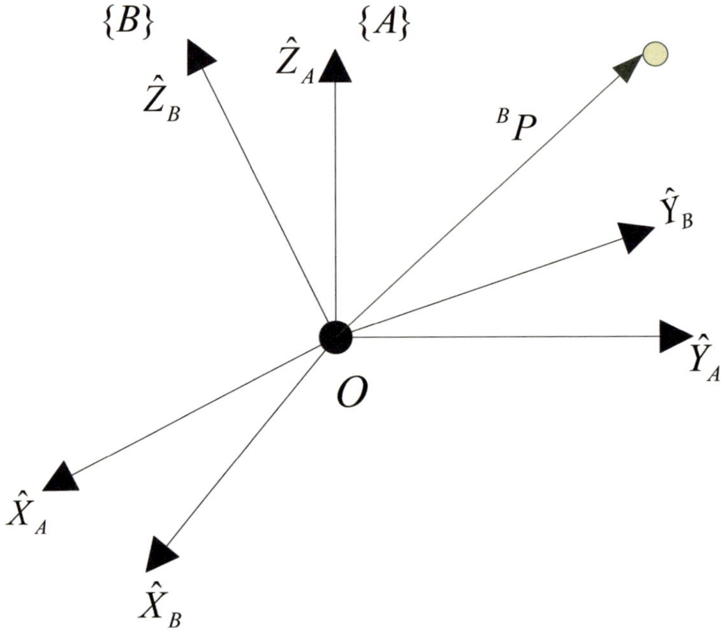

Fig. 5. rotating the description of a vector

In order to express (12) in terms of a rotation matrix multiplication, we note form (11) that the rows of ${}^{A}_{B}R$ are ${}^{B}\hat{X}_A$ ${}^{B}\hat{Y}_A$ and ${}^{B}\hat{Z}_A$. So (12) may be written compactly using a rotation matrix as

$$
{}^{A}P = {}^{A}_{B}R\,{}^{B}P \tag{13}
$$

Equation (13) implements a mapping—that is, it changes the description of a vector—from ${}^{B}P$, which description of the same point, but expressed relative to {A}.
We now see that out notation is of great help in keeping track of mappings and frames of reference. A helpful way of viewing the notation we have introduced is to imagine that leading subscripts cancel the leading superscripts of the following entity, for example the Bs in (13).

2.3 General rotation transformation

Mappings involving general frames

Very often we know the description of a vector with respect to some frame, {B}, and we would like to know its description with respect to another frame, {A}. We now consider the general case of mapping. Here the origin of frame {B} is not coincident with that of frame {A} but has a general vector offset. The vector that locates {B}'s origin is called ${}^{A}P_{BORG}$. Also {B} is rotated with respect to {A} as described by ${}^{A}_{B}R$. Given ${}^{B}P$, we wish to computer ${}^{A}P$, as in Fig.2.7.

We can first change ${}^{B}P$ to its description relative to an intermediate frame which has the same orientation as {A}, but whose origin is coincident with the origin of {B}. This is done by pre-multiplying by ${}^{A}_{B}R$ as in Section 2.3. We then account for the translation between origins by simple vector addition as in Section 2.3, yielding

$$ {}^{A}P = {}^{A}_{B}R\,{}^{B}P + {}^{A}P_{BORG} \tag{14} $$

Equation (2.17) describes a general transformation mapping of a vector from its description in one frame to a description in a second frame. Note the following interpretation of our notation as exemplified in (2.14): the B's cancel leaving all quantities as vectors written in terms of A, which may then be added.

The form of (2.14) is not as appealing as the conceptual form;

$$ {}^{A}P = {}^{A}_{B}T\,{}^{B}P \tag{15} $$

That is, we would like to think of a mapping from one frame to another as an operator in matrix form. This aids in writing compact equations as well as being conceptually clearer than (2.14). In order that we can write the mathematics given in (2.14) in the matrix operator form suggested by (2.15), we define a 4×4 matrix operator, and use 4×1 position vectors, so that (2.15) has the structure

$$ \begin{bmatrix} {}^{A}P \\ 1 \end{bmatrix} = \begin{bmatrix} & {}^{A}_{B}R & & {}^{A}P_{BORG} \\ 0 & 0 & 0 & 1 \end{bmatrix} \begin{bmatrix} {}^{B}P \\ 1 \end{bmatrix} \tag{16} $$

That is,

1□A "1"is added as the last element of the 4×1 vectors.

2□A row "$\begin{bmatrix} 0 & 0 & 0 \end{bmatrix}$"is added as the last row of the 4×4 matrix.

We adopt the convention that a position vector is 3×1 or 4×1 depending on whether it appears multiplied by a 3×3 matrix or by a 4×4 matrix. It is readily seen that (2.16) implements

$$ {}^{A}P = {}^{A}_{B}R\,{}^{B}P + {}^{A}P_{BORG} $$

$$ 1 = 1 \tag{17} $$

The 4×4 matrix in (2.16) is called a homogeneous transform. For our purposes it can be regarded purely as a construction used to cast the rotation and translation of the general transform into a single matrix form. In other fields of study it can be used to compute perspective and scaling operations (when the last row is other than" $\begin{bmatrix} 0 & 0 & 0 \end{bmatrix}$", or the rotation matrix is not orthonormal). The interested reader should see.

Often we will write equations like (2.15) without any notation indicating that this is a homogeneous representation, because it is obvious from context. Note that like while homogeneous transforms are useful in writing compact equations, a computer program to transform vectors would generally not use them because of time wasted multiplying ones and zeros. Thus, this representation is mainly for our convenience when thinking and writing equations down on paper.

Just as we used rotation matrices to specify an orientation, we will use transforms (usually in homogeneous representation) to specify a frame. Note that while we have introduced homogeneous transforms in the context of mappings, they also serve as descriptions of frames. The description of frame {B} relative to {A} is ${}^{A}_{B}T$.

2.4 Transformation matrix for links

Link description

A manipulator may be thought of as a set of bodies connected in a chain by joints. These bodies are called links. Joints form a connection between a neighboring pair of links. The term lower pair is used to describe the connection between a pair of bodies when the relative motion is characterized by tow surfaces sliding over one another.

Due to mechanical design considerations, manipulators are generally constructed from joints which exhibit just one degree of freedom. Most manipulators have revolute joints or have sliding joints called prismatic joints. In the rare case that a mechanism is built with a joint having n degrees of freedom, it can be modeled as n joints of one degree of freedom connected with n-1 links of zero length. Therefore, without loss of generality, we will consider only manipulators which have joints with a single degree of freedom.

The links are numbered starting from the immobile base of the arm, which might be called link 0. The first moving body is link 1, and so on, out to the free end of the arm, which is link n. In order to position an end—effector generally in 3-space, a minimum of six joints is required. Typical manipulators have five or six joints. Some robots may actually not be as simple as a single kinematic chain—they may have parallelogram linkages or other closed kinematic structures. We will consider one such manipulator later in this chapter.

A single link of a typical robot has many attributes which a mechanical designer had to consider during its design. These include the type of material used, the strength and stiffness of the link, the location and type of the joint bearings, the external shape, the weight and inertia, etc. However, for the purposes of obtaining the kinematic equations of the mechanism, a link is considered only as a rigid body which defines the relationship between two neighboring joint axes of a manipulator. Joint axes are defined by lines in space. Joint axis i is defined by a line in space, or a vector direction, about which link i rotates relative to link i-1. It turns out that for kinematic purpose, a link can be specified with two numbers which define the relative location of the two axes in space.

For any two axes in 3-space there exists a well—defined measure of distance between them. This distance is measured along a line which is mutually perpendicular to both axes. This distance is measured along line which is mutually perpendicular to both axes. This mutual perpendicular always exists and is unique except when both axes are parallel, in which case there are many mutual perpendiculars of equal length. Figure 3.2 shows link i-1 and the mutually perpendicular line along which the link length, a_{i-1}, is measured. Another way to visualize the link parameter a_{i-1} is to imagine an expanding cylinder whose axis is the joint i-1 axis—when it just touches joint axis is the joint i-1 axis—when it just touches joint axis i the radius of the cylinder is equal to a_{i-1}.

The second parameter need to define the relative location of the two axes is called the link twist. If we imagine a plane whose normal is the mutually perpendicular line just constructed, we can project both axes i-1 and i onto this plane and measure the angle between them. This angle is measured from axis i-1 to axis i in the right-hand sense about a_{i-1}. We will use this definition of the twist of link i-1, a_{i-1}. In Fig.3.2, a_{i-1} is indicated as the angle between axis i-1 and axis i (the lines with the triple hash marks are parallel). In the case of intersecting axes, twist is measured in the plane containing both axes, but the sense of a_{i-1} is lost. In this special case, one is free to assign the sign of a_{i-1} arbitrarily.

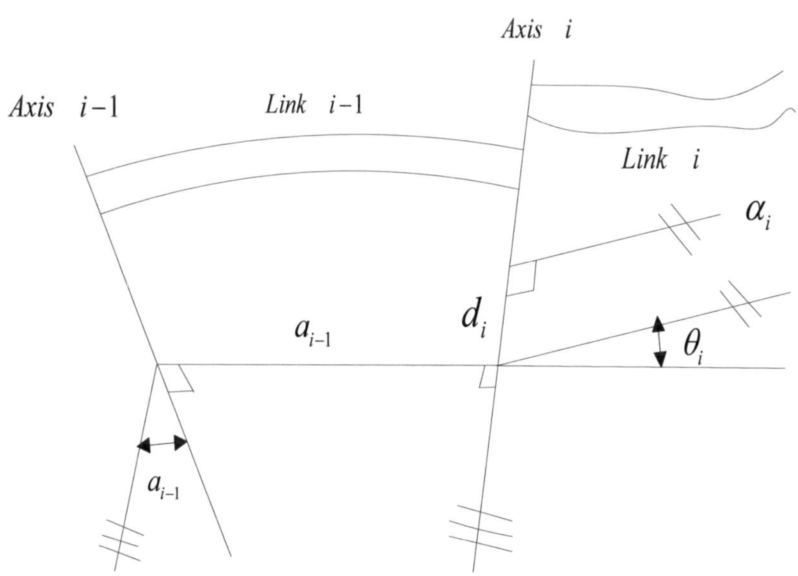

Fig. 6. The link offset, d, and the joint angle, θ, are two parameters which may be used to describe the nature of the connection between neighboring links.

2.5 Kinematics of robot

Robot kinematics is the study of the motion (kinematics) of robots. In a kinematic analysis the position, velocity and acceleration of all the links are calculated without considering the forces that cause this motion. The relationship between motion, and the associated forces and torques is studied in robot dynamics. One of the most active areas within robot kinematics is the screw theory.

Robot kinematics deals with aspects of redundancy, collision avoidance and singularity avoidance. While dealing with the kinematics used in the robots we deal each parts of the robot by assigning a frame of reference to it and hence robot with many parts may have many individual frames assigned to each movable parts. For simplicity we deal with the single manipulator arm of the robot. Each frames are named systematically with numbers, for example the immovable base part of the manipulator is numbered 0, and the first link joined to the base is numbered 1, and the next link 2 and similarly till n for the last nth link.

Robot kinematics is mainly of the following two types: forward kinematics and inverse kinematics. Forward kinematics is also known as direct kinematics. In forward kinematics, the length of each link and the angle of each join are given and we have to calculate the position of any point in the work volume of the robot. In inverse kinematics, the length of each link and position of the point in work volume is given and we have to calculate the angle of each joint.

Robot kinematics can be divided in serial manipulator kinematics, parallel manipulator kinematics, mobile robot kinematics and humanoid kinematics.

2.6 Reverse kinematics of robot

Direct kinematics consists in specifying the state vector of an articulated figure over time. This specification is usually done for a small set of "key-frames", while interpolation techniques are used to generate in-between positions. The main problems are the design of convenient key-frames, and the choice of adequate interpolation techniques. The latter problem, and in particular the way orientations can be represented and interpolated has been widely studied. Designing key positions is usually left onto the animator's hand, and the quality of resulting motions deeply depends on his skills. In many cases, available physical and biomechanical knowledge such as the characterization of motion phases for human walking, can help the animator to create relevant key-frames.

The exclusive use of direct kinematics makes it direct to add constraints to the motion, such as those specifying that the feet should not penetrate into the ground during the support phases. These constraints may be solved using inverse kinematic algorithms. Here, motion ΔX of the end link of a chain (ie. a foot) is specified by the animator in world coordinates. The system computes the variation $\Delta\theta$ of the state vector (ie. the orientations between intermediate links) that will meet the constraint. The relation between the _main task" ΔX and the angular displacements $\Delta\theta$ takes the form:

$$\Delta X = J \, \Delta\theta \tag{18}$$

where J is the Jacobian matrix of the system . J is not directly invertible, due to the direct dimensions of X and θ (ie. there is an infinity of angular positions at joints that lead to the same Cartesian position of a foot). So the most frequently used solution is:

$$\Delta\theta = J^{+}\Delta X + \alpha(I - J^{+}J)\Delta z \tag{19}$$

Where J^{+} is the pseudo-inverse of the Jacobian matrix J, α is a penalty constant, I is the identity matrix, and Δz is a constraint to minimize, called the secondary task. This secondary task is enforced on the null space of the main task. Thus, the second term does not affect the achievement of the main task, whatever the secondary task Δz is. Generally, Δz is used to account for joint angular limits or to minimize some energetic criteria.

3. Walking gait planning for humanoid robot

3.1 Walking pattern generation based on a inverted pendulum model

An inverted pendulum is a pendulum which has its mass above its pivot point. It is often implemented with the pivot point mounted on a cart that can move horizontally and may be called a cart and pole. Whereas a normal pendulum is stable when hanging downwards, an inverted pendulum is inherently unstable, and must be actively balanced in order to remain upright, either by applying a torque at the pivot point or by moving the pivot point horizontally as part of a feedback system.

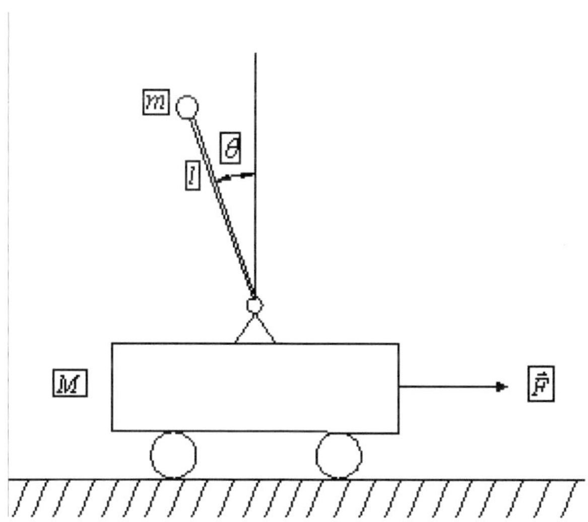

Fig. 7. a schematic drawing of the inverted pendulum on a cart. The rod is considered massless. The mass of the cart and the pointmass at the end of the rod are denoted by M and m. The rod has a length l.

The inverted pendulum is a classic problem in dynamics and control theory and widely used as benchmark for testing control algorithms (PID controllers, neural networks, fuzzy control, genetic algorithms, etc). Variations on this problem include multiple links, allowing the motion of the cart to be commanded while maintaining the pendulum, and balancing the cart-pendulum system on a see-saw. The inverted pendulum is related to rocket or missile guidance, where thrust is actuated at the bottom of a tall vehicle. The understanding of a similar problem is built in the technology of Segway, a self-balancing transportation device. The largest implemented uses are on huge lifting cranes move the box accordingly so that it never swings or sways. It always stays perfectly positioned under the operator even when moving or stopping quickly.

Another way that an inverted pendulum may be stabilized, without any feedback or control mechanism, is by oscillating the support rapidly up and down. If the oscillation is sufficiently strong (in terms of its acceleration and amplitude) then the inverted pendulum can recover from perturbations in a strikingly counterintuitive manner. If the driving point moves in simple harmonic motion, the pendulum's motion is described by the Mathieu equation.

In practice, the inverted pendulum is frequently made of an aluminum strip, mounted on a ball-bearing pivot; the oscillatory force is conveniently applied with a jigsaw.

Equations of motion

Stationary pivot point

The equation of motion is similar to that for an uninverted pendulum except that the sign of the angular position as measured from the vertical unstable equilibrium position:

$$\ddot{\theta} - \frac{g}{l}\sin\theta = 0 \tag{20}$$

When added to both sides, it will have the same sign as the angular acceleration term:

$$\ddot{\theta} = \frac{g}{l}\sin\theta \tag{21}$$

Thus, the inverted pendulum will accelerate away from the vertical unstable equilibrium in the direction initially displaced, and the acceleration is inversely proportional to the length. Tall pendulums fall more slowly than short ones.

Pendulum on a cart

The equations of motion can be derived easily using Lagrange's equations. Referring to the drawing where x (t) is the position of the cart, $\theta(t)$ is the angle of the pendulum with respect to the vertical direction and the acting forces are gravity and an external force in the x-direction, the lagrangian L = T - V, where T is the kinetic energy in the system and V the potential energy, so the written out expression for L is:

$$L = \frac{1}{2}Mv_1^2 + \frac{1}{2}mv_2^2 - mgl\cos\theta \tag{22}$$

Where v_1 is the velocity of the cart and v_2 is the velocity of the point mass m.

v_1 and v_2 can be expressed in terms of X and θ by writing the velocity as the first derivative of the position:

$$v_1^2 = \dot{x}^2 \tag{23}$$

$$v_2^2 = \left(\frac{d}{dt}(l\cos\theta)\right)^2 + \left(\frac{d}{dt}(x + l\sin\theta)\right)^2 \tag{24}$$

Simplifying the expression for v_2 leads to:

$$v_2^2 = \dot{x}^2 + 2\dot{x}l\dot{\theta} + l^2\dot{\theta}^2 \tag{25}$$

The Lagrangian is now given by:

$$L = \frac{1}{2}(M + m)\dot{x}^2 + ml\dot{x}\dot{\theta}\cos\theta + \frac{1}{2}ml^2\dot{\theta}^2 - mgl\cos\theta \tag{26}$$

and the equations of motion are

$$\frac{d}{dt}\frac{\partial L}{\partial \dot{x}} - \frac{\partial L}{\partial x} = F \tag{27}$$

$$\frac{d}{dt}\frac{\partial L}{\partial \dot{\theta}} - \frac{\partial L}{\partial \theta} = 0 \tag{28}$$

Substituting L in these equations and simplifying leads to the equations that describe the motion of the inverted pendulum:

$$(M + m)\ddot{x} + ml\ddot{\theta}\cos\theta - ml\dot{\theta}^2\sin\theta = F \tag{29}$$

$$ml\left(-g\sin\theta + \ddot{x}\cos\theta + l\ddot{\theta}\right) = 0 \tag{30}$$

These equations are nonlinear, but since the goal of a control system would be to keep the pendulum upright the equations can be linearized around $\theta \approx 0$.

Pendulum with oscillatory base

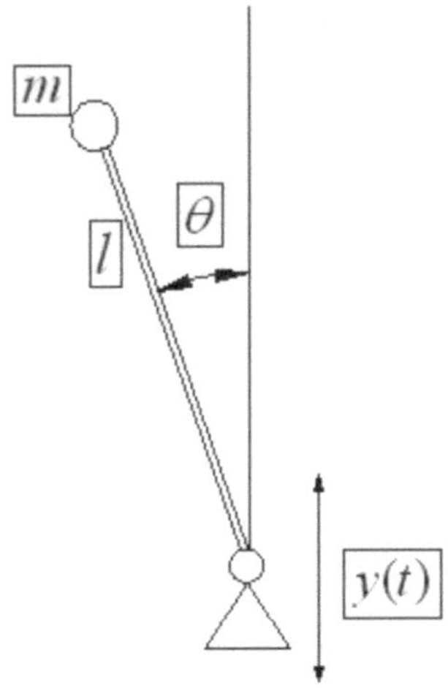

Fig. 8. a schematic drawing of the inverted pendulum on an oscillatory base. The rod is considered massless. The pointmass at the end of the road is demoted by m . The rod has a length l .

The equation of motion for a pendulum with an oscillatory base is derived the same way as with the pendulum on the cart, using the Lagrangian.
The position of the point mass is now given by:

$$\left(l\sin\theta, y + l\cos\theta\right) \tag{31}$$

And the velocity is found by taking the first derivative of the position:

$$v^2 = \dot{y}^2 - 2l\dot{\theta}\dot{y}\sin\theta + l^2\dot{\theta}^2 \tag{32}$$

The Lagrangian of this system can be written as:

$$L = \frac{1}{2}m\left(\dot{y}^2 - 2l\dot{\theta}\dot{y}\sin\theta + l^2\dot{\theta}^2\right) - mg(y + l\cos\theta) \tag{33}$$

and the equation of motion follows from:

$$\frac{d}{dt}\frac{\partial L}{\partial\dot{\theta}} - \frac{\partial L}{\partial\theta} = 0 \tag{34}$$

Resulting in:

$$l\ddot{\theta} - \ddot{y}\sin\theta = g\sin\theta \qquad (35)$$

If y represents a simple harmonic motion, $y = a\sin\omega t$, the following differential equation is:

$$\ddot{\theta} - \frac{g}{l}\sin\theta = -\frac{a}{l}\omega^2 \sin\omega t \sin\theta \qquad (36)$$

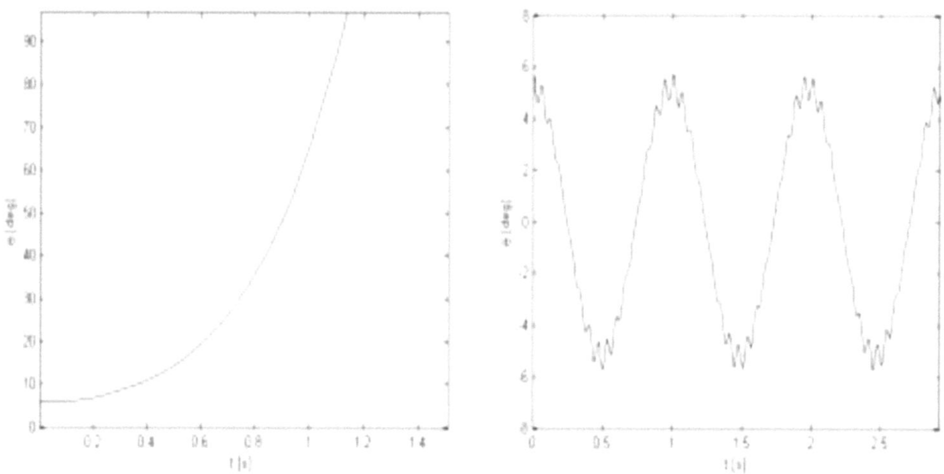

Fig. 9. Plots for the inverted pendulum on an oscillatory base. The first plot shows the response of the pendulum on a slow oscillation, the second the response on a fast oscillation

A solution for this equation will show that the pendulum stays upright for fast oscillations. The first plot shows that when y is a slow oscillation, the pendulum quickly falls over when disturbed from the upright position. The angle θ exceeds 90°after a short time, which means the pendulum has fallen on the ground.

If y is a fast oscillation the pendulum can be kept stable around the vertical position. The second plot shows that when disturbed from the vertical position, the pendulum now starts an oscillation around the vertical position ($\theta = 0$). The deviation from the vertical position stays small, and the pendulum doesn't fall over.

3.2 Gait planning of robot based on a seven-link model

In order to simplify research process we first discuss how to get ankle trajectory and hip trajectory. Then the knee trajectory could be got by kinematics. Here we take the left foot for example and the right foot is similar only with a delay of half cycle. The link model we used is shown in Figure 3.

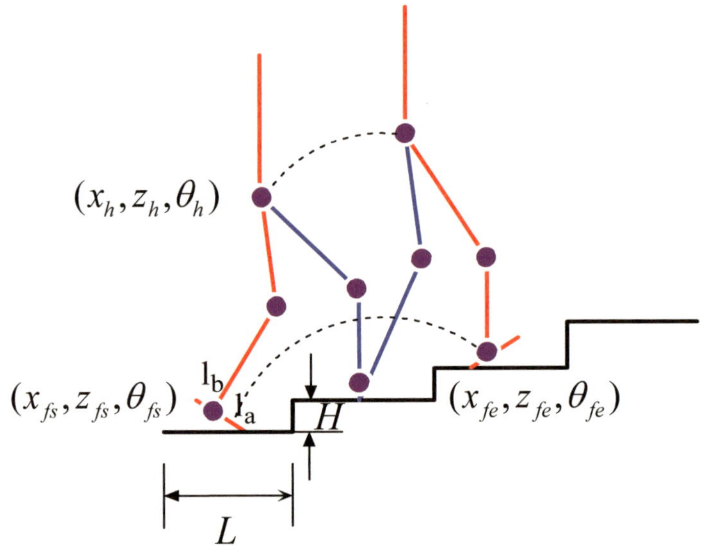

Fig. 10. The model of the humanoid robot going upstairs.

3.2.1 Gait Planning of Ankle

According to the walking procedure of human, we suppose that the walking cycle is T_c, $t = kT_c$ is the k th cycle begins with the moment when the left foot is just apart from the ground and ends with the left foot gets into contact with the ground; $kT_c < t \le kT_c + T_d$ is double support phase, during which the sole is rotated about toes, and the center of gravity moving forwards; the swing foot reaches the highest point when $t = kT_c + T_n$.
We get the key point $x_f(t), z_f(t)$ of ankle in plane XOZ as follows:

$$z_f = \begin{cases} z_{fs} + 2Hk + h_f & t = kT_c \\ z_{fs} + 2Hk + h_f \cos\theta_{fs} + l_a \left|\sin\theta_{fs}\right| & t = kT_c + T_d \\ z_{fs} + 2Hk + h_f + H & t = kT_c + T_n \\ z_{fs} + 2H(k+1) + h_f \cos\theta_{fe} + l_b \left|\sin\theta_{fe}\right| & t = (k+1)T_c \\ z_{fs} + 2H(k+1) + h_f & t = (k+1)T_c + T_d \\ z_{fs} + 2H(k+2) + h_f & t = (k+2)T_c \end{cases} \tag{37}$$

where l_a is the distance between tiptoe and the centre of gravity of sole; l_b is the distance between heel and the centre of gravity of sole; h_f is the height of heel and T_n is the time when the robot just walks through a step.
The key point of the angle between sole and ground can be denoted as follows:

$$\theta_f = \begin{cases} 0 & t = kT_c \\ \theta_{fs} & t = kT_c + T_d \\ 0 & t = kT_c + T_n \\ \theta_{fe} & t = (k+1)T_c \\ 0 & t = (k+1)T_c + T_d \end{cases} \tag{38}$$

Since the whole sole of the right foot is in contact with the ground at $t = kT_c$ and $t = (k+1)T_c + T_d$, the following derivative constraints must be satisfied.

$$\begin{cases} \dot{x}_f(kT_c) = 0 \\ \dot{x}_f((k+1)T_c + T_d) = 0 \end{cases} \tag{39}$$

$$\begin{cases} \dot{z}_f(kT_c) = 0 \\ \dot{z}_f((k+1)T_c + T_d) = 0 \end{cases} \tag{40}$$

$$\begin{cases} \dot{\theta}_f(kT_c) = 0 \\ \dot{\theta}_f((k+1)T_c + T_d) = 0 \end{cases} \tag{41}$$

3.2.2. Gait Planning of Hip

We assume that the robot is decelerated in double support phase and accelerated in single support phase and the acceleration in direction of x -axis and z -axis are a_{xh} and a_{zh} respectively. The distance between the hip and the ankle of supporting leg is x_s at the beginning of the double support phase and x_e at the end of the double support phase. The changes in the direction of z -axis are z_s and z_e at the beginning and end of the double support phase respectively. Then the trajectory of hip can be expressed like this:

$$x_h(t) = \begin{cases} x_h + 2Lk + x_s & t = kT_c \\ x_h + 2Lk + x_s + a_{xh}t & t = kT_c + T_d \\ x_h + 2Lk + L & t = kT_c + T_n \\ x_h + 2L(k+1) - x_e - a_{xh}t & t = (k+1)T_c \\ x_h + 2L(k+1) - x_e & t = (k+1)T_c + T_d \\ x_h + 2L(k+1) + x_s & t = (k+2)T_c \end{cases} \tag{42}$$

$$z_h(t)=\begin{cases} z_h+2Hk+z_s & t=kT_c \\ z_h+2Hk+z_s+a_{zh}t & t=kT_c+T_d \\ z_h+2Hk+H & t=kT_c+T_n \\ z_h+2H(k+1)-z_e-a_{zh}t & t=(k+1)T_c \\ z_h+2H(k+1)-z_e & t=(k+1)T_c+T_d \\ z_h+2H(k+1)+z_s & t=(k+2)T_c \end{cases} \quad (43)$$

It must satisfy the following constraints:

- The derivative constraints $\begin{cases}\dot{x}_h(kT_c)=\dot{x}_h(k+1)T_c \\ \ddot{x}_h(kT_c)=\ddot{x}_h(k+1)T_c\end{cases}$ and $\begin{cases}\dot{z}_h(kT_c)=\dot{z}_h(k+1)T_c \\ \ddot{z}_h(kT_c)=\ddot{z}_h(k+1)T_c\end{cases}$ must be satisfied.
- $z_h(t)\le h_{max}$, h_{max} is the maximum height of hip; $h_{max}=l_1+l_2+h_f$, l_1,l_2 are the length of thigh and shin respectively, h_f is the height of ankle.
- $z_h(t)\ge h_{min}$ h_{min} is the minimum height of hip and it's value can be set according to the process of human walking.
- $\{[x_h(t)-x_a(t)]^2+[z_h(t)-z_a(t)]^2\}^{1/2}\le l_1+l_2$

4. Stability control of humanoid robot

4.1 ZMP and FZMP concept

Zero moment point was introduced in January 1968 by Miomir Vukobratović at The Third All-Union Congress of Theoretical and Applied Mechanics in Moscow. In the following works and papers that were produced between 1970 and 1972 it would then be called zero moment point and would be spread around the world.

The zero moment point is a very important concept in the motion planning for biped robots. Since they have only two points of contact with the floor and they are supposed to walk, "run" or "jump" (in the motion context), their motion has to be planned concerning the dynamical stability of their whole body. This is not an easy task, especially because the upper body of the robot (torso) has larger mass and inertia than the legs which are supposed to support and move the robot. This can be compared to the problem of balancing an inverted pendulum.

The trajectory of a walking robot is planned using the angular momentum equation to ensure that the generated joint trajectories guarantee the dynamical postural stability of the robot, which usually is quantified by the distance of the zero moment point in the boundaries of a predefined stability region. The position of the zero moment point is affected by the referred mass and inertia of the robot's torso, since its motion generally requires large ankle torques to maintain a satisfactory dynamical postural stability.

One approach to solve this problem consists in using small trunk motions to stabilize the posture of the robot. However, some new planning methods are being developed to define the trajectories of the legs' links in such a way that the torso of the robot is naturally steered in order to reduce the ankle torque needed to compensate its motion. If the trajectory planning for the leg links is well succeeded, then the zero moment point won't move out of the predefined stability region and the motion of the robot will become smoother, mimicking a natural trajectory.

The resultant force of the inertia and gravity forces acting on a biped robot is expressed by the formula:

$$F^{gi} = mg - ma_G \tag{44}$$

Where m is the total mass of the robot, g is the acceleration of the gravity, G is the center of mass and a_G is the acceleration of the center of mass. The moment in any point X can be defined as:

$$M_X^{gi} = XG \times mg - XG \times ma_G - \dot{H}_G \tag{45}$$

where $\dot{H}_G$ is the rate of angular momentum at the center of mass. The Newton-Euler equations of the global motion of the biped robot can be written as:

$$F^c + mg = ma_G \tag{46}$$

$$M_X^c + XG \times mg = \dot{H}_G + XG \times ma_G \tag{47}$$

where F^c is the resultant of the contact forces at X and M_X^c is the moment related with contact forces about any point X.

The Newton-Euler equations can be rewritten as:

$$F^C + (mg - ma_G) = 0 \tag{48}$$

$$M_X^c + (XG \times mg - XG \times ma_G - \dot{H}_G) = 0 \tag{49}$$

So it's easier to see that we have:

$$F^c + F^{gi} = 0 \tag{50}$$

$$M_X^C + M_X^{gi} = 0 \tag{51}$$

These equations show that the biped robot is dynamically balanced if the contact forces and the inertia and gravity forces are strictly opposite.

If an axis Δ^{gi} is defined, where the moment is parallel to the normal vector n from the surface about every point of the axis, then the Zero Moment Point (ZMP) necessarily belongs to this axis, since it is by definition directed along the vector n. The ZMP will then be the intersection between the axis Δ^{gi} and the ground surface such that:

$$M_Z^{gi} = ZG \times mg - ZG \times ma_G - \dot{H}_G \tag{52}$$

with

$$M_Z^{gi} \times n = 0 \tag{53}$$

where Z represents the ZMP.
Because of the opposition between the gravity and inertia forces and the contact forces mentioned before, the Z point (ZMP) can be defined by:

$$PZ = \frac{n \times M_P^{gi}}{F^{gi} \cdot n} \tag{54}$$

where P is a point of the sole where is the normal projection of the ankle.
Fictitious zero moment point (FZMP) is an important expand of ZMP, it can be used in stability control. In order to evaluate dynamic stability, we use the ZMP principle. The ZMP is the point where the influence of all forces acting on the mechanism can be replaced by one single force. If the computed ZMP is the real ZMP, this means the computed ZMP inside the real support polygon, the biped robot can be stable. If the ZMP is not the real ZMP, this means the computed ZMP is on the boundary of the support polygon, the robot will fall down or have a trend of falling down. If the computed ZMP is outside the support polygon, then the robot will fall down and in this case, the computed ZMP is called fictitious ZMP. The link model of the humanoid robot is shown in Figure 3.4.

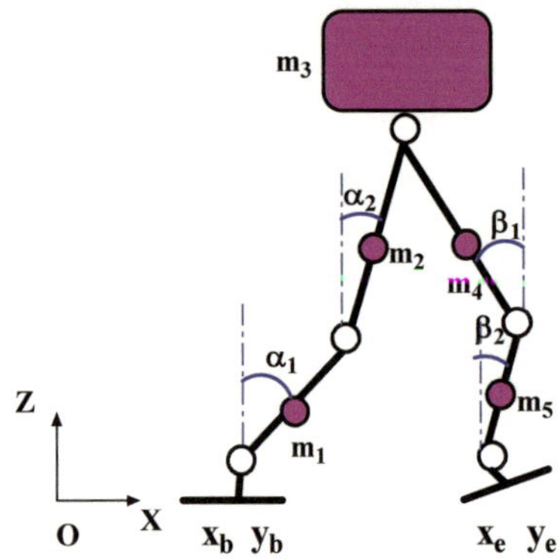

Fig. 11. The link model of the humanoid robot.

The projection of position vector of computed ZMP can be computed by the following equations:

$$x_{zmp} = \frac{\sum_{i=1}^{5} m_i(\ddot{z}_i + g)x_i - \sum_{i=1}^{5} m_i \ddot{x}_i z_i + \sum_{i=1}^{5} M_{iy}}{\sum_{i=1}^{5} m_i(\ddot{z}_i + g)} \tag{55}$$

$$y_{zmp} = \frac{\sum_{i=1}^{5} m_i(\ddot{z}_i + g)y_i - \sum_{i=1}^{5} m_i \ddot{y}_i z_i + \sum_{i=1}^{5} M_{ix}}{\sum_{i=1}^{5} m_i(\ddot{z}_i + g)} \tag{56}$$

where m_i is mass of every links, (x_i, y_i, z_i) is the coordinate of the mass center of the links, $(M_{ix}, M_{iy})^T$ is the moment vector.

If the ZMP is inside the support polygon and the minimum distance between the ZMP and the boundaries of support polygon is large, then the biped will be in high stable, and this distance is called the stability margin. We can know the situation of walking stability from the stability margin.

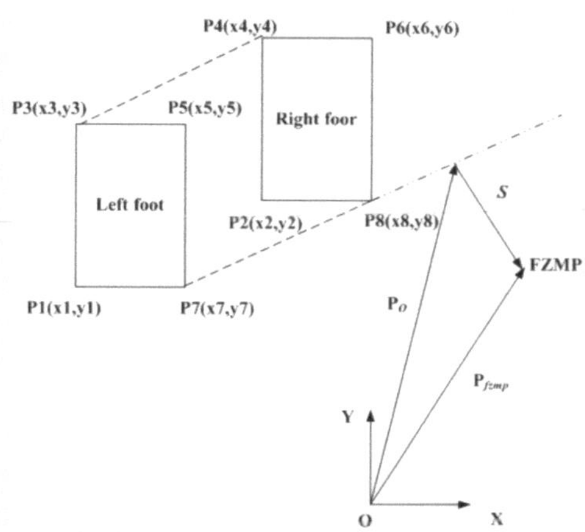

Fig. 12. The relationship between FZMP and support polygon.

As shown in Figure2, if the ZMP is outside the support polygon, i.e. FZMP, the norm of vector S represents the shortest distance between FZMP and the edges of the support polygon. This edge is called rotation edge. The direction of vector S is the rotation direction of the robot.

The importance of FZMP is:

- We can judge the falling down possibility by calculated the position of FZMP;
- According the position of FZMP, we can calculate the distance of rotate boundary and falling downing direction.

● When the robot in the stability situation, support polygon can be defined as the minimum distance between the boundaries of support polygon and the ZMP, this means robot stability margin; while in the instability situation, the minimum distance between the boundaries of support polygon and the ZMP, it is a measure of instability.

4.1.1 ZMP and stable walking

Apart from the realization of the relative motion of the mechanism's links, the most important task of a locomotion mechanism of humanoid robot during walking is to preserve its dynamic balance in contact with the ground. The foot relies freely on the upport and is realized via the friction force and vertical force of the ground reaction. The foot cannot be controlled directly but in an indirect way, by ensuring the appropriate dynamics of the mechanism above the foot. Thus, the overall indicator of the mechanism behavior is the point where the influence of all forces acting on the mechanism can be replaced by one single force. This point was termed the Zero-Moment Point (ZMP). ZMP is very important for humanoid robot as dynamic criterion of gait planning, stability and control. The ZMP principle can be generalized as follows.

● If ZMP is inside of the footprint of support foot in single support phase, or inside of support polygon in double support phase, then biped robot can keep its dynamic balance and the stable walking is possible.

● If ZMP is on the boundary of the footprint in single support phase or of support polygon in double support phase, then the robot will fall down or have a trend of falling down.

● If computed ZMP is outside of the footprint of support foot in single support phase, or without support polygon in double support phase, then the robot cannot be in the dynamic stable and will fall down. In this case, it should be called fictitious ZMP, shortly FZMP.

There are two different cases in which the ZMP plays a key role:

□(1) in determining the proper dynamics of the mechanism above the foot to ensure a desired ZMP position. This belongs to the task of gait synthesis.

□(2) in determining the ZMP position for the given mechanism motion. This refers to the gait control.

Biped walking is a periodic phenomenon. A complete walking cycle is composed of two phases: a double-support phase and a single-support phase. During the double-support phase, both feet are in contact with the ground. This phase begins with the heel of the forward foot touching the ground, and ends with the toe of the rear foot leaving the ground. During the single-support phase, one foot is stationary on the ground, the other foot swings from the rear to the front. The gait of walking robot can be generated by using ZMP principle.

4.1.2.FZMP and stability maintenance

For determination of dynamic equilibrium we have to consider the relationship between the computed position of reaction point P on the ground and the support polygon. If the position of point P is within the support polygon, the robot is in dynamic equilibrium. The computed position P is called traditional ZMP, if only one foot contacts with floor, the force acting at ZMP is real reaction force, if two feet contact with floor it is total force of all contact reaction forces. If the computed point P is located outside the support polygon, it can be called as a fictitious ZMP (FZMP). In this case, the humanoid robot would start to rotate about the edge and the robot would lose the stability. The real acting point of ground reaction force would be located in the edge. The calculated position of the point P outside the support polygon represents only fictitious locations. The FZMP is very useful to deal with the stability maintenance and control in the emergency case. For the stable walking of humanoid robot the ZMP must be kept within the support polygon. To maintain regularly the mechanism dynamic stable at the moment of the occurrence of an external disturbance an emergency-coping strategy based on FZMP concept can be applied. The importance of the FZMP to deal with the stability control and maintenance is mentioned by several authors. But how to fully utilize its property should be further researched. In this paper the FZMP is efficiently used to deal with the stability maintenance of humanoid robot under disturbance.

4.2 The determination of support polygon and stability margin

4.2.1 The determination of support polygon

If only one foot contacts with floor, the above mentioned support polygon is the region of the foot. But if the two feet contact with the floor, the situation would be sometime complex. In the current related researches the support polygon used to be expressed simply with the graphs. It is not convenient in stability analysis and control. In this paper, we present a computerization expression of support polygon.

We assume that the shape of the foot is rectangle. Then two feet contain eight edges all together. The support polygons are composed of some edges of the above mentioned eight edges and other two new edges. We call all edges that constitute the support polygon as valid connection edges (VCE). The candidates of the VCE are all connection edges of the eight corner points on two feet. In figure 3 the corner points in left foot are denoted with P_1, P_3, P_5, P_7, the right foot P_2 P_4, P_6, P_8. The line L_{ij} through two point P_i and P_j can be expressed as follows:

$$y = a_{ij}x + b_{ij}$$

$$a_{ij} = \frac{y_i - y_j}{x_i - x_j} \quad b_{ij} = \frac{x_j y_i - x_i y_j}{x_j - x_i} \tag{57}$$

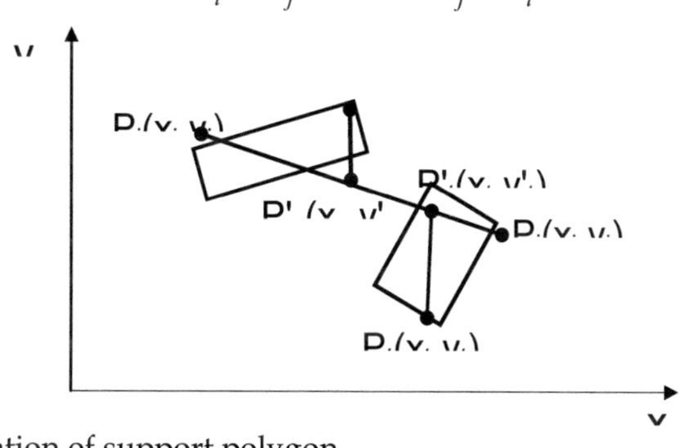

Fig. 13. The determination of support polygon

In order to determine whether the edge connecting point P_i and P_j is the VCE we have to consider the position relationship between L_{ij} and all eight corner points. If all eight corner points are the same side of the line L_{ij}, that is, satisfy (6), then edge E_{ij} according to line L_{ij} is the VCE. Otherwise E_{ij} is not VCE and should be ignored. Here, P_s, P_t is respectively the one of the eight corner points.

$$\begin{cases} y_s - y_s^{'} > 0 \\ y_t - y_t^{'} > 0 \end{cases} \text{ or } \begin{cases} y_s - y_s^{'} < 0 \\ y_t - y_t^{'} < 0 \end{cases} \quad (s,t = 1,...8.s \neq t) \tag{58}$$

Where $P_s^{'}$ and $P_t^{'}$ are the projection points of P_s and P_t on the

$$L_{ij} \, , \, y_s^{'} = a_{ij}x_s + b_{ij} \, , \, y_t^{'} = a_{ij}x_t + b_{ij} \, .$$

4.2.2 The Relationship Between FZMP and Support Polygon

We have to determine which is the rotation edge in all VCEs when robot lose stability. and in this case the distance from FZMP to the rotation edge can be calculated. In figure 4 the distance from FZMP to E_{ij} is expressed as follows:

$$\|s\| = \min_{i,j=np} \|p_{FZMP} - p_{ij}\| \tag{59}$$

Where p_{ij} is the position vector of vertical point from FZMP to E_{ij}.

$$\begin{aligned} x_{P_{ij}} &= \frac{a_{ij}b_{ij} - a_{ij}y_{fzmp} + x_{fzmp}}{1 - a_{ij}^2} \\ y_{P_{ij}} &= \frac{b_{ij} - a_{ij}^2 y_{fzmp} + a_{ij}x_{fzmp}}{1 - a_{ij}^2} \end{aligned} \tag{60}$$

We denote the point p_o as the position vector of the vertical point that satisfies (7). That is

$$\mathrm{s} = p_{fzmp} - p_o \tag{61}$$

If we know the position of FZMP the rotation edge can be determined according to (7), the distance from FZMP to the rotation edge and the direction of losing stability can be calculated by (9).Those two parameters play the key role in maintaining the robot stability.

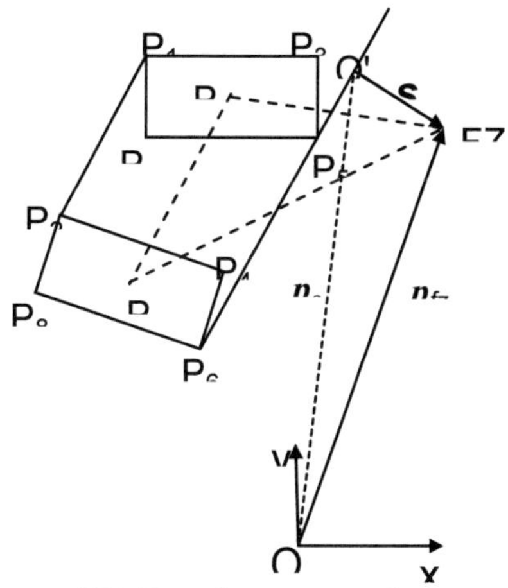

Fig. 14. The relationship between FZMP and support polygon

4.2.3 Control algorithm considering external environment

When the robot might lose stability because of the external disturbance, it must immediately react this situation and be controlled to keep in stable state.

The control approaches could be one of the several methods such as the movement of the upper body to change of the center of gravity, the enlargement or movement of the support polygon, the attachment of robot hand to the surrounding.

(1) The Enlargement of Support Polygon

We can enlarge the support polygon by modifying the prescribed landing position of the swing foot to maintain stability under disturbance. It is mentioned by some research, but it is not explained how to realize the enlargement of support polygon. It is not realistic to move parallel the rotation edge, which means to moving two feet at same time.

In figure 5 the moving foot should land the planed position expressed in dashed line if no external disturbance. But under disturbance the robot will be rotate about RB. In this case, the landing position should be changed to maintain the stability. The changed angle α_f^* of moving direction of the foot is determined by (10).

$$\alpha_f^* = \cos^{-1}\frac{e \cdot s}{\|e\| \cdot \|s\|} \tag{62}$$

Here *e* is the normal planed direction vector.

The foot moving distance l_f^* relative to planed landing position is determined by the formula (4.19)

$$l_f^* = \sin(\alpha^* - \beta)\left[l\cos(\gamma - \alpha^* - \beta) - b\right] + l\sin(\gamma - \alpha^* - \beta) + a \tag{63}$$

If the change of center of gravity due to the foot extra moving is ignored, then at new landing position the robot will be stable.

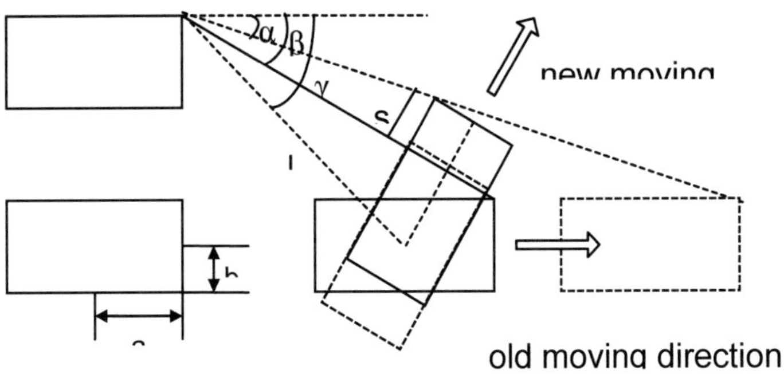

Fig. 15. The determination of foot landing position

(2) The Movement of Upper Body

We assume that the all links of upper body have same the displacement, the velocity and acceleration s_u, v_u, a_u respectively. The moving direction should be pointed to the FZMP. That is, $x_j = s_u \cos\alpha_f^*$, $y_j = s_u \sin\alpha_f^*$, $\dot{x}_j = v_u \cos\alpha_f^*$ $\dot{y}_j = v_u \sin\alpha_f^*$, $\ddot{x}_j = a_u \cos\alpha_f^*$, $\ddot{y}_j = a_u \sin\alpha_f^*$. We assume that $z_j = const, \ddot{z}_j = 0$ for all upper body j-th link. In this case, equation (3) is modified to equation (12).

$$
\begin{aligned}
&\bar{x}_{zmp}\left(\sum_{i=1}^{m} Q_{iz} + M\left(\ddot{z}_G + g\right)\right) = \Delta \dot{H}_{G_Y} + \left(Mx_G + M_u s_u \cos\alpha_f^*\right)\left(\ddot{z}_G + g\right) \\
&- z_G\left(M\ddot{x}_G + M_u a_u \cos\alpha_f^*\right) - \tau_y - \sum_{i=1}^{m}\left(x_{Q_i} Q_{iz} - z_{Q_i} Q_{ix}\right) \\
&\bar{y}_{zmp}\left(\sum_{i=1}^{m} Q_{iz} + M\left(\ddot{z}_G + g\right)\right) = \Delta \dot{H}_{G_x} + \left(My_G + M_u s_u \sin\alpha_f^*\right)\left(\ddot{z}_G + g\right) \\
&- z_G\left(M\ddot{y}_G + M_u a_u \sin\alpha_f^*\right) - \tau_x - \sum_{i=1}^{m}\left(y_{Q_i} Q_{iz} - z_{Q_i} Q_{iy}\right)
\end{aligned}
\tag{64}
$$

$$
\Delta H_{G_Y} = \sum_{i=1}^{n_1}\left(-\frac{M_u}{M} s_u \cos\alpha_f^* m_i \dot{z}_{G_i}\right) + \sum_{i=n_1+1}^{n_2}\left(\frac{M - M_u}{M} s_u m_i \cos\alpha_f^* \dot{z}_{G_i}\right)
$$

$$
\Delta H_{Gx} = \sum_{i=1}^{n_1}\left(-\frac{M_u}{M} s_u \sin\alpha_f^* m_i \dot{z}_{G_i}\right) + \sum_{i=n_1+1}^{n_2}\left(\frac{M - M_u}{M} s_u m_i \sin\alpha_f^* \dot{z}_{G_i}\right)
$$

where M_u is the total mass of upper body. $\bar{x}_{zmp}$ and $\bar{y}_{zmp}$ are the x- and y-projector of pre-designed ZMP. From this equation s_u^* and a_u^* can be calculated.

(3) The Attachment of the Hand to the Surrounding

Through the arm movement to attach with the surrounding to ensure additional support points the static equilibrium may be re-established and the dynamically balanced gait continued. This procedure of re-establishing dynamic equilibrium might be considered as a kind of total compliance procedure. The position of computed zero-moment point will be changed under the support reaction force as shown as following:

$$x'_c - x_c = \frac{u_{hy} + x_l Q_{lz} - z_l Q_{lx} + x_r Q_{rz} - z_r Q_{rx}}{\sum_{i=1}^{m} Q_{iz} + Q_{lz} + Q_{rz} + M\left(\ddot{z}_G + g\right)}$$

$$y'_c - y_c = \frac{u_{hx} + x_l Q_{lz} - z_l Q_{ly} + x_r Q_{rz} - z_r Q_{ry}}{\sum_{i=1}^{m} Q_{iz} + Q_{lz} + Q_{rz} + M\left(\ddot{z}_G + g\right)} \tag{65}$$

Here we assume $x_l = x_r = x_a$, $y_l = y_r = y_a$, $z_l = z_r = z_a$, $Q_{lx} = Q_{rx} = Q_x^* / 2$, $Q_{lz} = Q_{rz} = Q_z^* / 2$, $Q_{ly} = Q_{ry} = 0$, $\Delta x_{ch} = x'_c - x_c$, $\Delta y_{ch} = y'_c - y_c$, we can obtain:

$$Q_x^* = \frac{(\Delta y_{ch} - \Delta x_{ch}) x_a}{(\Delta y_{ch} - x_a) z_a} \left(\sum_{i=1}^{m} Q_{iz} + M(\ddot{z}_G + g \right)$$

$$Q_z^* = \frac{\Delta y_{ch}}{\Delta y_{ch} - x_a} \left(\sum_{i=1}^{m} Q_{iz} + M(\ddot{z}_G + g \right) \tag{66}$$

Because the hand can not only push but also pull the environment, the forces Q_x^* and Q_z^* can be positive or negative.

(4) The Optimization Control Strategy

The method above can be used to maintain the stability of the robot, but in some cases only one method is unrealistic because of the limitation of the time or foot stride etc.. we have to use the combination of the method above to maintain the stabilities of robot. In this case the stability maintenance can be considered the following dynamic optimization problem.

Objective function:

$$F(X) = \min_{X \in R^m} (\lambda \max_t (\left\| p_c(X) - p_{zmp}^* (\mathrm{k_f} X_f) \right\|_{p_c = p_{zmp}}) + \gamma \max_t (\left\| p_c(X) - p_o(X) \right\|_{p_c = p_{fzmp}})) \tag{67}$$

Subject to

$$x_c = \frac{\dot{H}_{Gy} + \left(Mx_G + M_u k_u s_u \cos\alpha_f^*\right)\left(\ddot{z}_G + g\right) - \left(M\ddot{x}_G + M_u k_u a_u \cos\alpha_f^*\right)z_G}{\sum_{i=1}^{m} Q_{iz} + k_a Q_z + M\left(\ddot{z}_G + g\right)}$$

$$- \frac{\tau_y + \sum_{i=1}^{m}\left(x_{Q_i} Q_{iz} - z_{Q_i} Q_{ix}\right) + k_a\left(x_a Q_z - z_a Q_x\right)}{\sum_{i=1}^{m} Q_{iz} + k_a Q_z + M\left(\ddot{z}_G + g\right)}$$

$$y_c = \frac{\dot{H}_{Gx} + \left(My_G + M_u k_u s_u \sin\alpha_f^*\right)\left(\ddot{z}_G + g\right) - \left(M\ddot{y}_G + M_u k_u a_u \sin\alpha_f^*\right)z_G}{\sum_{i=1}^{m} Q_{iz} + k_a Q_z + M\left(\ddot{z}_G + g\right)}$$

$$- \frac{\tau_x + \sum_{i=1}^{m}\left(y_{Q_i} Q_{iz} - z_{Q_i} Q_{iy}\right) + k_a y_a Q_z}{\sum_{i=1}^{m} Q_{iz} + k_a Q_z + M\left(\ddot{z}_G + g\right)}$$

$$\begin{aligned}
H_G &= \sum_{i=1}^{n} H_{G_i} + \sum_{i=1}^{n_1}\left(p_{G_i} - p_G - \frac{M_u}{M} k_u s_u I_u\right) \times m_i \dot{p}_{G_i} \\
&+ \sum_{i=n_1+1}^{n_2}\left(p_{G_i} - p_G + \frac{M - M_u}{M} k_u s_u I_u\right) \times m_i\left(\dot{p}_{G_i} + k_u v_u\right) \\
&= H_{G0} + \sum_{i=1}^{n_1}\left(-\frac{M_u}{M} k_u s_u I_u\right) \times m_i \dot{p}_{G_i} + \\
&\sum_{i=n_1+1}^{n_2}\left(\frac{M - M_u}{M} k_u s_u I_u\right) \times m_i\left(\dot{p}_{G_i} + k_u v_u\right) \\
&+ \sum_{i=n_1+1}^{n_2}\left(p_{G_i} - p_G\right) \times m_i\left(k_u v_u\right)
\end{aligned}$$

$$X_{\min} \le X \le X_{\max}$$

$p_o(X)$ is determined by (7)

Where $X = (x_1 \quad x_2 \quad x_3 \quad x_4 \quad x_5 \quad x_6 \quad x_7)^T = \left(X_f, X_u, X_a\right)^T$, $p_c(X) = (x_c \quad y_c \quad 0)^T$ is the optimization design variable. $X_f = \left(l_f, \alpha_f\right)^T$ is the movement vector of the foot, in which l_f, α_f is the moving distance and direction of the foot relative to planed landing position, respectively. $X_u = (s_u \quad v_u \quad a_u)^T$ is the movement parameter vector of upper body. $X_a = (Q_x, Q_z)^T$ is the hand support force parameters vector, and the force in y-direction is ignored. λ, γ are the weight coefficients, but they should be set with different numerical value. λ should be chose a relative small value, which depends on the optimization demand. The value γ expresses the influence extent of FZMP on the objective function. It is very important to choice right γ value. If computed value p_c is within the support polygon, γ =0. If outside the support polygon, γ should be chose the large value as the punishment. $k = (k_u \quad k_f \quad k_a)$ is the choice coefficient vector in which k_u , k_f , k_a equal 0 or 1. For example, $k = (1,1,0)$ means that the extra movement of upper body and foot are considered and the hand attachment does not exist.

This is a parameter optimization problem that means the normal gait pattern of robot before the external disturbance is introduced is known. From (15) we present a hierarchy control strategy as shown in figure 4.4.

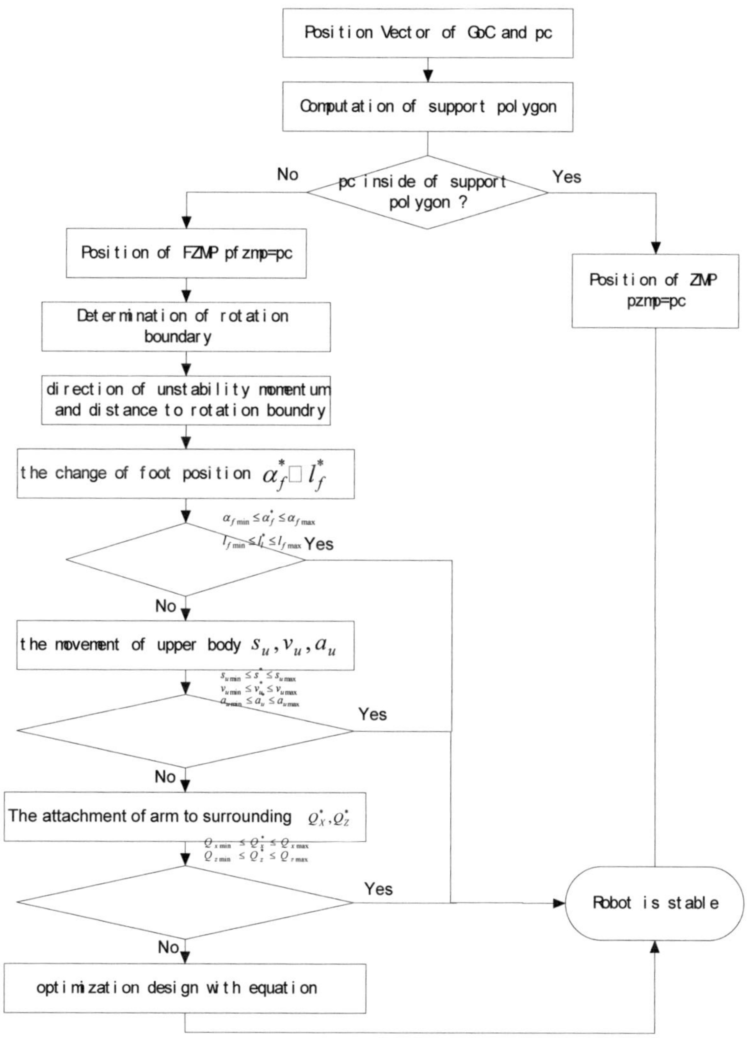

Fig. 16. The control strategy

4.3 Example:

With the method above we have constructed a simulator of humanoid robot by using dynamic analysis software package ADAMS. The total height of humanoid robot is 1650 mm. The walking speed is 2 km/h, the stride is 520 mm. The part of simulation parameters is shown in Table 1.

Components	Length(m)	Mass (kg)
Thigh	0.35	5.0
Shank	0.35	3.0
Ankle Height	0.10	0.5
Upper Arm	0.307	0.56
Lower Arm	0.241	0.58
Hand	0.178	20.0
Foot heel / Tiptoe	0.10 / 0.12	0.8

Table 1. Simulation Parameters

After building the model of humanoid robot we have made the several simulations considering the upper movement, enlargement of foot stride and their combination. In simulation the foot contact with the ground is emphasized. The figure 7 shows the normal walking process of humanoid robot.

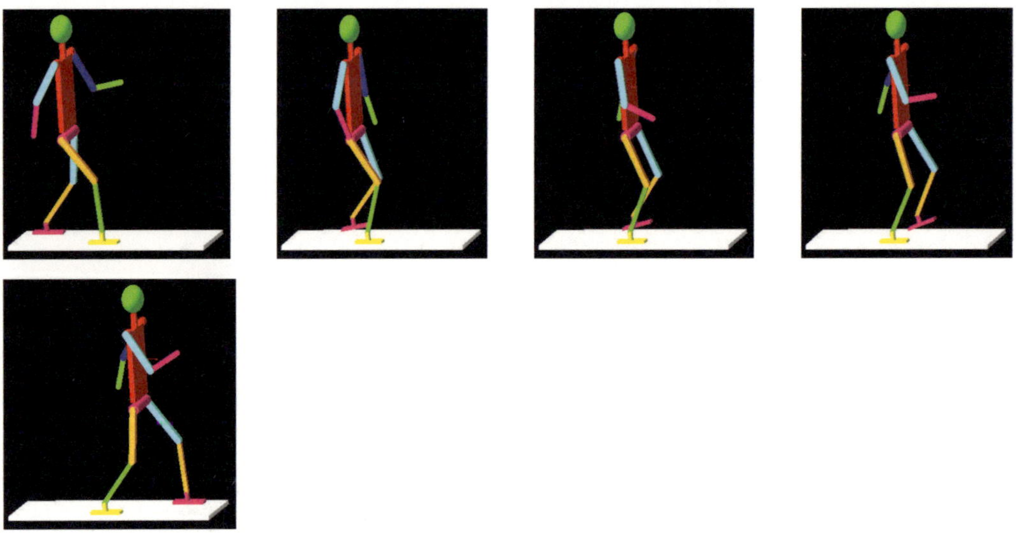

Fig. 17. The walking simulation

The x-coordinate and z-coordinate of center of gravity at normal walking and added upper body moving to maintain stability under external disturbance are shown in figure 8 and figure 9The stable walking under the reaction of external force can be kept also by the change of foot stride. In figure 10 the foot stride should be enlarged 120 mm compared the normal stride, where the walking direction was not changed. The optimization process is shown in figure 11.The simulations show that the stability under external disturbance can be kept with the described strategy.

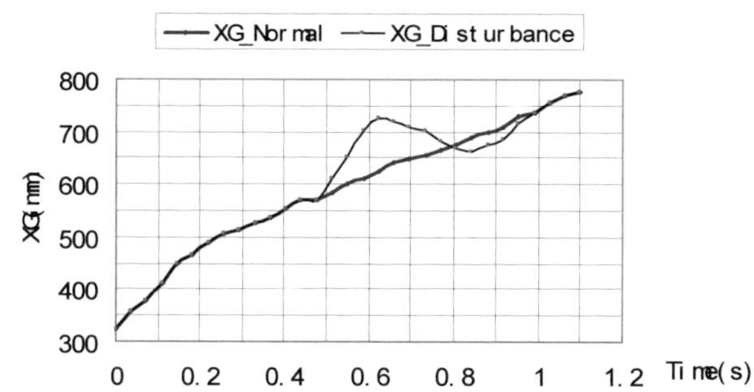

Fig. 18. The x-coordinate of center of gravity

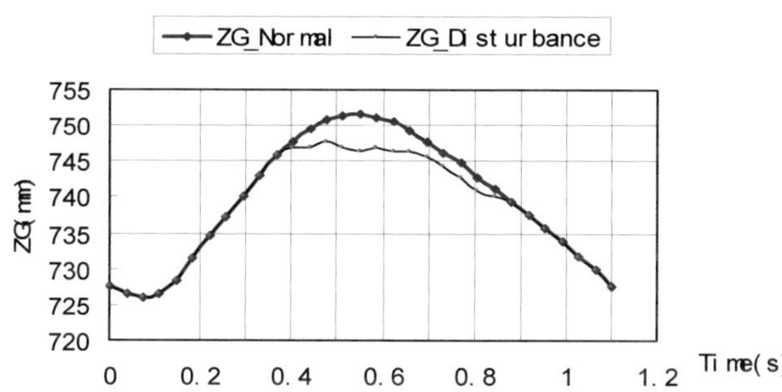

Fig. 19. The z-coordinate of center of gravity

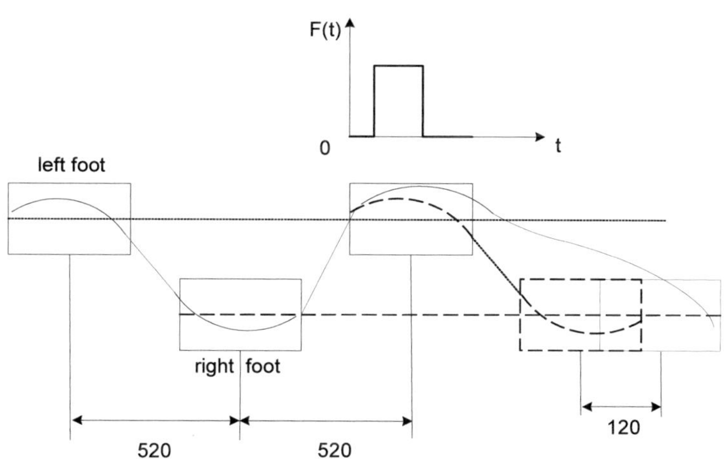

Fig. 20. The enlargement of foot stride (no change in direction)

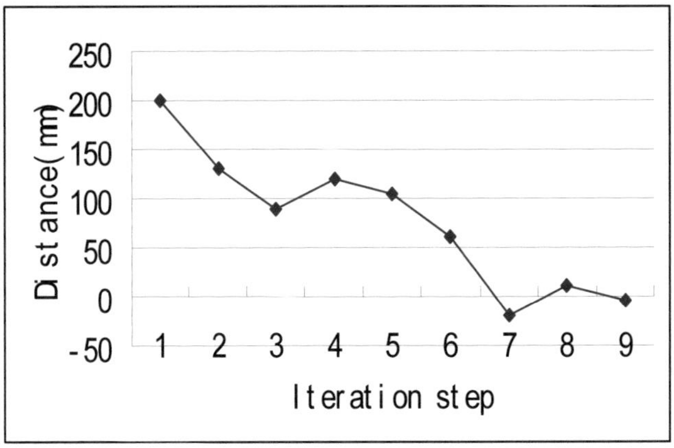

Fig. 21. The Distance between FZMP and center of support polygon

The stability maintenance is important issue in humanoid robot walking. The roles of FZMP are emphasized to maintain the robot stability. The support polygon is expressed with computerized form. According to the position of FZMP the rotation border can be automatic determined. The distance between the FZMP and the rotation border represents the strength and direction of losing the stability. The stability maintenance methods such as the movement of upper body, the change of foot landing position and the hand attachment with environment are discussed. The optimization control model considering different stability maintenance measures is proposed. The numerical simulation shows that the proposed method is effective and has the advantage of less calculation time.

5. Summary

The stability control is important issue in humanoid robot walking. A balance controller consisting of an off-line walk pattern planner and a real-time modification was proposed. If we can solve this problem, the robot can walk smoothly and adapt to unknown environments, and many functions will become true. Inverted pendulum model, ZMP and FZMP conception are effective methods for walking gait planning and the stability control.

6. References

K Kaneko, S Kajita, F Kanehiro, K Yokoi, K Fujiwara, H Hirukawa, T Kawasaki, M Hirata and T Isozumi, "Design of Advanced Leg Module for Humanoid Robotics Project of METI," Proceedings of the 2002 IEEE International Conference on Robotics & Automation, Washington, DC, May 2002

A Konno, "Design And Development of the Biped Prototype Robian," Proceedings of the 2002 IEEE International Conference on Robotics & Automation, Wasington,DC; May 2002.

F Pfeiffer, K Loeffer, M Gienger, "The Concept of Jogging JOHNNIE," Proceedings of the 2002

IEEE international conference on robotics & Automation, Washington, DC, May 2002

S Kajita, F Kanehiro, K Kaneko, K Fujiwara, K Harada, K Yokoi and H Hirukkawa, "Biped Walking Patern Generation By Using Preview Control of Zero-Moment Point," Proceedings of the 2003 IEEE International Conference on Robotics &Automation, Taipei, Taiwan, September 14-19,2003

R Stojic, C. Chevallereau, "On the Stability of Biped with Point Foot-Ground Contact," Proceedings of The 2000 IEEE International Conference on Robotics &Automation, San Francisco, CA, April 2000

Y Ogura, "Stretch Walking Pattern Generation for a Biped Humanoid Robot," Proceedings of 2003 International Conference in Intelligent Robots and Systems, Las Vegas, Nevada, October 2003.

M. Vukobratovic, "Biped Locomotion: Dynamics,Stability,Control And Application," Spring Verlag, Berlin,1990.

M Vukobratovic, B Borovac, "Zero-Moment Point – Thirty Five Years of Its Life," International Journal of Humanoid Robotics Vol. 1, No. 1 (2004), pp. 157–173

K Yoneda, S Hirose, "Tumble Stabilit Criterion of Integrated Locomotion and Manipulation," Proceedings of IEEE International Conference on Intelligent Robot and Systems, 1996, pp. 870-876

A Gowami, "Postural Stability of Biped Robots and the Foot-Rotation Indicator (FRI) Point," The International Journal of Robotics Research, vol.18, no. 6, June 1999, pp. 523-533

K Harada, S Kajita, K Kaneko and H Hirukawa, "Pushing Manipulation by Humanoid considering Two-kinds of ZMPs," Proceedings of the 2003 IEEE international conference on robotics & Automation, Taipei, Taiwan, September 14-19, 2003, pp. 1627-1632

A Goswami, V. Kallen, "Rate of changes of angular momentum and balance maintenance of biped robots," Proceedings of the 2004 IEEE international conference on robotics & Automation, New Orleans, LA, April 2004,pp. 3785-3790

S Kajita, F Kanehiro, K Kaneko etc., "Resolved Momentum Control: Humanoid Motion Planning Based on the Linear and Angular Momentum," Proceedings of IEEE International Conference on Intelligent Robot and Systems, Las Vegas, Nevada, October 2003, pp.1644-1650

Rüdiger Dillmann, Regine Becher and Peter Steinhaus, ARMAR II – A Learning and Cooperative Multimodal Humanoid Robot System, International Journal of Humanoid Robotics Vol. 1, No. 1 (2004) 143–155

Chenbo Yin□Qingmin Zhou and Le Xiao, walking stability of a humanoid robot based on fictitious zero-moment point, control, Automation, Robotics and Vision, 2006. ICARCV'06.9th International conference on, 5-8 Dec. 2006 On page(s): 1-6

Chenbo Yin, Albert Albers, Jens Ottnad and Pascal Häußler, stability maintenance of a humanoid robot under disturbance with fictitious zero-moment point, Intelligent Robots and Systems, 2005. (IROS 2005). 2005 IEEE/RSJ International Conference on Volume, Issue, 2-6 Aug. 2005 Page(s): 3149 - 3156

John J.Craig, Introduction to robotics: mechanics and control, Addison-Wesley Publishing

Company, 1989, 19-36.

Ming Tan, De Xu and Zengguang Hou ect., Advanced robot control(in Chinese), High education press, 2007.5, 20-56.

18

Towards Artificial Communication Partners with a Multiagent Mind Model Based on Mental Image Directed Semantic Theory

Masao Yokota
Fukuoka Institute of Technology
Japan

1. Introduction

In recent years, there have been developed various types of real or virtual robots as artificial communication partners. However, they are to play their roles according to programmed actions to stimuli and have not yet come to understand or imitate delicate mental functions of their human partners such as *Kansei*, one of the topics in this chapter. Kansei evaluates non-scientific matters such as art, music, natural scenery, etc. by Kansei words (e.g., 'heart-calming', 'fantastic', 'grotesque') (Fukuda et al, 1998; Sugita et al, 2003; Butz et al, 2005) and 'Artificial Kansei' namely 'Kansei for robots' is expected to play a part in 'artificial or robotic individuality' (Tauchi et al, 2006).

The author has proposed a human mind model consisting of Stimulus, Knowledge, Emotion and Response Processing Agents, intending the intelligent system IMAGES-M (Yokota, 2005a; Shiraishi et al, 2005) to understand and imitate miscellaneous human mental functions involving *emotion processing* as well as *knowledge processing* originally aimed at. For example, Kansei is defined and realized as tight collaboration of Knowledge and Emotion Processing Agents. This multiagent mind model is much simpler than Minsky's (Minsky, 1986) and its most remarkable feature is that the agents involved communicate with one another by exchanging and computing mental images represented in the formal language L_{md} developed for integrated representation and computation of multimedia information (Yokota, 2005, 2006) while other multiagent systems were designed to employ special programming languages for inter-agent communication (e.g., Labrou et al, 1999; Vieira et al, 2007).

IMAGES-M is originally intended for integrated multimedia understanding as *knowledge processing* for intuitive human-robot interaction such that may happen between ordinary (or non-expert) people and home robots (Yokota et al, 2008). Here, 'integrated multimedia understanding' means especially 'multimedia understanding based on such a knowledge representation common to multiple information media (i.e., natural language, picture, music, gesture, etc.) that can facilitate cross-media operations (Yokota et al, 1984; Eakins & Graham, 1999; Kherfi et al, 2004; Yokota & Capi, 2005a)'. For ordinary people, however, natural language is the most important because it can convey the exact intention of the sender to the receiver due to its syntax and semantics common to its users, which is not necessarily the case for another information medium such as gesture or so. Therefore, natural language can as well play the most crucial role in intuitive human-robot interaction and actually IMAGES-M is a language-centered multimedia understanding system where its attention mechanism is to be controlled efficiently in a top-down way according to people's suggestions in natural language (Yokota, 2007).

For integrated multimedia understanding intended here, it is essential to develop a systematically computable knowledge representation language (KRL) as well as representation-free technologies (Brooks, 1986) such as neural networks for unstructured sensory/motory data processing (i.e., *stimulus/response processing*). This type of language is indispensable to *knowledge-based* processing such as *understanding* sensory events, *planning* appropriate actions and *knowledgeable* communication with ordinary people in natural language, and therefore it needs to have at least a good capability of representing spatiotemporal events that correspond to humans'/robots' sensations and actions in the real world (Yokota & Capi, 2005b).

Most of conventional methods have provided robotic systems with such quasi-natural language expressions as 'move(*Velocity, Distance, Direction*)', 'find(*Object, Shape, Color*)', etc. for human instruction or suggestion, uniquely related to computer programs for deploying sensors/motors as their semantics (e.g., Coradeschi & Saffiotti, 2003; Drumwright et al, 2006). These expression schemas, however, are too linguistic or coarse to represent and compute sensory/motory events in such an integrated way as intuitive human-robot interaction intended here. This is also the case for AI planning ("action planning") which deals with the development of representation languages for planning problems and with the development of algorithms for plan construction (Wilkins & Myers, 1995).

In order to solve this problem, the author has employed the formal language so called 'Language for Mental-image Description (L_{md})' proposed in his original semantic theory 'Mental Image Directed Semantic Theory (MIDST)' (e.g., Yokota, 2005, 2006), the main topic in this chapter.

MIDST is based on the hypothesis that natural language understanding in humans is basically mental image processing and concerns:

(i) Modeling of the human mind as a society of agents;
(ii) Modeling of omnisensory mental image in humans;
(iii) Modeling of conceptualization in humans;
(iv) Designing the formal language L_{md} for mental image description;
(v) Formulation of word concepts as mental images in L_{md};
(vi) Mutual translation between expressions in miscellaneous information media (e.g., natural language, picture, robotic action, etc.) and those in L_{md};
(vii) Computation on L_{md} expressions;
(viii) Formalization of human mental competence and performance as a deductive system in L_{md};

The final goal of MIDST is to realize artificial communication partners with a good capability of intuitive interaction with ordinary people and the items (i)-(viii) above are assumed to be its subgoals. The key idea of MIDST is the model of human attention-guided (i.e., active) perception yielding omnisensory images that inevitably reflect certain movements of the focus of attention of the observer (FAO) scanning certain matters in the world, either inside or outside of the mind. More analytically, these omnisensory images are associated with spatiotemporal changes (or constancies) in certain attributes of the matters scanned by FAO and modeled as temporally parameterized "loci in attribute spaces", so called, to be formulated in the formal language L_{md}. This language has already been implemented on several types of computerized intelligent systems including IMAGES-M (e.g., Yokota et al, 1984; Oda, et al, 2001; Amano, et al, 2005; Yokota & Capi, 2005a). The most remarkable feature of L_{md} is its capability of formalizing spatiotemporal matter concepts grounded in human/robotic sensation while the other similar KRLs are designed to describe the logical relations among conceptual primitives represented by lexical tokens (e.g., Dorr & Bonnie, 1997; Zarri, 1997; Sowa, 2000). Moreover, in L_{md} expression are hinted what and how should be attended to in the world as analogy of human FAO movement and thereby the robotic attention can be controlled in a top-down way (Yokota, 2007), which is the author's answer to the essential issue in robotic imitation, namely, how to control robotic attention mechanism efficiently (e.g., Demiris & Khadhouri, 2006).

The remainder of this chapter is organized as follows. Section 2 presents MIDST, focusing on the multiagent mind model, the omnisensory mental image model and the formal language L_{md} with liguistic or pictorial manifestations for its validation. Section 3 details about grounding natural language expressions in mental images in view of natural language processing by computers. In Section 4, the mental fucntion Kansei is modeled as collaboration of **Kn** and **Em** viewed from artificial or robotic individuality. Section 5 presents a discussion on applications and further developments for the language L_{md} presented in this chapter. Conclusions and planned future work are given in the final section.

2. Mental Image Directed Semantic Theory

2.1 Multiagent mind model

Figure 1 shows the multiagent mind model proposed here, consisting of Stimulus, Knowledge, Emotion and Response Processing Agents. This is a functional model of human central nervous system consisting of the brain and the spine. These agents are to communicate with one another by exchanging and computing mental images represented in the formal language L_{md}. Their basic performances are as follows.

1) **Stimulus Processing Agent (St)** receives stimuli from the world (**W**) and encodes them into mental images (i.e. encoded sensations) such as *"I sensed something oily."* (if verbalized in English.)
2) **Knowledge Processing Agent (Kn)** evaluates mental images received from the other agents based on its memory (e.g. knowledge), producing other mental images such as *"It is false that the earth is flat."*
3) **Emotion Processing Agent (Em)** evaluates mental images received from the other agents based on its memory (e.g. instincts), producing other mental images such as *"I like the food."*
4) **Response Processing Agent (Re)** converts mental images (i.e. encoded actions such as *"I'll walk slowly."*) received from the other agents into real actions against **W**.

A *performance* **P** against a *stimulus* X with a *result* Y at each agent can be formalized as a function by the expression (1).

$$Y=\mathbf{P}(X) \tag{1}$$

where

P: a combination of ***Atomic Performances*** defined later in association with ***Attribute Spaces***;
X: a spatiotemporal distribution of stimuli from **W** to **St** or a mental image for another agent.
Y: a series of signals to drive actuators for **Re** or a mental image for another agent.

For example, all the agents are to work during understanding information media such as natural language, picture, music, gesture, etc., sometimes performing Kansei by tight collaboration of **Kn** and **Em** as detailed later while **St** and **Re** are exclusively to work during *reflection* so called.

A performance **P** is assumed as a function formed either consciously or unconsciously, or in other words, either with or without reasoning. In a conscious case, a set of atomic performances are to be chosen and combined according to *X* by a meta-function, so called, '***Performance Selector***' assumed as '***Conscience***'. On the contrary, in an unconscious case, such a performance as is associated most strongly with *X* is to be applied automatically as is in the case of reflection.

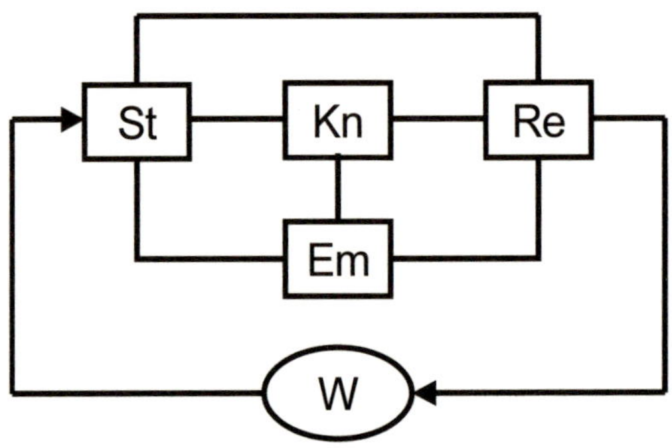

Multiagent model of human mind (St: Stimulus Processing Agent; Kn: Knowledge Processing Agent; Em: Emotion Processing Agent; Re: Response Processing Agent; W: World surrounding human mind, including his/her body).

2.2 Omnisensory mental image model and formal language L_{md}

Here are described the omnisensory mental image model and the syntax and semantics of L_{md} in association with the mental image model. In MIDST, word meanings are defined in association with mental images, not limited to visual but omnisensory, modeled as "Loci in Attribute Spaces" so called. See Fig.2-a and assume that the human is observing the phenomenon where the triangular gray object is moving in the sinusoidal trajectory and that its corresponding sensations (i.e., sensory images) are being caused in his/her mind. In this case, the moving triangular gray object is assumed to be perceived as the loci in the three attribute spaces, namely, those of 'Location', 'Color' and 'Shape' in the observer's mind. As easily imagined, attribute spaces correspond with human sensory systems and the loci represent certain sensations of the phenomena outside or inside human minds. From the viewpoint of artifact, an attribute space stands for a certain measuring instrument or sensor just like a chronograph, barometer, thermometer or so and the loci represent the movements of its indicator. The performance of an attribute space is the model of '***Atomic Performance***' introduced in Section 2.1.

These loci are to be articulated by "Atomic Locus" over a certain absolute time interval [t_i, t_f] as depicted in Fig.2-b and formulated as (2) in L_{md}, where the interval is suppressed because people are not aware of absolute time (nor always consult a chronograph).

$$L(x,y,p,q,a,g,k) \quad (2)$$

The expression (2) works as a formula in many-sorted predicate logic, where "L" is a predicate constant with five types of terms: "Matter" (at 'x' and 'y'), "Value" (at 'p' and 'q'), "Attribute" (at 'a'), "Event Type" (at 'g') and "Standard" (at 'k'). Conventionally, Matter variables are headed by 'x', 'y' and 'z'. This formula is called 'Atomic Locus Formula' whose first two arguments are sometimes referred to as 'Event Causer (EC)' and 'Attribute Carrier (AC)', respectively while ECs are often optional in natural concepts such as intransitive verbs. For simplicity, the syntax of L_{md} allows Matter terms (e.g., 'Tokyo' and 'Osaka' in (3) and (4)) to appear at Values or Standard in order to represent their values in each place at the time or over the time-interval. Moreover, when it is not so significant to discern ECs or Standards, anonymous variables, usually symbolized as '_', can be employed in their places (See (23) for example). A logical combination of atomic locus formulas defined as a well-formed formula (i.e., wff) in predicate logic is called simply 'Locus Formula'.

The intuitive interpretation of (2) is given as follows.

"Matter 'x' causes Attribute 'a' of Matter 'y' to keep* (p=q) *or change* (p ≠ q) *its values temporally* (g=Gt) *or spatially* (g=Gs) *over a certain absolute time-interval, where the values 'p' and 'q' are relative to the standard 'k'."

In (2), when $g=G_t$, the locus indicates monotonic change (or constancy) of the attribute in time domain, and when $g=G_s$, that in space domain. The former is called 'temporal event' and the latter, 'spatial event'. For example, the motion of the 'bus' represented by S1 is a temporal event and the ranging or extension of the 'road' by S2 is a spatial event whose meanings or concepts are formulated as (3) and (4), respectively, where 'A_{12}' denotes the attribute 'Physical Location'. These two formulas are different only at the term 'Event Type'.

(S1) The bus runs from Tokyo to Osaka.

(S2) The road runs from Tokyo to Osaka.

$$(\exists x,y,k)L(x,y,Tokyo,Osaka,A_{12},G_t,k)\wedge bus(y) \quad (3)$$

$$(\exists x,y,k)L(x,y,Tokyo,Osaka,A_{12},G_s,k)\wedge road(y) \quad (4)$$

It has been often argued that human active sensing processes may affect perception and in turn conceptualization and recognition of the physical world while such cognitive processes or products have seldom been formulated for computation (e.g., Leisi, 1961; Noton, 1970; Gardenfors, 2000; Langacker, 2005). The author has hypothesized that the difference between temporal and spatial event concepts can be attributed to the relationship between the Attribute Carrier (AC) and the Focus of the Attention of the Observer (FAO). To be brief, it is hypothesized that FAO is fixed on the whole AC in a temporal event but *runs* about on the AC in a spatial event. Consequently, as shown in Fig.3, the *bus* and FAO move together in the case of S1 while FAO solely moves along the *road* in the case of S2. That is, ***all loci in attribute spaces are assumed to correspond one to one with movements or, more generally, temporal events of FAO***.

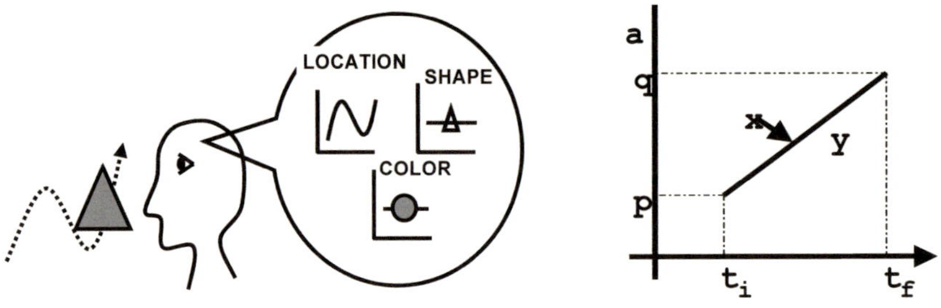

(a) (b)

Fig. 2. Mental image model (a) and Atomic Locus in Attribute Space (b)

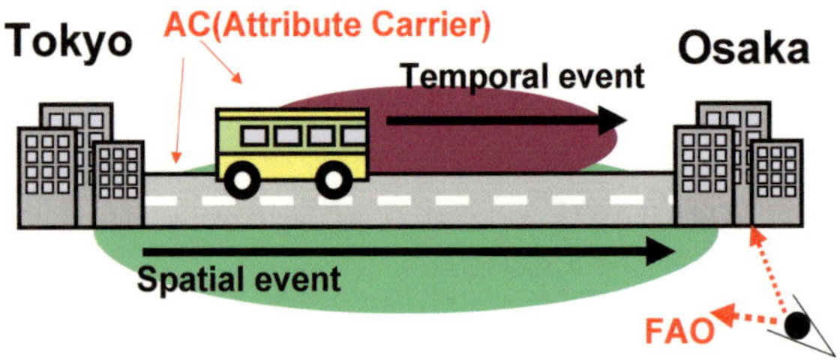

Fig. 3. FAO movements and Event types

2.3 Tempological connectives

The duration of a locus corresponds to an absolute time-interval over which FAO is put on the corresponding phenomenon outside or inside the mind. Such an absolute time-interval is suppressed in an atomic locus formula because it is assumed that people cannot measure the absolute time by any chronograph but a certain relative time (Actually, people do not always consult a chronograph even if they can). MIDST has employed 'tempo-logical connectives (TLCs)' denoting both logical and temporal relations between loci by themselves because these must be considered simultaneously in locus articulation.

A tempo-logical connective K_i is defined by (5), where τ_i, χ and K refer to one of the temporal relations indexed by an integer 'i', a locus, and an ordinary binary logical connective such as the conjunction '$\wedge$', respectively. The definition of τ_i is given in Table 1 from which the theorem (6) can be deduced. This table shows the complete list of topological relations between two intervals, where 13 types of relations are discriminated by τ_i ($-6 \leq i \leq 6$). This is in accordance with Allen's notation (Allen, 1984), which, to be strict, is exclusively for 'temporal conjunctions ($=\wedge_i$)' as introduced below.

$$\chi_1 \, K_i \, \chi_2 \Leftrightarrow (\chi_1 \, K \, \chi_2) \wedge \tau_i(\chi_1, \chi_2) \tag{5}$$

$$\tau_{-i}(\chi_2, \chi_1) \equiv \tau_i(\chi_1, \chi_2) \quad (\forall i \in \{0,\pm1,\pm2,\pm3,\pm4,\pm5, \pm6\}) \tag{6}$$

The TLCs used most frequently are 'SAND ($\wedge_0$)' and 'CAND ($\wedge_1$)', standing for 'Simultaneous AND' and 'Consecutive AND' and conventionally symbolized as 'Π' and '•', respectively. For example, the concepts of the English verbs 'carry' and 'return' are to be defined as (7) and (8), respectively. These can be depicted as Fig.4-a and b, respectively. The expression (9) is the definition of the English verb concept 'fetch' depicted as Fig.4-c. This implies such a temporal event that 'x' goes for 'y' and then comes back with it. In the same way, the English verb concept 'hand' or 'receive' depicted as Fig.4-d is defined equivalently as (10) or its abbreviation (10′) where ECs are merged into a set.

$$(\lambda x,y)\mathrm{carry}(x,y)\Leftrightarrow(\lambda x,y)(\exists p,q,k)L(x,x,p,q,A_{12},G_t,k)\Pi L(x,y,p,q,A_{12},G_t,k)\wedge x\neq y\wedge p\neq q \quad (7)$$

$$(\lambda x)\mathrm{return}(x)\Leftrightarrow(\lambda x)(\exists p,q,k)L(x,x,p,q,A_{12},G_t,k)\bullet L(x,x,p,q,A_{12},G_t,k)\wedge x\neq y\wedge p\neq q \quad (8)$$

$$(\lambda x,y)\mathrm{fetch}(x,y)\Leftrightarrow(\lambda x,y)\ (\exists p_1,p_2,k)\ L(x,x,p_1,p_2,A_{12},G_t,k)\bullet$$

$$((\underline{L(x,x,p_2,p_1,A_{12},G_t,k)\Pi L(x,y,p_2,p_1,A_{12},G_t,k)})\wedge x\neq y\wedge p_1\neq p_2 \quad (9)$$

$$(\lambda x,y,z)\mathrm{hand}(x,y,z).\equiv.(\lambda x,y,z)\mathrm{receive}(z,,y,x)$$

$$\Leftrightarrow(\lambda x,y,z)(\exists k)L(x,y,x,z,A_{12},G_t,k)\Pi L(z,y,x,z,A_{12},G_t,k)\wedge x\neq y\wedge y\neq z\wedge z\neq x \quad (10)$$

$$(.\equiv.\ (\lambda x,y,z)(\exists k)L(\{x,z\},y,x,z,A_{12},G_t,k)\wedge x\neq y\wedge y\neq z\wedge z\neq x) \quad (10')$$

Such locus formulas as correspond with natural event concepts are called 'Event Patterns' and about 40 kinds of event patterns have been found concerning the attribute 'Physical Location (A_{12})', for example, *start, stop, meet, separate, carry, return,* etc.

Temporal relations and definition of τ_i†			Allen's notation
χ_1 +·····················+ χ_2 +·····················+	$t_{11}=t_{21}$ $\wedge t_{12}=t_{22}$	$\tau_0(\chi_1,\chi_2)$	equals(χ_1,χ_2)
		$\tau_0(\chi_2,\chi_1)$	equals(χ_2,χ_1)
χ_1 +··········+ χ_2 +··········+	$t_{12}=t_{21}$	$\tau_1(\chi_1,\chi_2)$	meets(χ_1,χ_2)
		$\tau_{-1}(\chi_2,\chi_1)$	met-by(χ_2,χ_1)
χ_1 +··········+ χ_2 +·····················+	$t_{11}=t_{21}$ $\wedge t_{12}<t_{22}$	$\tau_2(\chi_1,\chi_2)$	starts(χ_1,χ_2)
		$\tau_{-2}(\chi_2,\chi_1)$	started-by(χ_2,χ_1)
χ_1 +··········+ χ_2 +·····················+	$t_{11}>t_{21}$ $\wedge t_{12}<t_{22}$	$\tau_3(\chi_1,\chi_2)$	during(χ_1,χ_2)
		$\tau_{-3}(\chi_2,\chi_1)$	contains(χ_2,χ_1)
χ_1 +··········+ χ_2 +·····················+	$t_{11}>t_{21}$ $\wedge t_{12}=t_{22}$	$\tau_4(\chi_1,\chi_2)$	finishes(χ_1,χ_2)
		$\tau_{-4}(\chi_2,\chi_1)$	finished-by(χ_2,χ_1)
χ_1 +······+ χ_2 +·········+	$t_{12}<t_{21}$	$\tau_5(\chi_1,\chi_2)$	before(χ_1,χ_2)
		$\tau_{-5}(\chi_2,\chi_1)$	after(χ_2,χ_1)
χ_1 +···············+ χ_2 +···············+	$t_{11}<t_{21}\wedge t_{21}<t_{12}$ $\wedge t_{12}<t_{22}$	$\tau_6(\chi_1,\chi_2)$	overlaps(χ_1,χ_2)
		$\tau_{-6}(\chi_2,\chi_1)$	overlapped-by(χ_2,χ_1)

Table 1. List of temporal relations (†χ_1 and χ_2 exist during $[t_{11}, t_{12}]$ and $[t_{21}, t_{22}]$, respectively)

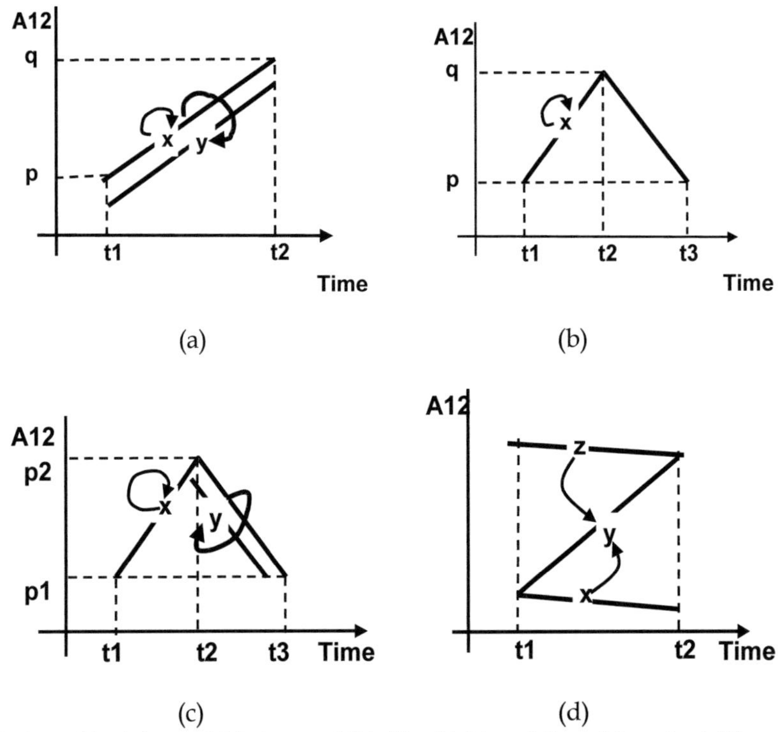

Fig. 4. Depiction of loci: 'carry' (a), 'return' (b), 'fetch' (c) and 'hand/receive' (d)

In order for explicit indication of time points, a very important concept called 'Empty Event (EE)' denoted by 'ε' is introduced. An EE stands only for absolute time elapsing and is explicitly defined as (11) with the attribute 'Time Point (A_{34})' and the standard 'K_{Ta}' denoting *absolute time*, where t_i and t_j are conventionally given as real numbers with the condition $t_i<t_j$. According to this scheme, the duration $[t_a, t_b]$ of an arbitrary locus χ can be expressed as (12).

$$\varepsilon([t_i,t_j]) \Leftrightarrow (\exists x,y,g)\, L(x,y,t_i,t_j,A_{34},g,K_{Ta}) \tag{11}$$

$$\chi \Pi \varepsilon([t_a,t_b]) \tag{12}$$

Any pair of loci temporally related in certain attribute spaces can be formulated as (13)-(17) in exclusive use of SANDs, CANDs and EEs. For example, the loci shown in Fig.5-a and b correspond to the formulas (14) and (17), respectively.

$$\chi_1 \wedge_2 \chi_2 .\equiv. (\chi_1 \bullet \varepsilon)\Pi\chi_2 \tag{13}$$

$$\chi_1 \wedge_3 \chi_2 .\equiv. (\varepsilon_1 \bullet \chi_1 \bullet \varepsilon_2)\Pi\chi_2 \tag{14}$$

$$\chi_1 \wedge_4 \chi_2 .\equiv. (\varepsilon \bullet \chi_1)\Pi\chi_2 \tag{15}$$

$$\chi_1 \wedge_5 \chi_2 .\equiv. \chi_1 \bullet \varepsilon \bullet \chi_2 \quad (16)$$

$$\chi_1 \wedge_6 \chi_2 .\equiv. (\chi_1 \bullet \varepsilon_3)\Pi(\varepsilon_1 \bullet \chi_2)\Pi(\varepsilon_1 \bullet \varepsilon_2 \bullet \varepsilon_3) \quad (17)$$

Employing TLCs, tempo-logical relationships between miscellaneous event concepts can be formulated without explicit indication of time intervals. For example, an event 'fetch(x,y)' is necessarily *finished by* an event 'carry(x,y)' as indicated by the underline at (9). This fact can be formulated as (18), where '$\supset_{-4}$' is the '*implication* ($\supset$)' furnished with the temporal relation '*finished-by* (τ_{-4})'. This kind of formula is not an axiom but a theorem deducible from the definitions of event concepts in the deductive system intended here.

$$(\forall x,y)(\text{fetch}(x,y) \supset_{-4} \text{carry}(x,y)) \quad (18)$$

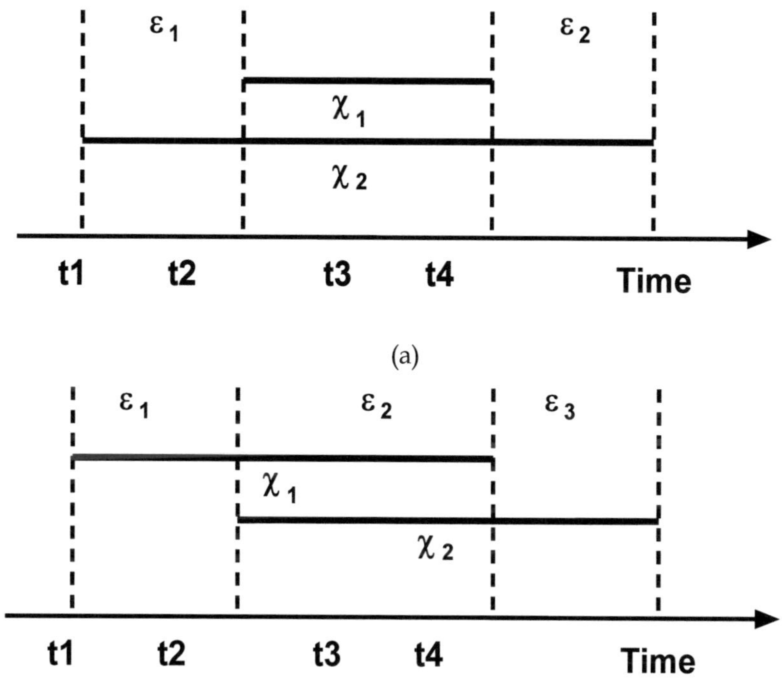

(a)

(b)

Fig. 5. Tempological relations: (a) during(χ_1, χ_2) and (b) overlaps(χ_1, χ_2)

2.4 Attributes and standards

The attribute spaces for humans correspond to the sensory receptive fields in their brains. At present, about 50 attributes concerning the physical world have been extracted as shown in Table 2 exclusively from Japanese and English words (e.g., Roget, 1975). They are associated with all of the 5 senses (i.e. sight, hearing, smell, taste and feeling) in our everyday life while those for information media other than languages correspond to limited senses. For example, those for pictorial media, marked with '*' in Table 2, associate limitedly with the sense 'sight' as a matter of course. The attributes of this sense occupy the greater part of all, which implies that the sight is essential for humans to conceptualize the external world by. And this kind of classification of attributes plays a very important role in our cross-media operating system (Yokota & Capi, 2005a).

Correspondingly, six categories of standards shown in Table 3 have been extracted after the conventional categorization (Leisi, 1961) that are assumed necessary for representing values of each attribute in Table 2. In general, the attribute values represented by words are relative to certain standards as explained briefly in Table 3. For example, (19) and (20) are different formulations of a locus due to the different standards 'k1' and 'k2' for scaling as shown in Fig.6-a and b, respectively. That is, whether the point (t_2, q) is significant or not, more generally, how to articulate a locus depends on the precisions or the granularities of these standards, which can be formulated as (21) and (22), so called, ***'Postulate of Arbitrariness in Locus Articulation'***. This postulate affects the process of conceptualization on a word based on its referents in the world.

Code	Attribute [Property]†	Linguistic expressions for attribute values
*A_{01}	PLACE OF EXISTE NCE [N]	The accident happened in Osaka.
*A_{02}	LENGTH [S]	The stick is 2 meters long.
	....	
*A_{11}	SHAPE [N]	The cake is round.
*A_{12}	PHYSICAL LOCATION [N]	Tom moved to Tokyo.
*A_{13}	DIRECTION [N]	The box is to the left of the chair.
*A_{14}	ORIENTATION [N]	The door faces to south.
*A_{15}	TRAJECTORY [N]	The plane circled in the sky.
*A_{16}	VELOCITY [S]	The boy runs very fast.
*A_{17}	MILEAGE [S]	The car ran ten miles.
A_{18}	STRENGTH OF EFFECT [S]	He is very strong.
A_{19}	DIRECTION OF EFFECT [N]	He pulled the door.
	....	
A_{28}	TEMPERATURE [S]	It is hot today.
A_{29}	TASTE [N]	The grapes here are very sour.
A_{30}	ODOUR [N]	The gas is pungent.
A_{31}	SOUND [N]	His voice is very loud.
*A_{32}	COLOR [N]	Tom painted the desk white.
A_{33}	INTERNAL SENSATION [N]	I am very tired.
A_{34}	TIME POINT [S]	It is ten o'clock.
A_{35}	DURATION [S]	He studies for two hours every day.
A_{36}	NUMBER [S]	Here are many people.
A_{37}	ORDER [S]	Tom sat next to Mary.
A_{38}	FREQUENCY [S]	He did it twice.
A_{39}	VITALITY [S]	The old man still alive.
*A_{44}	TOPOLOGY [N]	He is in the room.
*A_{45}	ANGULARITY [S]	The knife is dull.

Table 2. Examples of attributes (†S: 'scalar value', N : 'non-scalar value')

Categories of standards	Remarks
Rigid Standard	Objective standards such as denoted by measuring *units* (meter, gram, etc.).
Species Standard	The *attribute value ordinary* for a species. A *short train* is ordinarily longer than a *long pencil*.
Proportional Standard	'*Oblong*' means that the width is greater than the height at a physical object.
Individual Standard	*Much* money for one person can be too *little* for another.
Purposive Standard	One room large enough for a person's *sleeping* must be too small for his *jogging*.
Declarative Standard	Tom is taller *than Jim*. The origin of an order such as 'next' must be declared explicitly just as 'next *to him*'.

Table 3. List of standards

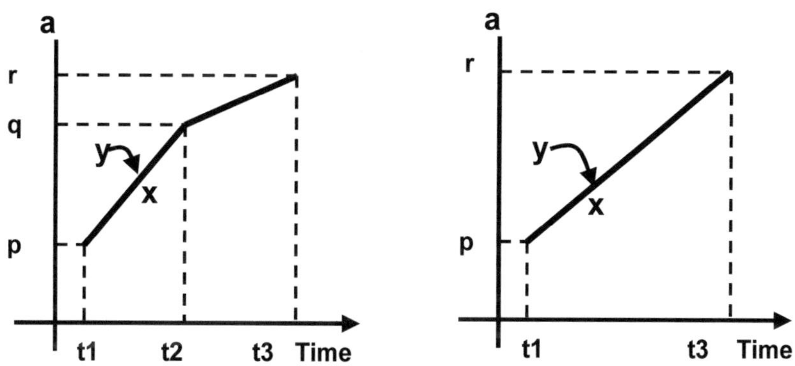

(a) (b)

Fig. 6. Arbitrariness in locus articulation due to standards: Standard k_1 (a) is finer than k_2 (b)

$$(L(y,x,p,q,a,g,k_1)\Pi\ \varepsilon([t_1,t_2]))\bullet(L(y,x,q,r,a,g,k_1)\Pi\ \varepsilon([t_2,t_3])) \tag{19}$$

$$L(y,x,p,r,a,g,k_2)\Pi\ \varepsilon([t_1,t_3]) \tag{20}$$

$$(\forall p,q,r,k)(L(y,x,p,q,a,g,k)\bullet L(y,x,q,r,a,g,k).\supset.(\exists k')L(y,x,p,r,a,g,k')\wedge k'\neq k) \tag{21}$$

$$(\forall p,r,k)(L(y,x,p,r,a,g,k).\supset.(\exists q,k')\ L(y,x,p,q,a,g,k')\bullet L(y,x,q,r,a,g,k')\wedge\ k'\neq k) \tag{22}$$

3. Mind model and natural language

3.1 Spatiotemporal expressions and perceptual processes

As already mentioned in Section 2, all loci in attribute spaces are assumed to correspond one to one with movements or, more generally, temporal events of FAO. Therefore, an event expressed in L_{md} is compared to a movie film recorded through a floating camera because it is necessarily grounded in FAO's movement over the event. And this is why S3 and S4 can refer to the same scene in spite of their appearances, where what 'sinks' or 'rises' is FAO as illustrated in Fig.7-a and whose conceptual descriptions are given as (23) and (24), respectively, where 'A_{13}', '↑' and '↓' refer to the attribute 'Direction' and its values 'upward' and 'downward', respectively.

(S3) The path *sinks to* the brook.

(S4) The path *rises from* the brook.

Such a fact is generalized as '***Postulate of Reversibility of a Spatial Event (PRS)***' that can be one of the principal inference rules belonging to people's common-sense knowledge about geography. This postulation is also valid for such a pair of S5 and S6 as interpreted approximately into (25) and (26), respectively. These pairs of conceptual descriptions are called equivalent in the PRS, and the paired sentences are treated as paraphrases each other.

(a) (b)

Fig. 7. FAO movements: 'slope' (a) and 'row' (b) as spatial events

$$(\exists y,p,z)L(\_,y,p,z,A_{12},G_{s},\_)\Pi L(\_,y,\downarrow,\downarrow,A_{13},G_{s},\_)\wedge path(y)\wedge brook(z)\wedge p\neq z \quad (23)$$

$$(\exists y,p,z)L(\_,y,z,p,A12,G_{s},\_)\Pi L(\_,y,\uparrow,\uparrow,A_{13},G_{s},\_)\wedge path(y)\wedge brook(z)\wedge p\neq z \quad (24)$$

(S5) Route A and Route B meet at the city.

$$(\exists p,y,q)L(\_,Route\_A,p,y,A_{12},G_{s},\_)\Pi L(\_,Route\_B,q,y,A_{12},G_{s},\_)\wedge city(y)\wedge p\neq q \quad (25)$$

(S6) Route A and Route B separate at the city.

$$(\exists p,y,q)L(\_,Route\_A,y,p,A_{12},G_{s},\_)\Pi L(\_,Route\_B,y,q,A_{12},G_{s},\_)\wedge city(y)\wedge p\neq q \quad (26)$$

For another example of spatial event, Fig.7-b concerns the perception of the formation of multiple objects, where FAO runs along an imaginary object so called '***Imaginary Space Region (ISR)***'. This spatial event can be verbalized as S7 using the preposition 'between' and formulated as (27) or (27'), corresponding also to such concepts as 'row', 'line-up', etc. Employing ISRs and the 9-intersection model (Egenhofer, 1991), all the topological relations between two objects can be formulated in such expressions as (28) or (28') for S8, and (29) for S9, where '*In*', '*Cont*' and '*Dis*' are the values 'inside', 'contains' and 'disjoint' of the attribute 'Topology (A_{44})' with the standard '9-intersection model (K_{9IM})', respectively. Practically, these topological values are given as 3×3 matrices with each element equal to 0 or 1 and therefore, for example, '*In*' and '*Cont*' are transpositional matrices each other.

(S7) The square is between the triangle and the circle.
(S8) Tom is in the room.
(S9) Tom exits the room.

$$(\exists y,x_1,x_2, x_3,p,q)\ (L(\_,y,x_1,x_2,A_{12},G_s,\_)\Pi L(\_,y,p,p,A_{13},G_s,\_))\bullet(L(\_,y,x_2,x_3,A_{12},G_s,\_)$$

$$\Pi L(\_,y,q,q,A_{13},G_s,\_))\wedge ISR(y)\wedge p=q\ \wedge triangle(x_1)\wedge square(x_2)\ \wedge circle(x_3) \qquad (27)$$

$$(\exists y,x_1,x_2, x_3,p)(L(\_,y,x_1,x_2,A_{12},G_s,\_)\bullet L(\_,y,x_2,x_3,A_{12},G_s,\_))\Pi L(\_,y,p,p,A_{13},G_s,\_)$$

$$\wedge ISR(y)\ \wedge triangle(x_1)\wedge square(x_2)\ \wedge circle(x_3) \qquad (27')$$

$$(\exists x,y)L(Tom,x,y,Tom,A_{12},G_s,\_)\Pi L(Tom,x,In,In,A_{44},G_t,K_{9IM})\ \wedge ISR(x)\ \wedge room(y) \qquad (28)$$

$$(\exists x,y)L(Tom,x,Tom,y,A_{12},G_s,\_)\Pi L(Tom,x,Cont,Cont,A_{44},G_t,K_{9IM})\wedge ISR(x)\ \wedge room(y) \qquad (28')$$

$$(\exists x,y,p,q)L(Tom,Tom,p,q,A_{12},G_t,\_)\Pi L(Tom,x,y,Tom,A_{12},G_s,\_)$$

$$\Pi L(Tom,x,In,Dis,A_{44},Gt,K_{9IM})\ \wedge ISR(x)\ \wedge room(y)\ \wedge p\neq q \qquad (29)$$

The rigid topology between two objects as in the 9-intersection model must be determined with the perfect knowledge of their insides, outsides and boundaries. Ordinary people, however, can comment on matters without knowing all about them. This is the very case when they encounter an unknown object too large to observe at a glance just like a road in a strange country. For example, Fig.8-a shows such a path viewed from the sky that is partly hidden by the woods. In this case, the topological relation between the path as a whole and the swamp/woods depends on how the path starts and ends in the woods, but people could utter such sentences as S10 and S11 about this scene. Actually, these sentences refer to such spatial events that reflect certain temporal changes in the topological relation between the swamp/woods and FAO running along the path. Therefore, their conceptual descriptions are to be given as (30) and (31), respectively. For another example, Fig.8-b shows a more complicated spatial event in topology that can be formulated as (32) and could be verbalized as S12.
(S10) The path enters the swamp/woods.
(S11) The path exits the swamp/woods.

$$(\exists x,y,z)L(\_,z,p,q,A_{12},G_s,\_)\Pi L(\_,x,y,z,A_{12},G_s,\_)\Pi$$

$$L(\_,x,Dis,In,A_{44},G_s,K_{9IM})\ \wedge ISR(x)\ \wedge\{swamp(y)/woods(y)\}\ \wedge path(z)\ \wedge p\neq q \qquad (30)$$

$$(\exists x,y,z)L(\_,z,p,q,A_{12},G_s,\_)\Pi L(\_,x,y,z,A_{12},G_s,\_)\Pi$$

$$L(\_,x,In,Dis,A_{44},G_s,K_{9IM})\ \wedge ISR(x)\ \wedge\{swamp(y)/woods(y)\}\ \wedge path(z)\ \wedge p\neq q \qquad (31)$$

(S12) The path cuts the swamp twice (as shown in Fig.8-b), passing p_1 outside, p_2 inside, p_3 outside, p_4 inside and p_5 outside the swamp on the way.

$$(\exists x,y,z,p_1,\ldots,p_5)L(\_,z,y,x,A_{12},G_s,\_)\Pi((L(\_,x,p_1,p_2,A_{12},G_s,\_)$$

$$\Pi(L(\_,z,Dis,In,A_{44},G_s,K_{9IM}))\bullet(L(\_,x,p_2,p_3,A_{12},G_s,\_)\Pi L(\_,z,In,Dis,A_{44},G_s,K_{9IM}))$$
$$\bullet(L(\_,x,p_3,p_4,A_{12},G_s,\_)\Pi L(\_,z,Dis,In,A_{44},G_s,K_{9IM}))\bullet(L(\_,x,p_4,p_5,A_{12},G_s,\_)\Pi$$

$$L(\_,z,In,Dis,A_{44},G_s,K_{9IM})))\wedge path(x)\wedge swamp(y)\wedge ISR(z) \quad (32)$$

Lastly, consider such somewhat complicated sentences as S13 and S14. The underlined parts are deemed to refer to some events neglected in time and in space, respectively. These events correspond with skipping of FAOs and are called 'Temporal Empty Event' and 'Spatial Empty Event', denoted by 'ε_t' and 'ε_s' as EEs with g=G_t and g=G_s at (11), respectively. The concepts of S13 and S14 are given by (33) and (34), where 'A_{15}' and 'A_{17}' represent the attribute 'Trajectory' and 'Mileage', respectively.

(S13) The *bus* runs 10km straight east from A to B, and <u>after a while</u>, at C it meets the street with the sidewalk.

(S14) The *road* runs 10km straight east from A to B, and <u>after a while</u>, at C it meets the street with the sidewalk.

$$(\exists x,y,z,p,q)(L(\_,x,A,B,A_{12},G_t,\_)\Pi L(\_,x,0,10km,A_{17},G_t,\_)\Pi$$

$$L(\_,x,Point,Line,A_{15},G_t,\_)\Pi L(\_,x,East,East,A_{13},G_t,\_))\bullet\varepsilon_t\bullet \quad (33)$$

$$(L(\_,x,p,C,A_{12},G_t,\_)\Pi L(\_,y,q,C,A_{12},G_s,\_)\Pi L(\_,z,y,y,A_{12},G_s,\_))$$

$$\wedge bus(x)\wedge street(y)\wedge sidewalk(z)\wedge p\neq q$$

$$(\exists x,y,z,p,q)(L(\_,x,A,B,A_{12},G_s,\_)\Pi L(\_,x,0,10km,A_{17},G_s,\_)\Pi$$

$$L(\_,x,Point,Line,A_{15},G_s,\_)\Pi L(\_,x,East,East,A_{13},G_s,\_))\bullet\varepsilon_s\bullet \quad (34)$$

$$(L(\_,x,p,C,A_{12},G_s,\_)\Pi L(\_,y,q,C,A_{12},G_s,\_)\Pi L(\_,z,y,y,A_{12},G_s,\_))$$
$$\wedge road(x)\wedge street(y)\wedge sidewalk(z)\wedge p\neq q$$

From the viewpoint of cross-media reference as integrated multimedia understanding, the formula (34) can refer to such a spatial event depicted as the still picture in Fig.9 while (33) can be interpreted into a motion picture.

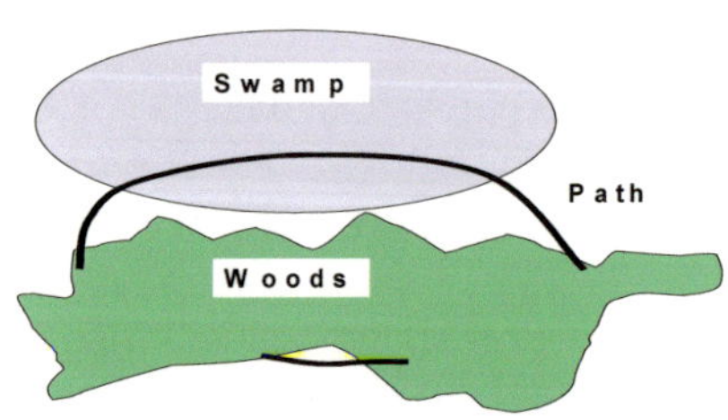

(a)

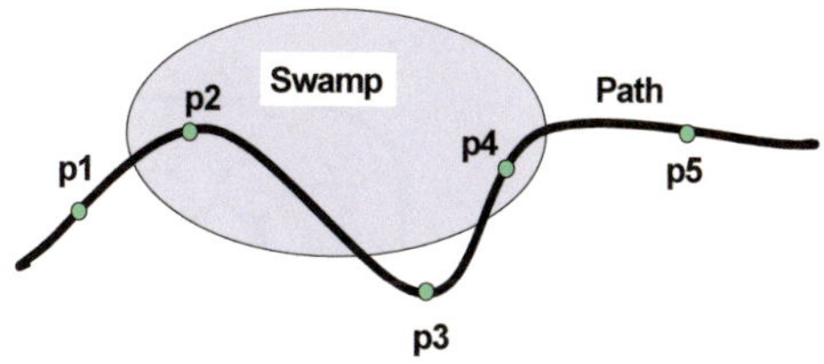

(b)

Fig. 8. Delicate topological relations: (a) the path partially hidden by the woods and (b) the path winding inside-outside-inside-outside of the swamp

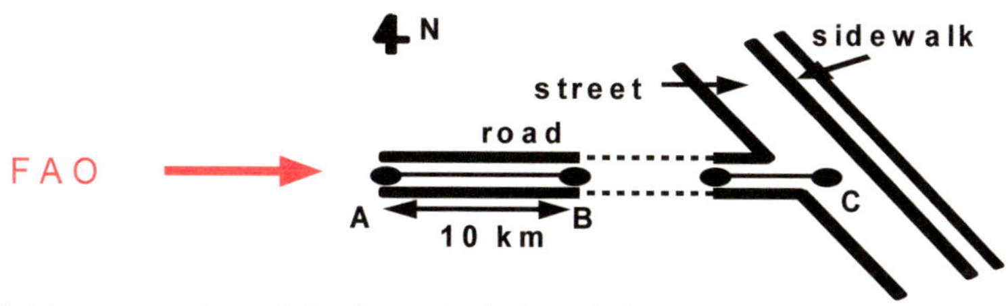

Fig. 9. Pictorial interpretation of the formula (34) and the movement of FAO involved

3.2 Conceptualization

It is well known that, from the cognitive viewpoint, there are two types of mental images, namely, (a) perceptual images and (b) conceptual images. The former are live images of the current world and the latter are recalled ones often in association with tokens such as words.

Ideally, a word concept should be associated with such a conceptual image that is abstract enough to represent the perceptual image of every matter referred to by the word. It is, however, practically impossible for an individual to obtain such a conceptual image because such instances or referents are usually too numerous for him/her to encounter and observe. In this sense, our conceptual image for a word is always imperfect or tentative to be sometimes updated by an exceptional instance just like a 'black swan'.

It is generally assumed that a word concept is an abstraction on properties and relations of the matters involved such as locations, shapes, colors, functions, potentialities, etc. In MIDST, a word concept is to be represented as an abstract locus formula resulted from generalization on the locus formulas of a number of matters referred to by the word.

Figure 10 illustrates the mental process of conceptualization on the word 'conveyance', where a set of its referents ***Sr*** (={*Matter_1*,..., *Matter_n*}) are generalized by abstraction and formulated as (35). The underlined part of this formula implies that matter 'z' includes two matters 'x' and 'y' in its 'Place of existence (A_{01})'. As easily imagined, the variable 'z' denotes a certain referent generalized so as to represent any member of ***Sr***.

This process consists of three stages as follows. Firstly, the attributes other than 'Place of existence (A_{01})' and 'Physical location (A_{12})' are discarded. Secondly, the concrete objects 'human', 'book', etc. and their concrete attribute values are replaced by the variables 'x', 'y', 'z', 'p', 'q', etc. And finally, the relationships '≠' and '=', the most essential for this concept, are placed among these variables. The equalities (or inequalities) in 'Physical location' are determined at the precision of the standard represented by the variable 'k'.

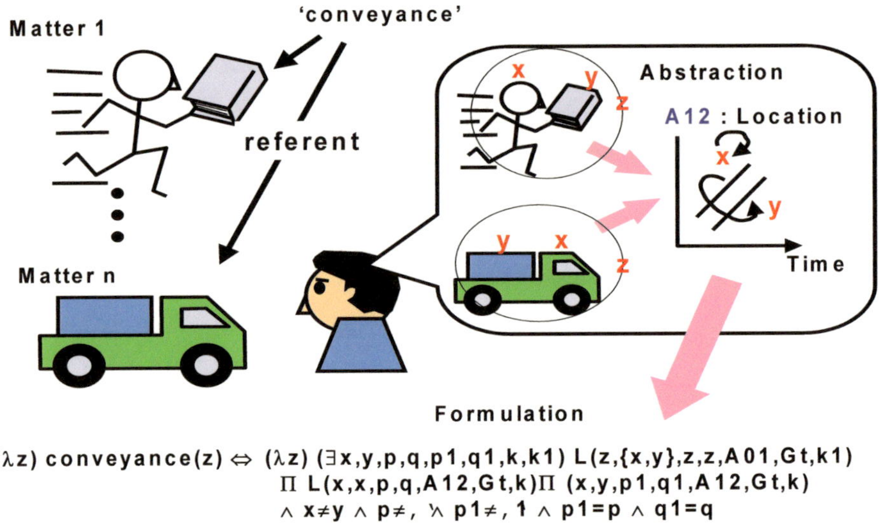

Fig.10. Conceptualization: the process of abstraction on referents of the word 'conveyance'.

$$(\lambda x)\mathrm{conveyance}(z) \Leftrightarrow (\lambda z)(\exists x,y,p,q,p_1,q_1,k,k_1)\underline{L(z,\{x,y\},z,z,A_{01},G_t,k_1)}$$

$$\Pi L(x,x,p,q,A_{12},G_t,k)\Pi L(x,y,p_1,q_1,A_{12},G_t,k)\wedge x\neq y\wedge p\neq q\wedge p_1\neq q_1\wedge p_1=p\wedge q_1=q$$

$$\Leftrightarrow (\lambda z)(\exists x,y,p,q,k,k_1)\underline{L(z,\{x,y\},z,z,A_{01},G_t,k_1)}$$

$$\Pi L(x,x,p,q,A_{12},G_t,k)\Pi L(x,y,p,q,A_{12},G_t,k)\wedge x\neq y\wedge p\neq q$$

$$\Leftrightarrow (\lambda z)(\exists x,y,p,q,k,k_1)\underline{L(z,\{x,y\},z,z,A_{01},G_t,k_1)}\Pi L(x,\{x,y\},p,q,A_{12},G_t,k) \wedge x\neq y \wedge p\neq q \quad (35)$$

For another example, the matter called 'snow' can be conceptualized as (36), where '_', as defined by (37), stands for the variable bound by the existential quantifier, reading 'Snow is powdered ice attracted from the sky by the earth, melts into water,…' (Hence forth, refer to Table 2 for undefined attributes.)

$$(\lambda x)\mathrm{snow}(x)\Leftrightarrow(\lambda x)(\exists x_1,x_2,\ldots)((L(\_,x,x_1,x_1,A_{41},G_t,\_)\Pi L(\mathrm{Earth},x,\mathrm{Sky},\mathrm{Earth},A_{12},G_t,\_))$$

$$\wedge L(\_,x,x_1,x_2,A_{41},G_t,\_) \wedge\mathrm{powder}(x_1)\wedge \mathrm{ice}(x_1)\wedge\mathrm{water}(x_2) \wedge\ldots) \quad (36)$$

$$L(\ldots,\_,\ldots) \leftrightarrow (\exists x)L(\ldots,x,\ldots) \quad (37)$$

For a more complicated example, the concept of 'umbrella' can be represented as (38), reading 'At raining, a human puts an umbrella in line between rain and himself/herself in order not to get wetter, …'

By the way, the concepts of 'rain' and 'wind' can be given as (39) and (40), reading 'Rain is water attracted from the sky by the earth, makes an object wetter, is pushed an umbrella to by a human,…,' and 'Wind is air, affects the direction of rain,… ,' respectively.

$$(\lambda x)\text{umbrella}(x) \Leftrightarrow (\lambda x)(\exists x,x_1,x_2,y,p,q_1,q_2,\ldots)(L(\_,x_1,p,p,A_{13},G_t,\_)\Pi$$
$$((L(\_,y,x_1,x,A_{12},G_s,\_)\bullet L(\_,y,x,x_2,A_{12},G_s,\_))\Pi\ L(\_,y,p,p,A_{13},G_s,\_)\Pi$$
$$L(x_1,x_2,q_1,q_2,A_{25},G_t,\_)\wedge ISR(y)\wedge \text{rain}(x_1)\wedge \text{human}(x_2)\wedge\sim(q_1<q_2)\ldots) \quad (38)$$

$$(\lambda x)\text{rain}(x) \Leftrightarrow (\lambda x)(\exists x_1,x_2,x_3,x_4,p,q,\ldots)L(\_,x,x_1,x_1,A_{41},G_t,\_)\Pi$$
$$L(\text{Earth},x,\text{Sky},\text{Earth},A_{12},G_t,\_)\Pi L(x,x_2,p,q,A_{25},G_t,\_)\Pi L(x_3,x_4,x,x,A_{19},G_t,x_3)$$
$$\wedge\text{water}(x_1)\wedge\text{object}(x_2)\wedge\text{human}(x_3)\wedge\text{umbrella}(x_4)\wedge(p<q)\ldots \quad (39)$$

$$(\lambda x)\text{wind}(x) \Leftrightarrow (\lambda x)(\exists x_1,x_2,p,q,\ldots)L(\_,x,x_1,x_1,A_{41},G_t,\_)\wedge\text{air}(x_1)\wedge$$
$$(L(x,x_2,p,q,A_{13},G_t,\_)\wedge\text{rain}(x_2)\ldots \quad (40)$$

3.3 Knowledge of word meanings

A word meaning M_W is defined as a pair of 'Concept Part (C_p)' and 'Unification Part (U_p)' and is formulated as (41).

$$M_W = [C_p{:}U_p] \quad (41)$$

The C_p of a word W is an $\boldsymbol{L_{md}}$ expression as its concept while its U_p is a set of operations for unifying the C_ps of W's syntactic governors or dependents. For example, the meaning of the English verb 'carry' can be given by (42).

$$[(\lambda x,y)(\exists p,q,k)L(x,\{x,y\},p,q,A_{12},G_t,k)\wedge x\neq y\wedge p\neq q{:}\ ARG(Dep.1,x);\ ARG(Dep.2,y);] \quad (42)$$

The U_p above consists of two operations to unify the first dependent (Dep.1) and the second dependent (Dep.2) of the current word with the variables x and y, respectively. Here, Dep.1 and Dep.2 are the 'subject' and the 'object' of 'carry', respectively. Therefore, the sentence '*Mary carries a book*' is translated into (43).

$$(\exists y,p,q,k)\ L(\text{Mary},\{\text{Mary},y\},p,q,A_{12},G_t,k)\wedge \text{Mary}\neq y\wedge p\neq q\wedge\text{book}(y) \quad (43)$$

Figure 11 shows the details of the conversion process of a surface structure (text) into a conceptual structure (text meaning) through a surface dependency structure.

For another example, the meaning description of the English preposition 'through' is also given by (44).

$$[(\lambda x,y)(\exists p_1,z,p_3,g,k,p_4,k_0)\ (\underline{L(x,y,p_1,z,A_{12},g,k)}\bullet L(x,y,z,p_3,A_{12},g,k))\Pi$$
$$L(x,y,p_4,p_4,A_{13},g,k_0)\wedge p_1\neq z\wedge z\neq p_3{:}\ ARG(Dep.1,z);$$

$$\text{IF(Gov=Verb)}\rightarrow\text{PAT(Gov,(1,1)); IF(Gov=Noun)}\rightarrow\text{ARG(Gov,y);]} \tag{44}$$

The U_p above is for unifying the C_ps of the very word, its governor (Gov, a verb or a noun) and its dependent (Dep.1, a noun). The second argument (1,1) of the command PAT indicates the underlined part of (44) and in general (i,j) refers to the partial formula covering from the ith to the jth atomic formula of the current C_p. This part is the pattern common to both the C_ps to be unified. This is called 'Unification Handle (U_h)' and when missing, the C_ps are to be combined simply with '$\wedge$'.

Therefore the sentences S15, S16 and S17 are interpreted as (45), (46) and (47), respectively. The underlined parts of these formulas are the results of PAT operations. The expression (48) is the C_p of the adjective 'long' implying 'there is some value greater than some standard of Length (A_{02}),' which is often simplified as (48').

(S15) The train runs through the tunnel.

$$(\exists x,y,p_1,z,p_3,k,p_4,k_0)\ (\underline{L(x,y,p_1,z,A_{12},G_t,k)}\bullet L(x,y,z,p_3,A_{12},G_t,k))$$
$$\Pi\ L(x,y,p_4,p_4,A_{13},G_t,k_0)\ \wedge p_1\neq z\ \wedge z\neq p_3\ \wedge \text{train}(y)\ \wedge \text{tunnel}(z) \tag{45}$$

(S16) The path runs through the forest.

$$(\exists x,y,p_1,z,p_3,k,p_4,k_0)\ (\underline{L(x,y,p_1,z,A_{12},G_s,k)}\bullet L(x,y,z,p_3,A_{12},G_s,k))\Pi$$
$$L(x,y,p_4,p_4,A_{13},G_s,k_0)\ \wedge p_1\neq z\ \wedge z\neq p_3\ \wedge \text{path}(y)\ \wedge \text{forest}(z) \tag{46}$$

(S17) The path through the forest is long.

$$(\exists x,y,p_1,z,p_3,x_1,k,q,k_1,p_4,k_0)\ (L(x,y,p_1,z,A_{12},G_s,k)\bullet L(x,y,z,p_3,A_{12},G_s,k))\Pi$$
$$L(x,y,p_4,p_4,A_{13},G_s,k_0)\Pi L(x_1,y,q,q,A_{02},G_t,k_1)\wedge p_1\neq z\ \wedge z\neq p_3\wedge q>k_1\wedge \text{path}(y)\wedge \text{forest}(z) \tag{47}$$

$$(\exists x_1,y_1,q,k_1)L(x_1,y_1,q,q,A_{02},G_t,k_1)\wedge q>k_1 \tag{48}$$

$$(\exists x_1,y_1,k_1)L(x_1,y_1,\text{Long},\text{Long},A_{02},G_t,k_1) \tag{48'}$$

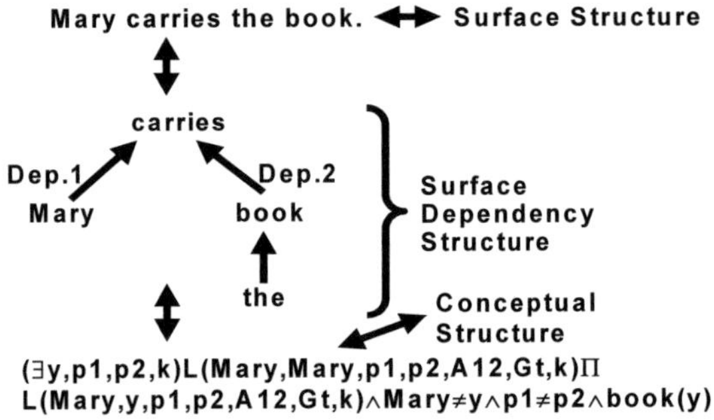

Fig. 11. Process of semantic interpretation of text

3.4 Fundamental semantic computations

Every version of the intelligent system IMAGES can perform text understanding based on word meanings as follows.

Firstly, a text is parsed into a surface dependency structure (or more than one if *syntactically* ambiguous). Secondly, each surface dependency structure is translated into a conceptual structure (or more than one if *semantically* ambiguous) based on word meanings. Finally, each conceptual structure is semantically evaluated.

The fundamental semantic computations on L_{md} expressions are performed to detect semantic anomalies, ambiguities and paraphrase relations in texts. Semantic anomaly detection is very important to cut off meaningless computations. For example, consider such a conceptual structure as (49), where 'A_{39}' is the attribute 'Vitality'. This locus formula can correspond to the English sentence 'The desk is alive', which is usually semantically anomalous because a 'desk' does never have vitality in the real world projected into the attribute spaces.

$$(\exists x)L(\_,x,Alive,Alive,A_{39},G_t,\_)\wedge desk(x) \qquad (49)$$

This kind of semantic anomaly can be detected in the following process. Firstly, assume the concept of 'desk' as (50), where 'A_{29}' refers to the attribute 'Taste'. The special symbols '*' and '/' are defined as (51) and (52) representing 'always' and 'no value', respectively.

$$(\lambda x)\ desk(x) \leftrightarrow (\lambda x)\ (\ldots L^*(\_,x,/,/,A_{29},G_t,\_) \wedge \ldots \wedge L^*(\_,x,/,/,A_{39},G_t,\_) \wedge \ldots) \qquad (50)$$

$$X^* \leftrightarrow (\forall[p,q])X\ \Pi\ \varepsilon([p,q]) \qquad (51)$$

$$L(\ldots,/,\ldots) \leftrightarrow \sim(\exists p)\ L(\ldots,p,\ldots) \qquad (52)$$

Secondly, the postulates (53) and (54) are utilized. The formula (53) means that ***<u>if one of two loci exists every time interval, then they can coexist</u>***. The formula (54) states that ***<u>a matter has never different values of an attribute with a standard at a time</u>***.

$$X \wedge Y^* .\supset. X\ \Pi\ Y \qquad (53)$$

$$L(x,y,p_1,q_1,a,g,k)\ \Pi\ L(z,y,p_2,q_2,a,g,k)\ .\supset.\ p_1=p_2 \wedge q_1=q_2 \qquad (54)$$

Lastly, the semantic anomaly of 'alive desk' is detected by using (50)-(54). That is, the formula (55) below is finally deduced from (49)-(54) and violates the commonsense given by (54), that is, " *Alive* $\neq$ / ".

$$(\exists x)L(\_,x,Alive,Alive,A_{39},G_t,\_)\ \Pi\ L(\_,x,/,/,A_{39},G_t,\_) \qquad (55)$$

This process above is also employed for dissolving such a syntactic ambiguity as found in S18. That is, the semantic anomaly of 'alive desk' is detected and eventually 'alive insect' is adopted as a plausible interpretation.

(S18) Look at the insect on the desk, which is still alive.

If a text has multiple plausible interpretations, it is semantically ambiguous. For example, S19 alone has two plausible interpretations (56) and (57) different at the underlined parts, implying 'Jack with the stick' and 'Tom with the stick', respectively.

(S19) Tom follows Jack with the stick.

$$(\exists x)(L(Jack,Jack,p,q,A_{12},G_t,\_)\underline{\Pi L(Jack,x,Jack,Jack,A_{12},G_t,\_)})\bullet L(Tom,Tom,p,q,A_{12},G_t,\_)\wedge p\neq q \tag{56}$$

$$(\exists x)L(Jack,Jack,p,q,A_{12},G_t,\_)\bullet(L(Tom,Tom,p,q,A_{12},G_t,\_)\underline{\Pi L(Tom,x,Tom,Tom,A_{12},G_t,\_)})\wedge p\neq q \wedge stick(x) \tag{57}$$

Among the fundamental semantic computations, detection of paraphrase relations is the most essential because it is for detecting equalities in semantic descriptions and the other two are for detecting inequalities in them. In the deductive system intended here, if two different texts are interpreted into the same locus formula, they are paraphrases of each other. For example, the sentence '*Mary goes with a book*' is interpreted into (58) which is proved to be equivalent to (43), the semantic description of '*Mary carries a book*'. In the process of this proof, the axioms (59) and (60) concerning the syntax of $\boldsymbol{L_{md}}$ are utilized.

$$(\exists y,p,q,k)L(Mary,Mary,p,q,A_{12},G_t,k)\Pi L(Mary,y,Mary,Mary,A_{12},G_t,k)\wedge p\neq q\wedge book(y) \tag{58}$$

$$(\forall x_1,x_2,x_3,x_4,p,q,a,g,k)\ (L(x_1,x_2,p,q,a,g,k)\Pi\ L(x_3,x_4,x_2,x_2,a,g,k) .\equiv.\ L(x_1,x_2,p,q,a,g,k)\Pi L(x_3,x_4,p,q,a,g,k)) \tag{59}$$

$$(\forall x_1,x_2,x_3,p,q,a,g,k)(L(x_1,x_2,p,q,a,g,k)\Pi L(x_1,x_3,p,q,a,g,k) .\equiv. L(x_1,\{x_2,x_3\},p,q,a,g,k)) \tag{60}$$

For another example, S14 and S20 below can be proved to be paraphrases each other by employing '***Postulate of Reversibility of a Spatial Event (PRS)***'.

(S20) The road separates at *C* from the street with the sidewalk and, after a while, runs 10km straight west from *B* to *A*.

The postulate PRS can be formulated as (61) using '$\equiv_0$', where χ and χ^R is a perceptual locus and its 'reversal' for a certain spatial event, respectively. These loci are substitutable with each other due to the property of '$\equiv_0$'.

$$\chi^R .\equiv_0. \chi \tag{61}$$

The recursive operations to transform χ into χ^R are defined by (62), where the reversed values p^R and q^R depend on the properties of the attribute concerned. For example, at (34), $p^R=p$, $q^R=q$ for A_{12}; $p^R=-p$, $q^R=-q$ for A_{13}.

$$(\chi_1\bullet\chi_2)^R \Leftrightarrow \chi_2^R \bullet\chi_1^R$$

$$(\chi_1\Pi\chi_2)^R \Leftrightarrow \chi_1^R\,\Pi\chi_2^R$$

$$(L(x,y,p,q,a,G_s,k))^R \Leftrightarrow L(x,y,q^R,p^R,a,G_s,k) \quad (62)$$

According to (62), (34) is transformed into (63) as its reversal and equivalent to the semantic interpretation of S20.

$$(\exists x,y,z,p,q)(L(\_,x,C,p,A_{12},G_s,\_)\Pi L(\_,y,C,q,A_{12},G_s,\_)\Pi L(\_,z,y,y,A_{12},G_s,\_))\bullet\varepsilon_s\bullet$$
$$(L(\_,x,B,A,A_{12},G_s,\_)\Pi L(\_,x,0,10km,A_{17},G_s,\_)\Pi L(\_,x,Point,Line,A_{15},G_s,\_)$$
$$\Pi L(\_,x,West,West,A_{13},G_s,\_)) \wedge road(x)\wedge street(y)\wedge sidewalk(z)\wedge p\neq q \quad (63)$$

3.5 Natural language understanding

Natural language, as already mentioned, is the most important for intuitive human-robot interaction. In order for comprehensible communication with humans, robots must understand natural language *semantically* and *pragmatically* well enough. As shown in Fig. 12, semantic understanding means associating symbols to conceptual images of matters (i.e., objects or events), and pragmatic understanding means anchoring symbols to real matters by unifying conceptual images with perceptual images. Robot manipulation by verbal suggestion here is defined as human-robot interaction where a human gives a robot a verbal expression of his/her intention and the robot behaviouralizes its conception, namely, semantic and pragmatic understanding of the suggestion, and maybe that repeatedly. As detailed above, semantic understanding is purely symbol manipulation for translation from verbal expression (text or speech) into L_{md} expression and fundamental semantic computations on L_{md} expressions by employing word meanings. On the other hand, pragmatic understanding is rather complicated because unstructured data processing for sensor-actor coordination is inevitably involved as well as pure symbol manipulation.

As also mentioned above, an event expressed in L_{md} is compared to a movie film recorded through a floating camera because it is necessarily grounded in FAO's movement over the event. ***This implies that L_{md} expression can suggest a robot what and how should be attended to in its environment.*** For example, consider such a suggestion as S21 presented to a robot by a human. In this case, unless the robot is aware of the existence of a certain box between the stool and the desk, such semantic understanding of the underlined part as (64) and such a semantic definition of the word 'box' as (65) are very helpful for it. The attributes A_{12} (Location), A_{13} (Direction), A_{32} (Color), A_{11} (Shape) and the spatial event on A_{12} in these L_{md} expressions indicate that ***the robot has only to activate its vision system in order to search for the box from the stool to the desk*** during the pragmatic understanding. That is, the robot can control its attention mechanism in the top-down way indicated in L_{md} to deploy its sensors or actuators during pragmatic understanding.

(S21) Avoid the green box between the stool and the desk.

$$(\exists x_1,x_2,x_3,x_4,p)(\underline{L(\_,x_4,x_1,x_2,A_{12},G_s,\_)\bullet L((\_,x_4,x_2,x_3,A_{12},G_s,\_))\Pi L(\_,x_4,p,p,A_{13},G_s,\_)}\Pi$$
$$L(\_,x_2,Green,Green,A_{32},G_t,\_)\wedge stool(x_1)\wedge box(x_2)\wedge desk(x_3)\wedge ISR(x_4) \quad (64)$$

$$(\lambda x)box(x) \leftrightarrow (\lambda x)L(\_,x,Hexahedron,Hexahedron,A_{11},G_{t},\_) \wedge container(x) \quad (65)$$

Conventionally, such quasi-natural language expressions as 'move(10meters)' and so on, uniquely related to computer programs, were employed for deploying sensors/motors in robotic systems as their semantics (e.g., Coradeschi & Saffiotti, 2003; Drumwright et al, 2006). These kinds of expression, however, were very specific to devices and apt to have miscellaneous syntactic variants among them such as 'move(10meters, quickly)', 'move(quickly, 10meters, leftward)', etc. for motors and 'find(object, red)', 'find(object, round, red)', etc. for sensors. This is also the case for spatial expressions such as 'left(x,y)' and 'left(x,y,z)', reading 'x is (z meters) to the left of y'. This fact is very inconvenient for communications especially between devices unknown to each other and, what is worse, they, unlike L_{md} expressions, are too linguistic or coarse to represent and compute sensory/motory events in such an integrated way as 'common coding approach to perception and action' (Prinz, 1990).

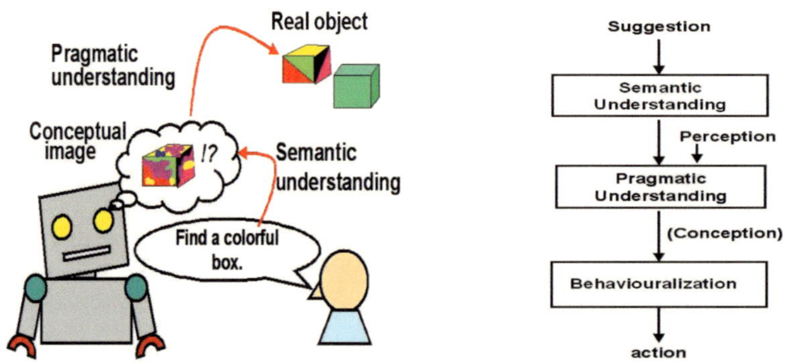

Fig. 12. Semantic and pragmatic understanding of natural language in robots

3.6 Formalization of communication

Here, a piece of communication is identified by a set of messages ($\boldsymbol{M}$) as is formalized in (66). In turn, a message (m) is defined by (67), where D, S, R and B mean the duration, sender(s), receiver(s) and the body of the message, respectively. The body (B) consists of the two elements defined by (68), where E and T mean the event referred to and the task requested or intended by the sender, respectively.

$$\boldsymbol{M}=\{m_1, m_2, \ldots, m_n\} \quad (66)$$

$$m=(D, S, R, B) \quad (67)$$

$$B=(E, T) \quad (68)$$

For example, each item of the message m_0: "Fetch me the book from the shelf, Tom" uttered by Jim during the time-interval [t_1, t_2] is as follows:
$m_o=(D_0, S_0, R_0, B_0)$, $B_0=(E_0, T_0)$, $D_0=[t_1, t_2]$, $S_0=$ *"Jim"*, $R_0=$ *"Tom"*, $E_0=$ *"Tom FETCH Jim BOOK FROM SHELF"*, and $T_0=$ *"realization of* E_0*"*.

In general, without any hint by language, it is not so easy for the receiver (R) to understand what task (*T*) is intended by the sender (*S*). Therefore, as mentioned in Section 1, IMAGES-M is designed to work as a language-centered (or language-driven) intelligent interface for human-robot multimedia communication. The author has adopted a simple assumuption that there are unique correspondences between the kinds of tasks and the types of sentences as shown in Table 4 while there were proposed more complicated and context-sensitive ones based on speech-act theory (Austin, 1962; Searle,1969).

Sentence type (Examples)	Task (*T*)
Declarative (It is ten o'clock now.)	To believe *E*.
Interrogative ([A] Is it ten o'clock now? [B] What time is it now?)	[A] To reply whether *E* is true or false. [B] To reply what makes *E* true.
Imperative (Show me your watch.)	To realize *E*.

Table 4. Sentence types and Tasks

4. Kansei processing as inter-agent communication

It is well known that emotion in a human can be affected by his/her world, namely, ***W*** in Fig.1. For example, a person's *evaluation* of live image of an object (i.e. image output from **St**) expressed by such words as 'favorite', 'beautiful', 'tasty', etc. can vary depending on his/her emotional bias such as 'hungry', 'depressed', etc. Kansei is one of mental functions with emotion involved but has a more complicated phase than pure emotion originated from instincts or imprinting. For example, sweet jam may be nice on toast but not on pizza for certain people knowledgeable about these foods. For another example, people can be affected on their evaluation of an art by its creator's name, for example, 'Picasso'. These are good examples of Kansei processing as emotional performance affected by knowledge in humans.

Therefore, Kansei can be defined as human emotion toward an object affected by its information for him/her, so called, 'concept', including his/her intellectual pursuits, traditions, cultures, etc. concerning it. In this sense, Kansei is assumed to be reasonable or intelligible among the people sharing such concepts unlike pure emotion. These hypothetical considerations are formalized as (69) and (70).

$$I_P(x) = \mathbf{P_E}(S(x)) \tag{69}$$

$$I_K(x) = \mathbf{P_E}(S(x) \wedge O(x)) = \mathbf{P_E}(S'(x)) \tag{70}$$

where

$\mathbf{P_E}(X)$: Performance of **Em** for mental image X;
$I_P(x)$: Mental image as pure emotion for object x;
$I_K(x)$: Mental image as the result of Kansei processing for object x;
S(x): Live image of object x sent from **St**;
O(x): Concept (i.e., conceptual image) of object x induced by S(x) from knowledge at **Kn**;
S′(x): Unified image of live image and concept of object x.

Figure 13 shows an example of Kansei processing in the mind model, where perceived, induced and inspired images correspond to S(x), S′(x) and $I_K(x)$, respectively, while Fig.14 is for pure emotion with $I_P(x)$ as the inspired image.

These two inspired images $I_K(x)$ and $I_P(x)$ are can be verbalized in **Re** as 'Appetizing!' and 'Fragrant!', as labeled in the figures, respectively. The essential difference between them is assumed to reside in whether or not they are affected by O(x), namely, the concept of 'chocolate cream bread', inferred by **Kn** from the shape, smell, etc. Whereas, pure emotion for an object can be a special case of Kansei processing without knowing or recognizing what it is.

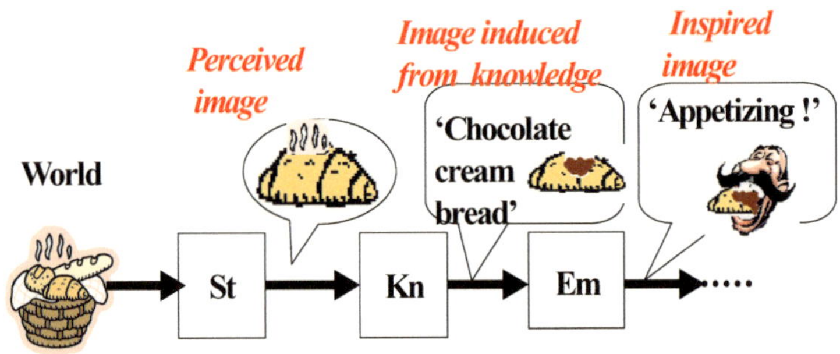

Fig. 13. Example of Kansei processing

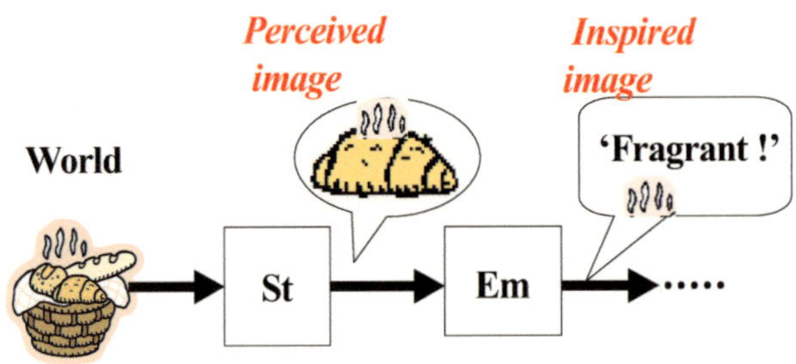

Fig. 14. Example of pure emotion

In MIDST, as already mentioned, the concept of an object *x* is given as an integrated omnisensory mental image of its properties and its relations with other objects involved. For example, the concept of 'chocolate cream bread' can be given by (71), reading that *x* is bread, sweet due to chocolate cream, fragrant of itself, etc., where A_{29} and A_{30} refer to 'Taste' and 'Odour', respectively.

$$(\lambda x)\text{chocolate\_cream\_bread}(x) \Leftrightarrow (\lambda x)(\exists y,k_1,k_2)L(y,x,Sweet,Sweet,A_{29},G_t,k_1)\Pi$$

$$L(x,x,Fragrant,Fragrant,A_{30},G_t,k_2) \wedge \text{bread}(x) \wedge \text{chocolate\_cream}(y) \wedge \ldots \quad (71)$$

Many psychologists have claimed that certain emotions are more basic than others (Ortony & Turner, 1990). The author has assumed that human emotion consists of four primitives such that represent the degrees of 1) Happiness, 2) Aggressiveness, 3) Surprise and 4) Superiority. For example, the degree of Happiness is measured by using such a word set as {anguish, distress, sorrow, gloom, content, joy, ecstasy}, whose each element is possibly ordered on a coordinate axis and fuzzified with a certain characteristic function.

Therefore, the author has assumed Kansei as a certain function to evaluate totally the loci in the attribute spaces of these primitives.

For comprehensible communication with humans, robots must understand natural language *semantically* and *pragmatically*. However, humans and robots can be equipped with sensors, actuators and brains of different competences and performances. Therefore, their vocabularies may well be grounded in quite different sensations, physical actions or mental actions and in turn such a situation may bring inevitably different kinds of semantics to them, so called, "Natural Semantics (NS)" for humans and "Artificial Semantics (AS)" for robots. For example, consider such a scenario as follows.

...***A human 'Kate' and a humanoid robot 'Robbie' encounter at the terrace in front of the room where a Christmas party is going on merrymaking. Kate says "Robbie, please fetch me some delicious food from the noisy room." Robbie replies "OK, Kate."....***

For a happy end of this dialog, Robbie must have a good knowledge of Kate's NS for Kansei and translate it into its AS appropriately enough to find out the real objects referred to by her words. In this case, Robbie needs at least to interpret Kate's statement as the expression (72) reading "If *Robbie* fetches *Kate* some food delicious ***for her*** from the room noisy ***for her*** (***E₁***), then consecutively it makes *Kate* happier (***E₂***)."

$$\mathbf{E}_1 \supset_1 \mathbf{E}_2, \tag{72}$$

$$\text{where } \mathbf{E}_1 = (\exists x_1,x_2,k_1,k_2,k_3,k_4)\,(L(\text{Robbie},\text{Robbie},\text{Kate},x_2,A_{12},G_t,k_1)\bullet$$

$$(L(\text{Robbie},\text{Robbie},x_2,\text{Kate},A_{12},G_t,k_1)\;\Pi\;L(\text{Robbie},x_1,x_2,\text{Kate},A_{12},G_t,k_1)))$$

$$\Pi(L(\text{Kate},x_1,\text{Delicious},\text{Delicious},B_{08},G_t,k_2)\;\Pi\;L(\text{Kate},x_2,\text{Noisy},\text{Noisy},B_{08},G_t,k_4)$$

$$\wedge\ \text{food}(x_1) \wedge \text{room}(x_2)$$

$$\text{and } \mathbf{E}_2 = (\exists v_1,v_2,k_7)\,L(\mathbf{E}_1,\text{Kate},v_1,v_2,B_{04},G_t,k_7) \wedge v_2>v_1$$

In (72), the special symbols and their meanings in the expressions are: '$X\supset_1 Y$' ='If X, then consecutively Y', B_{08}='Kansei' and B_{04}= 'Happiness'. As easily understood, 'Delicious' and 'Noisy' in Kansei (B_{08}) should be distiguished from 'Sweet', etc. in Taste(A_{29}) and 'Loud', etc. in Sound(A_{31}), respectively because the former result from total evaluation of the latter. Of course, the values in Kansei greatly depend on the standards (i.e., k_2 and k_4) given by the evaluator (i.e., Event Causer), which are most closely related to 'Individual' or 'Purposive' standard in Table 3.

By the way, Robbie's task is only to make $\boldsymbol{E_1}$ come true where each atomic locus formula is associated with his actuators/sensors. Of course, Robbie *believes* that he will become happier to make Kate happier as is given by (73) where 'B_{03}' is 'trueness (=degree of truth)' and 'K_B' is a certain standard of 'believability' for Robbie. That is *emotionally* to say, Robbie *likes* Kate as formulated as (74) reading 'Robbie is happy because of Kate' where 'K_H' is a certain standard of 'happy' for Robbie. And then, Robbie comes to believe that he will realize $\boldsymbol{E_1}$ in future (i.e., *decides* to make $\boldsymbol{E_1}$ take place) as is expressed by (75). Therefore, this example is also very significant for *intentional* sensing or action of a robot driven by logical description of its belief.

$$(\exists p)L(\text{Robbie},\mathbf{E},p,p,B_{03},G_t,K_B) \wedge p>K_B, \tag{73}$$

$$\text{where } \mathbf{E} = (\mathbf{E}_2 \supset_1 \mathbf{E}_3)$$

$$\text{and } \mathbf{E}_3 = (\exists v_3,v_4,k_8)\ L(\mathbf{E}_2,\text{Robbie},v_3,v_4,B_{04},G_t,k_8) \wedge v_4>v_3$$

$$(\exists p)\ L(\text{Kate},\text{Robbie},p,p,B_{04},G_t,K_H) \wedge p>K_H \quad (74)$$

$$(\exists p,q,t_1,t_2,k)\ L(\text{Robbie},\mathbf{E}_4,/,p,B_{03},G_t,K_B)\Pi\varepsilon([t_1,t_2]) \wedge p>K_B, \quad (75)$$

$$\text{where } \mathbf{E}_4 = \varepsilon([t_1,t_2]) \bullet L(\text{Robbie},\mathbf{E}_1,/,q,A_{01},G_t,k)$$

For realizing a plausible *Artificial Kansei,* it is most essential to find out functional features of **Em** and to deduce from them such laws that may rule $\mathbf{P}_E$. The author has already obtained a number of postulates concerning $\mathbf{P}_E$ such as Postulate 01-04 below and simulated human-robot dialog based on them (See Fig.17).

Postulate 01: "Some emotion can coexist with another in humans."
Postulate 02: "A desire for something is a belief of becoming happier with it."
Postulate 03: "Something delicious for someone makes him/her happier by its taste."
Postulate 04: "Something noisy for someone makes him/her less happy by its sounds."

5. IMAGES-M as an artificial communication partner

The methodology mentioned above has been being implemented on the intelligent system IMAGES-M shown in Fig.15-a. IMAGES-M is one kind of expert system consisting of Inference Engine (IE), Knowledge Base (KB) and five kinds of user interface: 1) Text Processing Unit (TPU), 2) Speech Processing Unit (SPU), 3) Picture Processing Unit (PPU), 4) Action Data Processing Unit (ADPU) and 5) Sensory Data Processing Unit (SDPU). The pair of IE and KB work as **Kn** and **Em** of the mind model, and the group of user interfaces as **St** and **Re**. As depicted in Fig.15-b, these user interfaces are intended to convert information media and L_{md} expressions mutually in collaboration with IE and KB in order to facilitate various types of cross-media operations such as language-to-picture translation.

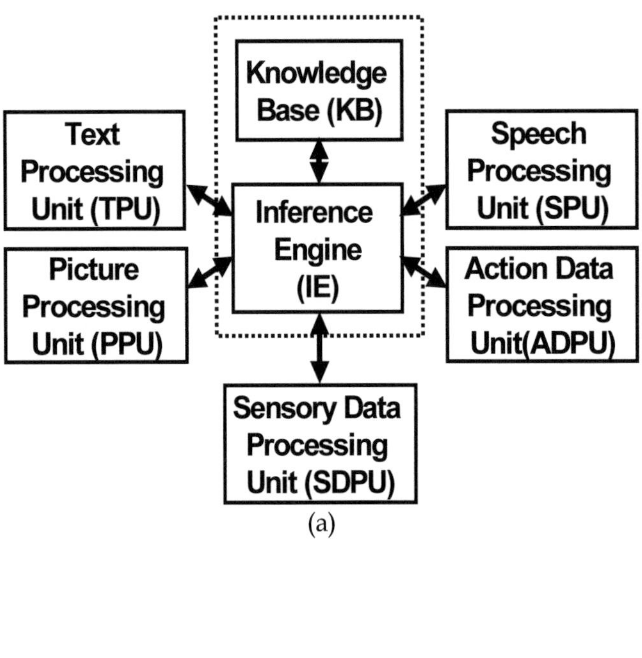

(a)

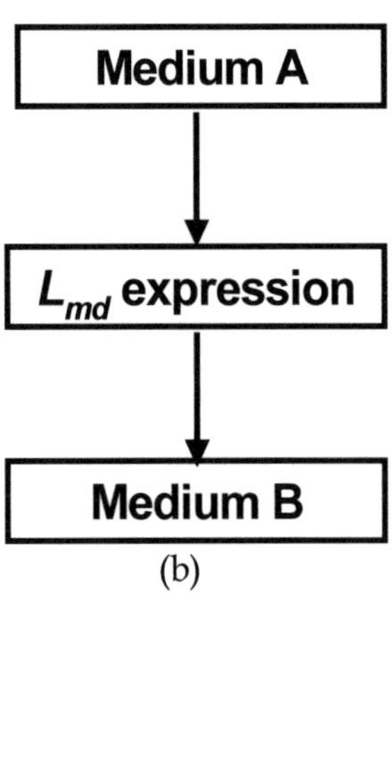

(b)

Fig. 15. (a) Configuration of IMAGES-M and (b) cross-media operation via L_{md} expression

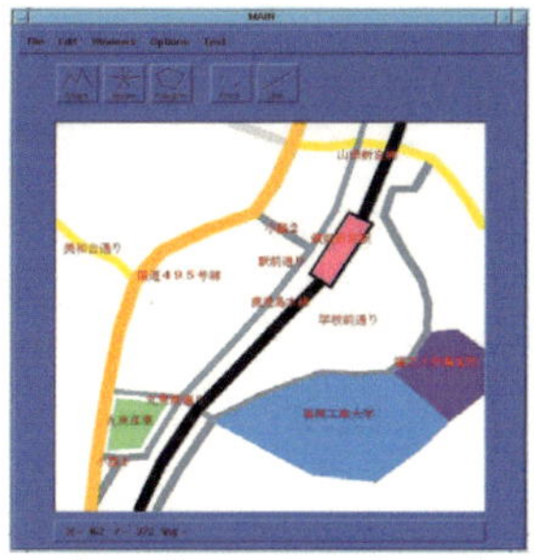

H: How does the national road run?
S: It extends between Pref. A and Pref. C via Pref. B.
H: Where does the bus go from the rail way station A?
S: It reaches the town D.
H: What is between the buildings A and B?
S: The railway D.
H: Where do the street A and the road B meet?
S: At the crossing C.
H: Where do the street A and the road B separate?
S: At the crossing C.

Fig. 16. Cross-media operations: A map reproduced from L_{md} expression and Q-A on the map ('H' and 'S' stand for 'Human' and 'IMAGES-M', respectively)

[Dialog 01] *Postulate 01: "Some emotion can coexist with another in humans."*
H: Tom loves Jane.
H: Whom does he like?
S: He likes Jane.
[Dialog 02] *Postulate 02: "A desire for something is a belief of becoming happier with it."*
H: Tom wants to go to Tokyo.
H: Does Tom believe to become happier if he goes to Tokyo?
S: Yes, he does.
[Dialog 03] *Postulate 03: "Something delicious for someone makes him/her happier by its taste."*
H: The food was delicious for Tom.
H: Did it make him happier by its taste?
S: Yes, it did.
[Dialog 04] *Postulate 04: "Something noisy for someone makes him/her less happy by its sounds."*
H: Mary evaluated the music noisy.
H: Did it make her less happy by its sounds?
S: Yes, it did.

Fig. 17. Examples of postulate-based dialog simulations ('H': human, 'S': IMAGES-M)

Fig. 18. AIBO (Sony) behaving to the command 'Walk forward and wave your left hand'

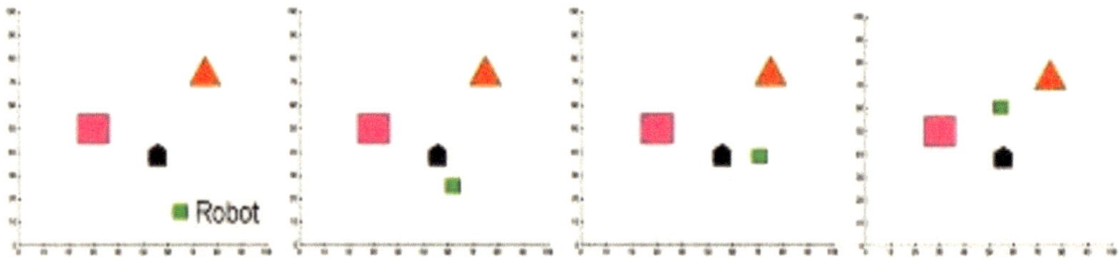

Fig. 19. Simulation in Matlab of the command 'Go to between the rectangle and the triangle, avoiding the pentagon'

Figure 16 presents a map reproduced from L_{md} expression and several transactions during Q-A on it, where IMAGES-M computed temporal and spatial events distinctively very well. In Fig.17 are shown several examples of postulate-based dialog between a human and IMAGES-M performing symbol manipulation, namely, deduction based on the postulates.

Figure 18 shows an example of langage-to-action translation by IMAGES-M. The real robot, AIBO (Sony), cannot perform the two actions (i.e., 'walking' and 'hand-waving') simultaneously and therefore it waved its hand after walking with the transit action (i.e., 'sitting-down') between them inserted, where each action was defined as an ordered set of values in 'Shape (A_{11})' (i.e., time-sequenced snapshots of the action) corresponding uniquely with the positions of their actuators determined by the rotations (i.e., changes in 'Angularity (A_{45})') of the joints (Yokota, 2006). In Fig.19 is presented another example of language-to-action translation for a virtual robot simulated in Matlab, where 'avoid' was defined as 'keep Topology (A_{44}) Disjoint'.

IMAGES-M has so far worked well enough as a language-centered intelligent system. To our best knowledge, there is no other system that can perform miscellaneous cross-media operations, including robot manipulation by natural language, in such a seamless way as ours. Most of computations on L_{md} are simply for unifying (or identifying) atomic loci and for evaluating arithmetic expressions such as 'p=q', and therefore the author believes that the methodology presented here can reduce the computational complexities of conventional ones when applied to the same kinds of problem described here. Recently, L_{md} has been enriched by a general theory of tempological connectives (Yokota, 2008), and another algorithm programmable in procedural languages has been developed in order for more efficient computation of L_{md} expressions (Yokota et al, 2008) while the earlier version of IMAGES-M was programmed in PROLOG with considerable redundancy. This advantage comes from the meaning representation scheme normalized by atomic locus formulas, which will also facilitate higher-order representation and computation such as described in Section 4 (although it is well known that expressions in higher-order predicate logic without any constraint like this are not computable).

6. Conlusion

The author presented an innovative concept of artificial communication partners with the multiagent mind model based on MIDST and its implementation on IMAGES-M. The competence and performance of the mind model are formalized as a deductive system in the formal language L_{md} employed for many-sorted predicate logic. This deductive system is one kind of appplied predicate logic (i.e., pure predicate logic with certain domain-specific constants (e.g., McDermott, 1982; Shoham, 1989; Kabanza, 1995; Haddawy, 1996)), but the domain-specificness in its syntax and semantics is exclusively related to atomic locus formulas and the essential part of its semantics is subject to their interpretation controled by the family of domain-specific constants, namely, Attributes, Values, Event Types and Standards intended to correspond well with human sensory systems. The author has found the implementation so far a success and come to have such a perspective that the scheme presented here is applicable to various mind models for humans or humanoid robots of different competences and performances simply by controling such a family.

The expressive power of L_{md} was demonstrated with linguistic or pictorial manifestations throughout this chapter. Its most remarkable point in comparison with other KRLs resides in that it can provide terms of physical matters such as *carry, snow,* etc. with precise semantic definitions that are normalized by atomic locus formulas and visualized as loci in attribute spaces in both temporal and spatial extents (i.e., temporal and spatial events), which leads to good computability and intuitive readability of L_{md} expressions. As easily imagined by the brief description in Section 4, this is also the case for terms of human mental matters such as *believe, like, desire,* etc. Among them, the semantic definitions of *belif, desire* and *intention* have been studied with special attention focused on inter-agent communication and computationally grounded in Belief-Desire-Intention logic (BDI logic) in conventional multiagent systems (MASs) (e.g., Vieira et al, 2007). These MASs, however, as well as other conventional deductive systems will have to employ another framework for dynamic (i.e., temporally or spatially continuous) processes in the attributes (or propeties) of matters such as "They *come to believe* it *gradually*", namely, *temporal event* of belief that can be easily formulated in L_{md}, needless to mention about *spatial event* that has seldom been considered for systematic formulation and computation by others.

The mind model proposed here is much simpler than Minsky's but it can work well enough to compute both physical and mental matters expressed in L_{md}. The most important problems remaining unsolved are how to provide each attribute space and how to build its corresponding atomic performance. These problems concern neuroscience (Brooks, 1986) as well as psychology (Yokota, 1988; Fukuda et al, 1998; Sugita et al, 2003) and therefore the author will consider employment of soft computing theories such as neural network, genetic algorithm, fuzzy logic, etc. for their self-organization in the near future.

At last of this chapter, the author would like to acknowledge that this work was partially funded by the Grants from Computer Science Laboratory, Fukuoka Institute of Technology and Ministry of Education, Culture, Sports, Science and Technology, Japanese Government, numbered 14580436 and 17500132.

7. References

Allen, J.F. (1984). Towards a general theory of action and time, *Artificial Intelligence*, Vol.23, No.2, pp.123-154

Amano,M., et al. (2005). Cross-media translation between human motions and texts based on Mental Image Directed Semantic Theory, *Proceedings of IEEE workshop on Multimedia Network Systems and Applications (ICDCS05)*, Columbus OH, pp.707-713

Austin, J. L. (1962). *How to Do Things with Words*, Oxford University Press, London

Brooks, R. A. (1986). A robust layered control system for a mobile robot, *IEEE Journal of Robotics and Automation*, Vol.RA-2, pp.14-23

Butz,A., Fisher,B.D., Krüger,A. & Olivier,P. (Eds.) (2005). *Smart Graphics, Proceedings of 5th International Symposium (SG 2005)*, Frauenwörth Cloister, Germany, August 22-24, 2005, Lecture Notes in Computer Science 3638 Springer

Coradeschi,S. & Saffiotti,A. (2003). An introduction to the anchoring problem, *Robotics and Autonomous Systems*, Vol.43, pp.85–96

Demiris,Y. & Khadhouri,B. (2006). Hierarchical attentive multiple models for execution and recognition of actions. *Robotics and Autonomous Systems*, Vol.54, pp.361–369

Dorr,B.& Bonnie,J. (1997). Large-Scale Dictionary Construction for Foreign Language Tutoring and Interlingual Machine Translation, *Machine Translation*, Vol.12, No.4, pp.271-322

Drumwright,E. Ng-Thow-Hing,V. & Mataric,M.J. (2006). Toward a vocabulary of primitive task programs for humanoid robots, *Proceedings of International Conference on Development and Learning (ICDL06)*, Bloomington IN, May 2006

Eakins, J.P. & Graham, M.E. (1999). Content-based Image Retrieval, *A report to the JISC Technology Applications Programme*, Institute for Image Data Research, University of Northumbria at Newcastle

Egenhofer,M. (1991). Point-set topological spatial relations, *Geographical Information Systems*, Vol.5, No.2, pp.161-174

Fukuda,M., Sugita,K.and Shibata, Y. (1998). Perceptional Retrieving Method for Distributed Design Image Database System, *Journal of Information Processing Society of Japan*, Vol.39, No.2, pp.158-169

Gardenfors,P. (2000). *Conceptual Space*, MIT Press

Haddawy,P. (1996). A logic of time, chance, and action for repesenting plans, *Artificial Intelligence*, Vol.80, No.2, pp.243-308

Kabanza,F.(1995). Synchronizing multiagent plans using temporal logic specifications, *Proceedings of 1st International Conf. on Multi-Agent Systems (ICMAS-95)*, pp.217-224, June 1995.

Kherfi,M.L., Ziou,D. & Bernardi,A. (2004). Image Retrieval from the World Wide Web : Issues, Techniques and Systems, *ACM Computer Surveys*, Vol.36, No.14, pp.35-67

Labrou,Y., Finin,T. & Peng,Y. (1999). The current landscape of agent communication languages, *Intelligent Systems*, 14 (2), pp.45-52

Langacker,R. (1991). *Concept, Image and Symbol*, Mouton de Gruyter, Berlin/New York

Langacker,R. (2005). Dynamicity, fictivity, and scanning: The imaginative basis of logic and linguistic meaning, In: *Grounding Cognition: The Role of Perception and Action in Memory, Language and Thinking*, Precher, D. & Rudolf,A.Z. (eds.), pp.164-197, Cambridge University Press, Cambridge

Leisi,E. (1961). *Der Waltinhalt-Seine Struktur im Deutschen und Englischen-*, Quelle & Meyer, Heidelberg

McDermott,D.V. (1982). A temporal logic for reasoning about processes and plans, *Cognitive Science*, Vol.6, pp.101-155

Miller,G.A.& Johnson-Laird,P.N. (1976). *Language and Perception*, Harvard University Press

Minsky, M. (1986). *The society of mind*, Simon and Schuster, New York

Noton,D. (1970). A Theory of Visual Pattern Perception, *IEEE Transaction on Systems Science and Cybernetcs*, Vol.SSC-6, No.4, pp.349-357

Oda,S., Oda,M.& Yokota,M. (2001). Conceptual Analysis Description of Words for Color and Lightness for Grounding them on Sensory Data, *Trans. of JSAI*, Vol.16, No.5-E, pp.436-444

Ortony,A. and Turner,T.J. (1990). What's basic about basic emotions?, *Psychological Review*, Vol.97, pp.315-331

Prinz, W. (1990).: A common coding approach to perception and action, In: *Relationships between perception and action*, Neumann, O. & Prinz, W. (Eds.), pp.167-201, Springer-Verlag

Roget,P. (1975). *Thesaurus of English Words and Phrases*, J.M.Dent & Sons Ltd., London

Searle, J. R. (1969). *Speech Acts: An Essay in the Philosophy of Language*, Cambridge University Press, Cambridge

Shiraishi,M, Capi,G.& Yokota,M. (2006). Human-robot communication based on a mind model, *Artificial Life and Robotics*, Vol.10, No.2, pp.136-140

Shoham,Y. (1989). Time for actions: on the relationship between time, knowledge, and action, *Proceedings of IJCAI89*, Detroit MI, pp.954-959

Sowa,J.F. (2000). *Knowledge Representation: Logical, Philosophical, and Computational Foundations*, Brooks Cole Publishing Co., Pacific Grove, CA

Sugita,K., Miyakawa, A. and Shibata, Y. (2003). Relation Between Kansei Words and the Room Space in Digital Traditional Japanese Crafting System, *Proceedings of International Conference on Advanced Information Networking and Applications (AINA03)*, pp.159-162, May 2003

Tauchi,E., Sugita,K., Capi,G. & Yokota,M. (2006). Towards Artificial Communication Partners with 'Kansei', *Proceedings of IEEE workshop on Network-based Virtual Reality and Tele-existence (INVITE06)*, Wien Austria, May 2006

Vieira,R., Moreira,A., Wooldridge,M. & Bordini,R.H. (2007). On the Formal Semantics of Speech-Act Based Communication in an Agent-Oriented Programming Language, *Journal of Artificial Intelligence Research*, Vol.29, pp.221–267

Wilkins, D.E. & Myers K.L. (1995). A common knowledge representation for plan generation and reactive execution, *Journal of Logic and Computation*, Vol.5, No.6, pp.731-761

Yokota,M., Taniguchi,R. & Kawaguchi,E. (1984). Language-Picture Question-Answering through Common Semantic Representation and Its Application to the World of Weather Report, In: *Natural Language Communication with Pictorial Information Systems*, Bolc,L.(Ed.), pp.203-254, Springer-Verlag

Yokota,M. (1988). A psychological experiment on human understanding process of natural language. *Transaction of IEICE Japan*, Vol.J71D, No.10 (in Japanese) pp.2120-2127

Yokota,M. (2005). An Approach to Natural Language Understanding Based on Mental Image Model. In: *Natural Language Understanding and Cognitive Science*, Sharp,B. (ed.), pp.22-31, INSTICC PRESS

Yokota,M.& Capi,G. (2005a). Cross-media Operations between Text and Picture Based on Mental Image Directed Semantic Theory, *WSEAS Transaction on Information Science and Applications*, Vol.10, No.2, pp.1541-1550

Yokota,M.& Capi,G. (2005b). Integrated Multimedia Understanding for Ubiquitous Intelligence Based on Mental Image Directed Semantic Theory, *Proceedings of IFIP EUC`2005 symposium (LNCS 3823)*, pp.538-546, Nagasaki Japan, Dec., 2005

Yokota,M. (2006). Towards a Universal Language for Distributed Intelligent Robot Networking, *Proceedings of International Conference on Systems, Man and Cybernetics (SMC06),* Taipei Taiwan, Oct. 2006

Yokota,M. (2007). A Theoretical Consideration on Artificial Imitation of Human Action Based on Mental Image Directed Semantic Theory, *Proceedings of IEEE workshop on Network-based Virtual Reality and Tele-existence (INVITE07)*, Niagara Canada, May 2007

Yokota,M., Shiraishi,M., Sugita,K. & Oka,T. (2008). Toward integrated multimedia understanding for intuitive human-system interaction, *Artificial Life and Robotics*, Vol.12, No.1-2, pp.188-193

Yokota,M., Abe,Y., Sugita,K. & Oka,T. (2008). Multimedia Description Language L_{md} and Its Application to Integrated Multimedia Understanding, *Proceedings of 4th International Conference on Soft Computing and Intelligent Systems and 9th International Symposium on Advanced Intelligent Systems (SCIS & ISIS 2008)*, Nagoya Japan, Sep. 2008

Yokota,M. (2008). A General Theory of Tempological Connectives and Its Application to Spatiotemporal Reasoning, *Proceedings of 4th International Conference on Soft Computing and Intelligent Systems and 9th International Symposium on Advanced Intelligent Systems (SCIS & ISIS 2008)*, Nagoya Japan, Sep. 2008

Zarri,G.P. (1997). NKRL, a Knowledge Representation Tool for Encoding the Meaning of Complex Narrative Texts, Natural Language Engineering, *Special Issue on Knowledge Representation for Natural Language Processing in Implemented Systems*, Vol.3, pp.231-253

19

New Approach of Neural Network for Controlling Locomotion and Reflex of Humanoid Robot

Zaier Riadh and Kanda Shinji
Fujitsu Laboratories Limited
Japan

1. Introduction

In recent years, several humanoid robotic platforms have been developed, and most of them were dealing with hardware problems to closely replicate the appearance and the motion ability of human beings. Along with this, motion control also has been advanced, and much work has been focused on the dynamics of the robot using the zero moment point (ZMP) approach (e.g. Vukobratovic & Juricic, 1969; Miura & Shimoyama, 1984; Kajita & Matsumoto, 2001; Huang et al., 2001). Huang et al. proposed a method for planning a walking pattern, where the reference trajectory is designed offline for given constraints on the foot and ground, and satisfying a particular ZMP constraint using third order spline functions. More recently, biologically inspired control strategies have been proposed to generate autonomously adaptable rhythmic movement. These are based on a neuronal network, termed a central pattern generator (CPG) (e.g., Griller, 1985; Taga et al., 1991; Taga, 1995) that is capable of generating a rhythmic pattern of motor activity in the absence of sensory input signals. Taga et al. demonstrated that bipedal locomotion can be realized as a global limit cycle generated through entrainment between a neural network consisting of a neural oscillator and the physical system.
On the other hand, toward a safe interaction of the humanoid robot with the environment, Morisawa et al. presented a method to generate an emergency stop motion based on the evaluation on the ZMP and the center of gravity (COG). Okada et al. presented a motion emergency system based on attractor design of the nonlinear dynamical system. Huang et al. proposed a feedback sensory reflex, which consists of ZMP reflex, landing-phase reflex, and a body-posture reflex. However, in the presence of large disturbances, the ZMP will have an arbitrary location that can be out of the stable region in spite of the stability of the robot's upper body.
In this chapter, we deal with both the motion patterns generation and reflexes against sudden events. Instead of using complex modeling and dealing with highly non-linear equations (Okada et al., 2005; Huang & Nakamura, 2005), our control strategy in generating rhythmic motion is simply based on piecewise linear oscillators and a small number of easily tunable parameters. The method needs not to satisfy constraints on robot's foot or ZMP stability margin as in (Huang et al., 2001). It simply uses piecewise-linear functions

and first order low-pass filters generated by an original recurrent neural network (RNN) (Zaier & Nagashima, 2006), where the "integrate and fire" neuron model (Gerstner, 1995) has been used. The method, therefore, provides much flexibility to the pattern generator so that combination of reflexes during locomotion can be realized without complexity or re-design of the system.

To deal with sudden events and large disturbance acting on the robot, most recent works have been approaching this problem by proposing reflex action for each event separately (Huang & Nakamura, 2005; Zaier & Nagashima, 2006). In contrast, our approach consists of presenting the reflex system in a unifying form with regards to four types of sudden events, where we use a simple recognizer based on Bayes rule to detect unexpected events. In this research framework, we 1) add normal feedback signals when a relatively small disturbance is detected by the gyro sensor; 2) change the walk parameters such as the gait, the stride, or the walk posture; 3) add an extra motion at some joints besides the motion generator outputs; 4) stop the walk motion in the case of large disturbance and generate a reflex based on the sensory data and an interpolation of four predefined postures. It should be noticed that the reflexes in 1) and 4) are decided according to the robot's upper body oscillation rather than the ZMP stability margin (Huang & Nakamura, 2005). Moreover, since the leg during swing phase is made very compliant, there will be neither tipping over when landing nor colliding with an obstacle.

In this chapter, we consider more focus on the reflex against sudden obstacle from the stability point of view. Indeed, the behavior of the robot when it encounters an obstacle depends on the timing and location of the collision with regards to the sole plate under the robot leg. In other words, and by considering the state space diagram, the stability of the robot will depend on the distance of the robot state from the equilibrium point and the ability of the robot to return to that point. In this research framework, therefore, the primitive reflex against sudden obstacle that has been proposed in (Zaier & Nagashima, 2006) will be modified by processing the force sensors' outputs such that the generated reflex movement can smoothly shift the robot state to the attraction domain of the equilibrium point.

Experiments using Fujitsu's humanoid robot HOAP-3 (Murase et al., 2001; Fujitsu Automation Ltd.) demonstrate that the reflex movement is successfully integrated with the rhythmic motion. The robot robustly walks on a number of different ground surfaces and exhibits an efficient natural looking reflex mode.

2. Neural network of rhythmic motion and reflex

Shaping an arbitrary motion pattern takes effort to think about the functions that fit the desired task with regards to given constraints. On the other hand, continuous piecewise linear functions have proved to be very powerful tools in modeling and analyzing nonlinear systems. For instance, to fit the real dynamics of the robot, the profile of the rolling motion of a humanoid robot can be approximated using (1).

$$\varepsilon_i \frac{da(t)}{dt} + a(t) = c(t)\,, \tag{1}$$

where $c(t)$ is the input signal described as a piecewise linear function in time t, $a(t)$ is the activation function. In fact, for small oscillations the passive behavior fits a sine-wave. But,

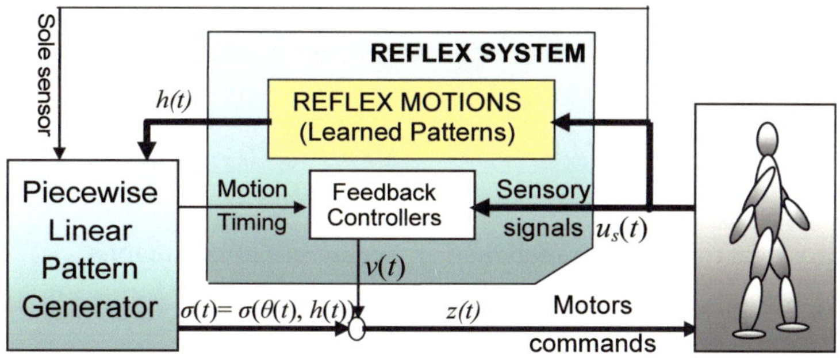

Fig. 1. Overall scheme of the reflex system and motion pattern generator

in general, the trajectories for the biped joints are not sine functions with only one frequency component. For such situations, a combination of more primitive functions will be much interesting. The input signal $c(t)$ can be expressed as time series of N piecewise-linear functions $u_i(t)$:

$$c(t) = \sum_{i=1}^{N} c_i u_i(t - t_i), \tag{2}$$

where $u_i(t) \in [0,1]$, $t_i \geq 0$ and c_i is a real number.

Let's write the equations of the proposed recurrent neural network (Fig.1) that controls the movement of the humanoid robot as follows;

$$\frac{d\theta_i(t)}{dt} = f_i(t, \theta_i) \tag{3}$$

$$f_i(t, \theta_i) = -\frac{1}{\varepsilon}\theta_i(t) + g_i(t) \tag{4}$$

where the index i represents either rolling or pitching, θ_i is the generated motion to the joints, $f_i(t,\theta_i)$ is a piecewise continuous function in time t, $g_i(t)$ is a piecewise function generating the walking as in (2). At the equilibrium points the robot is controlled by the following PD controller,

$$\begin{cases} \dfrac{dx_i}{dt} = -a_{ii}x_i + \sum\limits_{j=1, j\neq i}^{N} \dfrac{a_{ij}}{a_{ii}} x_j + \dfrac{b_i}{a_{ii}} u_s \\ v(t) = \sum\limits_{i=1}^{N} c_i x_i(t) \end{cases} \tag{5}$$

where $v(t)$ is the controller's output to the pitching or rolling joints, u_s is the sensory input, N is the order of controller. In this research framework we limit N=2 as it reduces the controller to two coupled neurons (Zaier & Nagashima, 2006). The parameters $a_{ii} > 0$, a_{ij}, and b_i can be easily designed, since optimal values are not required in the proposed design approach. Therefore, the motors' commands can be expressed as follows;

$$z_i(t) = v(t) + \sigma\big(\theta_i(t), h_i(t)\big), \tag{6}$$

where $h_i(t)$ represents the reflex motion, and $z_i(t)$ is the motor commands.

Note that the solution of (3) is $\theta_i(t) = \theta_{i0}(t) + \int_{t_0}^{t_1} f_i(s, \theta_i)ds$, and since $g_i(t)$ is a piecewise continuous in t and satisfies the Lipschitz condition over the interval $[t_0,t_1]$ (time interval

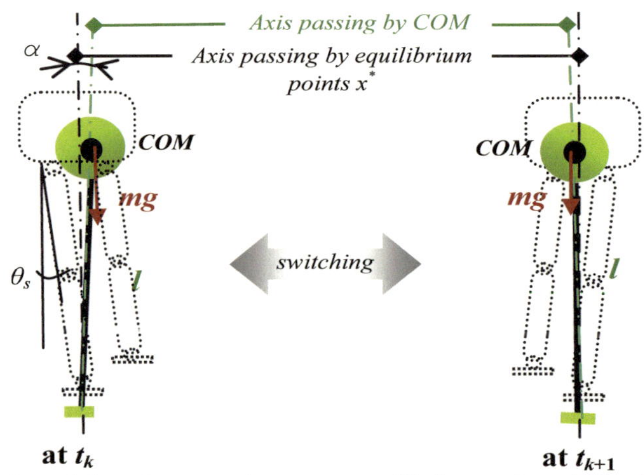

Fig. 2. Commutation of the robot state between equilibrium points defined at the single support phase

during commutation between single support phases), then $f_i(t,\theta_i)$ satisfies a local Lipschitz condition over the interval $[t_0,t_1]$ and has a unique solution (Khalil, 1996).

3. Generation of walking motion pattern

3.1 Stability at the single support phase

While the inverted pendulum model considerably simplifies the control of the humanoid robot, the inertia effect of distributed masses such as the arms may present a limitation to that approach. Our method, although inspired by the inverted pendulum, it considers not strictly the system dynamics. In other words, the method can generate a rhythmic movement without solving the nonlinear differential equations of the system dynamics. To illustrate the idea, consider Fig.2, where continuous commutations between single support phases take place. Let's model the system by a second order ODE with a feedback control input as follows;

$$\frac{d^2\alpha}{dt^2} + 2\varsigma\frac{d\alpha}{dt} - \mu\alpha = u_g(\alpha), \tag{7}$$

where α□$\boldsymbol{R}$ is the counter clockwise angle of the inverted pendulum from the vertical, ζ□$\boldsymbol{R}$ is damping ratio, $\mu=g/l$□$\boldsymbol{R}$, and u_g □$\boldsymbol{R}$ is the feedback control input stabilizing the system. Let $\alpha = \theta - \theta_s$, where θ is the angular position of the hip joint and θ_s is its value when the projection of the centre of mass (COM) is in the support polygon. Then (7) can be rewritten as follows;

$$\frac{d^2\theta}{dt^2} + 2\varsigma\frac{d\theta}{dt} - \mu\theta = u_g(\theta), \tag{8}$$

where $u_g(\theta)$ is the gyro feedback signal stabilizing the inverted pendulum around $\theta=\theta_s$. Then at each equilibrium point the system has the same eigenvalues as follows;

$$\lambda_{1,2} = -\varsigma \pm \sqrt{\varsigma^2 + \mu} \tag{9}$$

Since $\zeta<<\mu$, the equilibrium points (9) are saddle ones. Therefore, to generate periodic movement as explained in Fig. 2, we must first stabilize the equilibrium points by a feedback control $u_g(\theta)=k_1\theta+2k_2 d\theta/dt$ using the measured angular velocity of the inverted

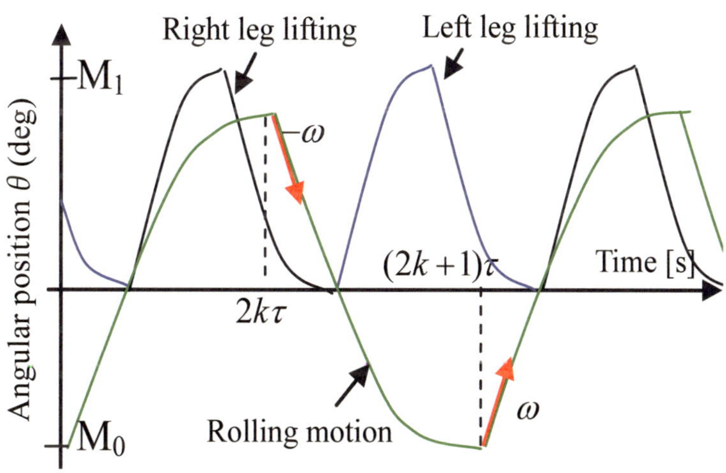

Fig. 3. The position commands to the joints; rolling commands to the ankle and hip, lifting commands to the ankle, knee and ankle joints, where the commands to the knee is twice of the ankle with opposed sign.

pendulum. It should be noted that our approach is based on rough model of the robot; there is no need to design optimal feedback gain, however, this gain has to be decided not to cause high vibration of the robot mechanism.

3.2 Rolling motion

The rhythmic motion is generated with regards to the rolling profile that is approximated as trapezoidal form, and smoothed using (1). The position command to the hip and ankle is illustrated by Fig.3. The control law at the switching times $t = k\tau$ is expressed as follows;

$$\begin{cases} \theta(t^-) = (-1)^{k+1}\theta_{\max}, & d\theta/dt\big|_{t=t^-} = 0 \\ \theta(t^+) = (-1)^{k+1}\theta_{\max}, & d\theta/dt\big|_{t=t^+} = (-1)^k \omega \end{cases} \quad k = 1,2,3\ldots \tag{10}$$

where k is the number of half walking cycle τ, the t^- and t^+ are the times just before and after commutation, respectively. For $t \neq k\tau$ the state trajectory will follow the desired rolling profile as described in Fig. 3, and the stability of the equilibrium points is guaranteed by the gyro feedback loop added to the system as in (8).

Let's formulate the rolling motion $\theta(t)$ during commutation from M_0 to M_1 as a function of time delay ε, joint angular velocity ω, walking cycle T, and the rolling amplitude a_s.

$$\theta = f(\varepsilon, \omega, T, \theta_s), \tag{11}$$

and define the operator $u(.)$ such that $u(t_i, \omega_i) = \omega_i (t - t_i)$ with $0 \le u(t_i, \omega_i) \le 1$. Using this operator, the periodic rolling motion can be expressed as follows;

$$\varepsilon \frac{d\theta(t)}{dt} + \theta(t) = \theta_s[u(t_{r0}, 2\omega_r) - 2u(t_{r0} + (n + f_1)T, \omega_r) + 2u(t_{r0} + (n + f_2)T, \omega_r)) - u(t_f, 2\omega_r)], \tag{12}$$

where t_{r0}, t_f are the switching times for the start and end of the rolling motion, respectively, and n is the number of walking steps. The f_1 and f_2 are the relative times with regards to the gait. By letting $p = t_{r1} - t_{r0} - 1/\omega_r$ be the time during which the robot stays at the maximum rolling and t_{r1} be the first switching time at the single support phase, we can write $f_1 = (1/\omega_r + p)/T$ and $f_2 = 2f_1 + 1/(\omega_r T)$.

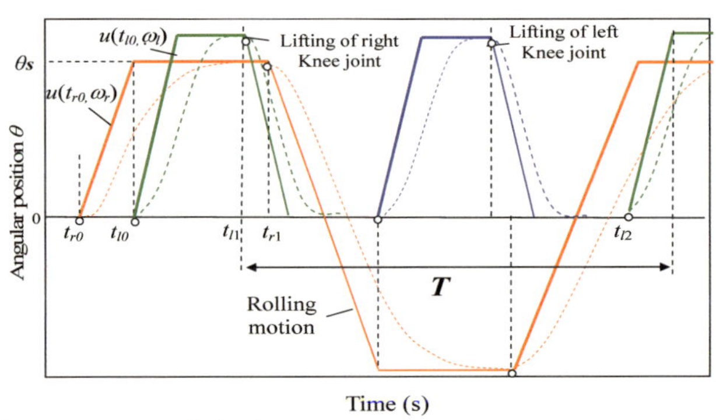

Fig. 4. Rolling motion pattern and design parameters.

The commands $\theta^r_{am}(t)$ and $\theta^r_{hm}(t)$ will be sent respectively to the hip and ankle joints, which are given by,

$$\theta^r_{am}(t) = -\theta^r_{hm}(t) = \theta(t) + \theta^r_{fb}(t)\,, \tag{13}$$

where $\theta^r_{fb}(t)$ is the feedback signal stabilizing (8).

Since the position command to the hip is the same as that to the ankle joint with opposed sign, the robot upper body orientation in the absence of disturbance remains unchanged, in other words the angular velocity around the rolling axis during locomotion is zero. Notice that the control law described by (10) is explicit of time. Instead, the state commutation can be constrained by angular velocity $d\theta/dt$=0, and the control law becomes independent of the cycle time as a pure feedback law, however, the commutation time will be no more constant.

3.3 Swing motion

The dynamic of the swing leg can be considered the same as that of a pendulum, and hence it is inherently stable without any compensator. This motion is generated in a similar fashion as that of the rolling motion, and according to Fig. 4 it is expressed by the following equation;

$$\begin{aligned}\varepsilon\frac{d\theta_l(t)}{dt}+\theta_l(t)=a_l[u(t_{l0},\omega_l)-\\ u(t_{l1}+nT,\omega_l)+u(t_{l2}+nT,\omega_l)-u(t_{lf},\omega_l)],\end{aligned} \tag{14}$$

where θ_l is the lifting motion and a_l is its amplitude. The t_{l0} and t_{l2} are the switching times of lifting for the first and second half walking cycle, t_{l1} and t_{lf} are the switching times of landing for the first and the last half cycle, and ω_l is the joints' angular velocity. It should be noticed that the proposed motion pattern generator generates dynamic walking motion as shown in Fig. 3, where the lifting time t_{l0} in Fig. 4 is very close to the rolling time t_{r0}. Similarly, the angular position θ_s generating the stride is expressed by the following equation;

$$\varepsilon\frac{d\theta_s(t)}{dt}+\theta_s(t)=a_s[u(t_{s1}+nT,\omega_s)-u(t_{s2}+nT,\omega_s)], \tag{15}$$

where a_s is the amplitude of the angular position generating the stride. The ω_s is the joints' angular velocity. The t_{s1} and t_{s2} are the times at the start and the end of stride, respectively.

On the other hand, we assume that the landing of each leg is accomplished with a flat foot on a flat ground, and the thigh and shank of the robot have the same length. Therefore, with respect to the angles defined by (Fujitsu Automation Ltd.), this condition can be satisfied as follows.

$$\begin{cases} \theta_{am}^{p}(t) = \theta_l(t) + \theta_{fb}^{l}(t) + \theta_s(t) \\ \theta_{km}(t) = -2\theta_l(t) \\ \theta_{hm}^{p}(t) = \theta_l(t) + \theta_{fb}^{l}(t) - \theta_s(t) \end{cases}, \tag{16}$$

where θ_{am}^{p}, θ_{km}, and θ_{hm}^{p} are the pitching motor commands to the ankle, the knee and the hip, respectively. The $\theta_{fb}^{l}(t)$ is the feedback signal to the ankle and the hip, satisfying the stability of inverted pendulum in the sagittal plane as in (7).
Moreover, to minimize the force of the collision of the landing leg with the ground, instead of using impact model, besides (17), we control the damping factor b_s and the spring stiffness k_s of the virtual damper-spring system (19) such that the leg is very compliant at the swing phase, and gradually get stiffer till it reaches the maximum stiffness at the single support phase.

$$m\frac{d^2 y_c(t)}{dt^2} + b_s\frac{dy_c(t)}{dt} + k_s y_c(t) = F_y, \tag{17}$$

where $y_c(t)$ is the displacement of the mass m along the vertical axis, and F_y is the external force acting on the supporting leg. The angular positions to the motors' commands becomes

$$\begin{cases} \theta_{am}^{p}(t) = \theta_l(t) + \theta_s(t) + \theta_{fb}^{l}(t) + \theta_c(t) \\ \theta_{km}(t) = -2\theta_l(t) - 2\theta_c(t) \\ \theta_{hm}^{p}(t) = \theta_l(t) - \theta_s(t) + \theta_{fb}^{l}(t) + \theta_c(t) \end{cases}, \tag{18}$$

where $\theta_c(t) = \arcsin(y_c(t)/L)$ is the angular position induced by the virtual damper spring system in (17), and L is the length of the thigh.

4. Reflex system

In order to prevent the falling down of a humanoid robot when it suffers a sudden deviation from its stable state, we have proposed a reflex system that can be triggered by sensory signals. In the previous result (Zaier & Nagashima, 2006), we studied three cases of sudden events; (1) when there is a sudden change in the ground level, (2) when the robot is pushed from arbitrary direction, and (3) when there is a sudden change in load.

4.1 Reflex against large disturbance

For the case of large disturbance, we defined a normalized angular velocity at the x-y plane termed gyro index "g_i", and it is given by,

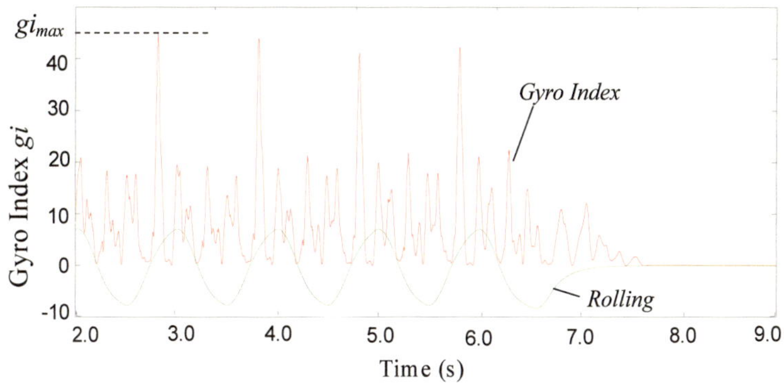

Fig. 5. Gyro index and rolling joint's output during walking when no large disturbance is present

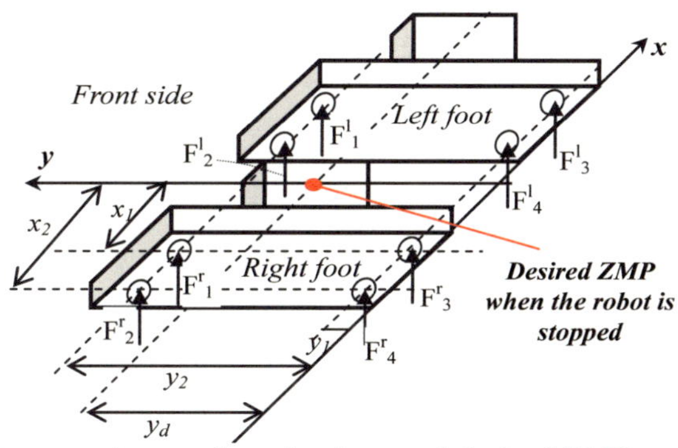

Fig. 6. Sole reaction forces acting on the sole plate and desired ZMP.

$$g_i = \sqrt{\mu^2 gyro_y^2 + gyro_x^2} \,, \tag{19}$$

where $gyro_x$ and $gyro_y$ are the components of the gyro sensor's output along the rolling and pitching axis, respectively. The parameter $\mu = L/w$ is the size ratio of the sole plate, where L and w are the length and the width of the robot sole plate, respectively. The decision about the presence of large disturbance is based on a threshold Tg_i, which we obtained experimentally by disturbing the robot to the level beyond which stability by PD controllers will not possible. Fig. 5 shows the experimental result of the *Gyro index* in the absence of large disturbance. Moreover, we defined four postures around the walking motion which we called learned postures. When a large perturbation occurs such that the threshold Tg_i is touched, the robot will stop walking and shifts its pose to one of learned postures selected according to the measured ZMP position. A gyro feedback controller will be activated stabilizing the final pose the humanoid robot has shifted to. In order to find the ZMP during locomotion, we simply use the sole reaction forces defined in Fig. 6, which is calculated using the force sensors' values as follows,

$$x_m = [x_1(F_1^r + F_3^r - F_2^l - F_4^l) + x_2(F_2^r + F_4^r - F_1^l - F_3^l)] / F_T \tag{20}$$

$$y_m = [y_1(F_3^r + F_4^r + F_3^l + F_4^l) + y_2(F_1^r + F_2^r + F_1^l + F_2^l)]/F_T \ , \tag{21}$$

where F_T is the sum of all forces acting on the legs. Notice that the stride effect on y_m is neglected since the duration of the double support phase is almost zero and is much smaller

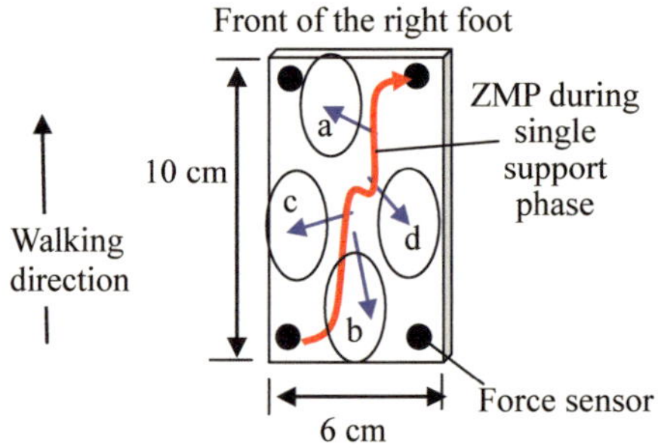

Fig. 7. Sole plate of the robot; the curve in red is the ZMP trajectory during the single support phase. The (a, b, c, and d) are the zones to which the ZMP will shift when disturbances occur

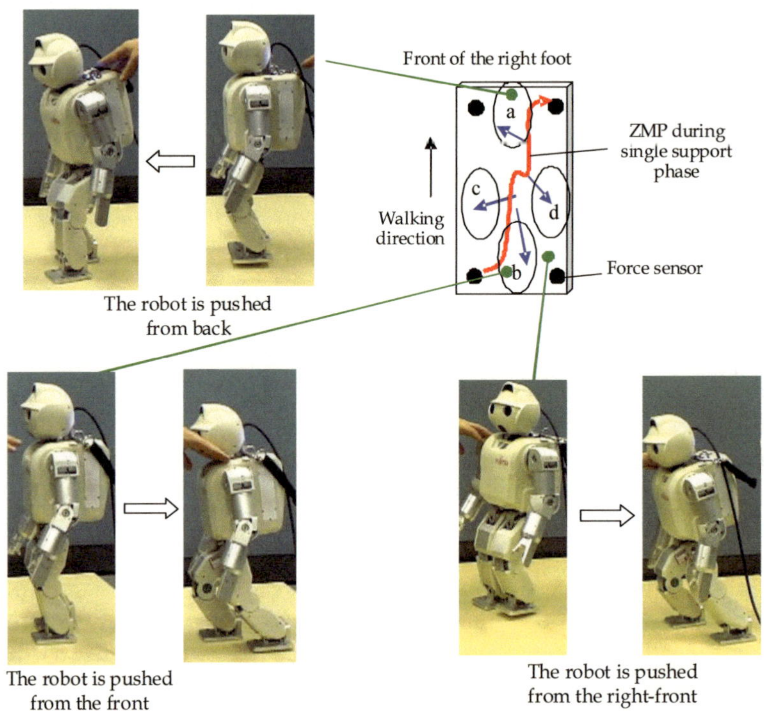

Fig. 8 Defined zones (a, b, c, and d) under the sole plate and reflex actions that take place when the large disturbance occurs. The reflex consists of switching the pose of the robot to the appropriate predefined posture according to the ZMP position. Notice that when the robot is pushed from the right-front, the ZMP position is between zone (b) and zone (d) shown in Fig 7, and consequently the left leg is stepped back to the left side.

than the time delay of the ZMP controller. The ZMP during walking and in the absence of large disturbance can be described by the curve in Fig.7. To generate a reflex motion when large disturbance acts on the robot's upper body, we define four zones under the sole plate and for each zone we define a posture such that when the ZMP is shifted to one zone in the presence of large disturbance, the robot will stop walking and modifies its pose to the appropriate posture, as described in Fig. 8. If the ZMP shifts between two of these four zones, linear interpolation will be considered between the two corresponding defined postures. Notice that a large disturbance is detected when the gyro index (19) touches the

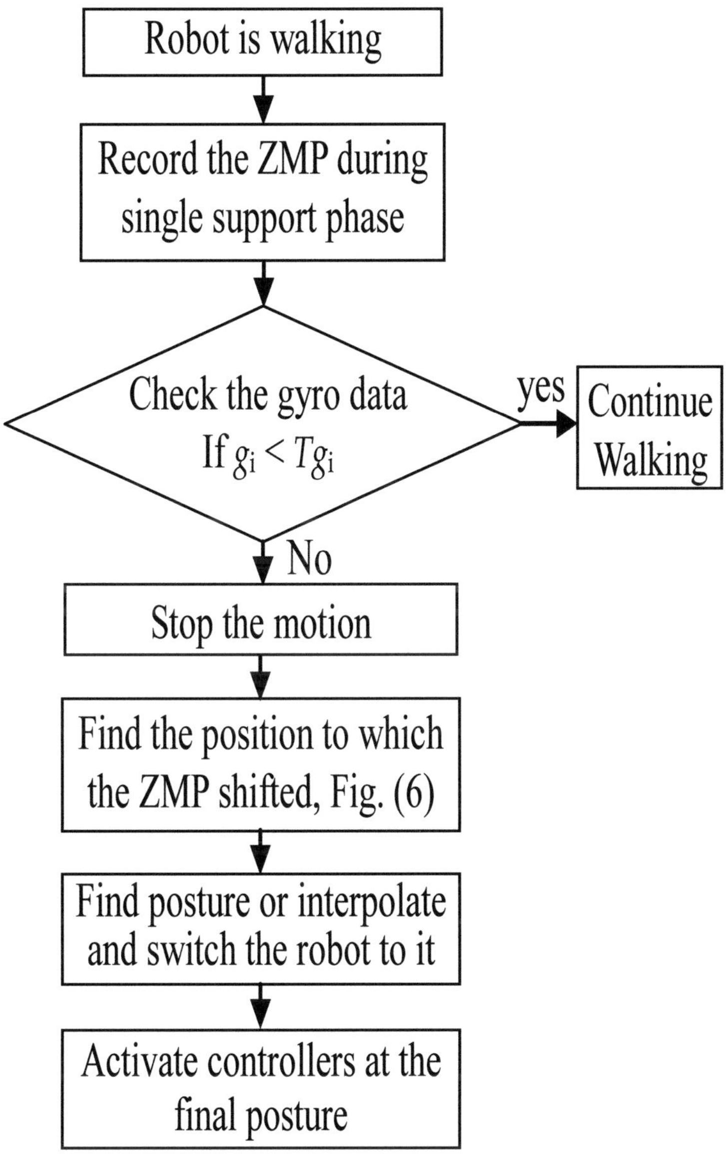

Fig. 9. Flowchart of reflex activation when a large disturbance is present

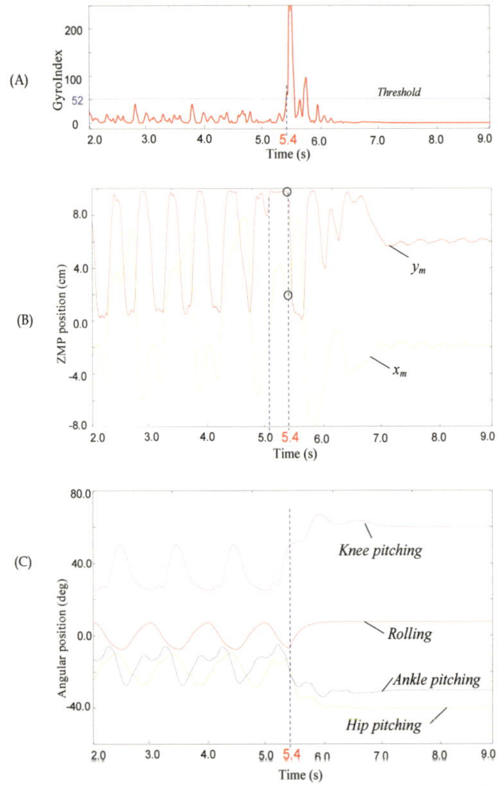

Fig. 10. Detailed results in the presence of large disturbance; (A) shows the gyro index that touches the threshold at 5.4 s. (B) shows the ZMP positions during walking defined within the support polygon, where x_m is varying between the right and left legs, while y_m is defined along the sole plate from 0 to 10 cm. At 5.4 s the large disturbance was detected and the ZMP has shifted to the circled points, which correspond to x_m =2.0cm and y_m =8.4cm. Finally, (C) shows the pitching and rolling joints' outputs of the left leg.

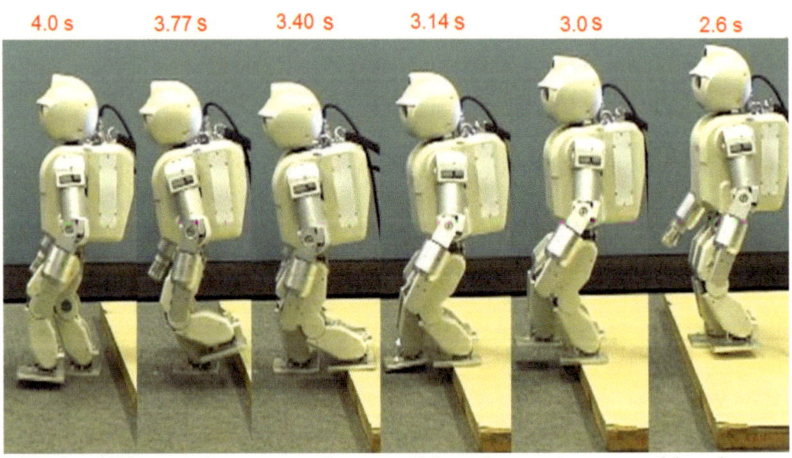

Fig. 11. HOAP-3 walking and encountering a sudden change in ground level

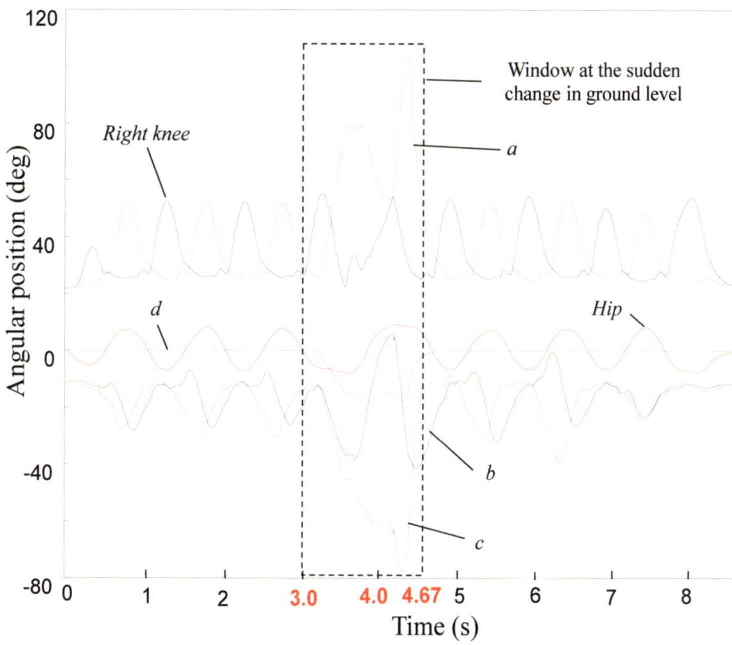

Fig. 12. Joints outputs at the sudden change in ground level. Pitching of the knee (*a*), ankle (*b*), and hip (*c*) joints of the left leg. (*d*): Extra motion added to the pitching joints of supporting leg

threshold Tg_i. The steps from the detection of large disturbance till the activation of a reflex motion are described by the flowchart in Fig. 9. The experimental results of the case when the robot is pushed from the back are detailed in Fig. 10.

4.2 Reflex against sudden change in ground level

Other type of sudden event we studied is the reflex against sudden change in ground level (Fig. 11) that is detected using a photo-interrupter attached to the front of the robot foot. The reflex movement will be triggered when the photo-interrupter, at the landing time of the leg, dose not detect a ground. The reflex movement consists of increasing the stride so that the foot collides not with the upper ground. Afterwards, the supporting leg will be contracted in height till the swing leg touches the lower surface. After landing, the walk parameters (gait and stride) will be set to their previous values. Notice that our proposed reflex needs not to know about the ground elevation. This parameter is detected automatically by the sole sensor, where we assume that the change in the ground level is within the hardware limit of the robot. The joints' outputs are shown in Fig. 12. The extra motion (*d*) is added to the supporting leg joints contracting the leg at height. The contraction is stopped when the swing leg touches the lower surface. The contraction phase of the supporting leg starts when the linear velocity of the upper body is almost zero. Moreover, the contraction speed is low enough to ensure that the contracting leg will not affect the stability of the robot. The walking cycle time is augmented by about 50% at the sudden change in ground level. The motion of the supporting leg is modified accordingly.

4.3 Reflex against sudden obstacle

This type of reflex is activated when a sudden obstacle touches the sole sensor of the leg at the swing phase. We assume here that the foot of the leg remains parallel to the ground during the swing motion, which is satisfied by (16).

The reflex process is abstracted as follows:

- Detect the sudden obstacle with a sole sensor.
- Stop the motion of the robot.
- Move the state of the robot to the attraction domain of the stable equilibrium point.
- Resume the walking motion with negative stride, then follow upper level control.

Let's write the equations of the reflex motion as follows,

$$\varepsilon_p \frac{dh_p(t)}{dt} + h_p(t) = u(t - t_{sw})(a_l - \theta_{swp}) \tag{22}$$

$$\varepsilon_r \frac{dh_r(t)}{dt} + h_r(t) = u(t - t_{sw})(\theta_s - \theta_{swr}) \tag{23}$$

where $u(t_i)$ is a unit step function starting at t_i. The $h_r(t)$ is the angle to the pitching joints of the leg touching the obstacle, which will be added as θ_l in (16), and $h_p(t)$ is the angle to the rolling joints of the ankles and hip of both legs. The ε_r and ε_p are neurons' time delays with much smaller values than ε in (4), θ_{swr} and θ_{swp} are the rolling and pitching values, respectively, at the time t_{sw}, when the sudden obstacle is detected, a_l is defined in (14), and θ_s is the rolling amplitude. Fig. 13 shows HOAP-3 detects a sudden obstacle as it walks. The joints' outputs are shown in Fig. 14. At 3.0 s, the sole sensor of the left leg touches the obstacle and the robot stops landing its leg and increases its rolling motion while keeping the gyro feedback controller active. The robot, then, retracts the left leg to its previous landing position. This is shown in Fig. 14 by the pitching motion of the hip, knee, and ankle within the time interval [3.0s, 4.8s]. The walking is resumed successfully at 6.8 s. Notice that the leg during swing motion is made very compliant, where the stiffness is controlled during the gait such that the leg at the middle of the supporting phase is very stiff, while it is very compliant during the swing phase.

4.4 Reflex against sudden change in load

We consider the reflex when the humanoid robot detects a sudden change in load as it walks, which is considered as a large disturbance but to the level that the ZMP remains

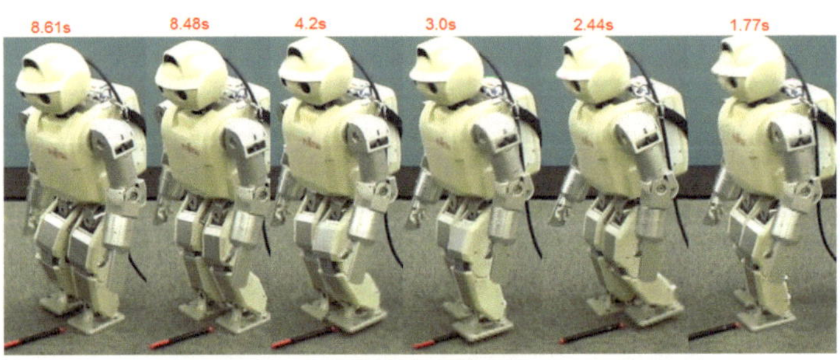

Fig. 13. Reflex against a sudden obstacle while walking.

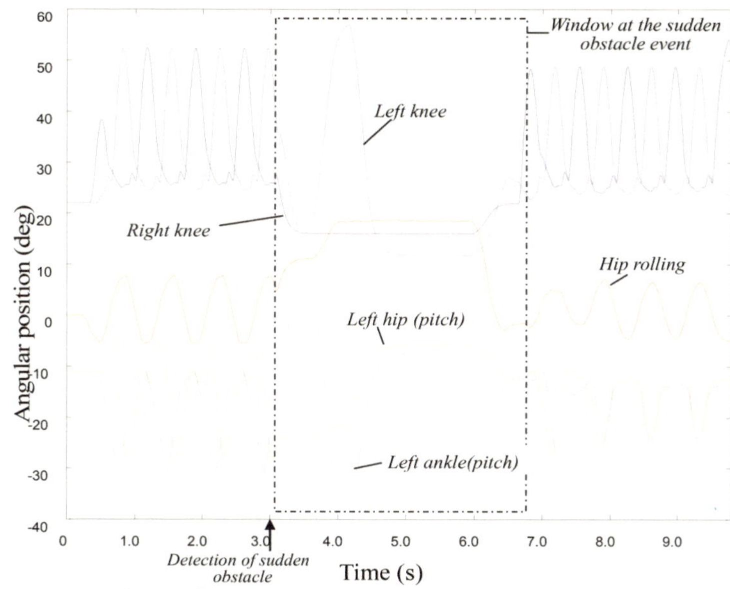

Fig. 14. Joints outputs when reflex against sudden obstacle is activated

inside the support polygon. The proposed reflex policy is shown in Figs. 15 and Fig. 16. The reflex motions θ_w and θ_a will be generated simultaneously at the ankles and waist joints, respectively (Fig. 15 c). The amplitudes of these angles are set according to the deviation angle of the upper body. During this phase only, the gain of the gyro feedback controller, which works against oscillation of the robot's upper body, is reduced by more than half. After this phase, the gain of the gyro feedback controller is set to its previous value (Fig 15 d). Fig. 17 shows HOAP-3 carrying a relatively heavy box. As it walked, the box was taken away and the robot could walk stably and compensate for the change by adjusting its walking posture at the ankle and waist joints according to the algorithm in Fig. 16. The positions of the ankles and their change during walking are shown in Fig. 18. The reflex motion is generated within the time interval [6.0s, 7.0s]. It should be noticed here that the reflex is triggered by the gyro sensor when its value exceed a given threshold, while the reflex motion is selected according to the recorded ZMP at the disturbance detection time.

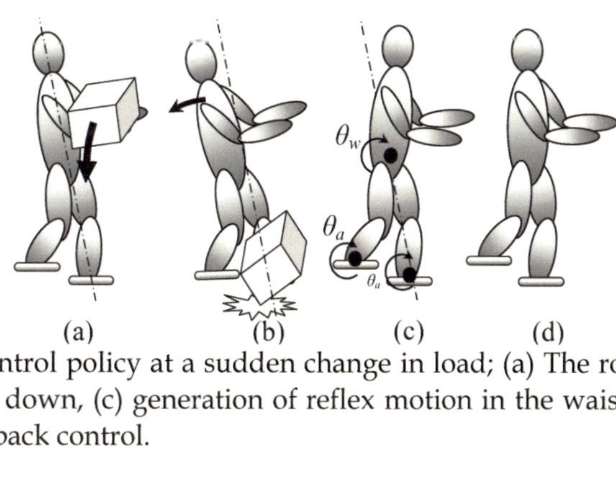

Fig. 15. Proposed control policy at a sudden change in load; (a) The robot is carrying a box, (b) The box is fallen down, (c) generation of reflex motion in the waist and ankle joints, (d) ZMP and Gyro feedback control.

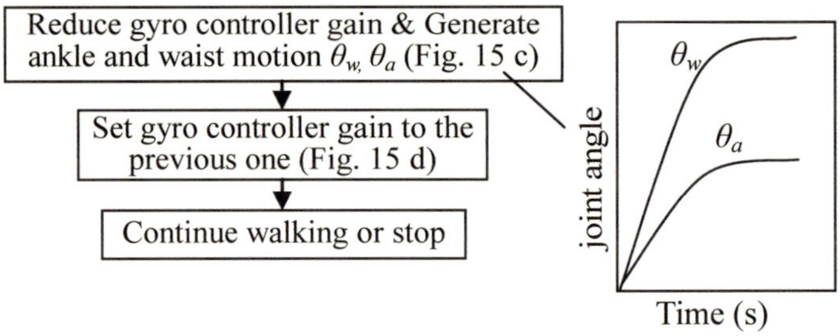

Fig. 16. Flowchart of reflex at sudden change in load.

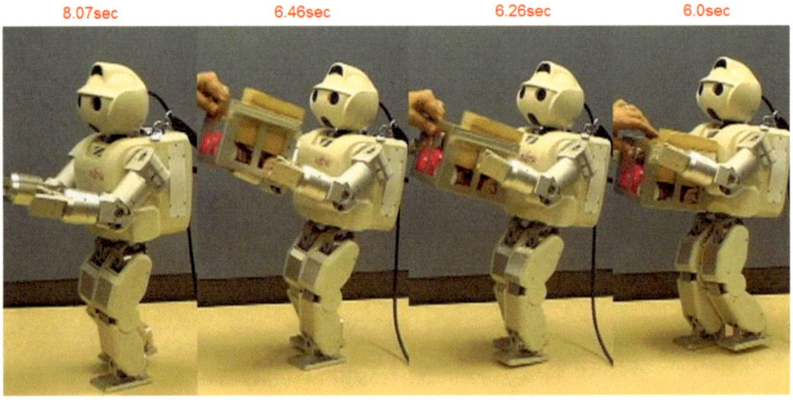

Fig. 17. Reflex against sudden change in load while walking

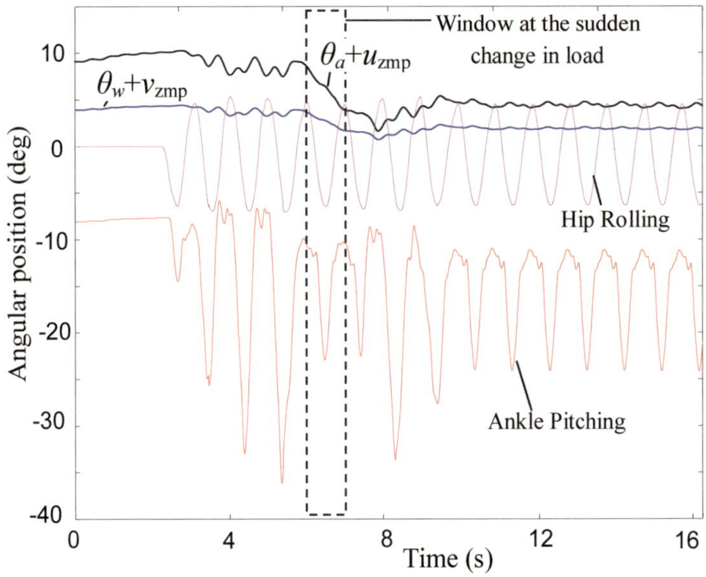

Fig. 18. Joints outputs at the sudden change in load. As shown in Fig. 16, θ_a, and θ_w are the angles added to the ankle and waist joints, respectively. u_{zmp}, and v_{zmp} are the ZMP feedback controls to the ankles and waist joints.

4.5 Unified reflex system

To deal simultaneously with several types of undesired events, the reflex system has to be designed with a structure that can be extended to cover, to some extent (e.g., hardware), several types of sudden events that may act on the robot. If we consider that prior knowledge about the type of sudden event is provided, and by defining a discriminant function $d_i(s)$ based on the Bayes rule, the general structure of the recognizer can be represented as shown in Fig. 19. In this research frame work, the recognizer is limited to four types of sudden events that we studied in the previous subsection. To build up the recognizer, sensory data and motion related parameters are collected. When extracting the appropriate features required for each type of sudden event, we realized that the recognizer can be simplified into two dichotomizers (Fig. 20). The first dichotomizer consists of reflex actions against large disturbances. That is, when an input from the gyro sensor s_1 exceeds a given threshold, discriminant functions d_1 and d_2 will use that feature but the decision will depend on input's feature s_2. For instance, when the ZMP remains inside the support polygon, the disturbance is considered to be of type one. If the ZMP leaves the support polygon, then a disturbance will be of type two. For example, pushing a robot as it is walking can be classified as a large disturbance of type one if the ZMP remains inside the supporting polygon, otherwise, it is of type two. As for the second dichotomizer, $d_3(s)$ activates the reflex action against a sudden change in ground level when the photo-interrupter that is attached to the front of the leg is enabled at landing time, while $d_4(s)$ activates the reflex against a sudden obstacle when the sole sensor of the swing leg touches an obstacle. Fig.21 shows the overall scheme of the reflex system and motion pattern generator.

4.6 Adaptive reflex system

In the previous subsection, although the reflex was proposed in a unified fashion with respect to four types of sudden events, adaptation of the reflex was not considered. For instance, the reflex against sudden obstacle, in some situation, may fail to prevent the falling

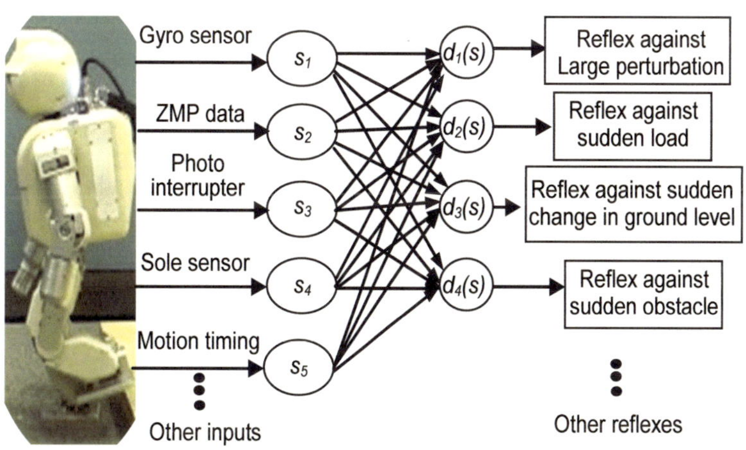

Fig. 19. General structure of the unexpected pattern recognizer, where s_i represents the normalized input signal and $d_i(s)$ is the discriminant function.

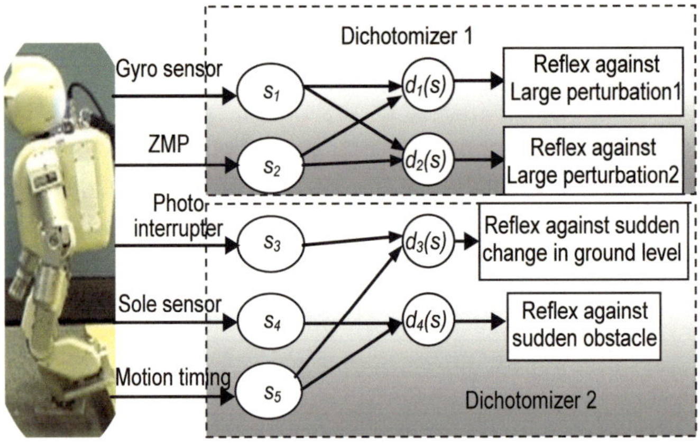

Fig. 20. Structure of the unexpected pattern recognizer after features extraction.

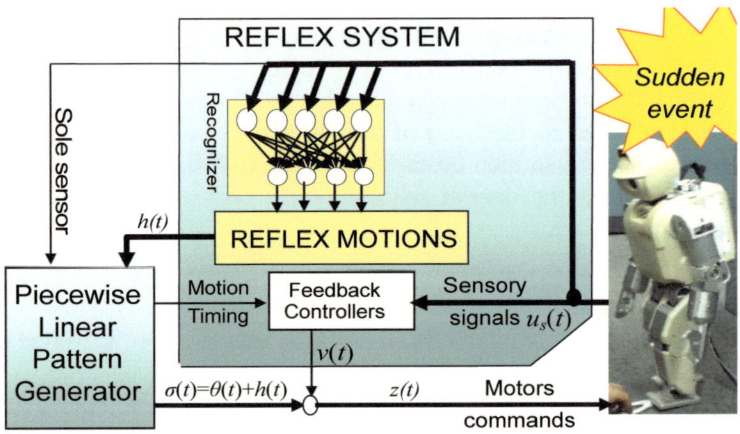

Fig. 21. Overall scheme of the reflex system and motion pattern generator; $\sigma(t)$ is the sum of reflex $h(t)$ and the motion pattern generator output $\theta(t)$

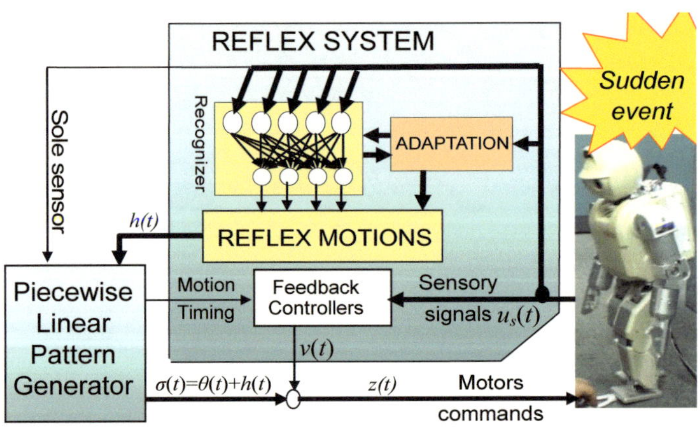

Fig. 22. Overall scheme of the adaptive reflex system and motion pattern generator

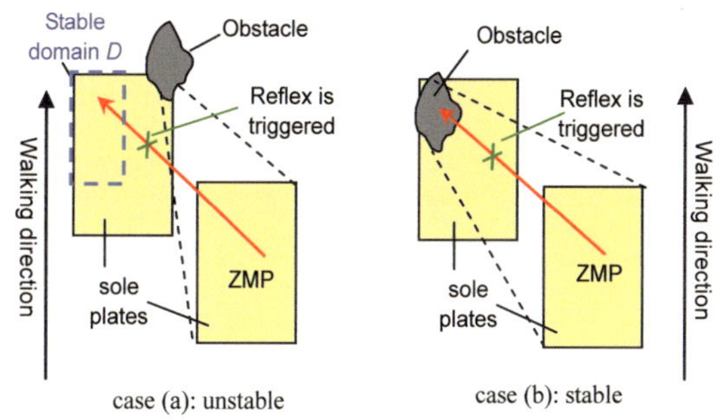

Fig. 23. The state of the robot depends on the relative position of the sudden obstacle with regards to the sole plate. Notice that the commutation of the state between equilibrium points is realised by gravity

of the robot. In this section, therefore, we propose a new adaptation algorithm as shown in Fig. 22, which will adapt the reflex motion and the recognizer as well. In this research framework we limit the adaptation to the reflex against sudden obstacle.

Let consider our previously proposed reflex (22) and (23) as a primitive one, and abstracted as follows:

- Detect the sudden obstacle with a sole sensor.
- Stop the motion of the robot.
- Move the state of the robot to the attraction domain of the previous equilibrium point.

However, if the obstacle does not provide a stable supporting surface, it will be hard for the robot to step forward and stand up stably on the obstacle. For instance, when an obstacle touches the obstacle in the half sole plate adjacent with the other leg, the robot state will not be able to return to the equilibrium point defined by the other leg.

Due to the dynamic walking pattern (i.e., the commutation of the ZMP from one leg to the other is realized by a free fall of the humanoid robot) and the presence of time delay in triggering the reflex, the ZMP may become out of the support polygon formed by the leg's sole plate and the obstacle (Fig. 23a), which make the primitive reflex unable to prevent the falling down of the robot. To face such a problem there are two solutions; either to increase the support polygon area or to change the robot posture so that the ZMP becomes inside the stability zone. Intuitively, it is easier to consider the first possibility rather than the second one, by rolling or pitching the ankle joint of the leg touching the obstacle. Moreover, using

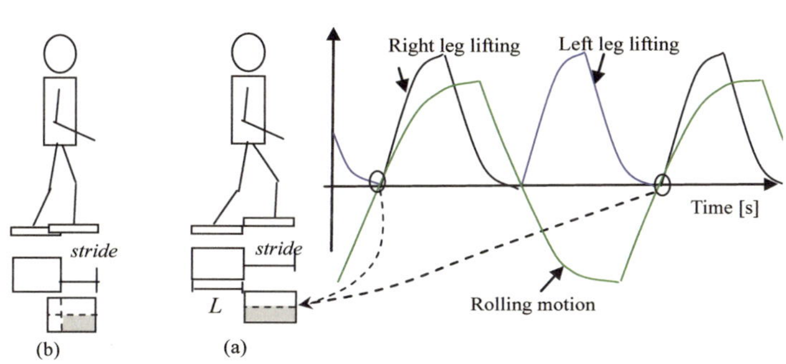

Fig. 24. The stability zone D is coloured in gray; it corresponds to the zone from which the landing and rolling to the other leg start

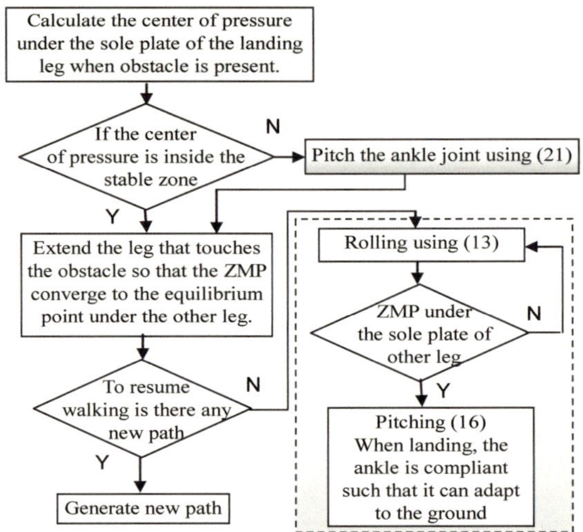

Fig. 25. Flowchart of the improved reflex against sudden obstacle, where the reflex is triggered when the sole plate touches the obstacle at the swing phase

dynamics calculation, it is possible to show that this reflex corresponds to the minimum energy the robot may spend to achieve stability. In some cases, however, environmental constraints may impose the second solution to be adopted, for instance, by lowering the COM, or even by relying on a fixed object nearby the robot. These intuitive actions can be considered as hints to adapt the neural network with additional reflexes covering cases similar to that of Fig.23a, which depend on the relative position of the obstacle with respect to sole plate and the environment surrounding the robot as well. Therefore, the augmented reflex is proposed as follows;

$$\begin{cases} \varepsilon_r \dfrac{ds_a}{dt} + s_a = s_a(t)u(t_{obs}) + k(u(t_{obs}) - u(t_{cont})) \\ \dfrac{dx_i}{dt} = -a_{ii}x_i + \sum\limits_{j=1, j\neq i}^{N} \dfrac{a_{ij}}{a_{ii}} x_j + \dfrac{b_i}{a_{ii}} e_j(t) \\ r_a(t) = s_a(t) + \sum\limits_{i=1}^{N} c_i x_i \end{cases}, \qquad (24)$$

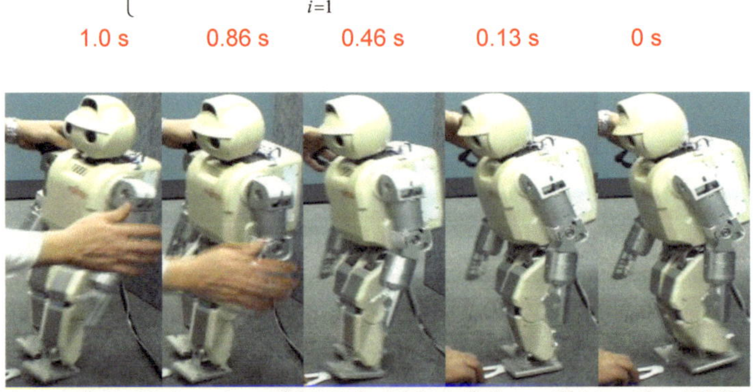

Fig. 26. Primitive reflex against a sudden obstacle while walking; the ZMP is outside the support polygon formed by the right leg and the obstacle.

where $u(t_i)$ is as defined in (22), $\varepsilon_r > 0$, e_j is the distance of the ZMP from the center of the sole plate, k is the integration constant, $r_a(t)$ is the reflex motion to the ankle pitching joint. $s_a(t)$ is an integrated angular position from the time when the obstacle is detected (t_{obst}) till the time when the sole plate touches the ground making a stable support polygon (t_{cont}). The $a_{ii} > 0$, a_{ij} and b_i are parameters of the PD controller defined as in (5). The stability zone D is described in Fig. 24, which depends on the walking stride and is defined by the timing when the leg touches the ground (for dynamic walking the rolling at this timing is zero). Fig.25 shows the proposed adaptive reflex algorithm. Notice that the whole control diagram is shown in Fig. 22, where the adaptation part consists of (24) and some check regarding the stability domain D at the collision time. The first step consists of finding the center of pressure (or position of the obstacle). Then, if this position belongs to the stability zone D, the leg will expands in length by inverting the sign of sensor feedback output in (16) so that the state of the robot reaches the attraction domain of the equilibrium point under the supporting leg. Otherwise, the reflex (24) will be generated, which consists of augmenting the support polygon by pitching the ankle joint. Once the robot reaches the stable state it follows upper level commands, either by walking over the obstacle or stepping back and finding a new path. In the case of walking over the obstacle, the locomotion controller will generate a static walking; this is by shifting the time t_{l0} from t_{r0} (Fig.4), so that the lifting starts only when the projection of COM is under the other leg. This walking style is also adopted by human whenever he walks on rough or unstable terrain.

Remark: The advantage of (24) over the pitching expression (22) is that only the ankle of the leg that collides with the obstacle is pitched by an angle such that the support polygon covers the ZMP at the collision time. Moreover, although we used a ZMP feedback loop besides (22), the response is not fast enough to prevent the falling down of the robot.

To show the effectiveness of our newly developed adaptive reflex we conducted two experiments. Fig. 26 shows the experiment results when HOAP-3 detects a sudden obstacle as it walks, using the primitive reflex. Since the collision position of obstacle with the sole plate is out of the stability zone D (Fig. 24), the robot was not able to get stable by the primitive reflex (22) and (23), and therefore it has fallen down. Fig.27 shows the case when the proposed reflex was implemented, where the joints' outputs are shown in Fig. 28.

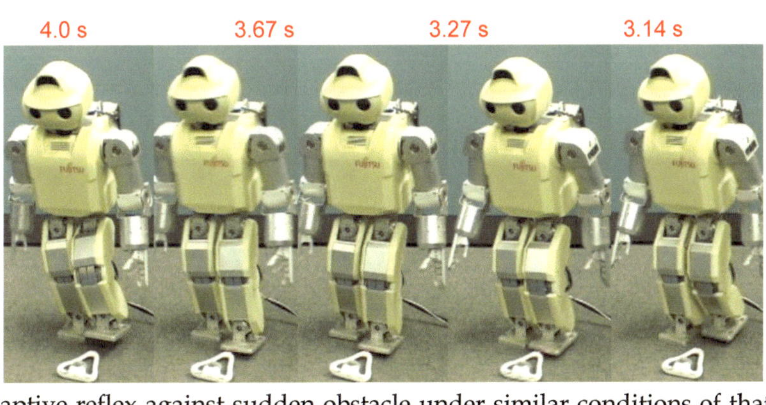

Fig. 27. Adaptive reflex against sudden obstacle under similar conditions of that in Fig. 10; the ankle of the left leg is pitched at the detection time of the obstacle, then it is controlled

such that the center of pressure is in the middle of the sole plate. At the same time the leg is extended so that the state of the robot reaches the equilibrium point under the right leg.

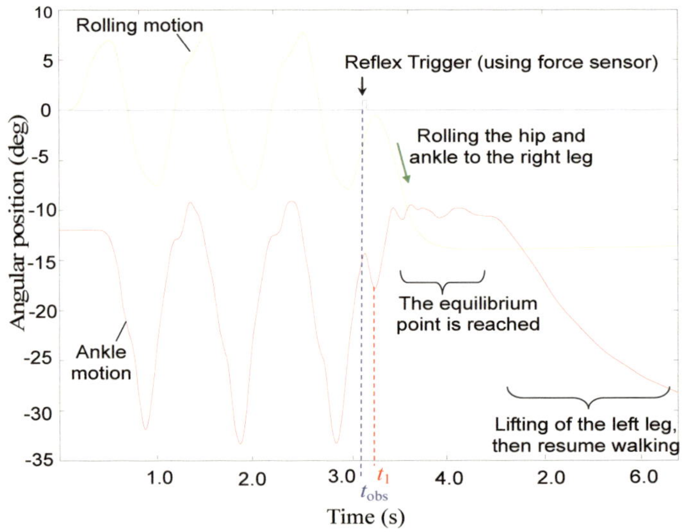

Fig. 28. Ankle joints' outputs when the proposed reflex against sudden obstacle is activated

At 3.14 s, the sole sensor of the left leg touches the obstacle and the pitching joint of the ankle was controlled according to (24) by $s_u(t)$ (fast response), then by $r_a(t)$ at 3.27 s, which is delayed by $1/a_{ii}$. Fig. 29 shows the ZMP position, where the location of obstacle is as described in Fig. 23a. The state of the robot reached the equilibrium point at about 4.0s, where the rolling joints get the static value (the projection of COM is under the right leg). Notice that the mode to resume motion by walking on obstacle is still under research.

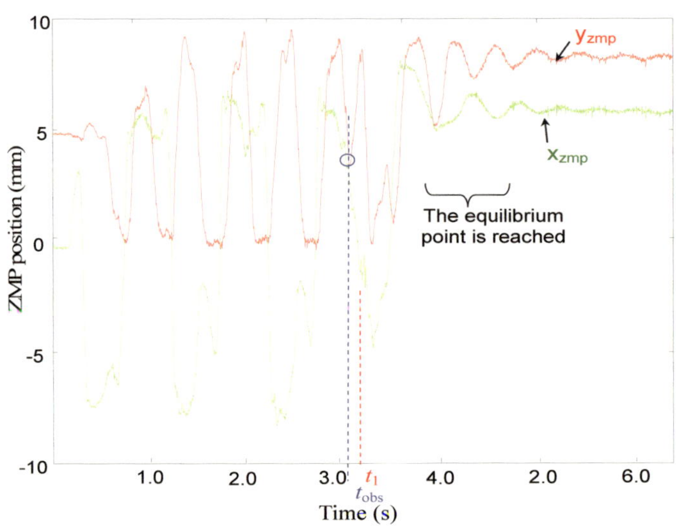

Fig. 29. ZMP position when reflex against sudden obstacle is activated

5. Conclusion

The reflex system was combined with the motion pattern generator to improve the robustness against lateral and frontal disturbances. On the other hand, the reflexes that highly adapt and control the movement of the humanoid in the presence of large disturbance acting on the robot's upper body were considered in a unified form. It was shown that a reflex movement was successfully enabled and integrated with the rhythmic motion in the cases of a sudden change in the floor level and a sudden obstacle, and in the presence of a large disturbance. Moreover, the proposed reflex against sudden obstacle was further investigated, and it was shown that when obstacle collides with the sole plate of the robot leg in a location out of a defined stable domain, the robot could not reach stability and fall down. On the other hand, an adaptive reflex was proposed that consisted of increasing the support polygon by controlling the ankle joint of the leg touching the obstacle. In order to demonstrate the effectiveness of the proposed system, we used the humanoid robot HOAP-3 of Fujitsu. It was shown that a reflex movement was successfully enabled and integrated with the rhythmic motion. As a future work, we will extend the reflex system by considering the interaction between the upper and lower body of the humanoid robots.

6. References

Fujitsu Automation. http://jp.fujitsu.com/group/automation/en/

Gerstner, W. (1995). Time structure of the activity in neural network models, *Phys.* E. 51, pp.738-758, *Rev*, 1995

Grillner, S. (1985). Neurobiological bases of rhythmic motor acts in vertebrates, *Science*, no. 228, pp143-149, 1985

Huang, Q.; & Nakamura, Y. (2005). Sensory Reflex Control for Humanoid Walking, *IEEE Trans. Robotics and Automation*, vol. 21, no.5, pp. 977-984, Oct. 2005

Huang, Q.; Yokoi, K.; Kajita, S.; Kaneko, K.; Koyachi, N.; Arai, H.; & Tanie, K. (2001). Planning walking patterns for a biped robot, *IEEE Trans. Robotics and Automation*, vol. 17, no. 3, pp. 280–289, Jun 2001

Kajita, S.; & Matsumoto, O. (2001). Real-time 3D walking pattern generation for a biped robot with telescopic legs, *Proceeding of the IEEE International Conference on Robotics & Automation*, pp.2299-2306, 2001

Khalil, H. (1996). *Nonlinear Systems*, Printice Hall, ISBN 0-13-067389-7, New York

Miura, H., & Shimoyama, I. (1984). Dynamic walk of a biped, *The International Journal of Robotics Research*, vol. 3 no. 2, pp.60-74, 1984

Morisawa, M.; Kajita, S.; Harada, K.; & Fujiwara, K. Emergency Stop Algorithm for Walking Humanoid Robots, *IEEE/RSJ International Conference on Intelligent Robots and Systems*, pp. 2109- 2115, 2005

Murase, Y.; Yasukawa, Y.; Sakai, K.; & Ueki, M.(2001). Design of Compact Humanoid Robot as a Platform", *19th Annual Conf. of the Robotics Society of Japan*, pp.789-790, 2001

Okada, M.; Osato, K.; & Nakamura, Y. (2005). Motion Emergency of Humanoid Robots by an Attractor Design of a Nonlinear Dynamics, *IEEE International Conference on Robotics and Automation*, pp. 18- 23, 2005

Taga, G. (1995). A model of the neuro-musculo-skeletal system for human locomotion, I. Emergence of basic gait, *Boil. Cybern*, no.73, pp.97-111. 1995

Taga, G.; Yamaguchi, Y.; & Shimizu, H. (1991). Self organized control of bipedal locomotion by neural oscillators in unpredictable environment, *Biological Cybernetics*, vol. 65, pp.147-159, 1991

Vukobratovic, M.; & Juricic, D. (1969). Contributuion to the Synthesis of Biped Gait, *IEEE Trans. On Biomedical engineering*, vol BME-16, No. 1, pp. 1-6, 1969

Zaier, R. & Nagashima, F. (2004). Motion Generation of Humanoid Robot based on Polynomials Generated by Recurrent Neural Network, *Proceedings of the First Asia International Symposium on Mechatronics*, p.659-664, 2004

Zaier, R.; & Nagashima, F. (2006). Motion Pattern Generator and Reflex System for Humanoid Robots, *Proceeding of IEEE/RSJ International Conference on Intelligent Robots and Systems*, 2006